OCEAN SCIENCE

Readings from

SCIENTIFIC AMERICAN

OCEAN SCIENCE

With Introductions by

H. W. Menard

Institute of Marine Resources

and

Scripps Institution of Oceanography

W. H. Freeman and Company
San Francisco

Cover: The research submersible *Alvin* seen against the bizarre scenery of the ocean floor along the Mid-Atlantic Rift. This illustration is taken from "The Floor of the Mid-Atlantic Rift," by J. R. Heirtzler and W. B. Bryan, which begins on p. 65.

Most of the *Scientific American* articles in *Ocean Science* are available as separate Offprints. For a complete list of more than 950 articles now available as Offprints, write to W. H. Freeman and Company, 660 Market Street, San Francisco, California 94104.

Library of Congress Cataloging in Publication Data

Main entry under title:

Ocean science.

Bibliography: p. 297
Includes index.
1. Oceanography—Addresses, essays, lectures.
2. Marine resources—Addresses, essays, lectures.
I. Scientific American.
GC26.023 1977 551.4′6′008 77-23465
ISBN 0-7167-0014-X
ISBN 0-7167-0013-1 pbk.

Printed in the United States of America

9 8 7 6 5 4 3 2 1

PREFACE

The science of the sea is as boundless, complex, and changing as the sea itself. It embraces the broadest spectrum of specialties in mathematics, all the sciences, engineering, and resource development and management. Thus, the chief problem of the oceanographer, whether student or active scientist, is staying in communication with other oceanographers with peripheral specialties that may or may not be overlapping. Isolation brings its perils. Knowledge continually advances in all the special fields of science—albeit understanding does not proceed everywhere at the same pace. You never know when a breakthrough in an apparently unrelated subject will suddenly illuminate your ongoing research.

How to keep up? How to communicate? At a large oceanographic institution it is fairly easy, at least in principle. Perhaps ease of communication is why the invention of the social organization we call a research laboratory has been such an outstanding success. To begin with, each specialist is in a group with closely overlapping interests and they keep each other informed about the matters that most concern them. In addition, almost every day some scientist, resident or visiting, presents a seminar intended to inform people in other fields about some development in specialized research. It is in principle easy to keep up, but in fact oceanographers, like everyone else, tend to stick to their small groups and skip the general seminars.

If communication is not easy even in circumstances ideal for it, how can anyone hope to become or keep informed about the full breadth of ocean science? This reader offers one of the best ways to do so. It presents twenty-five articles selected from *Scientific American* and written by specialists, mostly discussing their own research. Many of the articles appeared just after some significant breakthrough had occurred that merited the interest of a more diversified audience than usual. Thus they are full of the excitement of current work, the thrill of discovery, the quiet pride of achievement, the dawning realization that the work may be important to the future of mankind—all those emotions that make science a human endeavor.

April, 1977 *H. W. Menard*

CONTENTS

V PHYSICAL MARINE RESOURCES

Note on cross-references to SCIENTIFIC AMERICAN *articles:* Articles included in this book are referred to by title and page number; articles not included in this book but available as Offprints are referred to by title and offprint number; articles not included in this book and not available as Offprints are referred to by title and date of publication.

OCEAN SCIENCE

I

THE HISTORY OF OCEANOGRAPHY

THE HISTORY OF OCEANOGRAPHY

I

INTRODUCTION

Modern oceanography began with the *Challenger* expedition, which was sent out by the royal Society in 1872 and spent the next three and a half years sounding and sampling the world ocean. Thus, it is fitting that these readings begin with "The Voyage of the *Challenger*" by Herbert S. Bailey, Jr., who describes the scientists, the ship, the cruise, and the principal results from a historical perspective. He remarks that "far more handsomely outfitted expeditions . . . are [now] filling in details and retouching parts of a picture which in its broad outlines has remained essentially unchanged." This was an entirely appropriate viewpoint in 1953, when the article was published, but by this time oceanography has changed so much that such an attitude itself has some historical interest.

Oceanography changes so fast that it may be usefully analyzed more from the viewpoint of a sociologist than from that of a historian. The size, activity, and growth of a scientific field can be evaluated from the size of the population of scientists who work in it or from the number of scientific papers that they publish to describe their research results. By these measures, science as a whole doubles every fifteen years. It follows that at the end of a professional career of about 45 years an aging scientist finds that there are eight times as many scientists and scientific books and journals as when he was a student. In business terms, science is a dynamic growth industry.

Oceanography and a few other fields, such as molecular biology and particle physics, are even more dynamic. They double every four or five years. During a professional career there are ten doublings, which equals a thousandfold expansion. The oceanographer is "always climbing up the climbing wave," in Tennyson's phrase, just trying to keep his head above a flood of scientific literature. The doubling time is so short that it is equal to the normal tenure of a student in graduate school. Thus, I tell the incoming students at Scripps Institution to go to the library, look around, and remember that for every book that they see, another will be published before they receive their degrees. Fortunately, there are compensations for this hectic pace. A scientist may be considered to be at the beginning of a career when he enters graduate school; in oceanography, by the time the student completes his training half the people in the field will look upon him as a newcomer, but, by the same token, half will view him as a pioneer who began before them.

The enormous expansion in oceanography cannot continue forever, but it does not yet appear to be slackening. This is because it is constantly refueled by revolutionary scientific discoveries and by the need for additional information as an exploding human population exploits the sea. We shall consider the expansion from three aspects, namely, the ability to do research at sea, the institutional support ashore, and the harvest of new understanding.

In 1953 most of the instruments were mechanical devices like those of *Challenger.* We used wires instead of rope, but the things on the ends of the wire were about the same and it still took more than an hour for them to reach bottom. We had an electronic echo sounder which was about 100 times faster than the sounding line used by *Challenger.* However, it was only about a tenth as reliable, and, as a consequence, bathymetric charts were full of spurious soundings and were steadily getting more fanciful. At that time there was a saying to the effect that the optimum number of vacuum tubes in an oceanographic instrument was one less than one.

Without electronics, oceanographers could not understand the data they were collecting while they were still at sea. Thus, the style of oceanographic exploration was unchanged from a century before. A program of sampling was laid out before the ship left port, scientists or (more likely) technicians collected the data, and scientists ashore analyzed it months afterwards. The recording echo sounder, reliable or not, showed what could be done when data were available instantly on a ship. When a new undersea mountain was discovered, the course of the ship could be modified instantly to make a survey. Progress in bathymetric exploration was correspondingly rapid. Gradually, other programs of research were modified to take advantage of the ability to survey bathymetry instantly—in what is now called "real time." Instead of taking a bottom sample at a predetermined place or time, the sedimentologist would take one on a hill or in a valley. This gave him the opportunity to discover that in some regions the sediment on hills was not like that on the surrounding plains—which turned out to be a very important result.

Physical oceanographers, marine biologists, and the other specialists that we lump together as oceanographers also began to take advantage of electronics and real-time oceanography. Formerly, oceanographers had measured the temperature and salinity of sea water at widely spaced stations by lowering instruments on (it was hoped) vertical wires suspended from (it was hoped) motionless ships. They now began to trail electronic temperature sensors behind the ship, so they could detect the boundaries between ocean currents as they crossed them. Next, they began to zig-zag, and map the boundaries from place to place. Then they used many ships on the same project and mapped the temporal changes in the boundaries in real time.

A similar sequence of events has occurred in most research at sea during the past quarter of a century. In contrast, the ships themselves have not changed very much except that they have become less uncomfortable and they cost ten times as much to operate. However, these are short-term changes. Compared to *Challenger,* the changes have been about the reverse. The great expedition of a century ago was probably the most expensive one ever launched: taking into account the steady decline in the value of money, the cost per day was not very different from the cost for a ship now; however, by measures such as the earning power of a factory worker or the fraction spent of the gross national product, *Challenger* was much more costly. Of course there was only one *Challenger,* and it went to sea only once. Now we have scores of ships, and they sail until they rust through their plates. As for amenity, the gentlemen on *Challenger* used to take advantage of the fact that mud dredged from the sea floor is almost as cold as ice. They would plunge bottles of champagne into the mud to chill them. Now we have refrigerators, but their contents consist increasingly of tasteless convenience foods; certainly, champagne is not a staple.

The changes in oceanography have come about because of the enormous expansion of instrument capabilities and the invention of special ships. Electronic instruments with vacuum tubes gave us the ability to pursue our discoveries as soon as we made them. However, they were crude and liable to break down, especially in the hot tropics. Gradually, we improved them by putting in devices for controlling frequency and by air conditioning the laboratories to control temperature. Ironically, much of the need for air conditioning ceased just as we obtained it. Transistors and other solid-state electronic instruments, which are not so sensitive to the environment, began to replace vacuum tubes. Of course we still had the same scientists, and they didn't mind the air conditioning, since it happened to be available.

With solid-state electronics we could do many kinds of research that were previously impossible, but we still could not analyze some kinds of data in real time, and we could not map small features because there was no way to locate a ship accurately when it was out of sight of land. What is, at this instant, called "modern" oceanography is a development of computer technology and the space age. Ships are now located by a navigation system that depends on the existence of a special group of artificial satellites. The system requires an on-board computer, which is also used to process many kinds of marine data that formerly could be analyzed only after return to port. Minute by minute and hour by hour the ships collect data and feed them into the computer. If you go to the computer output and type "SHIP" it will print out the ship's position, its speed, and such other information as you desire. The computer has such capacity that if no one asks it any questions for some time, it begins to feel neglected and it types out "WOULD ANYONE LIKE TO PLAY TIC TAC TOE?" Of course, it is hard to beat.

Oceanography is also being done not only with ships of the ordinary kind but with space ships. The sensors on artificial satellites are capable of measuring the temperature of the ocean surface, thereby mapping ocean currents almost instantly over enormous areas. Many other uses for space technology are being developed at the present time; they will surely revolutionize some kinds of oceanography and the utilization of the sea.

In the space program we have gained the ability to study some features of the sea by getting far away from it. We have also greatly expanded our abilities in the opposite direction, namely by descending into the sea itself with SCUBA gear and research submarines. These and other aspects of new technology are described in "Technology and the Ocean," by Willard Bascom. In the early 1950s, Jacques Cousteau's invention of the Aqua-Lung opened shallow water to ready access by scientists. Within a few years, marine biologists and geologists made tens of thousands of dives to personally examine at close range the phenomena about which they had previously speculated. Their branches of science were so completely revolutionized that now no student could expect to work in them without using SCUBA gear.

The use of research submersibles in very deep water came about more slowly. These devices, such as the *Bathyscaphe*, were expensive to build and more so to operate. Moreover, they could do little more than go down and up and allow people to view the sea floor through portholes. Much the same results could have been achieved with a television camera. Thus, deep submersibles were used initially more for geographical exploration than for scientific research. Men went to the sea floor just as later they went to the moon. Gradually, the submersibles have changed and become more useful. They can now move at reasonable speeds for some distance, so that one can follow something when it is discovered. Moreover, the technology of remote handling developed for the atomic-energy program has provided strong, sensitive arms for manipulating the environment outside the submersibles. These improvements are opening new lines of research, such as the detailed mapping of the Mid-Atlantic Rift, and many more new uses may be expected to follow.

Other remarkable advances have resulted from the invention of special types of research vessels, some of which cannot easily be called ships. Foremost among these is the *Glomar Challenger*, an actual ship but one with a hole in the middle and a conventional oil-drilling rig above the hole. This *Challenger*, like the first one, is unique: it can drill holes in the deep-sea floor. Another remarkable vessel is *FLIP*, a very long barge that can be towed into deep water and then partially flooded so it stands on end. In this attitude it is almost motionless in normal ocean waves and thus can be used to measure the motions of the sea. Some oceanographers become sick at sea; others are immune to motion sickness, but after a long time at sea they become queasy when they first set foot ashore. *FLIP* is so stable that the latter type of oceanographer has been known to become sick when stepping aboard from a pitching boat.

Let us turn now to the institutions that were developed ashore as oceanography expanded and the use of the ocean increased. All the institutions are interrelated, so it is impossible to say in any meaningful way which logically came first or was most important. The main points I would like to make here are related to the interdependencies. You can't run ships without docks; you can't produce scientists without schools; you can't produce scientists indefinitely without jobs; you can't do anything complex without coordinating committees; and you can't do anything at all without money.

The money came initially (in the early 1950s) from the Office of Naval Research (ONR) because, of all rich organizations, the Navy was the first to realize that it needed to know much more about the sea. The money went to the few oceanographic institutions that existed at the time, of which the principal ones were Scripps Institution of Oceanography and the Woods Hole Oceanographic Institution. With the influx of money, these and other institutions began to expand. They built new laboratories, outfitted new ships, and hired new scientists to do research and teach. As the ferment in oceanography became known and the scholarships grew more abundant, an increasing number of students entered the field.

Within a decade, the National Science Foundation and the Atomic Energy Commission began to funnel money into oceanography, and the expansion continued. New oceanographic institutions were established in most coastal states, and departments of oceanography were formed in major universities throughout the country. At first, the graduates of the older oceanographic institutions staffed the new departments, but they, in turn, produced graduates who went to the even newer departments. This was all in marked contrast to the few original institutions themselves, which were initially staffed by physicists, chemists, geologists, and biologists—because there were no oceanographers.

As the university and research institutions expanded, many oceanographers were needed to staff the federal agencies that were distributing money. In addition, some agencies, such as the Navy, the Bureau of Commercial Fisheries, and the Coast and Geodetic Survey, began to appreciate the need for applied research that would answer questions of direct and immediate importance to their missions. Thus, they too began to hire oceanographers and to establish federal oceanographic laboratories and centers.

Meanwhile, various coordinating organizations came into existence in the hope of making all these activities more efficient. They operated at all levels: within the federal government, within the nation, and internationally. The function of these organizations can be illustrated by the example of ship use. In the initial flush of expansion, when no goal seemed impossible, because none had been, it was assumed that each small new laboratory would grow large and acquire a diversified fleet of ships. At some point this dream collapsed. It was realized that the nation had no need, let alone funds, for a continuous wall of giant oceanographic institutions along the coastline. Thus, the small labs would remain small and would not have fleets. Nonetheless, the scientists at the small labs might need different ships and had as much right to them as the scientists at the big labs. The solution was the formation of an organization to schedule the use of all the ships on a nationwide basis. The Deep Sea Drilling Program posed a similar problem. Only one deep-sea drilling ship could be funded, and yet it would produce enough data to overwhelm the scientists of any single laboratory. As a consequence, the program was initially put under the joint administration of a group of oceanographic laboratories. Now it is funded and administered internationally.

As Parkinson has taught us, governmental and academic bureaucracies tend to grow whether or not there is any continuing need for them. However, it seems likely that the expansion of oceanography could not have continued for so long had it not been so successful. It may have seemed in 1953 that oceanographers were just filling in the details of the great picture sketched by the *Challenger* expedition long ago, but that is not what happened. Those were daggers, not brushes, and the oceanographers have cut the old picture of the world into shreds. In the remaining sections of this reader we shall see what the last quarter-century of oceanography has achieved and how old ideas have given way to new. As we read of the latest miracles of modern science we should always remember that many of them will appear quaint a quarter-century hence.

1 The Voyage of the *Challenger*

by Herbert S. Bailey, Jr.
May 1953

From 1872 to 1876 a doughty little ship sailed the seven seas and gathered an unprecedented amount of information about them, thereby founding the science of oceanography

JUST 77 years ago this month a spar-decked little ship of 2,300 tons sailed into the harbor of Spithead, England. She was home from a voyage of three and a half years and 68,890 miles over the seven seas. Her expedition had been a bold attack upon the unknown in the tradition of the great sea explorations of the 15th and 16th centuries. The unknown she had explored was the sea bottom. When she had left England, the ocean deeps were an almost unfathomed mystery. When she returned, she had sounded the depths of every ocean except the Arctic and laid the foundation for the modern science of oceanography.

The ship was called the *Challenger*. Her name and voyage are already covered with the dust of time, but her story is worth reviving today, when far more handsomely outfitted expeditions are once more exploring the sea deeps. They are filling in details and retouching parts of a picture which in its broad outlines has remained essentially unchanged since that pioneering voyage. It was the *Challenger*, rigged with crude but ingenious sounding equipment, that charted what is still our basic map of the world under the oceans.

Before the *Challenger*, only a few isolated soundings had been taken in the deep seas. Magellan is believed to have made the first. During his voyage around the globe in 1521 he lowered hand lines to a depth of perhaps 200 fathoms (1,200 feet) in the Pacific; failing to reach bottom, he concluded that he was over the deepest part of the ocean. (Actually the water where he took his soundings is 12,000 feet deep, far from the deepest bottom in the Pacific.) After Magellan no deep-sea soundings were taken for about 300 years. In the 19th century a few sea captains and layers of telegraph cables began to plumb deep waters, some of them getting their lines down as deep as two miles or more.

One of the first men to take a scientific interest in the ocean depths was Edward Forbes, professor of natural philosophy at the University of Edinburgh. He did some dredging in the Aegean Sea, studying the distribution of flora and fauna and their relation to depths, temperatures and other factors. Forbes never dredged deeper than about 1,200 feet, and he acquired some curious notions, including a belief that nothing lived in the sea below 1,500 feet. But his pioneering work led the way for the *Challenger* expedition.

H.M.S. Challenger, *as depicted in the official* Challenger *report*

THE MAN WHO organized the expedition was Charles Wyville Thomson, Forbes's successor as professor of natural philosophy at Edinburgh. Thomson first made some summer dredging cruises in ships borrowed from the British Admiralty, and the results were so interesting that they prompted Thomson and the Royal Society to approach the Admiralty with a much more ambitious project. They asked for a vessel that could carry out an investigation of the "conditions of the Deep Sea throughout all the Great Oceanic Basins." The naval authorities, now fully awake to the importance of oceanic research, provided H.M.S. *Challenger*, a corvette fitted with auxiliary steam power in addition to her sails. A naval crew under Captain George S. Nares was assigned to the mission, and Thomson selected a staff of scientists and other civilians to assist him.

They proceeded to adapt or improvise the necessary scientific equipment and to fit out laboratories on the ship. To make room for their gear they removed all but two of the warship's 18 guns. Their equipment included instruments for taking soundings, bottom samples and undersea temperatures; winches and a donkey engine; 144 miles of sounding rope and 12.5 miles of sounding wire; sinkers, nets, dredges, a small library, hundreds of miscellaneous scientific instruments and "spirits of wine" for preserving specimens.

The expedition, coming after Charles Darwin's famous voyage in the *Beagle* and in the midst of the great uproar

Dredging and sounding apparatus on the deck of the Challenger

over his new theory of evolution, naturally attracted public attention. Even *Punch* gave the *Challenger* a send-off:

Her task's to sound Ocean, smooth
humours or rough in,
To examine old Nep's deep-sea
bed. . . .
In a word, all her secrets from Na-
ture to wheedle,
And the great freight of facts
homeward bear.

The *Challenger* sailed from Portsmouth on December 21, 1872. She immediately ran into a howling storm at sea. Thomson found this no evil omen, pointing out that the gale "brought all our weak points to light" and increased confidence in the arrangements. His staff spent the first leg of the voyage, as far as Bermuda, in training and practice on their work: sounding, dredging, trawling and making measurements. Holding the ship steady with her steam engines, the civil and naval crews each time took a standard series of observations: the total depth of water, the temperatures at various depths, the atmospheric and meteorological conditions, the direction and rate of the current on the ocean surface and occasionally of the currents at different depths. They also dredged up samples of the bottom, including its plant and animal life, and dipped up samples of the water and of the sea life at various levels. They found they had to make their soundings with the hemp rope, because the wire tended to kink and break. Attached to the line were sinkers, thermometers and water bottles; when the sinkers hit the bottom they were automatically detached. By the time they had finished their voyage, they had made such observations at 360 stations scattered over the 140 million square miles of the ocean floor.

The routine was long and laborious. In really deep water it took more than an hour and a half to reach bottom—and much longer to haul the line back. "Dredging," wrote one of the naval officers, "was our *bête noire*. The romance of deep-water trawling or dredging in the *Challenger*, when repeated several hundred times, was regarded from two points of view: the one was the naval officer's, who had to stand for 10 or 12 hours at a stretch carrying on the work . . . the other was the naturalist's . . . to whom some new worm, coral, or echinoderm is a joy forever, who retires to a comfortable cabin to describe with enthusiasm this new animal, which we, without much enthusiasm, and with much weariness of spirit, to the rumbling tune of the donkey engine only, had dragged up for him from the bottom of the sea."

THE TRIP was not, however, all drudgery. There was romance and adventure enough to inspire the officers and scientists, almost to a man, to produce memoirs, logs and other accounts for an eagerly waiting public at the end of the voyage. One of the first diversions occurred in the South Atlantic. Putting in at the tiny colony of Tristan Island, the *Challenger's* crew learned from the inhabitants that two brothers named Stoltenhoff, seeking their fortune at seal-hunting, had marooned themselves nearly two years earlier on aptly named Inaccessible Island. The ship diverted its course to rescue the brothers. They had kept themselves alive on a diet of penguins' eggs and wild pigs, but had had no luck catching seals. The ship also stopped at nearby Nightingale Island, a rookery for hundreds of thousands of penguins, and found it so covered with eggs that the shore party could hardly walk without stepping on them. The penguins defended their nests furiously and pecked one of the ship's dogs to death.

Beyond Cape Horn, on Marion Island, they saw multitudes of white albatross, but, heeding the warning of the Ancient Mariner, they killed none. This was a rare exception, for it was their practice to collect specimens of indigenous flora and fauna wherever they touched land.

Exploring in the southernmost Indian Ocean, the *Challenger* became the first steamship to cross the Antarctic Circle. The scientists were tremendously interested in the icebergs and even fired a cannon at one to break off a chunk. In an unsuccessful attempt to find the "Termination Land" reported by the U. S. explorer Charles Wilkes, the *Challenger* ran into a sudden antarctic storm while traversing a pack of icebergs. With the wind at 42 miles per hour and night coming on, the ship took refuge in the lee of a large berg, holding position close beside it with the steam engine. During an unexpected lull in the wind,

The zoological laboratory of the Challenger (above), *and the principle of its deep-sea dredges* (below)

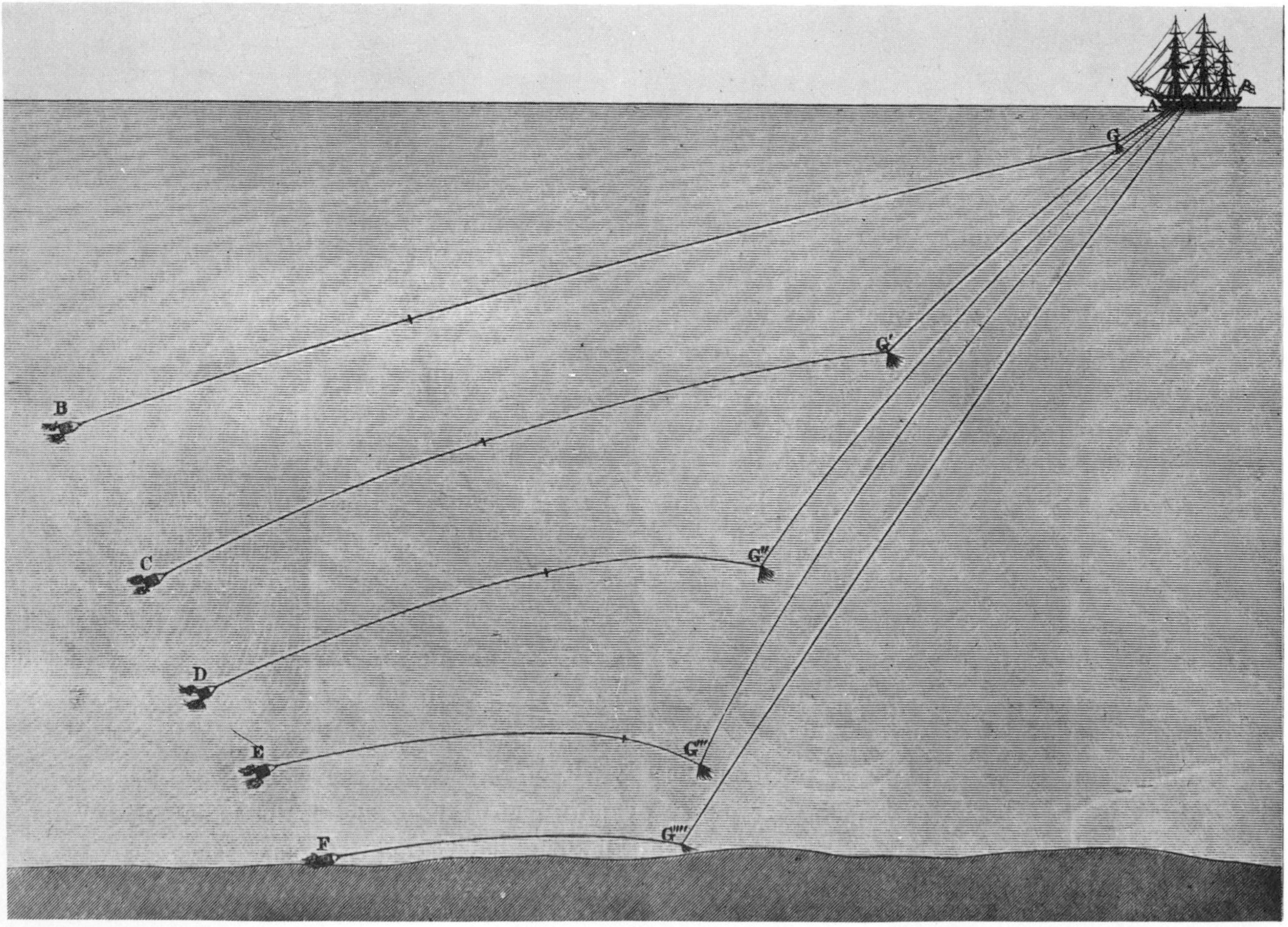

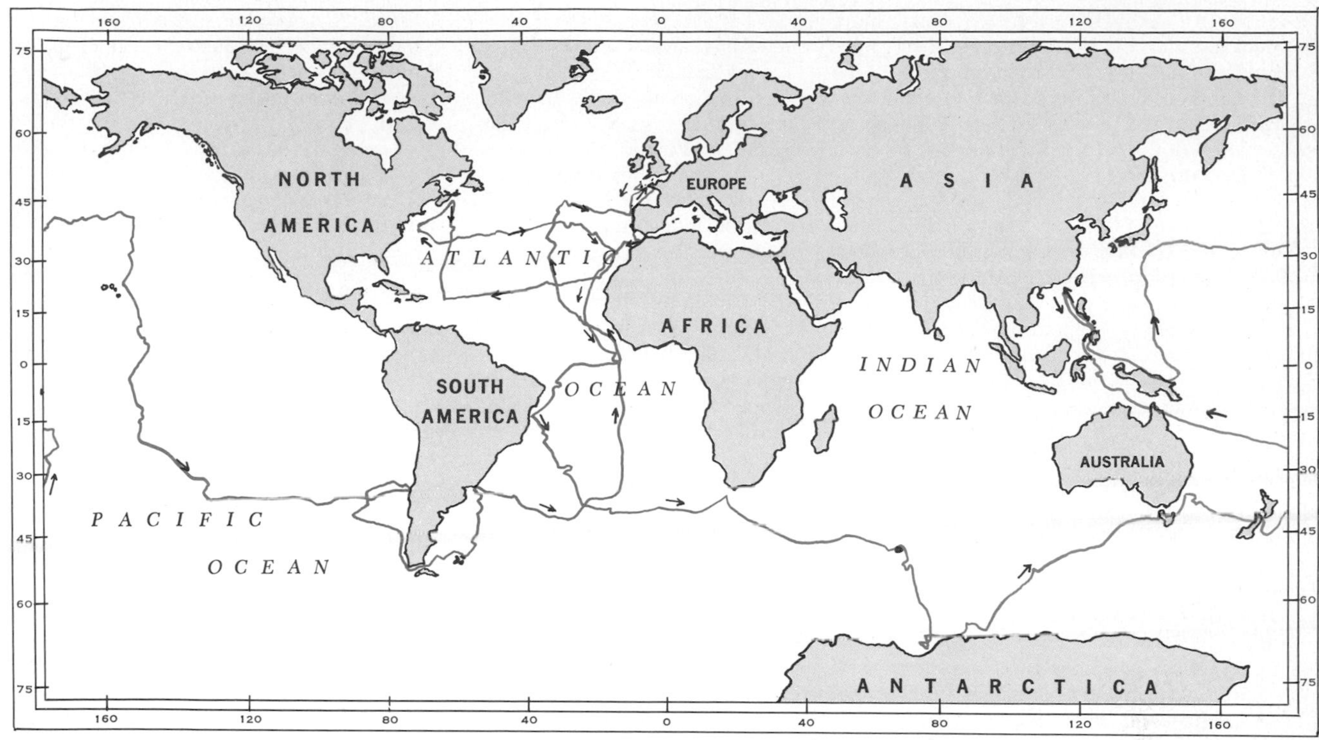

***A general outline of the voyage of the* Challenger**

the ship rammed the berg before the engine could be reversed and lost its jib boom and other rigging. The damage was not serious and the rigging was recovered, but the company spent an anxious night steaming back and forth in a dense snowstorm between two large icebergs.

The ship next went on to Australia, New Zealand and the Pacific islands. In the Fijis they interviewed King Thackombau, a converted Christian, who had earlier cut out a prisoner's tongue and eaten it in his sight—before eating the prisoner himself. The ship called at the Philippines, Japan, China and the Admiralties. On March 23, 1875, off the Marianas Islands, the explorers hit their deepest sounding—26,850 feet. This was not very far from the deepest of all time: the record to date is a sounding of 34,440 feet, made in the Mindanao Trench off the Philippines by a U. S. Navy vessel in 1950.

The *Challenger* zigzagged across the Pacific, stopping at the Hawaiian Islands and Tahiti, and then rounded South America through the Strait of Magellan. It swung north through the South and North Atlantic and finally arrived home in England on May 24, 1876.

Of the *Challenger's* crew of some 240 men, seven died during the trip: two by drowning, one of yellow fever, the others of accidents and miscellaneous causes. Several of the crew jumped ship in Australia. The remainder returned to a joyful welcome—and to the long, hard task of organizing the vast amount of data accumulated on the voyage.

A COMMISSION was set up in Edinburgh to assess the results of the voyage, which were eventually published in an official report of 50 volumes. Two volumes contain a "summary of scientific results"; two a "narrative of the voyage." The other 46 are monographs written by some of the leading scientists of the day, among them T. H. Huxley,

Cladodactyla crocea, ***taken in the Falkland Islands***

Japetella prismatica

Alexander Agassiz, H. N. Moseley and the great German biologist Ernst Haeckel. Most famous of the official reports are Haeckel's monographs on certain sea organisms that had previously been relatively little known. One of the most interesting is on the radiolaria, of which the expedition collected 3,508 new species to add to the 600 then known.

To see the *Challenger's* scientific results in proper perspective one must remember that the voyage took place at a time when every new discovery was an exciting prize to be fitted into the evolutionary table. The *Challenger* discovered 715 new genera and 4,417 species of living things, thus demonstrating that the oceans were teeming with unknown life waiting to be classified. It proved beyond question that life existed at great depths in the sea. The voyage opened the great descriptive era of oceanography, which was followed by the analytic oceanography of our own century.

The summary volumes were written by Sir John Murray, who became head of the commission after Thomson, exhausted by the voyage, died in 1882. Murray's comments and theories have had an important influence on oceanography. He strongly put forward the view, for example, that at equivalent latitudes both the Arctic and the Antarctic have similar marine organisms, and that these are not to be found in the more temperate zones. This "bipolarity" theory has now been discarded, but for a time it stimulated much investigation. Murray also asserted that, contrary to what had been hoped and expected, the deep sea did not yield a widespread fauna of great antiquity, though some very ancient species were found. He believed that under about 600 feet below sea level the bottom deposits and fauna become more uniform with increasing depth until a point is reached at which conditions are almost the same in all parts of the world. He added, "When once animals have accommodated themselves to deep-sea conditions there are few barriers to further vertical or horizontal migration." Such suggestions, based on the *Challenger's* observations, gave direction to further investigation.

The expedition washed out of existence a form of living matter that had been "observed" by Huxley and described by Haeckel. On the basis of preserved specimens dredged during earlier expeditions, Haeckel had decided that the entire ocean floor, or at least a major part of it, was covered with a thin layer of almost structureless living slime which he named "Bathybius." At a time when Darwin's theories were still under severe debate, "Bathybius" had been hailed as a living example of the primordial protoplasm. The scientists of the *Challenger* looked for it in vain, and finally discovered the answer to the puzzle. The alcohol and sea water in which the sea-bottom specimens were preserved had combined to form an amorphous precipitate of sulfate of lime. This was Haeckel's "Bathybius."

THE *Challenger* expedition made thousands of other contributions to various sciences—meteorology, hydrography, the physics and chemistry of sea water, geology, petrology, botany, zoology, geography. Murray's map of the world-wide sampling of oozes and other bottom deposits collected by the expedition has not been changed much by the many subsequent explorations. The *Challenger's* crew perfected the method of "swinging the compass" to get accurate magnetic readings. The voyage established the main contour lines of the ocean basins and disproved the myth of the lost continent Atlantis. It yielded the first systematic plot of currents and temperatures in the oceans, and showed that the temperature in each zone was fairly constant in all seasons.

The achievement of the *Challenger* was tremendous: a barrier had been broken and the world of the depths explored. In a sense the *Challenger* had answered the question that had echoed down the ages in the words of Ecclesiastes: "That which is far off and exceeding deep, who can find it out?"

Technology and the Ocean

2

by Willard Bascom
September 1969

The materials, machines and techniques that can be employed in the ocean have advanced greatly during the past decade. Major developments include superships and deep-sea drilling

Without technology, meaning knowledge fortified by machinery and tools, men would be ineffective against the sea. During the past decade the technology that can be brought to bear in the oceans has improved enormously and in many ways. The improvements have not only increased knowledge of the oceans but also speeded the flow of commerce while decreasing its cost, brought new mineral provinces within reach and made food from the sea more readily available.

With today's technology it is possible, given a sufficient investment of time and money, to design and build marine hardware that can do almost anything. The problem is to decide whether it is sensible to make a given investment. Industry decides on the basis of whether a proposed technological step will solve a specific problem and improve the firm's competitive position. Government has more latitude: it does not need to show a prompt return on investment, and it can better afford the high risk of developing expensive and exotic devices for which there may be no immediate or clearly defined need. The gains in ocean technology have resulted from the largely independent efforts of both industry and government.

This article will deal broadly with the progress in ocean technology over the past decade, concentrating on developments that seem to be the most important at present. I shall begin by making my own selection of the five most important advances. The main criterion in this selection is that each advance represents an order-of-magnitude improvement: in one way or another it is a tenfold step forward since the beginning of the decade. I have also given weight to the social and economic significance of these developments and to the degree of engineering imagination and perseverance that each one required.

The first development is the supership. Not long ago a "supertanker" carried 35,000 deadweight tons. Now a fleet of ships with nearly 10 times that capacity is coming into being. For these vessels the Panama Canal and the Suez Canal are obsolete. By the same token the ships are making large new demands on the technology that provides the terminal facilities.

Second is the deep-diving submarine. Man can now go to the deep-ocean bottom in an "underwater balloon" submersible such as the *Trieste*, which reached a depth of 36,000 feet in the Mariana Trench 200 miles southwest of Guam. Somewhat more conveniently he can go to a depth of about 6,000 feet in any of several small submarines. This rapidly developing technology still has a long way to go, but it has certainly improved by an order of magnitude in the past decade. Several techniques have been employed to solve the problem of how to make a submarine hull that is strong enough to resist great pressure and still light enough to return to the surface.

The third development is the ability to drill in deep water. This category includes both the drilling that is done in very deep water for scientific purposes and the use of full-scale drilling equipment on a floating platform to obtain oil from the continental shelf. The first deep-ocean drilling, which was carried out eight years ago by the National Academy of Sciences in water 12,000 feet deep near Guadalupe Island off the west coast of Mexico, improved on four previous records by an order of magnitude: the ship held its position at sea for a month without anchors, drilled in water 20 times deeper than that at earlier marine drilling sites, penetrated 600 feet of the deep-sea floor and lifted weights of 40 tons from the bottom. These records have since been improved on even more. In fact, virtually all floating drilling equipment, including semisubmersible platforms and self-propelled vessels, has been designed and built in the past decade.

Fourth is the ability to navigate precisely. A ship in mid-ocean has rarely known its position within a mile; indeed, five miles is probably closer to the truth, notwithstanding assertions to the contrary. Now a ship 1,000 miles from land can fix its position within .1 mile. If the vessel is within 500 miles of land, the position can be ascertained within .01 mile. The position of a ship within 10 miles of land can be fixed to an accuracy of 10 feet. The techniques for these determinations include orbiting satellites, inertial guidance systems and a number of electronic devices that compare phases of radio waves.

Finally I would cite the ability to examine the ocean bottom in detail from the surface by means of television and side-looking sonar. These techniques, together with their recording devices and the capacity for precise navigation, have made it possible to inspect the sea floor much as land areas have been examined by aerial photography. New television tubes that amplify light by a factor of 30,000 make it possible to eliminate artificial lighting, thereby eliminating also the backscatter of light by small particles in the water.

The supership and the improvement in drilling are mainly industrial developments. The evolution of navigation technology has resulted largely from government efforts. Both industry and government have figured prominently

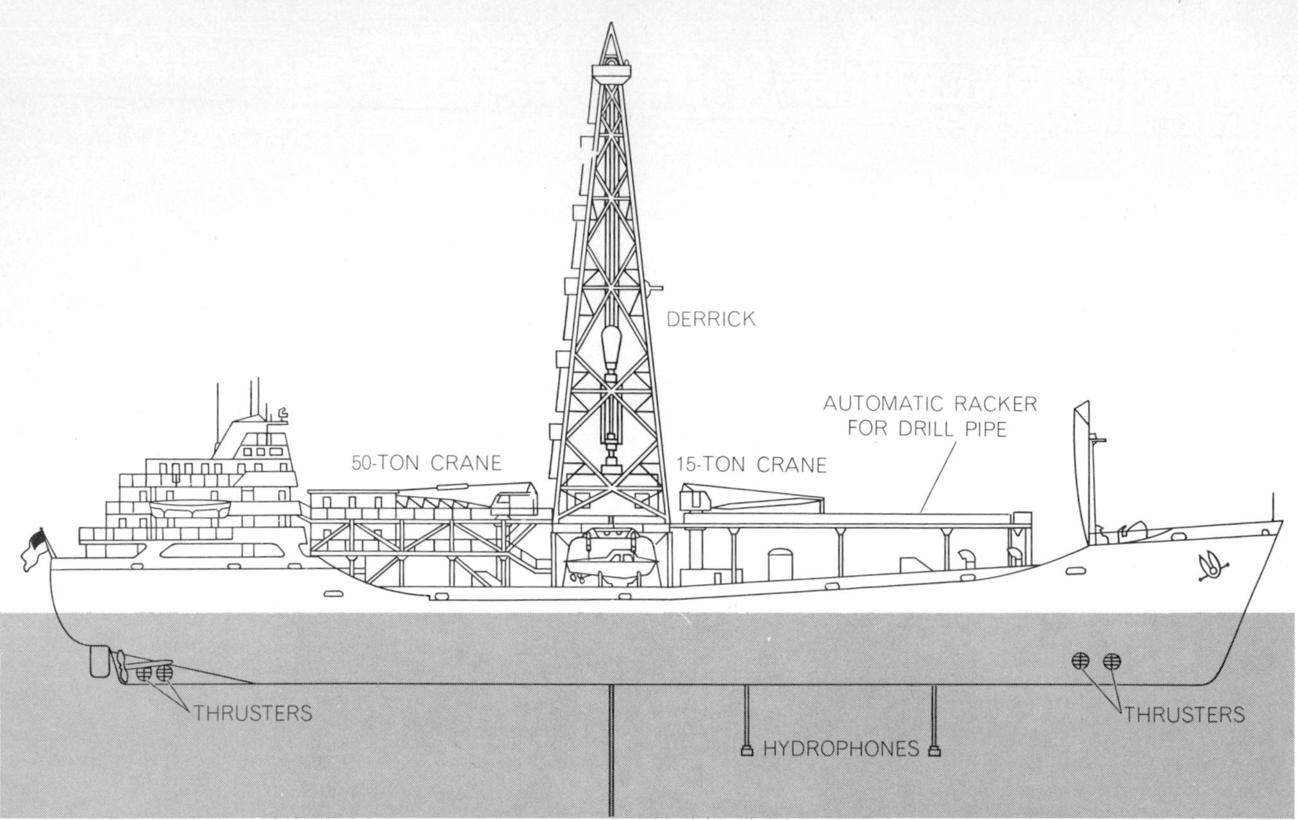

GLOMAR CHALLENGER has a 142-foot derrick as her most conspicuous feature. Her automatic pipe racker can hold 23,000 feet of drill pipe. Positioning equipment includes two tunnel thrusters at the bow and two near the stern to provide for sidewise maneuvers. When the ship is on station (*above*), four hydrophones are extended under the hull to receive signals from a sonar beacon on the ocean floor. The signals are fed into a computer that controls the thrusters to maintain the ship's position over the drill hole. At the sea bottom (*below*), as much as four miles under the ship, the drill penetrates as much as 2,500 feet of sediment and basement rock.

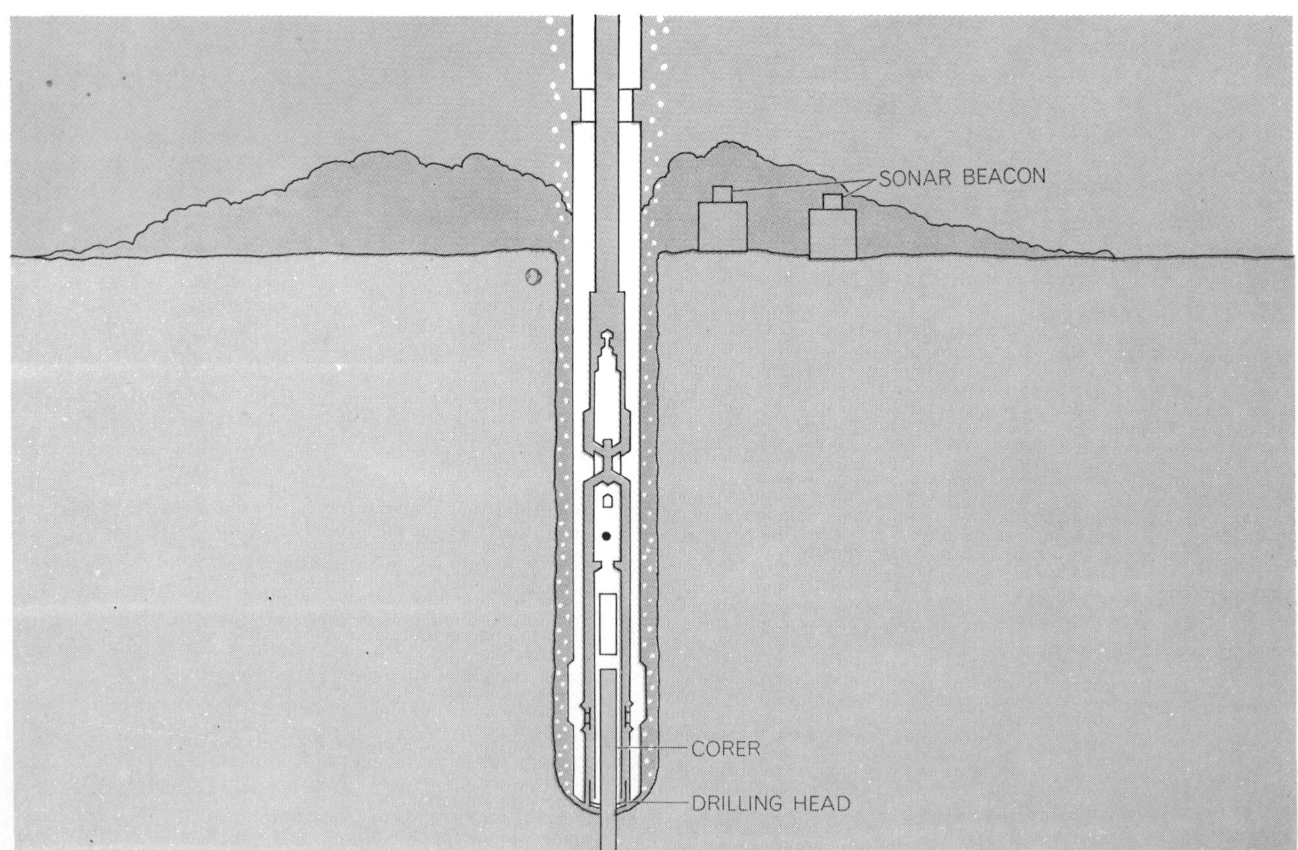

in the development of deep-diving submarines and techniques for examining the bottom with television and sonar.

In considering the application of these and other techniques one might classify them according to who uses them. For example, scientific investigators use research ships and submarines, instruments, buoys, samplers and computers. Industry constantly seeks better methods for mining, fishing, salvage and the production of oil. Waterborne commerce needs better ships, better cargo-handling methods and better port facilities. Exploration becomes more efficient as improved navigational systems, vehicles, geophysical tools and communications equipment become available. Adventure and recreation offer new toys such as air-cushion vehicles and scuba equipment.

The entire area of military technology, which is the most sophisticated of all, must be outside the scope of this article. The best of modern seaborne military technology is done in secrecy, with budgets far in excess of those spent for any of the other areas. Thus we shall not go into such matters as the duel between the submarine builder, who endlessly tries to make submarines go deeper, faster and quieter, and the antisubmarine expert, who tries to detect, identify and destroy the steadily improving submarines.

In any case, the classification of marine technology according to users is somewhat impractical because there is so much overlap. For example, certain kinds of diving and television equipment might be used by all the groups. Therefore I shall discuss marine technology in terms of materials, vehicles, instruments and systems.

What characteristics should a marine material have? It should be light, strong, easy to form and connect, rigid or flexible as desired and inexpensive. The difficulty ocean design engineers have in finding a material that meets most of the requirements for a given task has led them to speak whimsically of an ideal material called "nonobtainium." The problem is that characteristics such as lightness and strength are relative. Nonetheless, engineers and manufacturers recall that not many years ago fiber glass, Dacron and titanium were not obtainable, and so they are optimistic about the development of materials that come ever closer to the qualities of nonobtainium.

In the past decade the steel available for marine purposes has improved substantially under the spur of demands for submarines that can withstand the pressure of great depth, drill pipe (unsupported by the hole wall that pipe in a land well has) that must survive high bending stresses, and great lengths of oceanographic cable that must not twist. For example, a new kind of maraging steel with a high nickel content is tougher, more resistant to notching and less subject to corrosion fatigue than the steel formerly available. The minimum yield strength of conveniently available steel shell plate has risen from 80,000 pounds per square inch to 130,000 pounds per square inch and more for the shells of deep-diving submarines. Steel in wire form now attains a strength of 350,000 pounds per square inch. Steel is becoming more uniform and reliable as the processes of mixing and rolling are subjected to better quality control. Indeed, some metallurgists believe nearly any metal requirement can be met by properly alloyed steel.

Also available for marine purposes are new, high-strength aluminum alloys, such as 5456 (a designation indicating the mix of metals in the alloy), that have a strength of more than 30,000 pounds per square inch after welding and are resistant to corrosion. They can also be cut with ordinary power saws instead of torches and welded by a technique that is easily taught. With this material small boats, ships up to 2,000 tons and superstructures for much larger ships can be built, as can a number of other structures where lightness and flexibility are important.

Titanium is becoming more readily available. When special properties of lightness, strength (as high as 120,000 pounds per square inch) and good resistance to corrosion are required, its relatively high cost becomes acceptable.

Glass, fiber glass and plastics are the glamorous materials of oceanography. They are virtually free of the problems of corrosion and electrolysis that have afflicted most materials in a marine environment, and they are easily formed

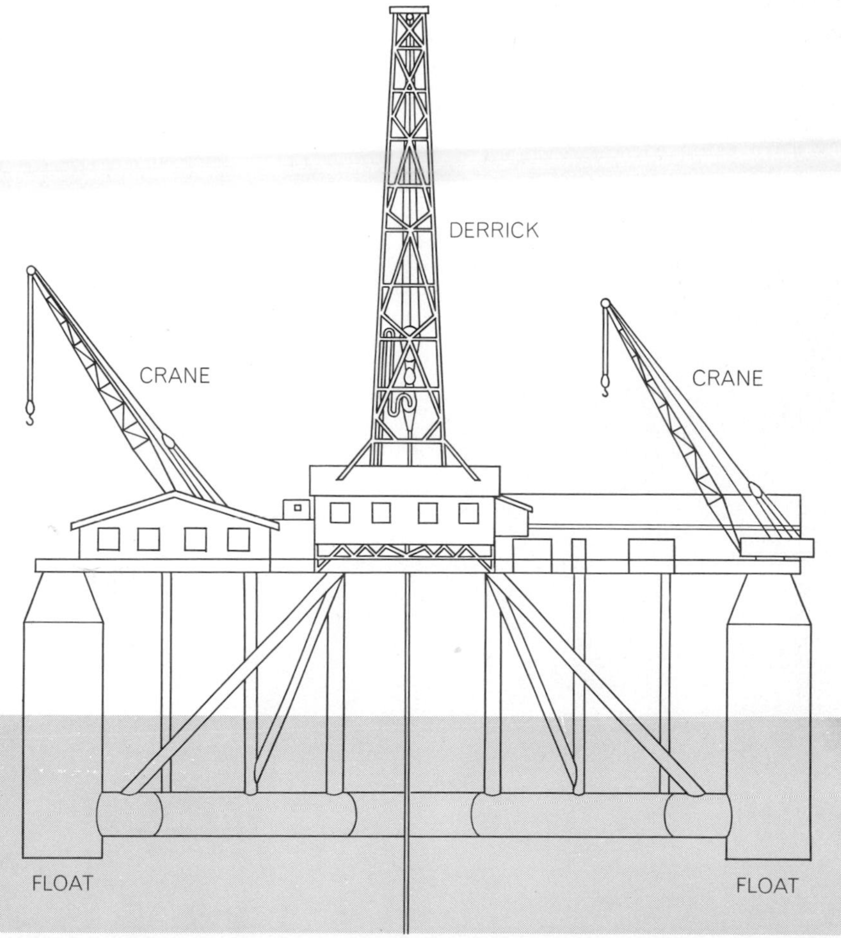

SEMISUBMERSIBLE PLATFORM, *Blue Water 3*, is now drilling for oil off Trinidad. When the platform has been towed to its position, water is drawn into the four corner cylinders to make the structure submerge enough so that wave motions have little effect on it. The platform, which is 220 feet by 198 feet, was designed for work in the open ocean.

into complex shapes. Constant research is improving the strength and versatility of these materials.

Glass is less fragile than most people think and has excellent properties in compression. It is finding increasing favor among the designers of small submarines, who want a glass-bubble pressure hull that is also a superwindow. Glass microspheres, which do the same thing as a submarine hull but on a microscopic scale, are packaged in blocks of epoxy and used to furnish incompressible flotation at depths of as much as 20,000 feet. The best such material to withstand the pressure at that depth so far weighs about 40 pounds per cubic foot; in seawater at 64 pounds per cubic foot the material therefore has a net buoyancy of 24 pounds per cubic foot.

The many remarkable characteristics

SUPERTANKER *UNIVERSE IRELAND* is seen at her loading berth in the Persian Gulf. The vessel, which is 1,135 feet in length, carries 312,000 deadweight tons of oil from the Persian Gulf to Ireland, going around Africa at an average speed of 15 knots.

of fiber glass are widely known. Its outstanding virtue as a marine material is that it enables precisely shaped hulls with complex lines to be reproduced easily. The result has been a revolution in the construction of small craft over the past decade. Fiber-glass hulls, which are light and strong without a rib structure, are a major contribution to marine technology and a boon to the small-boat owners, whose maintenance problems are reduced accordingly.

Among the plastics polyvinyl chloride has found use in marine pipelines subject to severe internal corrosion. It is light, inexpensive and easily joined. Nylon and polypropylene for cordage and fishing nets and Dacron for sails are appreciated by all sailors because the materials are light and elastic and do not rot. Ship-bottom paints, designed to reduce fouling by marine organisms, have been greatly improved. Inorganic zinc underpaints promise to decrease substantially the pitting of hulls and decks, which should increase the life of ships and the time between dry-dockings.

A remarkable collection of marine vehicles and equipment has made its appearance in the past 10 years. Ships now exist that can go up, down and sideways and can flip. They skim, fly and dive. Some of them are amphibious and some go through ice, over ice or under it.

This versatility is important; a ship cannot be efficient unless it has been designed to do exactly what the user wants. Widely varying requirements mean very different sea vehicles. A distinct place on the spectrum is occupied by the superships I mentioned earlier. They are bigger than anyone dreamed of only a few years ago; in fact, they are almost the largest man-made structures. The largest vessel now in service is the *Universe Ireland*, a tanker with a capacity of 312,000 deadweight tons. The ship is 1,135 feet long and delivers 37,400 shaft horsepower. It plies between the Persian Gulf and Ireland, going around Africa at 15 knots and pushing a 12-foot breaking wave ahead of it. Even the enormous vessels of the *Universe Ireland* class will soon be surpassed in size by ships being built in Japan and West Germany. They will be so large they will not be able to dry-dock in any yard except where they were built, and they will not be able to enter any ordinary harbor because they will draw up to 80 feet of water.

A variation in the tanker field is the conversion of the comparatively small (114,000 deadweight-ton capacity) *Man-*

***ALUMINAUT*, a mobile submersible capable of carrying two crew and four passengers to depths of 15,000 feet and of probing the bottom or moving heavy objects with manipulator arms attached to the hull, is photographed during a dive. The craft is 51 feet in length.**

MANIPULATOR ARM of *Aluminaut* explores bottom off Bimini at a depth of nearly 1,800 feet. Numbered sample boxes are nearby. Thin layer of sand is rippled by a current moving it over a rock base; the dark areas are debris that are caught in filamentous organisms attached to rock. Photograph was made by A. Conrad Neumann of the University of Miami.

hattan to a supericebreaker. The purpose is to move the petroleum from the large new oil fields on the northern slope of Alaska to more moderate climates. A fleet of such ships may be able to keep open a northwest passage from the U.S. East Coast to Alaska. From ships of the *Manhattan* class it is only a small step conceptually to a ship five times larger that could cross the Arctic Ocean at will, treating the ice, which averages eight feet in thickness, as an annoying scum.

The ships that go up include both the ground-effect machine, which can rise a few feet above the surface of the sea on a cushion of air, and the hydrofoil, which has a hull that flies above the surface at high speed with the support of small, precisely shaped underwater foils. The newest versions of these "flying boats" represent substantial technical achievements, and yet neither vehicle seems likely to become a very important factor in marine affairs because each has basic problems, such as the danger to the hydrofoil of hitting heavy flotsam and the inability of the ground-effect machine to carry large loads or to operate in high waves. The ground-effect machine does have a potential, not much exploit-

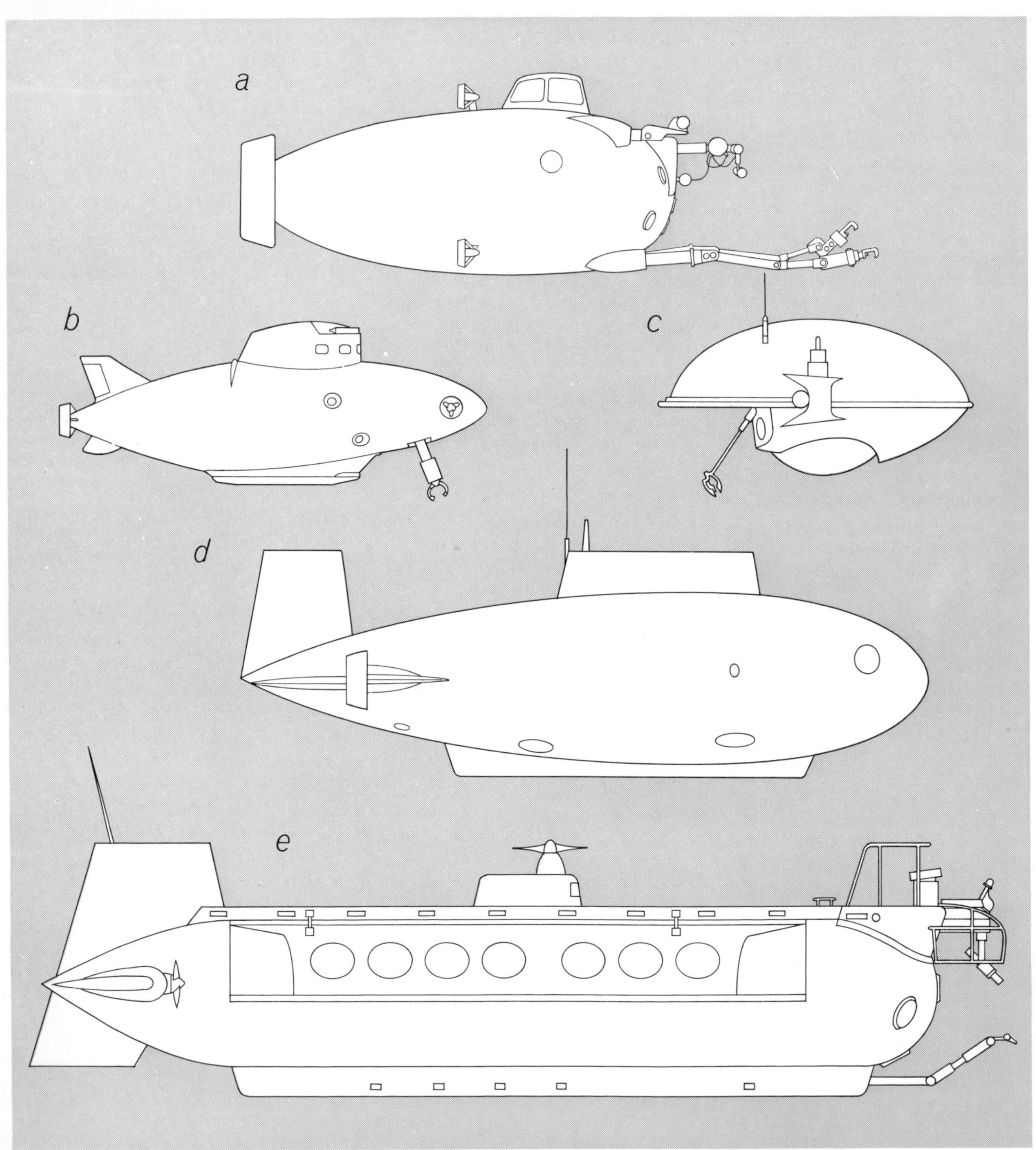

MANNED RESEARCH SUBMARINES are designed for deep diving. They include (*a*) *Beaver IV*, which can dive to 2,000 feet; (*b*) *Star III*, to 2,000 feet; (*c*) *Deepstar IV*, to 4,000 feet; (*d*) *Deep Quest*, to 8,000 feet, and (*e*) *Aluminaut*, which is designed to go to 15,000 feet with a staff of six. Vessels are drawn to scale. *Aluminaut* is made primarily of aluminum; the others are steel craft.

ed as yet, stemming from its ability to run up a beach and cross mud flats, ice and smooth land surfaces.

The ships that go down are of course submarines. Nuclear power in military submarines dates back further than the decade under discussion, but large advances in nuclear propulsion have been made during the decade. The circumnavigation of the earth without surfacing and the trip under ice to the North Pole were both made possible by nuclear power and highly developed life-support systems for keeping the crews alive and well on the long missions.

Quite a number of small, deep-diving submarines are in existence. Of them the *Aluminaut,* designed to go to 15,000 feet while carrying six people, has accomplished the most. (Because of the problem of obtaining life insurance for its crew its deepest dive has been about 6,000 feet.) Among the many other small submarines are the ones of the *Alvin* class, which can dive to 6,000 feet; *Deep Quest,* to 8,000 feet; *Deepstar IV,* to 4,000 feet; *Beaver IV* and *Star III,* to 2,000 feet, and *Deep Diver,* to 1,000 feet. There is therefore a considerable choice of vehicles, instruments and sup-

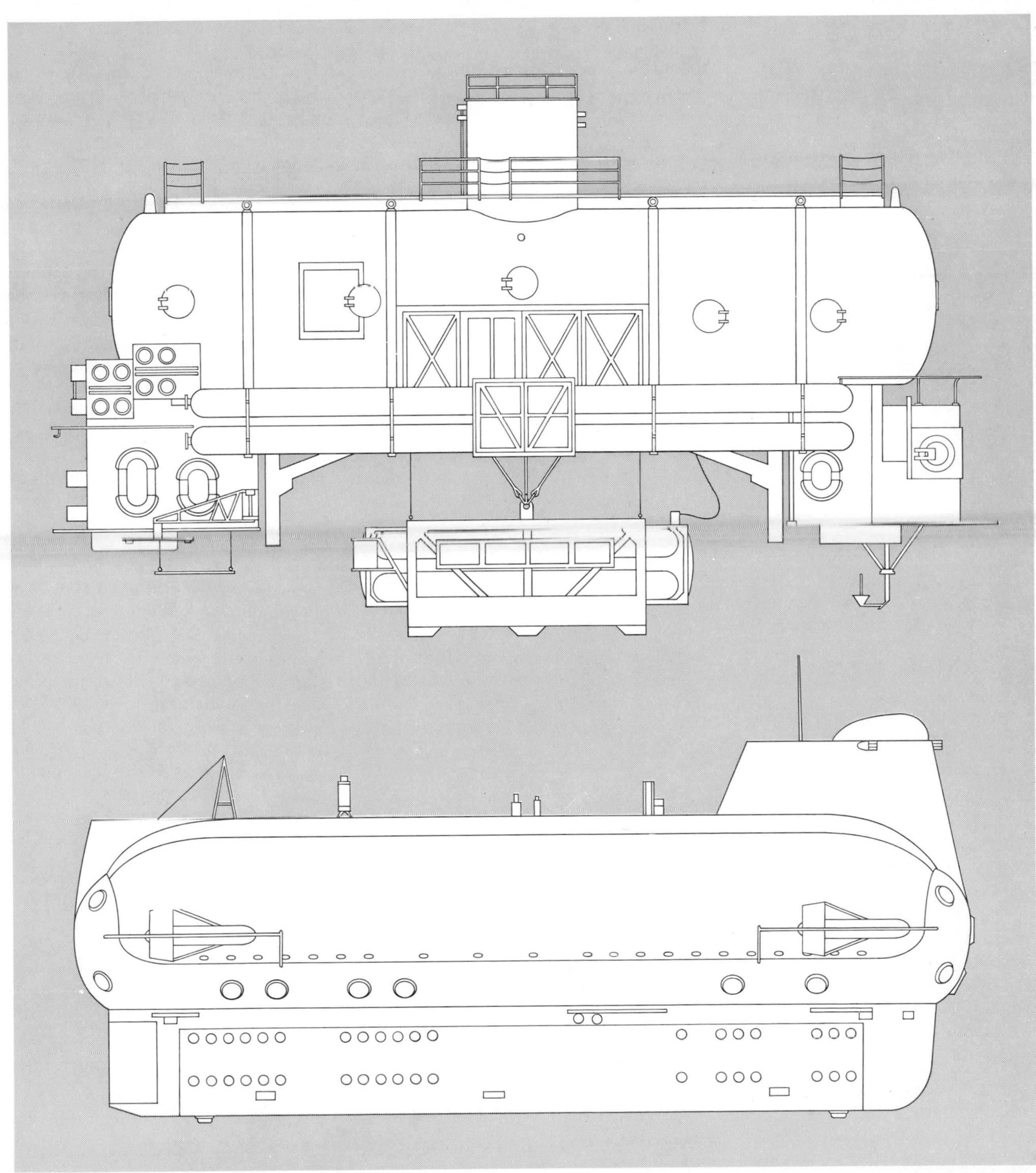

UNDERWATER LABORATORIES include the Navy's *Sealab III* (*top*) and the Grumman-Piccard submersible *Ben Franklin* (*bottom*). *Sealab* is designed to operate on the sea bottom, where it will provide living quarters for divers who will venture forth periodically in heated diving suits to explore the bottom. The first mission of *Ben Franklin* was a submerged drift up the Gulf Stream.

porting facilities. The problem is that there are few customers with the inclination to employ these vehicles at $1,000 or more per hour of diving.

The ships that move sideways are those with trainable propellers or vertical-axis propellers or tunnel thrusters. (A tunnel thruster enables a pilot to move the ship's bow sideways.) Vessels so equipped have found use in self-docking situations and in such waterways as the St. Lawrence Seaway. Dynamic positioning, which means holding position without anchors, is possible with ships that have precise local-navigation systems and central control of several maneuvering propellers.

The ship that flips is *FLIP*, operated by the Scripps Institution of Oceanography. It has two positions of stability. While it is under tow it lies on the surface and looks like a barge made from a big piece of pipe. On station it ballasts itself so as to float on end, much like a big, habitable buoy. In this position *FLIP*, because of its size, is detuned from the motion of the sea surface: it does not move vertically under the influence of ordinary waves and swell. As a result it is an excellent platform for making underwater sound measurements.

Among instruments and tools the now venerable sonar, the sound-ranging device, still figures prominently. It has been improved substantially. Frequencies have risen steadily, making it possible to narrow the beam width to searchlight dimensions, with the result that the distance to (or depth of) discrete areas of the ocean bottom can be measured more accurately. Sonars employing the Doppler effect, which is the change of pitch of a sound resulting from relative motion between the source and the observer, make possible the direct measurement of a ship's speed over the bottom—a measurement that is essential to the high-quality navigation required for such purposes as determining gravity at sea by means of a shipboard gravity meter. Frequency-scanning sonars are now available that better match the signal with the reflector.

Hydrophone arrays, sometimes a mile long, make it possible to use the low-frequency sound created by a series of gas explosions to examine the rocks under the sea bottom in great detail. The result is a continuous picture of a vertical geologic section. Such continuous-reflection seismic profiles have revealed folds and faults in sub-bottom rocks to depths of as much as 15,000 feet and have found many new undersea oil deposits.

Satellites are valuable ocean instruments. They are the essential elements in the system that makes it possible to determine a ship's position accurately wherever it may be. Other satellites transmit photographs of weather patterns, cloud cover and the state of the sea. By combining the picture with other weather data, meteorologists can produce accurate charts with reliable and up-to-date information. The information is useful in routing ships and forecasting waves.

Buoys moored in the deep ocean hold instruments for measuring, recording and transmitting sea and weather conditions. A number of buoys are already in use, producing an abundance of hitherto unavailable information at minimal expense. It seems likely that hundreds of additional buoys will be put to work in the next few years.

Shipboard computers are becoming an accepted convenience. Such a computer plots the ship's position continuously and matches it with accumulating data of other kinds so that investigators aboard the ship have an information system describing the pulse of the sea below them.

Occasionally a single device can revolutionize an industry. Such a device is the Puretic power block, which handles fishing nets. It is, like many good inventions, basically simple: it is a wide-mouthed, rubber-lined pulley driven by a small hydraulic motor. During the past decade it has been adopted by many fishing fleets and now accounts for some 40 percent of the world's catch. With the block it is possible to handle much larger nets with fewer men. One result is that the tuna industry has shifted almost entirely from line fishing to net fishing.

Another trend in fishing has been to put fish-processing equipment on boats. The equipment includes automatic filleting machines, quick-freeze boxes and even packaging machines so that a finished frozen product can be delivered at dockside. Scallops, for example, can now be shucked and eviscerated on shipboard, so that the scalloper can remain at sea for a week at a time and return with a cargo of ready-to-eat scallops.

Barge-mounted cranes capable of lift-

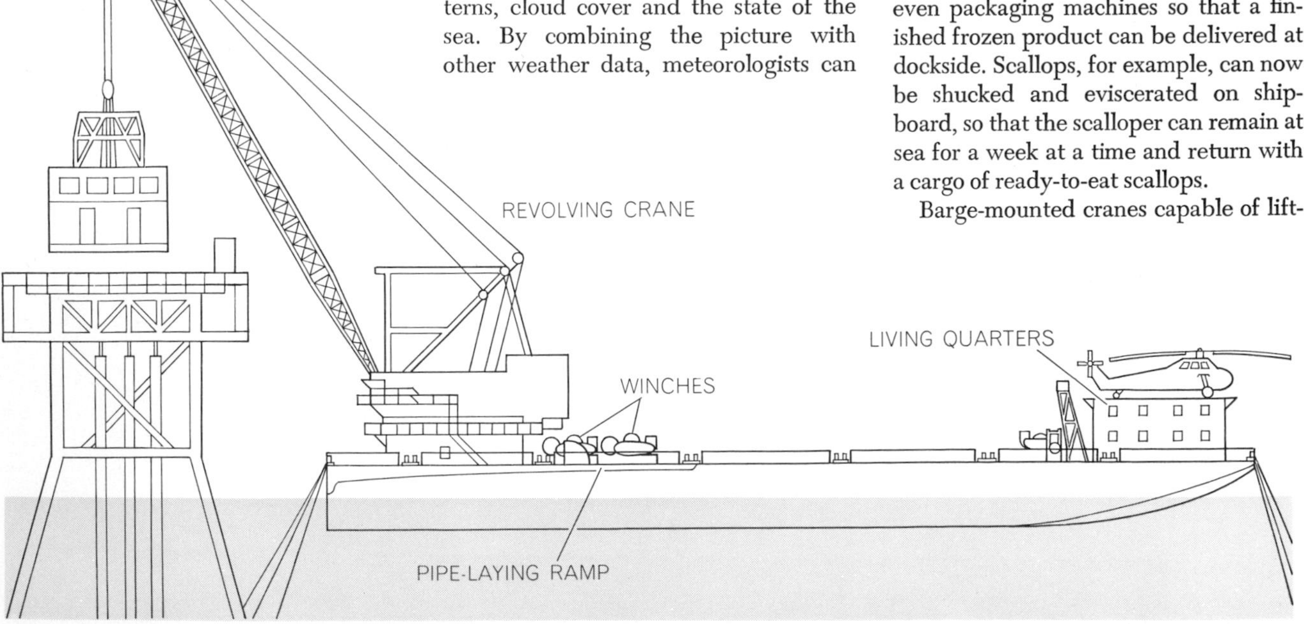

CONSTRUCTION BARGE, the *William Denny*, is about to be put in ocean operation by Raymond International Inc. It is 350 feet long and 25 feet deep and has a 100-foot beam. Its revolving crane, which has a 250-foot boom, can lift 500 tons at a 70-foot radius and 100 tons at a 215-foot radius. The craft can lift 750 tons over the stern. It can build structures, drive piles and lay pipelines.

ing 600-ton loads as much as 200 feet above the water are now available along many coasts. The result is a change in construction techniques. For example, a bridge can be built in large sections, which are then hoisted into place by the crane.

Shipyards are using elevators called Syncrolifts to lift and launch ships weighing as much as 6,000 tons. The machines are replacing dry docks and marine railways. The Syncrolift is simply a big platform that can be lowered below keel depth. A ship is then floated in, and a dozen or more synchronized winches hoist it up to the level of a transfer railway, which moves it to a position in the yard. In this way the yard can work on several ships simultaneously.

The first undersea dredge has just made its appearance. The machine moves along the bottom on crawler tracks in depths of as much as 200 feet. Hence it is not affected by the wave action that makes life on floating dredges hard. The machine was designed to replace eroding beaches with sand from offshore: the dredge has a 700-horsepower pump that moves the sand slurry a mile to shore.

It is fashionable now to speak of the "systems approach," which is a way of expressing the obvious idea that all the elements in the solution of a problem should fit together and be headed toward the same goal. All the ships and instru-

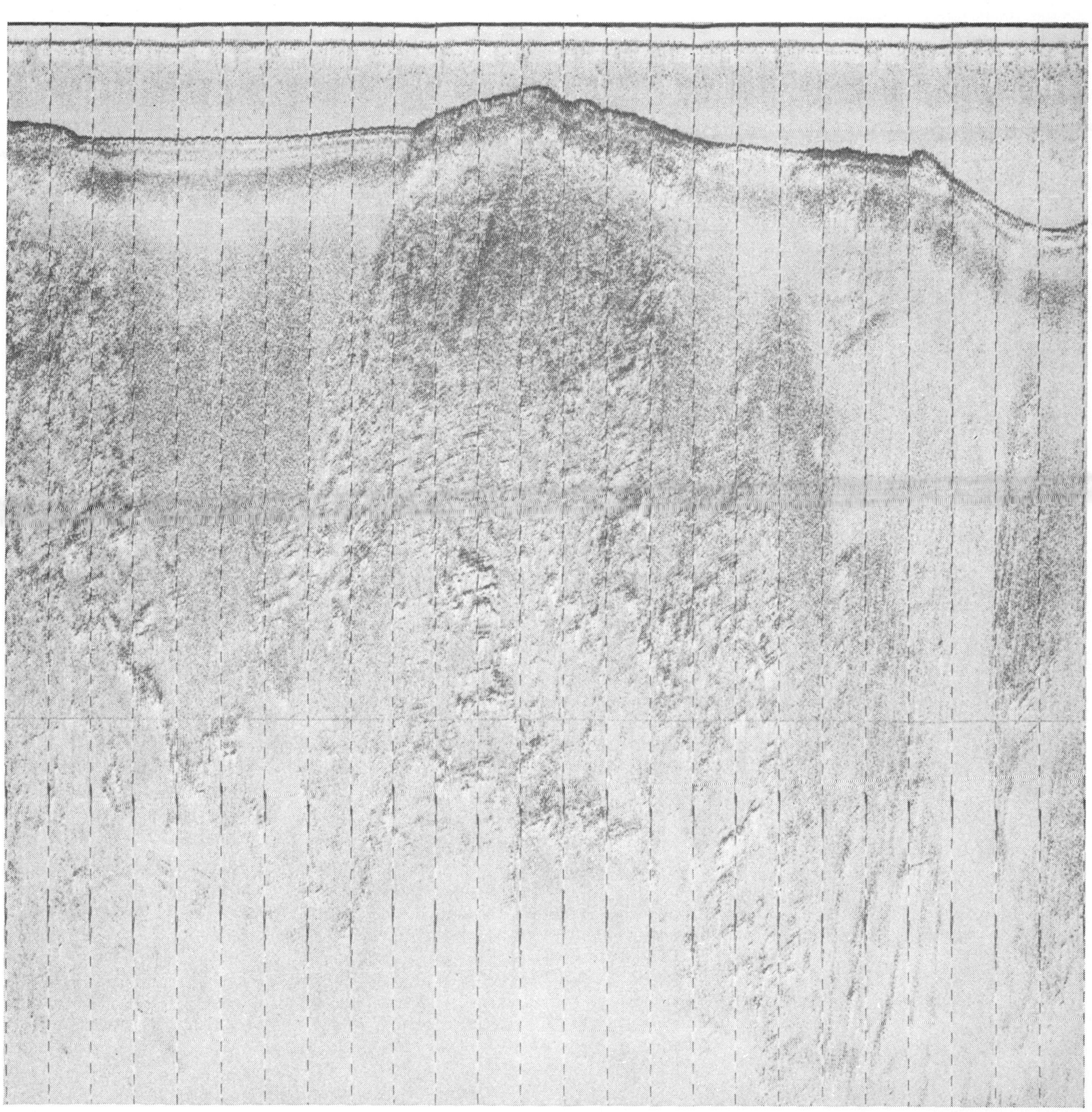

SIDE-LOOKING SONAR produced this view of the ocean bottom on the continental shelf northeast of Boston. From right to left the record covers a distance of about two kilometers along the ship's track. Broken lines show one-minute intervals. Irregular line near top is a profile view of the sea bottom as it appeared at the instrument's horizon. Irregular portions of the photograph are bedrock; darker flat areas are sand waves and gravel; lighter flat areas are smooth sand. Bottom was 60 to 140 meters below the ship. Record was made by John E. Sanders of Barnard College and K. O. Emery and Elazar Uchupi of Woods Hole Oceanographic Institution.

ments I have described are employed as parts of systems. There are, however, several integrated combinations of technology that are best described under the heading of systems. They include containerized cargo-handling, desalination of seawater, deep-ocean drilling and deep diving.

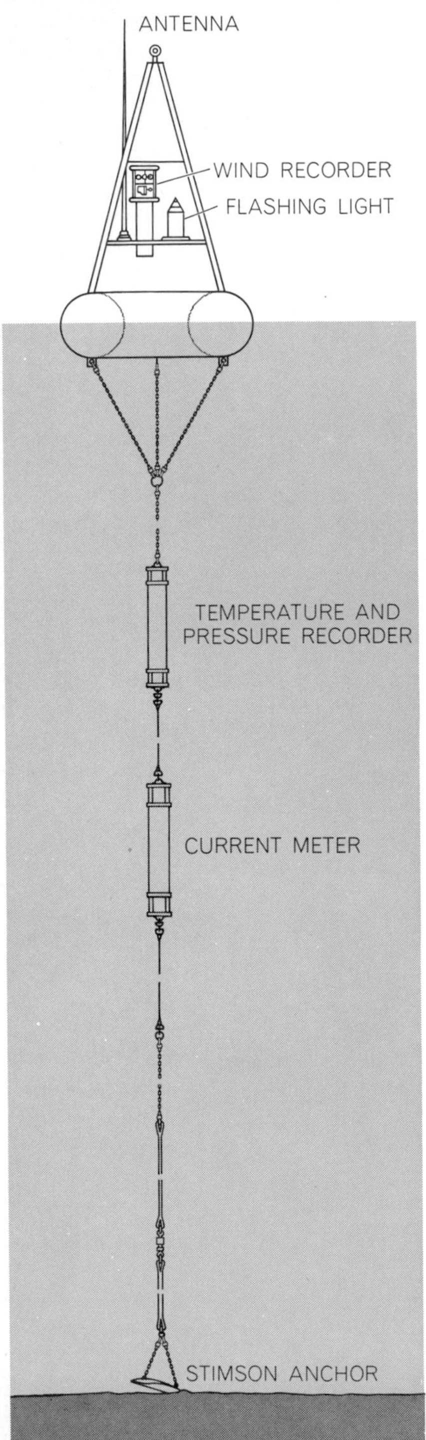

BIG RESEARCH BUOY employed by the Woods Hole Oceanographic Institution gathers and transmits data from the surface to the sea floor. At the surface it records wind speed and direction; below the surface it measures temperature, pressure and current.

Containerization has become a magic word on the waterfront. The basic idea is that a shipper can move his goods from his inland manufacturing point to an inland customer overseas in a private container. A container moves by train or truck to a marshaling yard on the waterfront. There it is picked up by a straddle truck and moved within reach of a gantry crane, which sets it in slots on a container ship. The contents are safe from pilferage, weather and damage. A harbor facility, dealing with containers of standardized size, can semiautomatically unload and reload a large cargo ship in less than 24 hours. Labor cost is lowered; the ship spends more time at sea and less alongside a dock, and the freight moves faster and more cheaply than on a break-bulk cargo ship [see "Cargo-handling," by Roger H. Gilman; SCIENTIFIC AMERICAN, October, 1968].

Methods for desalting water have been improved substantially. The worldwide use of desalted water from the sea is now almost 100 million gallons per day. Most of the water is obtained by various distillation processes; the average cost is estimated to be about 75 cents per 1,000 gallons. Other means of desalination, such as vacuum freezing and reverse osmosis, are being developed. Major nuclear plants that would produce both fresh water and electricity are under study. Most of the desalted water now obtained or in prospect is for household and industrial purposes. The day of cheap irrigation water in large quantities is still far away.

Offshore drilling from floating platforms is less than a decade old and has evolved rapidly. The self-propelled drilling ship and the semisubmersible platform, both of which drill while anchored, represent the two ends of the spectrum. The ship emphasizes speed of movement to the drilling site; the platform provides more steadiness and room for working. A semisubmersible platform is towed to its drilling site, where it takes on enough water ballast to submerge its lower portions. In that position it floats on large cylindrical columns. The arrangement is such that the platform is little affected by waves or other motions of the sea.

The unanchored deep-ocean drilling system, which consists of a drilling rig in a ship hull, has so far been used only for scientific work. It has improved substantially the ability of geologists and other investigators to explore the strata under the deep ocean. The technology dates from 1961 and includes dynamic positioning, the control of stress in a long and unsupported drill pipe, placement of conductor pipe (leading the drill through the soft sea bottom to bedrock) in deep water and the use of a seawater turbodrill to drill hard rock in more than 12,000 feet of water.

Later developments have led to the system employed on the *Glomar Challenger,* which is operated by the Scripps Institution of Oceanography in a National Science Foundation program involving the coring of deep-ocean sediments in water depths of up to 20,000 feet [*see illustration on page 14*]. The developments include acoustical position-sensing equipment and automatic control of the propulsion units, so that dynamic positioning is much more reliable. The ship has successfully drilled several dozen holes in water depths to 17,000 feet, penetrating as much as 2,500 feet of the sea bottom. The cores thus obtained have yielded much valuable information. Moreover, the discovery of hydrocarbons on Sigsbee Knolls deep in the Gulf of Mexico has done much to modify geological thinking about the possibility of oil in deep water.

Offshore oil production is moving steadily into deeper and rougher water and more remote areas. If it is to be profitable, several producing wells must be established in each cluster and the capital cost should not exceed the present cost of producing oil in 200 feet of water. Probably it is possible to build stationary platforms that would resemble existing ones for depths of up to 600 feet. The cost would be high, however, particularly for a system that involved completing the well atop the platform (installing the pipes and valves and related equipment needed to put the well into production after drilling has reached oil). The current trend is toward the use of floating drilling platforms such as the semisubmersible ones, with completion of the well being made on the sea floor. A system of this kind would include remotely controlled valves and flow lines to central collecting points. In depths of 1,000 feet or more submarine work chambers analogous to pressure-resistant elevators will lower workmen to the bottom; while they remain inside at normal surface pressure they will be able to remove and replace heavy components, make flow-line and electrical connections and inspect the machinery.

Deep diving is receiving increasing attention in ocean technology. Men go into increasingly deeper water by means of systems that grow ever more complicated, involving a variety of chambers, hoists, gases and instruments.

There are two competing methods: the bounce dive and the saturation dive. In a bounce dive the diver goes from atmospheric pressure to the required depth in a chamber, breathing gaseous mixtures that change in accordance with depth and physiological requirements. He works for a few minutes (perhaps 10) and then returns to the surface in a fully pressurized chamber for slow decompression on the deck of the mother ship. The saturation-dive system makes possible multiple dives. In it the diver's body is saturated with inert gases while he lives in a pressure chamber on shipboard. When it is time to dive, he moves into a similarly pressurized capsule that is lowered to the bottom. Since his body is already prepared for the pressure, he can immediately go to work. He can work much longer than the bounce diver

HOISTING DEVICE called the Syncrolift allows the Canadian submarine *Ojibwa* to be lifted out of the water (*above*) and pulled ashore on rails (*below*) for repair in a dockyard at Halifax, Nova Scotia. The platform is lowered under the keel of a vessel, the ship is floated in and the platform is raised by the array of winches visible in the photograph. The device has a capacity of 6,000 tons.

DIVING GEOLOGISTS of the Shell Oil Company probe the ocean floor at a depth of about 20 feet on the Bahama Banks. The pipe and hose slanting across the center are parts of an air-lift apparatus that they are using to obtain information helpful in the search for oil.

before he returns to the capsule and thence to the chamber on deck. Since he lives on the surface but at the pressure of the bottom, the procedure can be repeated for many days. Then the diver takes a slow decompression to atmospheric pressure. Divers using each of these systems have reached 1,000 feet.

Another scheme has divers living on the bottom in shallow water at ambient pressures in undersea chambers. Examples are the experiments that have been carried out in such chamber systems as the U.S. Navy's *Sealab II*, the University of Miami's *Tektite* and the French *Conshelf*.

Doubtless oceanographers can point to elements of ocean technology I have overlooked. The developments have been too rapid and profuse for one man to be familiar with them all. Sometimes one feels that the first question to be asked on hearing about a new oceanographic device is: "Is it obsolete yet?" The answer should be: "We're working on that problem and it soon will be."

II

MARINE GEOLOGY

II MARINE GEOLOGY

INTRODUCTION

During the last quarter-century, geologists and geophysicists have been blessed with a science in ferment, full of local revolts and overwhelmed by a conceptual revolution—continental drift—whose effects continue to spread. I say "blessed" because scientists are trained to believe that facts count more than faith and that, in principal, a student with new data may overthrow the hypotheses of the greatest authority. These are not ordinary or even easy things to believe, especially if as a student you have learned all the evidence that supports a hypothesis, and you then spend years collecting more evidence that you interpret as supporting the hypothesis. A scientist can usually ignore new data and new hypotheses on the grounds that they have nothing to do with his own specialty. Sometimes a real crunch comes, however, because the new data in his specialty simply are incompatible with what he believes. Then a scientist, somehow, shrugs his shoulders, shakes his head, mutters "Easy come, easy go," or its equivalent, and sets about relearning his business. Not everyone survives the crunch—some find it easier to give up a profession than to change an idea—but most people make the transition. Students, of course, learn the new ideas in school, just as their professors learned the old ideas. The students think "What was all the fuss about?" and they may never experience the pleasure of meeting the crunch and finding out whether they really are scientists.

Let me classify the changes that have occurred in geology and, particularly, marine geology. I propose two classifications: internal revolutions that completely change ideas within a science, and external revolutions that radically change several sciences and affect the cultural heritage of mankind. In the 1950s there was an internal revolution in geology when new data changed the understanding of erosion and deposition at the boundary between continents and ocean basins. In the 1960s it was proved that the sea floor spreads, continents drift, and the earth's crust is broken into gigantic ridid plates. This produced an external revolution that still continues. In the 1970s another internal revolution began as deep-sea drilling started to reveal the complex geological history of the deep-sea floor. The articles in this reader deal with all these great changes, but the emphasis is on the external revolution.

Let us take the changes in chronological order. Early in this century little was known about the origin and distribution of either modern or ancient sediments. Observations by oceanographers were few, and land geologists were compelled to explain the origin and history of ancient sedimentary rocks according to what logically ought to have happened. In the general picture that emerged, the submarine world was quiescent. Geology then was in the same state as biology before the *Challenger* expedition. Forbes had said that there was no life below 540 meters, but *Challenger's* trawl proved otherwise. Geologists said that there was no water motion that affected rocks or sediment below a level called "wave base," at a depth of a few scores of meters. Thus, any sedimentary rock that contained ripple marks or current scours must have been produced in shallow water. The same argument applied to well-sorted sands, because they could have been moved into place only in shallow water, where water had sufficient energy for transportation. The whole distribution of sediment on the sea floor could be hypothesized in terms of energy and distance from sediment sources. Near the mouths of rivers and beaches the sea floor would be covered by sand. Farther out in shallow water would be silt; beyond that, clay. The deep-sea floor, being below wave base, would be covered with layer on layer of pelagic oozes consisting of the finest clay and the skeletons of the marine life that rained down from the productive surface waters; it was thereafter eternally at rest. The real world was much more interesting.

Practically the only seagoing marine geologist in the United States in the 1920s and one of the few in the 1930s and 1940s was Francis P. Shepard. He worked where ships could take him, near shore, and found that the sediment on the continental shelves was patchy and quite unlike the distribution implied by logic. He also surveyed and dredged the great submarine canyons incised in the continental slope, and he reached the conclusion that they were cut by rivers during a lower stand of sea level, probably during the ice ages, when water was stored in glaciers.

Reginald Daly proposed, on the contrary, that the canyons were cut by currents of dense turbid water that could flow under the sea just as they flowed under lakes. His proposal was supported by Philip Kuenen, who conducted experiments with such turbidity currents. In 1951, Shepard, Kuenen, and a few lucky onlookers including myself were on the research ship *E. W. Scripps*, anchored athwart the Scripps Submarine Canyon off (naturally) Scripps Institution of Oceanography. Shepard had lowered current meters into the canyon. Divers had buried explosive charges in the loose sediment in the head of the canyon. At a signal, they were exploded, and we all waited to see if the instruments would report the passage of a fast turbidity current down the canyon under the ship. Nature is a worthy opponent. All we observed was a slow current moving up the canyon. Shepard is still studying these currents in canyons, and he has established that they are related to the tides.

Other approaches to the turbidity-current problem were more fruitful. Within a few years it was proved that enormous sedimentary fans lie at the mouths of even the deepest submarine canyons and that these fans are crossed by leveed, meandering channels. Downslope the fans and channels lead to almost flat abyssal plains. Clearly, river-like flows occur at every depth in the ocean. This conclusion was confirmed by sediment samples. In the fans and plains the sediment in many places contains coarse sand and the remains of organisms that live only on the bottom in shallow water. In contrast, the hills that stick up from the plains typically are covered with pelagic oozes that have rained down from the overlying waters. The flows are turbidity currents too thin to cover the hills. Thus we know that enormous numbers of turbidity currents surge through submarine canyons, but even now we cannot be sure that they have cut all the canyons.

The concept of "wave base" was demolished even before the existence of turbidity currents was proved. The first deep-sea expeditions that had suitable cameras photographed ripple marks and current scours on normal sea floor. The assumptions that geologists had used to reconstruct ancient environments were in need of revision.

Soon ancient beach and lagoon deposits were discovered on the continental shelves; it was evident that the sediment distribution had been influenced by repeated changes in sea level and migration of the shoreline across the shelf.

The chronology of these sea-level changes was more or less established after it was discovered that fossil shells can be dated by the radiocarbon they contain.

The origin and history of the rocks under the continental shelves remained controversial until new techniques of sub-bottom profiling were introduced and oil companies began to drill very expensive holes into the rocks. What we now know about the continental shelves is discussed in "The Continental Shelves," by K. O. Emery, who continues to be one of the principal investigators of continental margins. As he predicted, the continental shelves have become increasingly important sources of oil and gas. All around the world, American drilling ships and platforms are exploring for petroleum. Everywhere, that is, except off some of the coasts of the chief energy hog, the United States. However, even that has been changing. Within a few decades the last parts of the continental shelf will be explored and drilled as the land resources are depleted.

So much for the internal revolution of the 1950s; the revolution of the 1960s was much broader and its ramifications may not all be known for decades. The discovery that the earth's crust is broken into enormous, drifting, rigid plates brought about a great conceptual revolution, but even it was not comparable to some revolutions of the past. This is chiefly because no element of religion is related to continental drift. In contrast, the Copernican revolution proposed ideas that were opposed by religious dogma and flew in the face of common sense, namely, that the earth moves although we are unable to feel it do so. Almost anyone who believes that the earth moves about the sun should hardly cavil at continents drifting about the earth. Nor was the new geological revolution as far-reaching in its implications as the theory of evolution, which is still contested by fundamentalist religious groups a century after it was accepted by scientists. Has anyone denounced continental drift from the pulpit?

The discovery that the sea floor spreads, continents drift, and tectonic plates are rigid shocked only scientists, who, as we know, are conditioned for such shocks. Indeed, many geologists in the Southern Hemisphere, Great Britain, and Western Europe had believed in continental drift for decades since it was first elaborated by Alfred Wegener. The evidence, they thought, was clear. The Atlantic coasts of Africa and South America looked as though they had been split apart from each other, and much evidence from geology and paleontology could be explained most easily by the hypothesis that a great southern continent had broken into fragments.

The data collected by marine geologists in the 1950s did not at first seem related to continental drift, but they did show that the sea floor was tectonically much more active than anyone had thought. Gigantic fault scarps were discovered and it was estimated that the deep basins were pimpled with 20,000 volcanoes. In the late 1950s, however, Bruce Heezen and Maurice Ewing discovered in the center of the Mid-Atlantic Ridge a rift that did seem to pertain to continental drift. They identified the ridge as a submarine mountain range reaching around the earth, and they realized that the central rift was seismically active.

Thereafter progress was rapid, and within a few years general tectonic syntheses were being offered to explain many of the new discoveries. In 1962 Harry Hess proposed what ultimately turned out to be the correct explanation, namely that the sea floor spreads apart and the gap is filled with new crust. Despite rapid and vigorous support from Robert Dietz, this idea did not gain widespread acceptance. The reasons for resistance were two: other explanations were available; and Hess' idea, as proposed, flew in the face of some facts that appeared important.

The concept of sea-floor spreading gained acceptance only gradually as it withstood every test and offered explanations for new discoveries. One by one, different specialists began to find that their data were compatible with spreading, but still they were not sure it was right. For them the final proof came at different times and in various ways. For some African paleontologists there probably never was any question. For myself the critical proof came in 1966 with the discovery of symmetrical magnetic anomalies by Tuzo Wilson and Fred Vine. Even after reading their papers I was not wholly convinced by the small illustrations, but as soon as I looked at the original data on file at Scripps I was sure they were right.

Some marine scientists were convinced earlier, others later; but before long, all were convinced and a renaissance followed in 1966–1967 as old data were reinterpreted with new understanding. If anyone had had time then to think about it, he probably would have believed that we had gone about as far as we could go for a long time. Yet by the end of 1967 Dan McKenzie and Robert Parker, and Jason Morgan, had discovered plate tectonics. On a large scale, the crust does not deform plasticly, as earth scientists thought; instead it was rigid. The concept of sea-floor spreading had introduced a revolutionary qualitative model of the world. The theory of plate tectonics used mathematics to understand the model and to make testable quantitative predictions from it. From then on it was all wine and roses.

Sir Edward Bullard, the author of "The Origin of the Oceans," learned the mathematics that is now used in plate tectonics as a schoolboy; unlike most people, he did not forget it. Thus, when he wanted to use a computer to make a best-fit match of the facing coastlines of Africa and South America he programmed it with Euler's theorem about the movement of concentric spherical shells. The computer produced a marvellously impressive match, which converted many mathematically minded specialists to continental drift who would have been skeptical about the significance of a mere glance at a map. Properly so, because flat maps are distortions.

Bullard's article, and mine ("The Deep-Ocean Floor"), which follows it, were published in 1969, and thus they are high-speed photographs of a conceptual explosion. Bullard says in his last paragraph, "What we have is a sketch of the outlines of a history." However, the concepts that he describes have remained valid, and in many ways they are easier to follow when unencumbered by the detail now available.

Marine geology at the time these two articles were written was a delight, because everything was falling into place. The mathematics of plate tectonics made it possible to integrate scattered and previously unrelated data, and it was as if we suddenly knew ten times more than before. Formerly isolated facts, such as the general slope of the sea floor, suddenly became obvious consequences of the plate-tectonic model. For example, the depth of the sea floor depends on its age, and upon this fact was based the preliminary explanation for the distribution of pelagic sedimentation in space and time. This age-depth hypothesis underwent a remarkable metamorphosis and expansion a few years later when John Sclater and a few co-workers developed a quantitative model that explained age, depth, and heat flow by cooling of the lithosphere. This begins as soon as the lithosphere solidifies at a ridge crest and continues until it disappears down a subduction zone. (A jargon has developed since these articles were written in 1969.)

Another idea that has been transformed by the mathematics of plate tectonics concerns the origin of the linear chains of volcanic islands that are so characteristic of the Pacific. It had long been known that there is an active volcano at one end of most chains and much older atolls at the other end. Tuzo Wilson proposed that the chains form as the lithosphere drifts over hot spots in the mantle. Jason Morgan showed that, to a reasonable approximation, the hot spots appear to be in a rigid framework; that is, they are either immobile or all drifting together. This hypothesis has stimulated a wide range of marine research, which is still continuing. If the hypothesis is correct, or to the extent that it is correct, marine geologists can reconstruct the movement of plates relative to latitude and longitude rather than just relative to other plates.

In "The Floor of the Mid-Atlantic Rift," by J. R. Heirtzler and W. B. Bryan, we see the results of applying the new technology of manned research submersibles and deep-towed instruments to the investigation of the spreading sea floor. For the first time we can really know what is happening in a spreading center, as we examine the incredible forms that the lava assumes when it cools in the abyss. Think how much more we shall learn when we have examined other parts of the sea floor in the same detail.

The last two articles in this section build upon plate tectonics, but only in a broad sense. Because plate tectonics brought about a major external revolution, it can be applied to many otherwise unrelated aspects of science. Later articles in the reader discuss the distribution of marine life and the origin of ore deposits in terms of moving plates. The article on "The Evolution of the Pacific" and "When the Mediterranean Dried Up" are geological equivalents to the later articles. They are concerned with an internal revolution in geology that is being based on the Deep Sea Drilling Program (DSDP).

The *Glomar Challenger* first put to sea in 1968, to begin this highly successful drilling program, which still continues, although it has changed its financing and changed its name to International Program of Ocean Drilling (IPOD). Initially, much of the drilling effort was expended on testing the concept of plate tectonics. You may wonder why anyone thought it necessary to test the concept as late as 1968–1969. Hadn't it been proved two years before? The drilling program demonstrates that not everyone was convinced. In those transitional years there were still experts and specialists who had not yet had to face the crunch in their own specialties. The fossil-laden sediment cored by the DSDP contained the kind of evidence that many stratigraphers, micropaleontologists, and land geologists required in order to be convinced.

In addition to convincing these specialists, the early deep drilling added a priceless ingredient to the quantification of geological history: it dated the older magnetic anomalies. As Bullard indicates in his article, in 1969 the reversals of the earth's magnetic field could be dated for a mere 4 million years. The age of earlier reversals and magnetic anomalies could only be inferred, by assuming that the sea floor tended to spread at a constant rate. A comparison of the magnetic anomalies in different oceans demonstrated that the rate of spreading had varied in some places, but which ones? James Heirtzler and his colleagues made the inspired assumption (or correct analysis, if you prefer) that spreading was constant in the South Atlantic. Continental geology indicated that Africa and South America began to separate about 70 million years ago, and this appeared to confirm the assumption of constant spreading.

Appearance is not necessarily reality, however, and many scientists had deep reservations about constructing a quantitative geological history on a questionable assumption. Consequently, the DSDP sampling in the South Atlantic was watched with great expectation. The results proved that Heirtzler had indeed been inspired.

Increasingly, deep drilling turned to the problems of the geological history of ocean basins; in part, the results were used to analyze in elegant fashion the early inferences of plate tectonics about marine geology. Such an analysis is described in "The Evolution of the Pacific," by Bruce Heezen and Ian MacGregor. Most of the history of the Pacific could have been, and was, inferred from the geometry of plate motions, the fixed hot-spot framework, known rates of pelagic sedimentation, and the existence of a belt of rapid sedimentation along the equator. Inferred, yes; proved no. No method except deep drilling was capable of testing these inferences. Progress in this exciting field of plate stratigraphy is still very rapid. The actual sedimentation rates measured in the DSDP holes have been integrated into a computer program by my colleagues Wolf Berger and Edward Winterer. By combining these sedimentation rates with the rate of drift of the Pacific plate and its rate of subsidence they can predict the composition of a sedimentary column almost anywhere in the ocean basin.

The last article, "When the Mediterranean Dried Up," by Kenneth Hsü, reports confirmation of an event that had been inferred but already largely rejected—namely, that the Mediterranean once dried up and then was refilled by a gigantic deluge. Hsü describes the new evidence: the sediment in the bottom of the Mediterranean contains layers of salts that could have formed only by complete evaporation of the overlying water. The evidence that had been suppressed, ignored or scorned is that deep gorges and submarine canyons indicate that the water level in the basin had once been very low. Thus, the last article harks back to the first in this section. One of the facts that convinced Francis Shepard that submarine canyons were cut by subaerial rivers was the existence of gorges in the hard rocky submarine slopes of the Mediterranean island of Corsica.

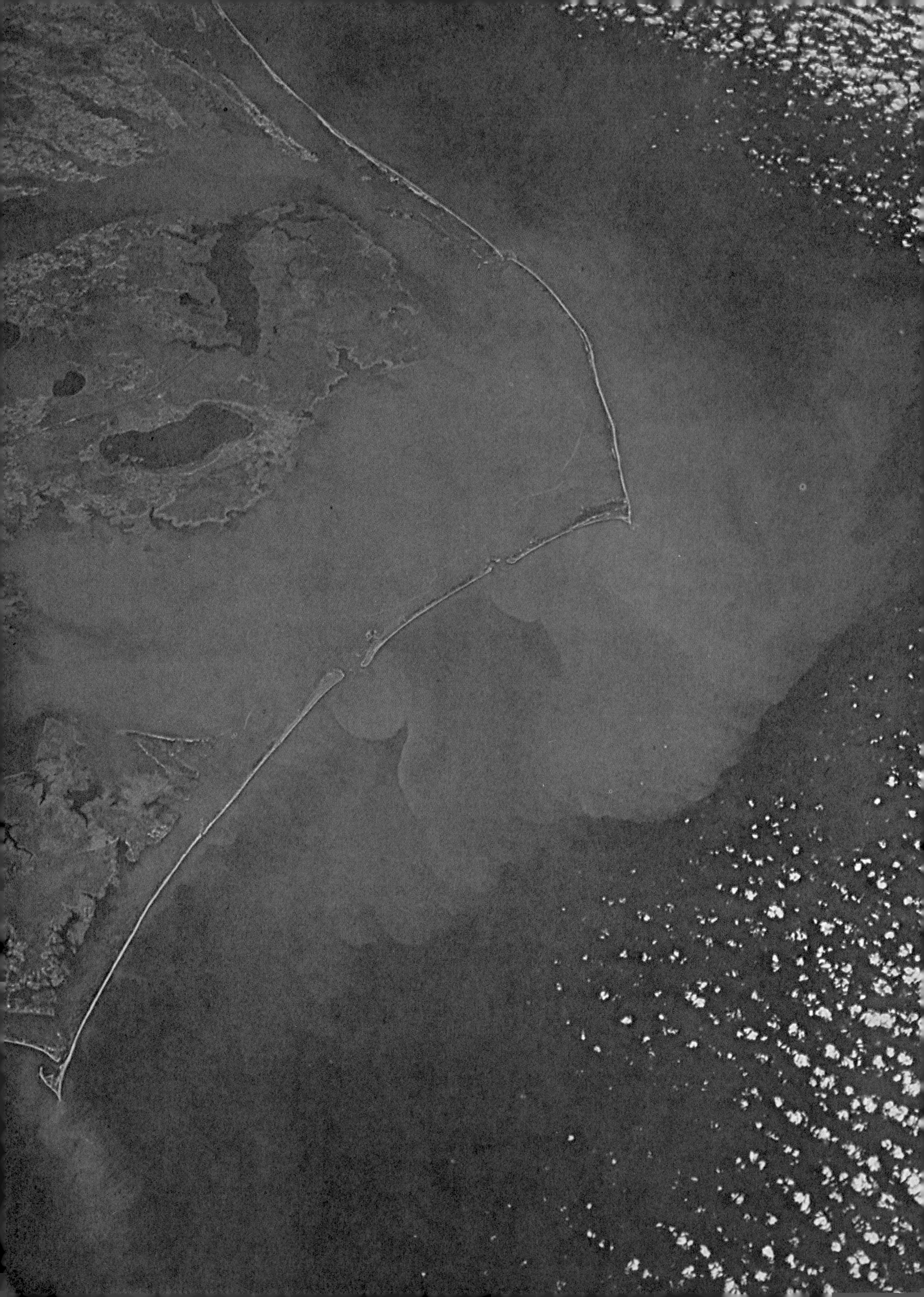

The Continental Shelves 3

by K. O. Emery
September 1969

The shallow regions adjacent to the continents are equal in extent to 18 percent of the earth's total land area. They are alternately exposed and drowned as the continental glaciers advance and retreat

The continental shelves were the first part of the sea floor that was studied by man, chiefly as an aid to navigation and fishing. Perhaps the earliest recorded observation was one made by Herodotus about 450 B.C. "The nature of the land of Egypt is such," he wrote, "that when a ship is approaching it and is yet one day's sail from the shore, if a man try the sounding, he will bring up mud even at a depth of 11 fathoms." A more recent example is found in the diary of a 19th-century seaman: "An old captain once told me to take a cast of the lead at 4 a.m. We were bound to Hull from the Baltic. He came on deck before breakfast and on showing him the arming of the lead, which consisted of sand and small pebbles, I was surprised to see him take a small pebble and put it in his mouth. He tried to break it with his teeth. I was very curious and asked him why he did so. He told me that the small pebbles were called Yorkshire beans, and if you could break them you were toward the westward of the Dogger Bank; if you could not, you were toward the eastward."

Fishing success often depends on knowledge of the kind of bottom frequented by particular fish and on the avoidance of rocky areas that can catch and tear nets. As a result governmental agencies routinely chart bottom topography and materials to aid the fishing industry, but the successful fisherman generally keeps much additional information to himself. Similarly, the production of oil and gas from the continental shelves during the past two decades has led governmental and international agencies to make broad geological surveys, which help to guide the oil industry to the areas of greatest economic promise. The oil companies make studies that are much more detailed and so expensive that the results are considered proprietary, at least until after exploitation rights are secured.

During World War II submarines took a large toll of the ships that crossed the continental shelves, mainly at the approaches of ports. The effectiveness of acoustical detection equipment on submarine hunting ships was much increased by a knowledge of bottom materials and their effects: long ranges over sand, short ones over mud, confusing echoes over rock or coral. Accordingly charts of bottom sediments were compiled by the American and German navies for many areas of the world. This problem had not arisen earlier because both submarines and the search gear of surface ships were too primitive during World War I, and it may not be important in the future owing to the greatly increased sophistication of submarines and to the different role they may play in any future war.

The conflict between disseminating information and keeping it secret is about what one would expect in an environment that is both economically and militarily important. The recent political interest in the sovereignty of the ocean is also to be expected, considering the way the economic potential of the shelves has often been exaggerated in recent years. Thus it is not surprising that there have been a number of proposals to redefine the continental shelf so as to extend it seaward to whatever distance and to whatever depth are necessary to give a nation access to the resources presumably lying or hidden there. In 1953, before the world developed its present large appetite for seafood and minerals, an international commission defined the continental shelf, shelf edge and continental borderland as: "The zone around the continent extending from the low-water line to the depth at which there is a marked increase of slope to greater depth. Where this increase occurs the term shelf edge is appropriate. Conventionally the edge is taken at 100 fathoms (or 200 meters), but instances are known where the increase of slope occurs at more than 200 or less than 65 fathoms. When the zone below the low-water line is highly irregular and includes depths well in excess of those typical of continental shelves, the term continental borderland is appropriate."

Somewhat similar, but shorter, definitions are presented by most textbooks of geology. Where a depth limit is given it is 100 fathoms, an inheritance from the time when navigational charts had only three depth contours: 10, 100 and 1,000 fathoms. On a global basis the edge of the continental shelf ranges in depth from 20 to 550 meters, with an average of 133 meters; the shelf ranges in width from zero to 1,500 kilometers, with an average of 78 kilometers [*see illustration on next two pages*].

CONTINENTAL SHELF in the Atlantic Ocean off Cape Hatteras is delineated by puffy clouds that show where cold surface water on the eastern edge of the shelf meets warmer surface water. The boundary is near the edge of the shelf, which at this point averages about 120 meters in depth. The picture also shows turbid water moving from Pamlico Sound into the ocean, where it is carried northward by a fringe of the Gulf Stream that lies atop the shelf. The photograph was taken from *Apollo 9* on March 12, 1969, at an altitude of 134 statute miles; the view extends 175 miles in the north-south direction. The astronauts on this mission were James A. McDivitt, David R. Scott and Russell L. Schweikart.

The continental shelves underlie only 7.5 percent of the total area of the oceans, but they are equal to 18 percent of the earth's total land area. A geological understanding of this huge region requires a knowledge of its topography, sediments, rocks and geologic structure. For nearly all shelves there is some information about topography; for perhaps a fourth of them something can be said about the surface sediments, but the rocks and geologic structure are known for less than 10 percent. Detailed knowledge is far less available. The best-known large areas are the ones off the U.S., eastern Canada, western Europe and Japan—in short, the shelves next to countries where scientific knowledge is well developed and freely disseminated. Smaller areas of knowledge are found where oil companies have worked (such as parts of northern South America, parts of Australia and the Persian Gulf) and where oceanographic institutions have conducted repeated operations (northwestern Alaska, the Gulf of California, northwestern Africa, the shelf off Argentina, the Red Sea and the Yellow Sea). Recently some developing countries have effectively closed their shelves to foreign scientific studies; these areas are fated to remain unknown and unexploited for the foreseeable future.

The information that is most costly and most difficult to obtain concerns the underlying rocks and the structural geology of the continental shelves. This information is essential for understanding the origin and most of the history of the shelves. Data about the surface topography and sediments, which are readily accessible, tell only the late history.

Samples of bedrock have been dredged from the surface of many shelves, mainly from the top of small projecting hills and the sides and heads of submarine canyons that incise the shelves. Additional rock samples have come from the top of the adjacent continental slopes. Care must be taken in deciding whether the rock samples are from outcrops, whether they are loose pieces that were deposited by ancient streams or glaciers, or whether they were rafted to their present location by ice, kelp, marine animals or man. The decision is usually based on the size of the piece, on the presence or absence of fresh fractures, on the similarity of lithologic types within a given dredging area or between adjacent dredging areas, and on the amount of tension of the dredge cable. It is helpful to have submarine photographs, which may reveal rock outcrops in the dredging area. Rock also can be sampled by coring: by dropping or forcing a heavily weighted pipe into the bottom. This method can show the dip of the strata, and it can sample rock that is covered by sediments if the sediment is thin or if it is first removed by hydraulic jetting. A better but more expensive method of rock sampling is provided by well-drilling methods. Many holes have been drilled for geological information in shelf areas of structural interest; they provide good information on the sequence and depth of strata, the date of original deposition and geologic structure. In addition several thousand oil wells have been sunk into shelves, but most have been drilled in abnormal geologic structures such as salt domes and folds.

Geophysical methods provide excellent, although indirect, data from which geological cross sections can be constructed. These methods include seismic reflection and refraction, as well as measurements of geomagnetism and gravity. Each method has its advantages, but the most generally successful one is seismic reflection. In practice a ship traverses the shelf and produces a loud acoustical signal in the water at intervals of a few seconds. The chief source of sound energy a few years ago was a chemical explosive, usually dynamite; other sources are now preferred because they are

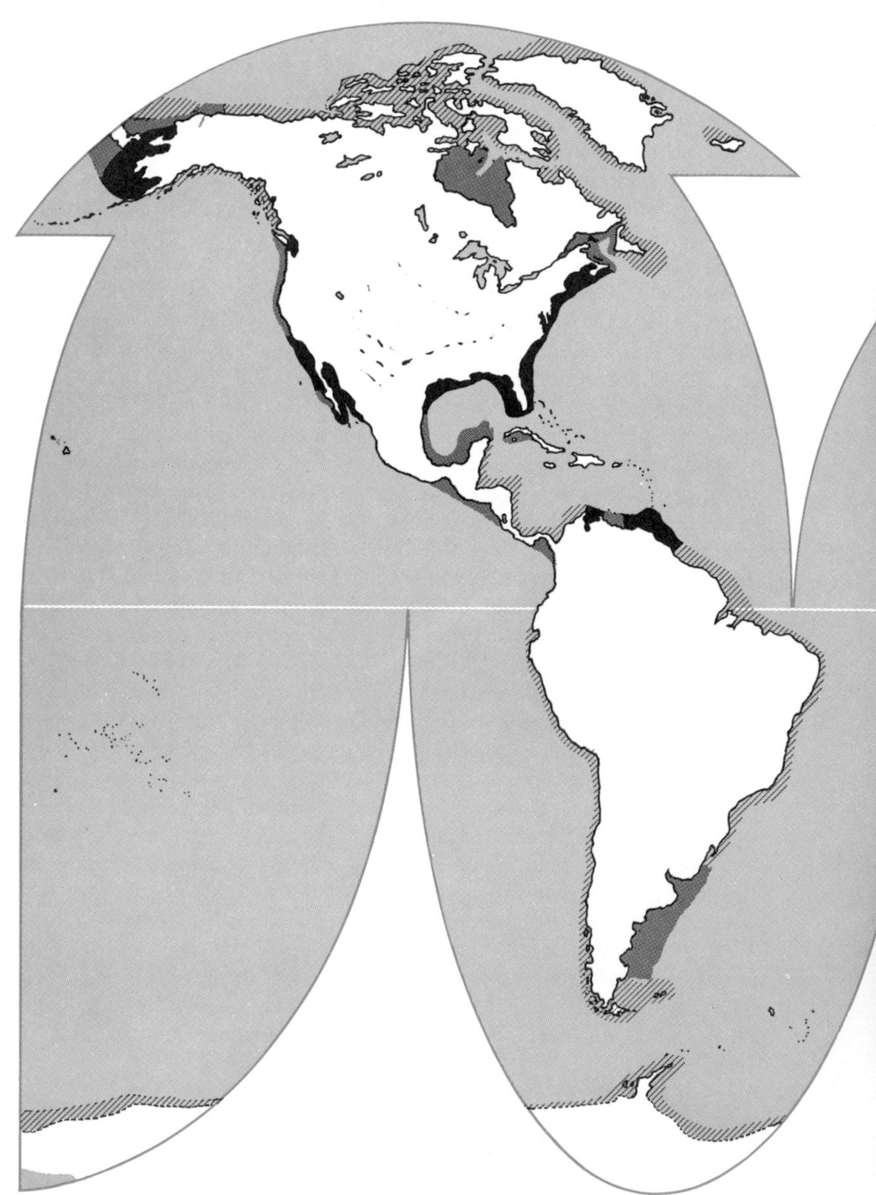

CONTINENTAL SHELVES underlie about 7.5 percent of the total ocean; all together they occupy an area roughly equal to that of Europe and South America combined, or some 10 million square miles. The shelf is defined as the zone around a continent extending from

cheaper, easier to trigger accurately and greatly reduce the danger both to the operators on the ship and to the fish in the sea. These newer energy sources include electric spark, compressed air and propane gas. Although part of the sound energy is reflected from the sea floor, much of it enters the bottom to be reflected upward from various layers of rock under the bottom. The reflected energy is received by hydrophones trailed behind the ship; the signal is amplified, filtered and recorded on continuously moving paper tapes. This method, termed continuous seismic-reflection profiling, is rapid and can yield information from depths of several kilometers under the sea floor, making it possible to construct geological cross sections. When the interpretations are supplemented by dredging or drilling, they provide the best information now available about the structure of continental shelves.

When existing geological and geophysical information is assembled on a worldwide basis, it shows that continental shelves can be classified into two main types by composition: those that are underlain by sedimentary strata and those underlain by igneous and metamorphic rocks. A large majority of the world's continental shelves mark the top surface of long, thick prisms of sedimentary strata [*see illustration on next two pages*]. Many of the prisms are held in position against the continents by long, narrow fault blocks. Such is true of almost the entire perimeter of the Pacific Ocean, where tectonic activity has also produced deep trenches that are parallel to the base of the continental slope. In some areas, such as the West Coast of the U.S., a single geologic dam is known to have extended for thousands of kilometers along the coast. Locally part of the dam rises above sea level to form the granitic Farallon Islands that lie immediately off San Francisco. These rocks are some 100 million years old, but they were thrust up to form the dam only about 25 million years ago. Elsewhere, as in the Yellow Sea of Asia, half a dozen such fault dams or fold dams

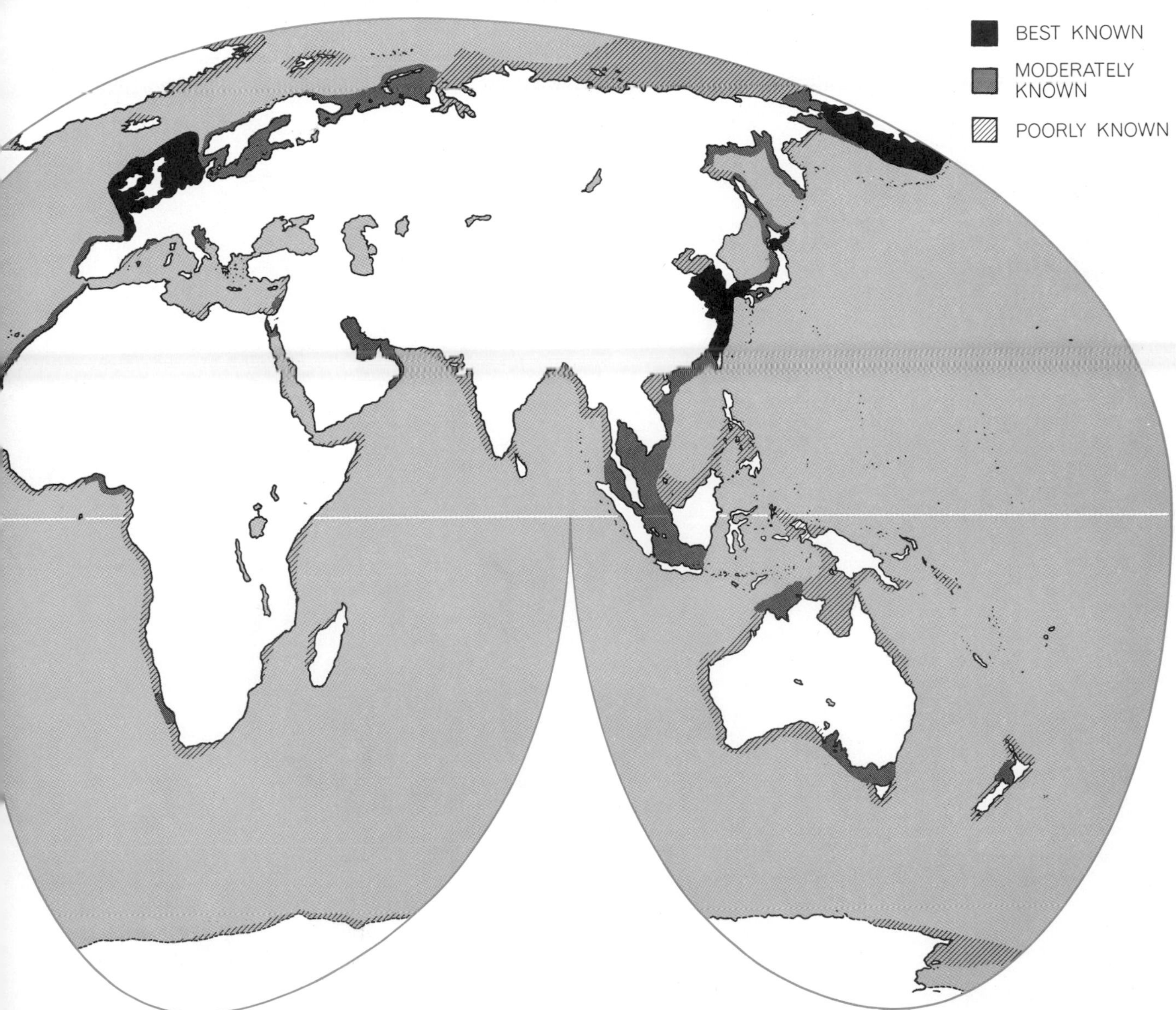

the low-water line to the depth at which the ocean bottom slopes markedly downward. Conventionally the edge of the shelf is taken to lie at 100 fathoms, or 200 meters, but a more accurate average value for all continents is about 130 meters. Worldwide the shelf has an average width of 78 kilometers. The illustration indicates the present state of knowledge of the continental shelves of the world.

have risen in the last 500 million years or so, each dam in turn causing the ponding of sediments from the land. There was a similar dam off the entire length of the East Coast of the U.S. between 270 and 60 million years ago; in time the trench on its landward side was filled with sediments that subsequently spilled over the former dam to build a continental slope that is held in place only by the angle of rest of the sediments. That this angle is unstable is shown by numerous landslides and erosion features recorded in seismic profiles across the continental slope and rise.

The continental shelf off the western part of the Gulf Coast of the U.S. is held in place by a diapir dam: a dam formed by the upward movement of salt from a bed that is buried several kilometers deep and is about 150 million years old. Seismic profiles and dredgings show the presence of still another kind of dam in the eastern Gulf of Mexico and off the southeastern coast of the U.S. This is an algal reef that dates from 130 million years ago and was succeeded by a coral reef off Florida at some time before 25 million years ago; even today the Florida keys are bordered by a living coral reef. Similar tectonic and biogenic dams elsewhere in the world have trapped huge quantities of sediments in the geological past.

Shelf areas underlain by igneous and metamorphic rocks are found on top of the tectonic dams. Off Maine, however, glacial erosion has removed the sedimentary rocks that once covered such a dam [*see illustration on pages 38 and 39*]. Other shelves underlain by igneous and metamorphic rocks are known, but most of them appear to be at high latitudes where glacial erosion has been effective. Nevertheless, even at high altitudes probably most of the shelves are underlain by sedimentary rock. In a sense we can consider the shelves whose shape is chiefly due to glacial or wave erosion as youthful ones (or rejuvenated ones); the shelves that are mainly depositional, with a thick prism of sediments on top of igneous and metamorphic rocks, can be regarded as mature ones. The sediments have built the shelves upward during the concurrent sinking of the edges of the continents. Perhaps more important from the viewpoint of real estate, these sediments have increased the size of the continents, widening them as much as 800 kilometers in areas where rivers have brought much sediment from the land and where tectonic or other dams effectively prevent the escape of the sediment to the ocean basins. This dammed sediment, as well as the sediment held only by its angle of rest, has an estimated average thickness of about two kilometers, yielding a total

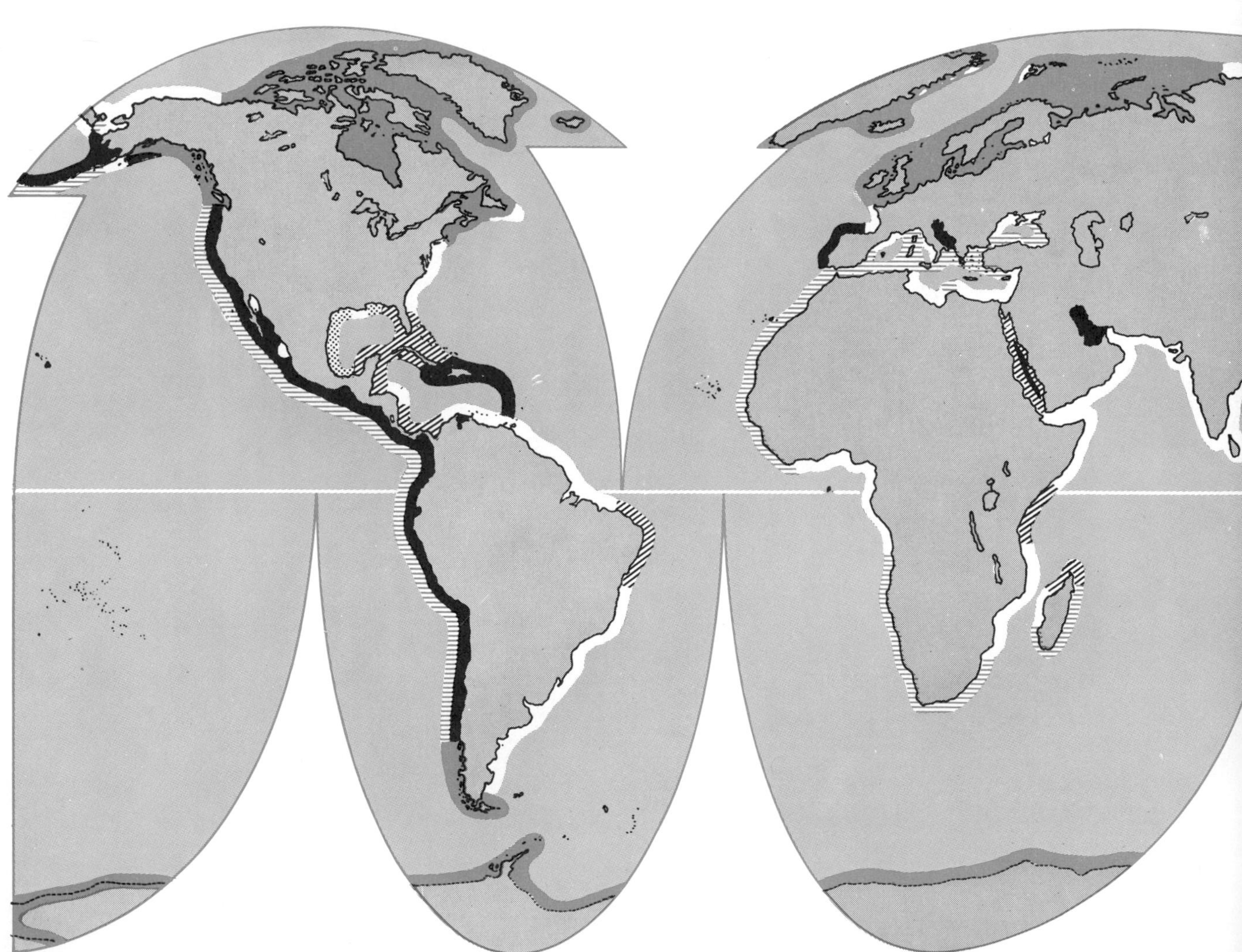

CHARACTER OF CONTINENTAL SHELF depends largely on how the shelf was formed. Six types of shelf, classified by origin, are indicated in this worldwide map. Many shelves are deposited behind three kinds of dams: tectonic dams, formed by geological uplift or upwelling of lava; reef dams, created by marine organisms, and diapir dams, which are pushed upward by salt domes.

volume of sedimentary strata under continental shelves of about 50 million cubic kilometers.

Perhaps the most dramatic period in the history of the continental shelves was the million-year passage of the Pleistocene epoch when the sea level changed in response to the waxing and waning of the continental glaciers. At their maximum the glaciers appear to have been so extensive as to have stored in the form of ice enough water to lower the surface of the ocean nearly 150 meters below the present level. Four major lowerings of the sea level were produced by the four main glaciations, with minor lowerings caused by secondary fluctuations of climate and ice volume. Limited investigations with special seismic equipment off the East Coast of the U.S. show four or five somewhat irregular acoustical reflecting surfaces near the top of the shelf sediments. These reflecting surfaces probably can be explained by erosion and sand deposition at stages of low sea level. Cores from these beds probably would provide much interesting information about Pleistocene climates and Pleistocene chronology. For the present, however, our data for glacial effects on the continental shelf are restricted largely to surface sediments and topography.

About 50 years ago most textbooks of geology led the reader to believe that sediments became progressively finer in texture with distance from shore: gravel and sand at the shore, coarse sand grading to fine sand across the shelf, and finally silt and clay (the "mud line") at the shelf edge. Bottom-sediment charts compiled during World War II, however, showed that this simple pattern is very rare, and that the size of sediment grains is unrelated to the distance from shore. The examination of actual samples showed that most of the shelves are floored with coarse sands that commonly are stained by iron and contain the empty shells of mollusks that live only close to shore in shallow depths. Broken shells or shell sand are particularly abundant at the outer edge of the shelf and on small submerged hills that are relatively inaccessible to detrital sediments. Some of these same areas contain glauconite and phosphorite, minerals that are precipitated from seawater, but so slowly that they are obscured or diluted beyond recognition where detrital minerals are present. The only areas that exhibit a consistent seaward decrease of grain size are those between the shore and depths of 10 or 20 meters—in short, whatever areas are shallow enough to be ruled by the waves. At greater depths the sediments are too deep to be reached by new supplies of sand. These sections

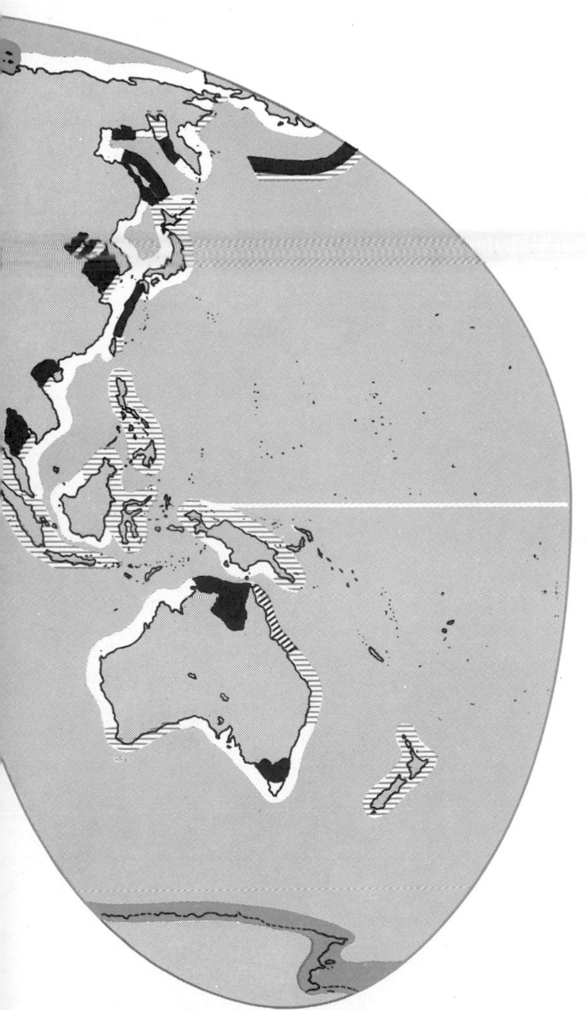

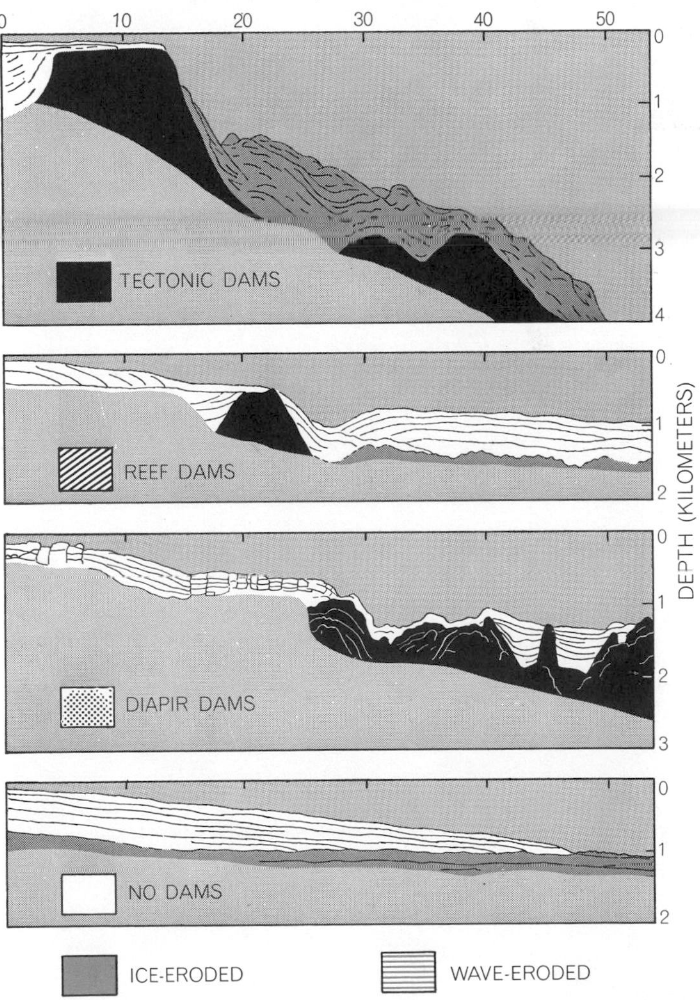

These three kinds of dams are shown in black in the typical shelf cross sections at the right; a simple damless shelf is also depicted. The vertical scale is exaggerated six times. Sediment deposited before formation of the dam is shaded gray; sediment deposited subsequently is unshaded. All four kinds of shelf structure may be eroded by waves (*hatched color on map*) or by ice (*solid color*).

of the shelves are also bypassed by contemporary silts and clays that remain in suspension en route to deeper or quieter waters.

The sediments on about 70 percent of the world's continental-shelf area have been laid down in the past 15,000 years, since the last glacial lowering of the sea level. The rest of the shelf is floored by silts at the mouths of large rivers, in quiet waters behind barriers and in shelf basins, by recent shell debris and by chemically deposited minerals. This means that when the sea level was low, the entire shelf was exposed and the rivers deposited sands on the then broader coastal plain and transported their silts and clays to the ocean. At that time ocean waves, with no shelf to reduce their height, were probably higher at the shore than they are today, with the result that shore sediments were probably coarser. The broad expanse of lowland favored the development of ponds and marshes, which were partly filled with debris from the forests and meadows that extended unbroken from the inland areas across what is now the sea floor. Freshwater peat now submerged in the ocean has been sampled at 10 sites off the eastern U.S. and at many other sites on the shore; similar peats have been found off Europe, Japan and elsewhere.

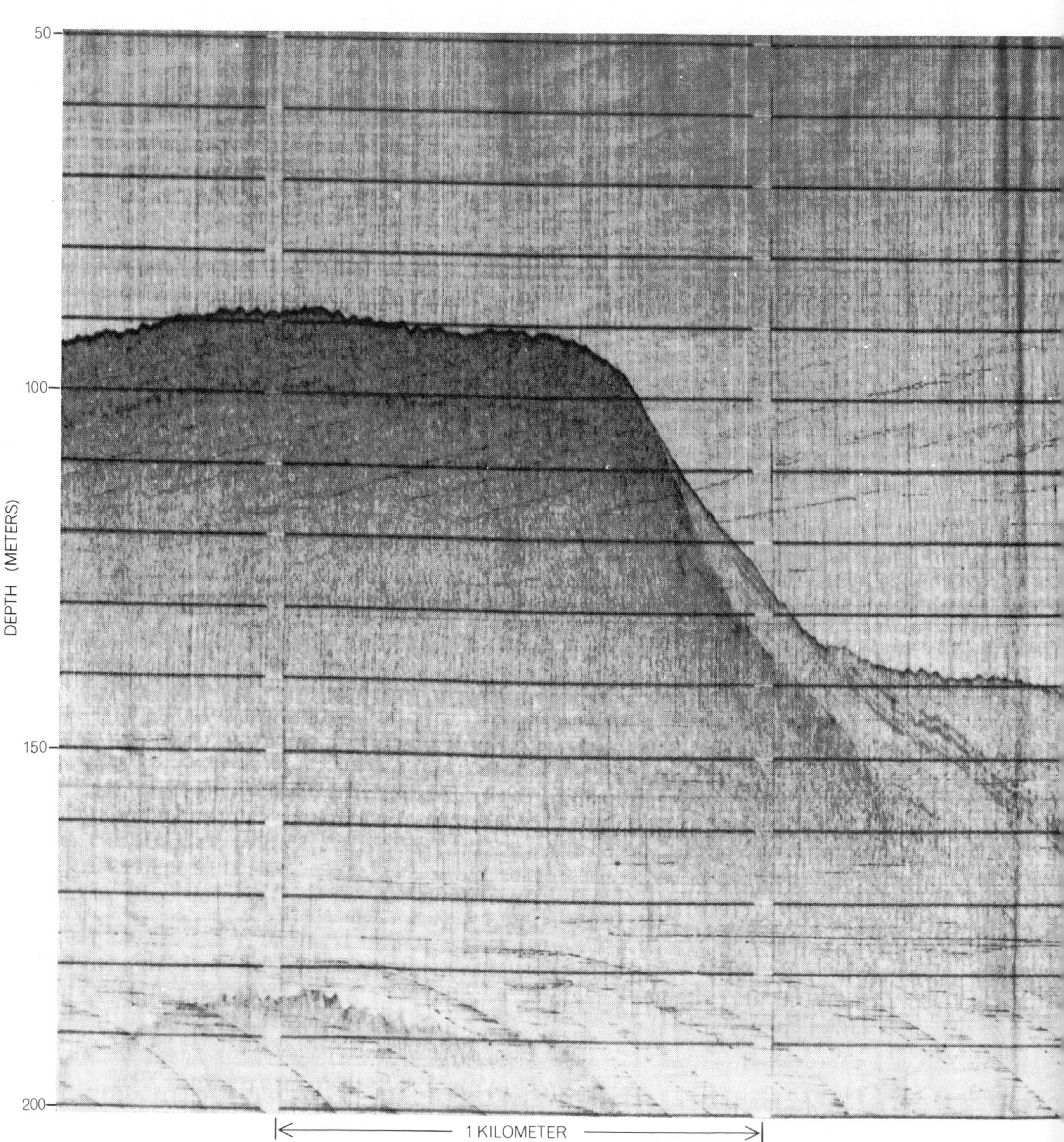

ICE-ERODED SHELF about 100 kilometers off the coast of Maine, landward of Georges Bank, is shown in this seismic-reflection record. **The deep trough was gouged out of the basement rock by ice some 15,000 years ago during the last glaciation. Subsequently the trough**

Pollen analysis shows a succession from tundra to boreal spruce and pine some 12,000 years ago, followed by oak and other Temperate Zone deciduous trees about 8,500 years ago; the deciduous trees flourished until the site was submerged. Birds once flew among the trees in many areas where fish now swim.

The vegetation attracted many animals, but only their heavier bones are preserved or are readily detected by dredging. Nearly 50 teeth of mammoths and mastodons have been collected off the East Coast of the U.S., along with the bones of the musk ox, giant moose, horse, tapir and giant ground sloth. Similar finds have been reported off Europe and Japan.

Carbon-14 dates have been obtained for more than 50 samples of shallow-water material from the shelf off the East Coast of the U.S. The materials include salt-marsh peat, oölites (concentrically banded calcium carbonate pellets that typically form only in warm, shallow, agitated seawater) and the shells of oysters and other mollusks (which live in only a few meters of water but whose empty shells are found as deep as 130 meters). The dates and depths make it possible to draw a curve showing the changes of sea level in an-

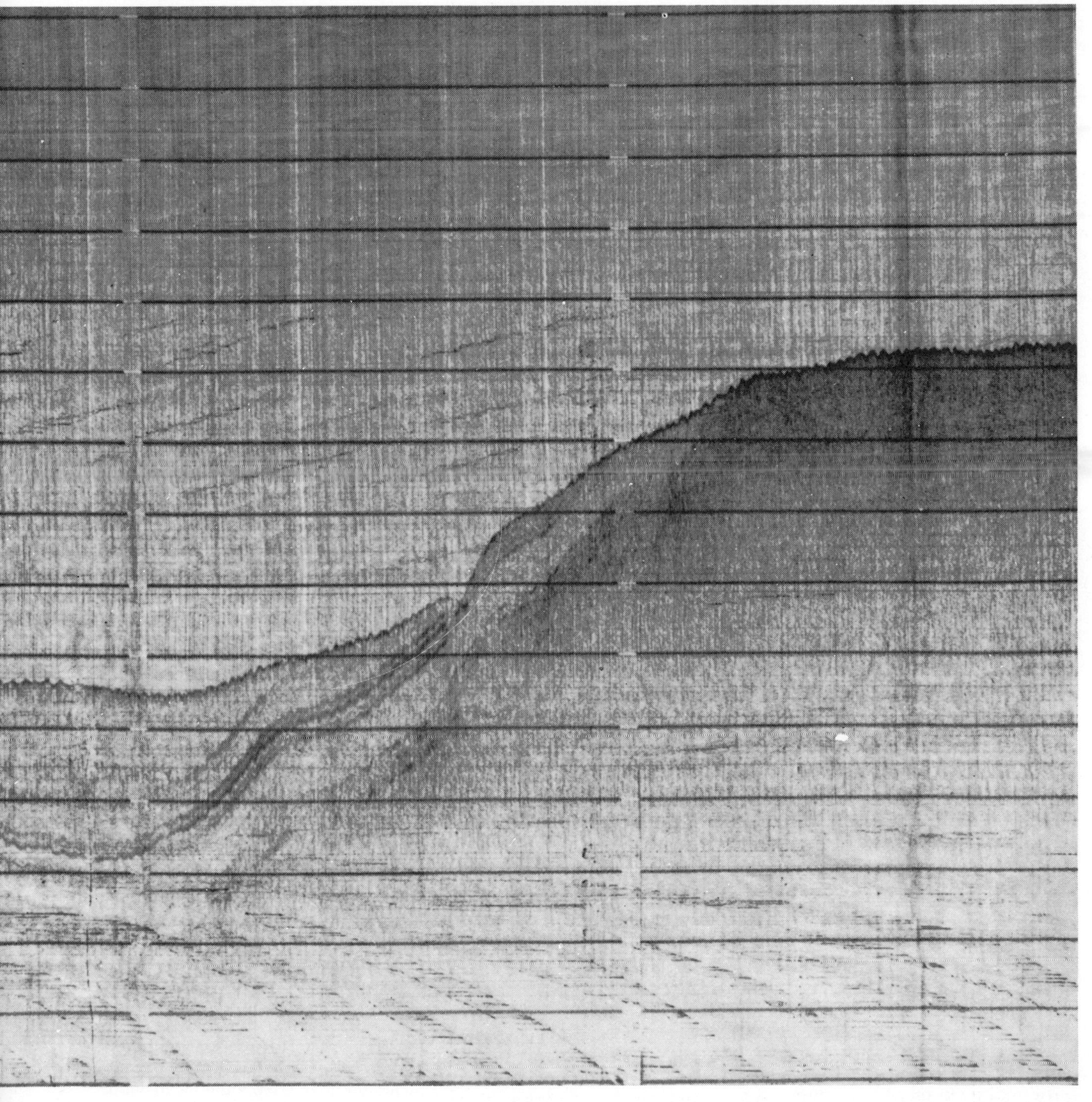

became filled with sediments about 30 meters thick. The recording was made this past July with high-frequency seismic equipment aboard the *Dolphin*, a vessel operated by the U.S. Geological Survey in cooperation with the Woods Hole Oceanographic Institution.

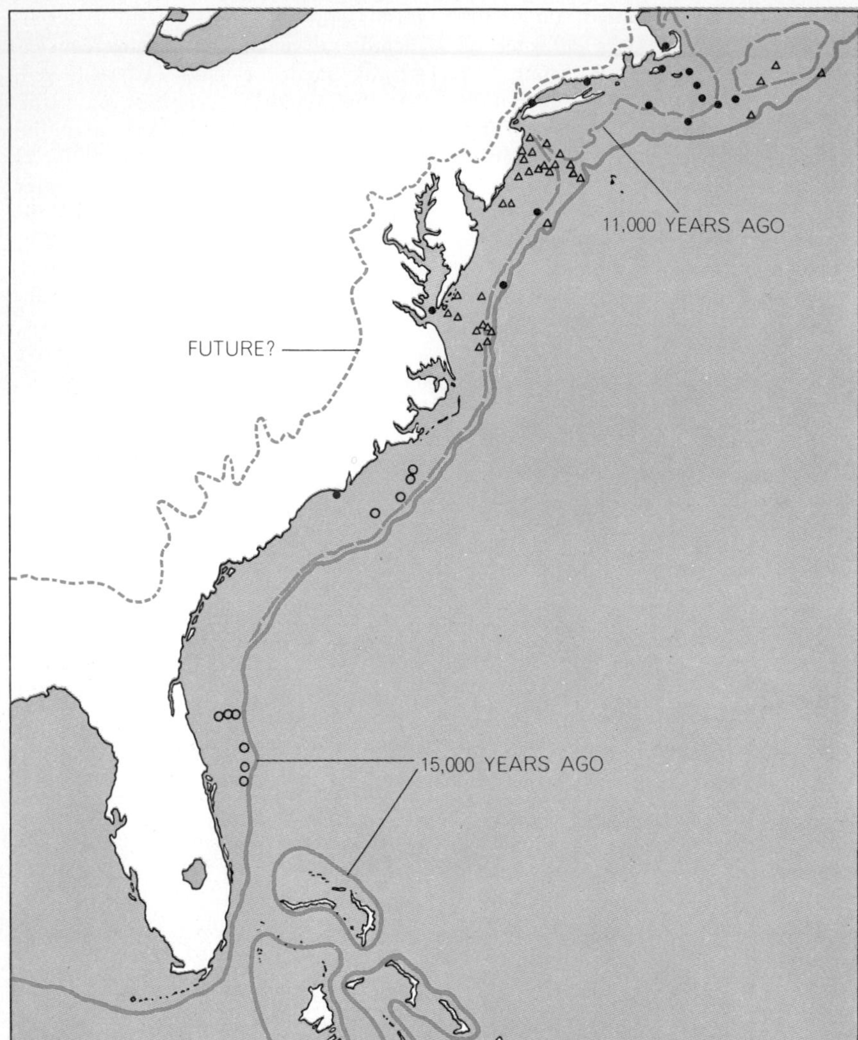

ATLANTIC COAST SHORELINE has varied greatly in the past and will undoubtedly continue to in the future. This illustration compares the shoreline of 15,000 and 11,000 years ago with the probable shoreline if all the ice at the poles were to melt. Confirmation that the continental shelf was once laid bare is found in discoveries of elephant teeth (*triangles*), freshwater peat (*dots*) and the shallow-water formations called oölites (*circles*).

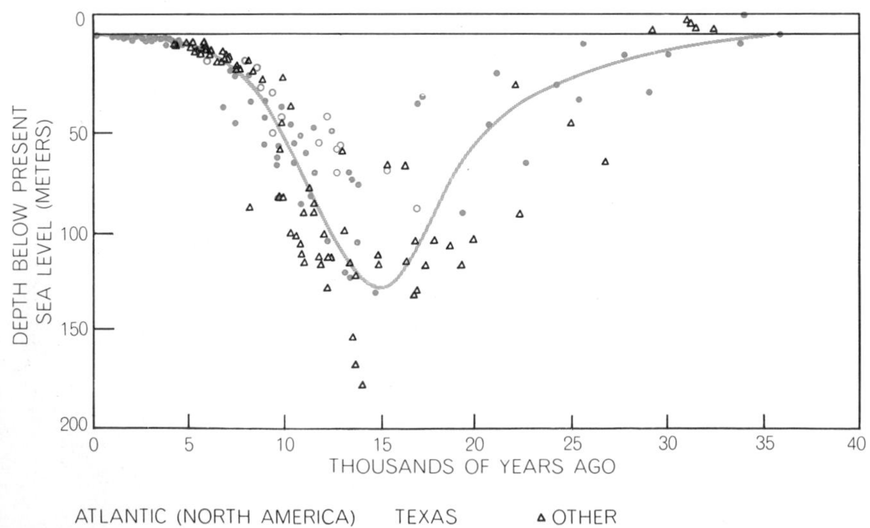

WORLDWIDE CHANGES IN SEA LEVEL can be inferred from the radiocarbon ages of shallow-water marine organisms and the depth at which they were recovered. Samples are from the Atlantic shelf of North America, the Texas shelf and other parts of the world. The depth inconsistency of the Texas samples implies that the shelf there has been uplifted.

cient times [*see bottom illustration at left*].

Apparently the sea was near its present level about 35,000 years ago and began to recede about 30,000 years ago. The level dropped by 130 meters or more 15,000 years ago; then it rose rather rapidly to within about five meters of the present level 5,000 years ago. The slow rise during the past 5,000 years has been documented by perhaps 100 carbon-14 dates for peat under existing salt marshes. Less complete sequences of dates for similar samples from elsewhere in the world show a sea-level curve resembling the one for the East Coast. Only the samples from the shelf in the western Gulf of Mexico provide a different curve, which suggests that this part of the gulf shelf was uplifted about 40 meters during the past 10,000 years.

Early men of the Clovis culture (characterized by fluted stone projectile points) appeared in North America some 12,000 years ago, when the sea level was still very low. What is more reasonable than to suppose such men ranged over the forested lowland that is now continental shelf? Game, fish and oysters were abundant. How were they to know or care that in a few thousand years the area was to be drowned by the advancing sea, any more than New Yorkers know or care that when the remaining glaciers melt, the ocean will rise to the 20th story of tall buildings? [*see top illustration at left*]. The search for traces of early man far out on the shelf began with the discovery of what may be the remains of an oyster dinner on a former beach off Chesapeake Bay, a site that is now 43 meters below sea level. This discovery was made from the Woods Hole Oceanographic Institution's research submarine *Alvin;* many similar discoveries will probably be made during the next decade.

Submerged barrier beaches are common on the continental shelf, but they are easily confused with the sand waves that are formed by strong currents. More spectacular and of certain origin are the submerged sea cliffs and terraces that mark the temporary stillstands of the sea level. Most of the shelves that have been studied have four to six such terraces, but the recognition of the terraces depends on their width and sharpness. On gently sloping shelves the terraces are almost imperceptible; on steep shelves they are narrow or absent; on shelves receiving a large supply of sediment they are buried. Variation in depth is to be expected in view of the large variation in depth of the most prominent terrace of all—the edge of the shelf. Passing through the terraces are channels cut by streams that flowed across the shelf when the sea level was low. Most of these channels have been filled by sediment; they can be recognized only by seismic profiling and by drilling on the shore at the mouths of stream valleys. Probably hundreds of channels cross the continental shelves of the U.S., but only a dozen are known. One channel, the one cut by the Hudson River off New York, is so large it is not yet filled with sediment.

At the seaward end of the channels, near the edge of the shelf, the channels are replaced by the heads of submarine canyons that continue down the continental slope to depths of several kilometers. The continuation of the submarine canyons to depths far below the maximum probable lowering of the sea level means that the canyons must have been formed by some process that operates under the ocean surface. Although the matter is still the subject of debate, most of the evidence favors the view that the canyons were excavated by turbidity currents: currents that arise when sediment slips down a slope and becomes mixed with overlying water, thereby increasing its density so that it continues down the slope, often at high speed. Today the shelf off the East Coast of the U.S. is only slightly modified by submarine canyons; only the heads of the canyons indent the shelf edge. When the sea level was at its lowest, the canyons were probably important factors in sedimentation. The shelf off the West Coast of the U.S. is so narrow that the heads of many canyons reach almost to the shore. In those areas the canyons serve to trap and divert sand that is moved along in the shore zone under the influence of wind-driven waves and their associated currents. As a result the sand that is brought to the shore by streams and cliff erosion is only temporarily added to the beaches; eventually it moves seaward through the canyons in the form of slow sand glaciers or rapid turbidity currents.

The water above the continental shelves is complex in composition and movement because it is shallow and close to the land. Large rivers contribute so much fresh water that they dilute the ocean, but they also increase the local concentration of calcium, phosphate, silica and nitrate—precisely those elements and compounds that elsewhere in the ocean have been reduced to low concentration by incorporation into marine plants and animals. Continental-shelf waters that are distant from river mouths are sometimes saltier than the open sea because their rate of evaporation is high. Local variations in salinity (and therefore density) control the direction of currents on the shelf. For example, the low salinity at the mouth of a river means a higher sea level near the shore than farther out on the shelf, leading to a flow toward the right (when one is facing the ocean in the Northern Hemisphere).

Just at the shore, however, the longshore currents are mainly controlled by the angle at which waves intersect the beach, which in turn is a function of the wind direction. As a result the cur-

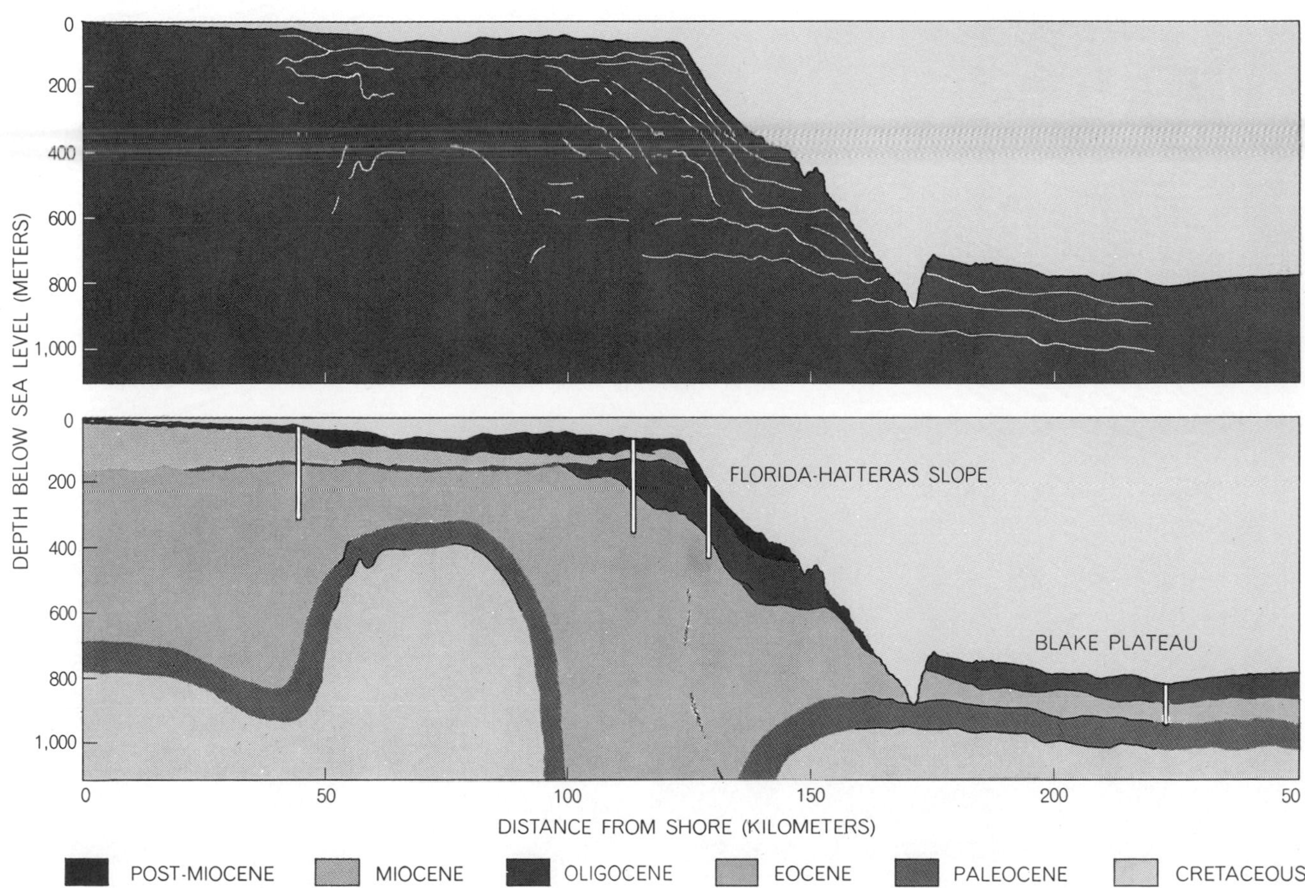

SHELF OFF JACKSONVILLE, FLA., has been studied by two geophysical methods: seismic reflection (*top*) and drilling (*bottom*). Seismic studies can show only the general nature of the stratigraphy. Cores obtained by the JOIDES project, a drilling study conducted by a consortium of institutions, made it possible to map the stratigraphy in considerable detail. The vertical scale is exaggerated 67 times. The approximate termination dates for the various geologic periods are as follows: Miocene, 10 million years ago; Oligocene, 25 million years ago; Eocene, 40 million years ago; Paleocene, 55 million years ago; Cretaceous, 65 million years ago.

rent in the wave zone may be northward, the current on the inner half of the shelf may be southward and the current on the outer shelf may be northward again (for example where an oceanic current such as the Gulf Stream runs along the edge of the shelf). Where rivers bring water to the ocean there must be a general current component toward the ocean at the surface; this induces a return flow toward the land at the bottom [*see top illustration on page 43*]. Thus the sediment on the sea floor may be moved landward often working its way into the mouths of estuaries. This means estuaries are truly ephemeral features, receiving sediments from both rivers and the open shelf.

Temperature zones on land are mainly a function of latitude, with secondary modifications resulting from winds whose direction may change seasonally or may be controlled by topography. Similarly, ocean water is cooled at high

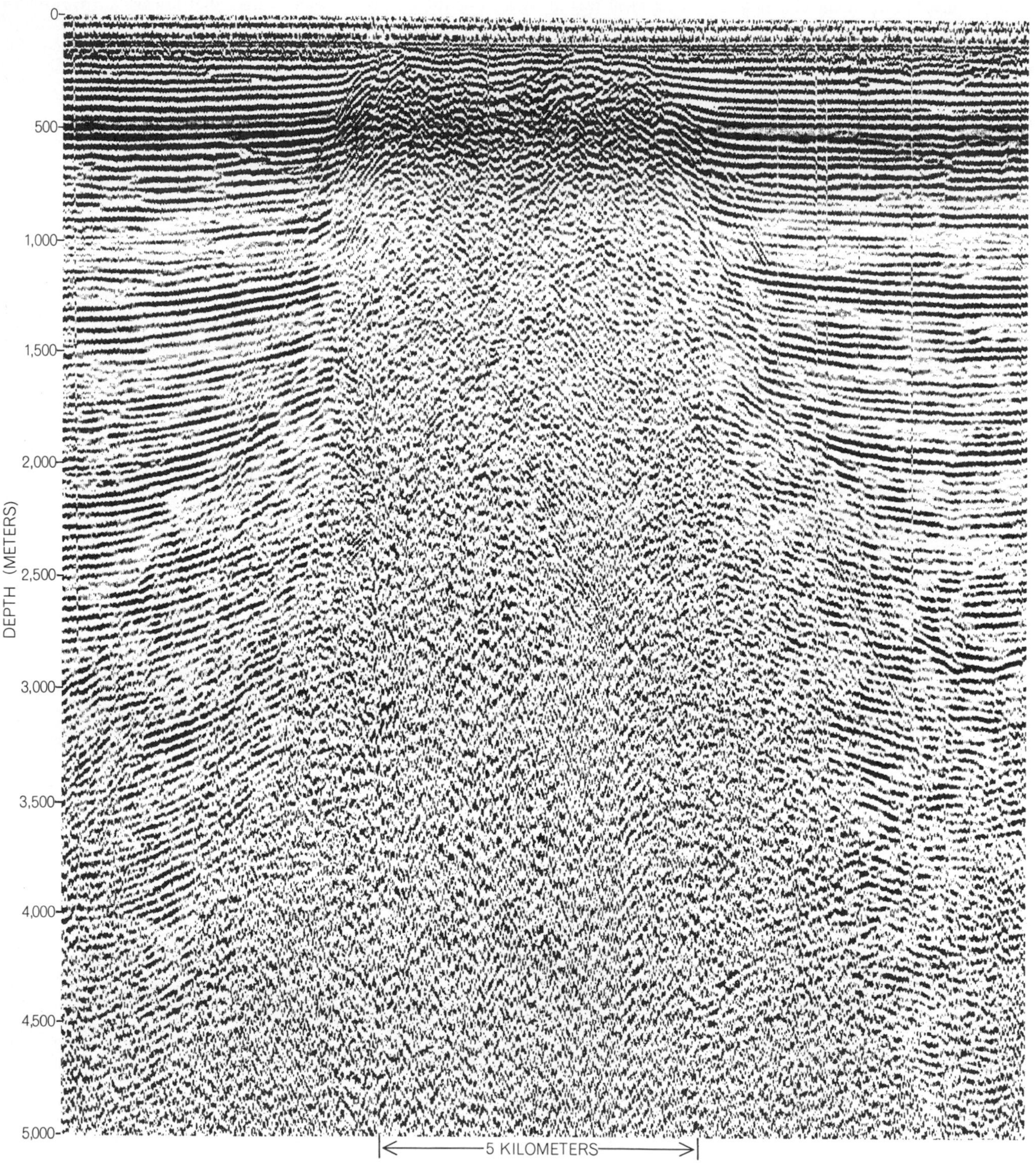

SALT DOME bulging upward into the continental shelf about 10 miles south of Galveston in the Gulf of Mexico is shown in this seismic record. The water is so shallow (between 10 and 20 meters) that the reflection from the surface of the shelf is virtually at the top of the recording. Geologists can discern significant features in such a record down to a depth of about 3,000 meters. The record was made by the Teledyne Exploration Company. The salt dome was subsequently drilled and was found to contain hydrocarbons.

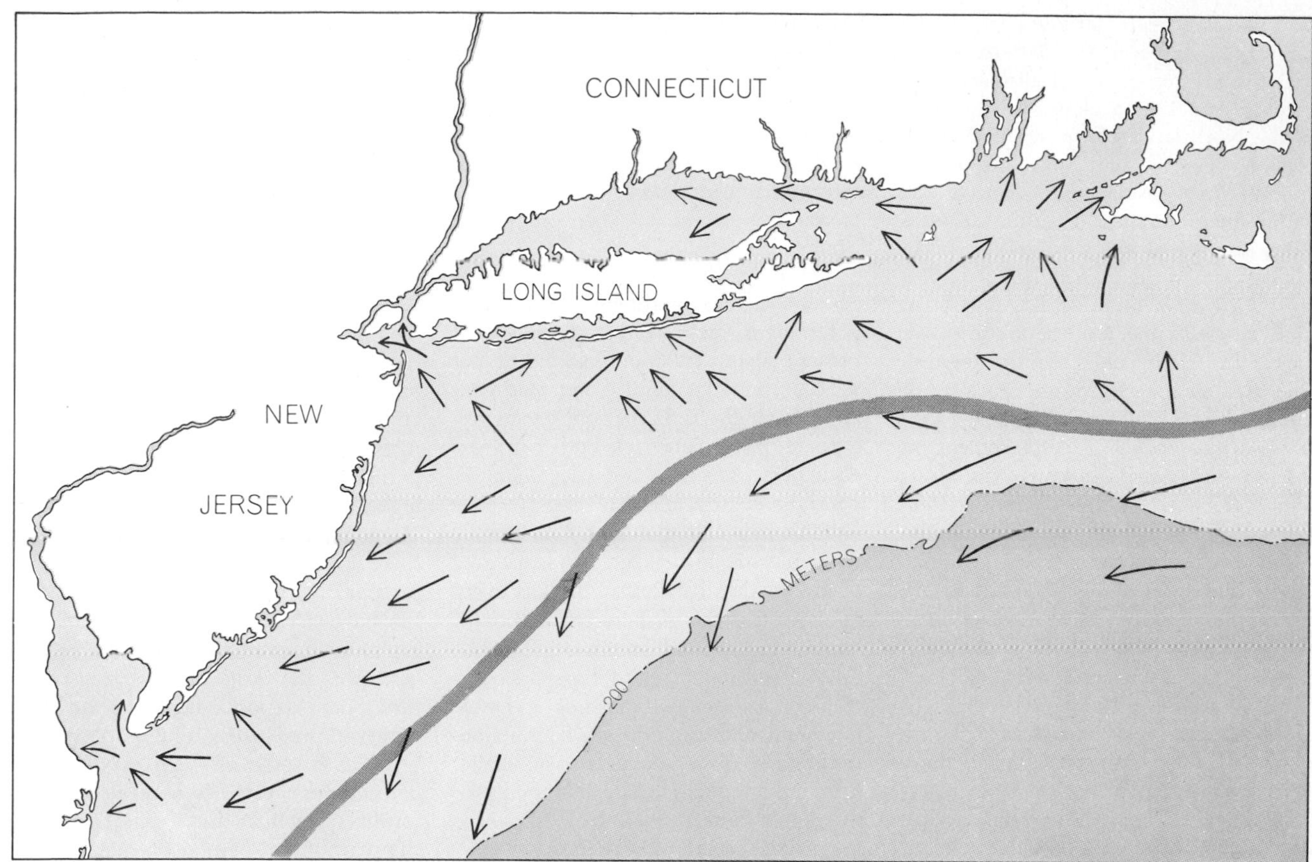

BOTTOM CURRENTS, indicated by arrows, can be traced with the help of simple plastic devices called bottom-drifters. The gray band marks the boundary between landward flow and seaward flow. The broken line represents the edge of the continental shelf.

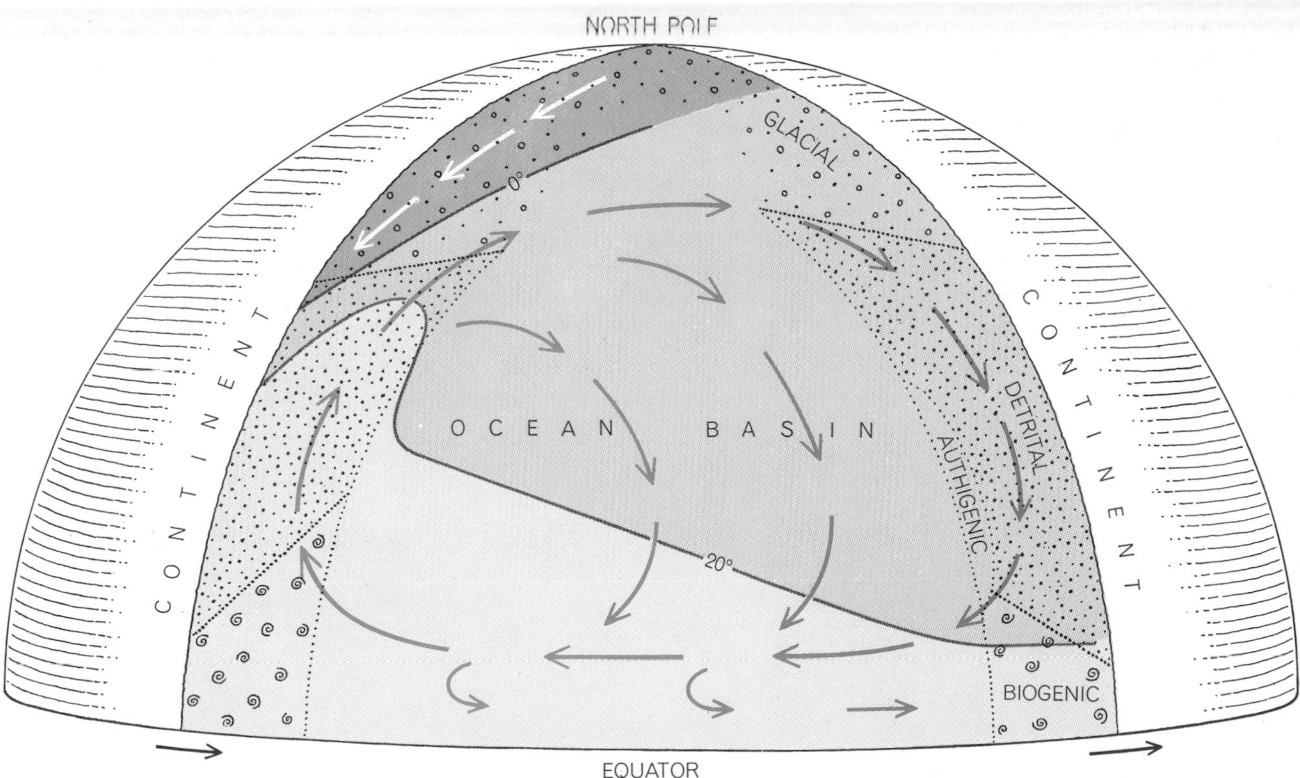

CHARACTER OF SHELF SEDIMENTS around an ocean basin, shown here schematically, is heavily influenced by oceanic currents. The rotation of the earth produces a clockwise flow in the Northern Hemisphere, so that the western edge of the basin, up to a certain latitude, is warmer than the eastern edge. Thus biogenic sediments extend farther north on the west than on the east. The effect would be greater except for a counterflow of arctic water on the western side of the basin. Detrital sediments are the typical outwash of continents. Authigenic sediments are minerals that come out of solution under suitable conditions and fall to the ocean floor.

latitudes and heated at low ones. At the same time, however, the pattern of currents in the open ocean displaces the climatic zones in a clockwise direction in the Northern Hemisphere and counterclockwise in the Southern Hemisphere. The displacement causes the water above the shelf at middle latitudes to be warmer on the western side of an ocean than on the eastern side [*see bottom illustration on preceding page*]. At high latitudes the flow of arctic water makes the shelf colder on the western side than on the eastern one. As a result the correlation of animal species with shelf latitude shows a displacement on the opposite sides of oceans. Moreover, temperature zones are compressed on the western side and expanded on the eastern. The movements of currents, waves and tides above the shelves are so complex that they have received little study compared with those of the deep ocean. Much fieldwork is needed.

The present great interest in exploring the world's continental shelves flows from their potential economic exploitation. About 90 percent of the world's marine food resources, now extracted at the rate of $8 billion per year, comes from the shelves and adjacent bays [see the article "The Food Resources of the Ocean," by S. J. Holt, beginning on page 210]. Most of this is fish for human and animal consumption; the remainder is largely used for fertilizer.

Second in economic importance is petroleum and natural gas from the shelf; their present annual value is about $4 billion, representing nearly a fifth of the total world production of these substances [see the article "The Physical Resources of the Ocean," by Edward Wenk, Jr., beginning on page 257]. Currently about $1 billion worth of oil and gas a year is extracted from the shelves off the U.S., and much of the rest was developed by American companies with interests abroad. It is safe to predict that the future production from the world's continental shelves will increase at a greater rate than production from wells drilled on land.

The third marine resource in terms of present annual production and future potential is lowly sand and gravel. At present about $200 million worth per year is mined for landfill and road construction in the U.S., for concrete aggregate in Britain and for both purposes elsewhere. As cities and megalopolises continue to grow and show a preference for the coastal regions, and as readily available stream deposits are exhausted or are overlain by houses, there is every prospect that the offshore production of sand and gravel will increase greatly.

We read much about the possibility of economic exploitation of valuable heavy minerals from the sea floor, namely ilmenite, rutile, zircon, tin, monazite, iron, gold and diamonds. The total production of these minerals from below the sea is now less than $50 million per year. Production may increase, particularly in the case of tin, but it is decreasing for iron. Prospects for gold are not very hopeful, and diamonds have never been mined profitably from the sea floor. The basic problem is that economic placer deposits of tin and gold are found only within a few kilometers of the original igneous sources, and few continental shelves contain metalliferous igneous sources. Similarly, ilmenite, rutile, zircon and monazite require the high-energy wave environment of beaches to form deposits that are concentrated enough and large enough to be mined at a profit. When ancient beach deposits are submerged, even if they are not buried under worthless sediment or mixed with it, the cost of mining increases substantially. They will probably be mined in the future but not until they are economically competitive with shoreline deposits. This could come about either through a rise in prices, resulting from a diminution of known deposits on land, or when more efficient mining and separation methods are devised for the marine environment.

Phosphorite is present in large quantities on shelves off southern California, Peru, southeastern Africa, northeastern Africa and Florida. It can be mined off the U.S., but it has to compete with high-grade land deposits in Florida, Montana, Idaho and Wyoming (where there is about a 1,000-year supply at present rates of mining). Most investigators have concluded that the cost of mining at sea exceeds the cost of mining on land plus the costs of land transportation. Some deposits far from the U.S., however, may justify mining, particularly because some of them may be near places where there is a great need for fertilizer, such as India. Unfortunately the distribution of phosphorite in these areas is poorly known, and little or no effort is currently being expended on their investigation.

The would-be exploiter of the ocean will do well to remember the words of the old Newfoundland skipper, "We don't be *takin'* nothin' from the sea. We has to sneak up on what we wants and wiggle it away." Nevertheless, the continental shelves, when they are properly investigated, promise to greatly increase our knowledge of the earth's history and to become a steadily more important source of food and raw materials.

The Origin of the Oceans **4**

by Sir Edward Bullard
September 1969

In recent years it has become increasingly apparent that the floor of the deep ocean is remarkably young. It is growing outward from mid-ocean ridges, pushing most of the continents apart as it does

The earth is uniquely favored among the planets: it has rain, rivers and seas. The large planets (Jupiter, Saturn, Uranus and Neptune) have only a small solid core, presumably overlain by gases liquefied by pressure; they are also surrounded by enormous atmospheres. The inner planets are more like the earth. Mercury, however, has practically no atmosphere and the side of the planet facing the sun is hot enough to melt lead. Venus has a thick atmosphere containing little water and a surface that, according to recent measurements, may be even hotter than the surface of Mercury. Mars and the moon appear to show us their primeval surfaces, affected only by craters formed by the impact of meteorites, and perhaps by volcanoes. Only on the earth has the repetition of erosion and sedimentation —"the colossal hour glass of rock destruction and rock formation"—run its course cycle after cycle and produced the diverse surface that we see. The mountains are raised and then worn away by falling and running water; the debris is carried onto the lowlands and then out to the ocean. Geologically speaking, the process is rapid. The great plateau of Africa is reduced by a foot in a few thousand years, and in a few million years it will be near sea level, like the Precambrian rocks of Canada and Finland. All trace of the original surface of the earth has been removed, but as far back as one can see there is evidence in rounded, water-worn pebbles for the existence of running water and therefore, presumably, of an ocean and of dry land.

The obvious things that no one comments on are often the most remarkable; one of them is the constancy of the total volume of water through the ages. The level of the sea, of course, has varied from time to time. During the ice ages, when much water was locked up in ice sheets on the continents, the level of the sea was lower than it is at present, and the continental shelves of Europe and North America were laid bare. Often the sea has advanced over the coastal plains, but never has it covered all the land or even most of it. The mechanism of this equilibrium is unknown; it might have been expected that water would be expelled gradually from the interior of the earth and that the seas would grow steadily larger, or that water would be dissociated into hydrogen and oxygen in the upper atmosphere and that the hydrogen would escape, leading to a gradual drying up of the seas. These things either do not happen or they balance each other.

The mystery is deepened by the almost complete loss of neon from the earth; in the sun and the stars neon is only a little rarer than oxygen. The neon was presumably lost when the earth was built up from dust and solid grains because neon normally does not form compounds, but if that is so, why was the water not lost too? Water has a molecular weight of 18, which is less than the atomic weight of neon, and thus should escape more easily. It looks as if the water must have been tied up in compounds, perhaps hydrated silicates, until the earth had formed and the neon had escaped. Water must then have been released as a liquid sometime during the first billion years of the earth's history, for which we have no geological record. The planet Mercury and the moon would have been too small to retain water after it was released. Mars seems to have been able to retain a trace, not enough to make oceans but enough to be detectable by spectroscopy.

These speculations about the early history of the earth are open to many doubts. The evidence is almost nonexistent, and all one can say is, "It might have been that...." The great increase in understanding of the present state and recent history of the ocean basins that we have gained in the past 20 years is something quite different. For the first time the geology of the oceans has been studied with energy and resources commensurate with the tremendous task. It turns out that the main processes of geology can be understood only when the oceans have been studied; no amount of effort on land could have told us what we now know. The study of marine geology has unlocked the history of the oceans, and it seems likely to make intelligible the history of the continents as well. We are in the middle of a rejuvenating process in geology comparable to the one that physics experienced in the 1890's and to the one that is now in progress in molecular biology.

The critical step was the realization that the oceans are quite different from the continents. The mountains of the oceans are nothing like the Alps or the Rockies, which are largely built from folded sediments. There is a world-encircling mountain range—the mid-ocean ridge—on the sea bottom, but it is built entirely of igneous rocks, of basalts that have emerged from the interior of the earth. Although the undersea mountains have a covering of sediments in

⟶

RED SEA and the Gulf of Aden represent two of the newest seaways created by the worldwide spreading of the ocean floor. In this photograph, taken at an altitude of 390 miles from the spacecraft *Gemini 11* in September, 1966, the Red Sea separates Ethiopia (*at left*) from the Arabian peninsula (*at right*). The Gulf of Aden lies between the southern shore of Arabia and Somalia. The excellent fit between drifting land masses is depicted in the illustrations on page 47.

many places, they are not made of sediments, they are not folded and they have not been compressed.

A cracklike valley runs along the crest of the mid-ocean ridge for most of its length, and it is here that new ocean floor is being formed today [*see illustration on next two pages*]. From a study of the numerous earthquakes along this crack it is clear that the two sides are moving apart and that the crack would continually widen if it were not being filled with material from below. As the rocks on the two sides move away and new rock solidifies in the crack, the events are recorded by a kind of geological tape recorder: the newly solidified rock is magnetized in the direction of the earth's magnetic field. For at least the past 10,000 years, and possibly for as long as 700,000 years, the north magnetic pole has been close to its present location, so that the magnetic field is to the north and downward in the Northern Hemisphere, and to the north and upward in the Southern Hemisphere. As the cracking and the spreading of the ocean floor go on, a strip of magnetized rock is produced. Then one day, or rather in the course of several thousand years, the earth's field reverses, the next effusion of lava is magnetized in the reverse direction and a strip of reversely magnetized rocks is built up between the two split halves of the earlier strip. The reversals succeed one another at widely varying intervals; sometimes the change comes after 50,000 years, often there is no change for a million years and occasionally, as during the Permian period, there is no reversal for 20 million years. The sequence of reversals and the progress of spreading is recorded in all the oceans by the magnetization of the rocks of the ocean floor. The message can be read by a magnetometer towed behind a ship.

We now have enough examples of these magnetic messages to leave no doubt about what is happening. It is a truly remarkable fact that the results of magnetic surveys in the South Pacific can be explained—indeed predicted—from the sequence of reversals of the direction of the earth's magnetic field known from magnetic and age measurements, made quite independently on lavas in California, Africa and elsewhere. The only adjustable factor in the calculation is the rate of spreading. Such worldwide theoretical ideas and such detailed agreement between calculation and theory are rare in geology, where theories are usually qualitative, local and of little predictive value.

The speed of spreading on each side of a mid-ocean ridge varies from less than a centimeter per year to as much as eight centimeters. The fastest rate is the one from the East Pacific Rise and the slowest rates are those from the Mid-Atlantic Ridge and from the Carlsberg Ridge of the northwest Indian Ocean. The rate of production of new terrestrial crust at the central valley of a ridge is the sum of the rates of spreading on the two sides. Since the rates on the two sides are commonly almost equal, this sum is twice the rate on each side and may be as much as 16 centimeters (six inches) per year. Such rates are, geologically speaking, fast. At 16 centimeters per year the entire floor of the Pacific Ocean, which is about 15,000 kilometers (10,000 miles) wide, could be produced in 100 million years.

When the mid-ocean ridges are examined in more detail, they are found not to be continuous but to be cut into sections by "fracture zones" [*see top illustration on page 50*]. A study of the earthquakes on these fracture zones shows that the separate pieces of ridge crest on the two sides of a fracture zone are not moving apart, as might seem likely on first consideration. The two pieces of ridge remain fixed with respect to each other while on each side a plate of the crust moves away as a rigid body; such a fracture is called a transform fault. The earthquakes occur only on the piece of the fracture zone between the two ridge crests; there is no relative motion along the parts outside this section.

If two rigid plates on a sphere are spreading out on each side of a ridge that is crossed by fracture zones, the relative motion of the two plates must be a rotation around some point, termed the pole of spreading. The "axis of spreading," around which the rotation takes place, passes through this pole and the center of the earth. The existence of a pole of spreading and an axis of spreading is geometrically necessary, as was shown by Leonhard Euler in the 18th century. If the only motion on the fracture zones is the sliding of the two plates past each other, then the fracture zones must lie along circles of latitude with respect to the pole of spreading, and the rates of spreading at any point on the ridge must be proportional to the perpendicular distance from the point to the axis of spreading [*see bottom illustration on page 50*].

All of this is well verified for the spreading that is going on today. The rates of spreading can be obtained from the magnetic patterns and the dates of the reversals. The poles of spreading can be found from the directions of the frac-

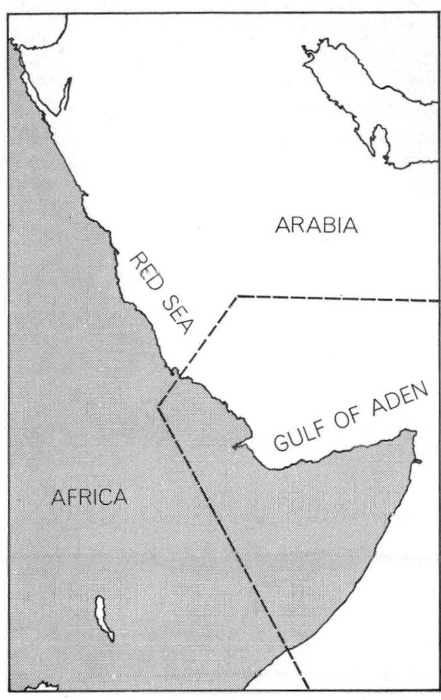

RUPTURE OF MIDDLE EAST is being caused by the widening of the Red Sea and the Gulf of Aden. Some 20 million years ago the Arabian peninsula was joined to Africa, as evidenced by the remarkable fit between shorelines (*see illustration below*). The area within the *Gemini 11* photograph on page 46 is shown by the broken lines.

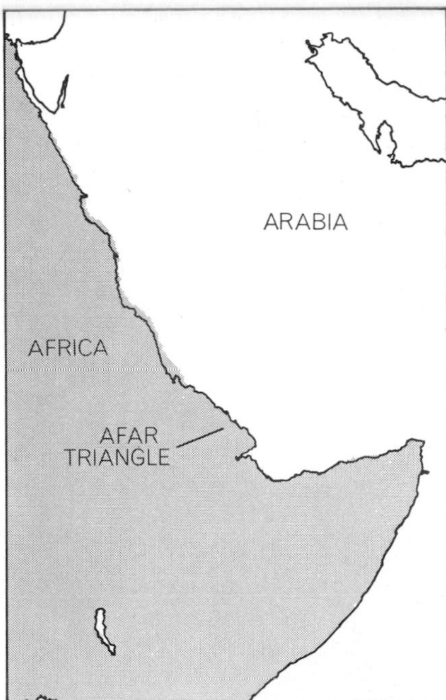

FIT OF SHORELINES of Arabia and Africa works out most successfully if the African coast (*black*) is left intact and if the Arabian coast (*color*) is superposed in two separate sections. In the reconstruction a corner of Arabia overlaps the "Afar triangle" in northern Ethiopia, an area that now has some of the characteristics of an ocean floor.

ture zones and checked by the direction of earthquake motions. It turns out that the ridge axes and the magnetic pattern are usually almost at right angles to the fracture zones. This is not a geometrical necessity, but when it does happen it means that the lines of the ridge axes and of the magnetic pattern must, if they are extrapolated, go through the pole of spreading. If the ridge consists of a number of offset sections at right angles to the fracture zones, the axes of these sections will converge on the pole of spreading. It is one of the surprises of the work at sea that this rather simple geometry embraces so large a part of the facts. It seems that marine geology is truly simpler than continental geology and that this is not merely an illusion based on our lesser knowledge of the oceans.

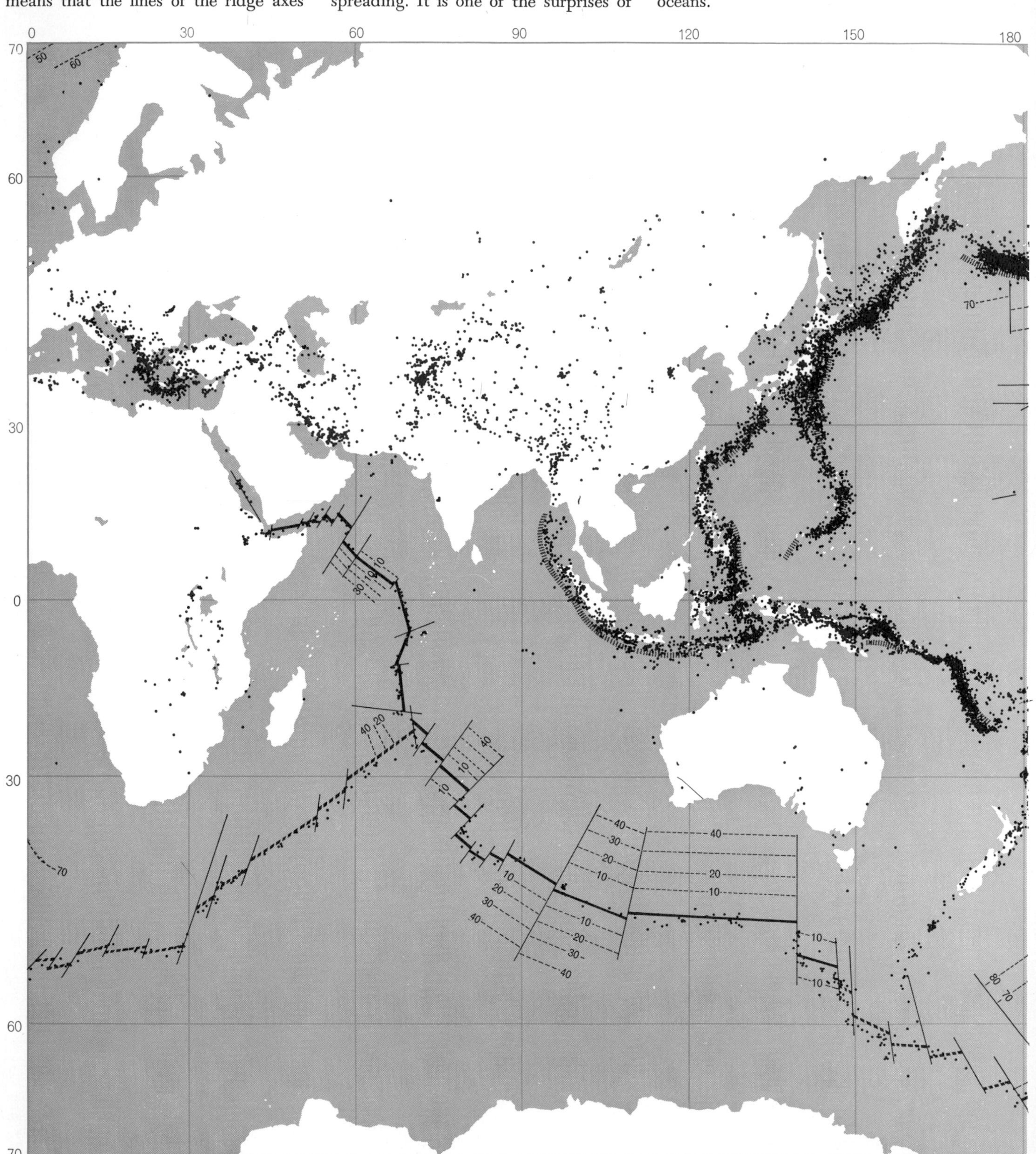

OCEANIC GEOLOGY has turned out to be much simpler than the geology of the continents. New ocean bottom is continuously being extruded along the crest of a worldwide system of ridges (*thick black lines*). The present position of material extruded at intervals of 10 million years, as determined by magnetic studies, appears as broken lines parallel to the ridge system, which is offset by fracture

The regularity of the magnetic pattern suggests that the ocean floor can move as a rigid plate over areas several thousand kilometers across. The thickness of the rigid moving plate is quite uncertain, but a value between 70 and 100 kilometers seems likely. If this is so, the greater part of the plate will be made of the same material as the upper part of the earth's mantle—probably of peridotite, a rock largely composed of olivine, a silicate of magnesium and iron, $(Mg, Fe)_2SiO_4$. The basaltic rocks of the oceanic crust will form the upper five kilometers or so of the plate, with a veneer of sediments on top.

What happens at the boundary of an ocean and a continent? Sometimes, as in the South Atlantic, nothing happens; there are no earthquakes, no distortion, nothing to indicate relative motion be-

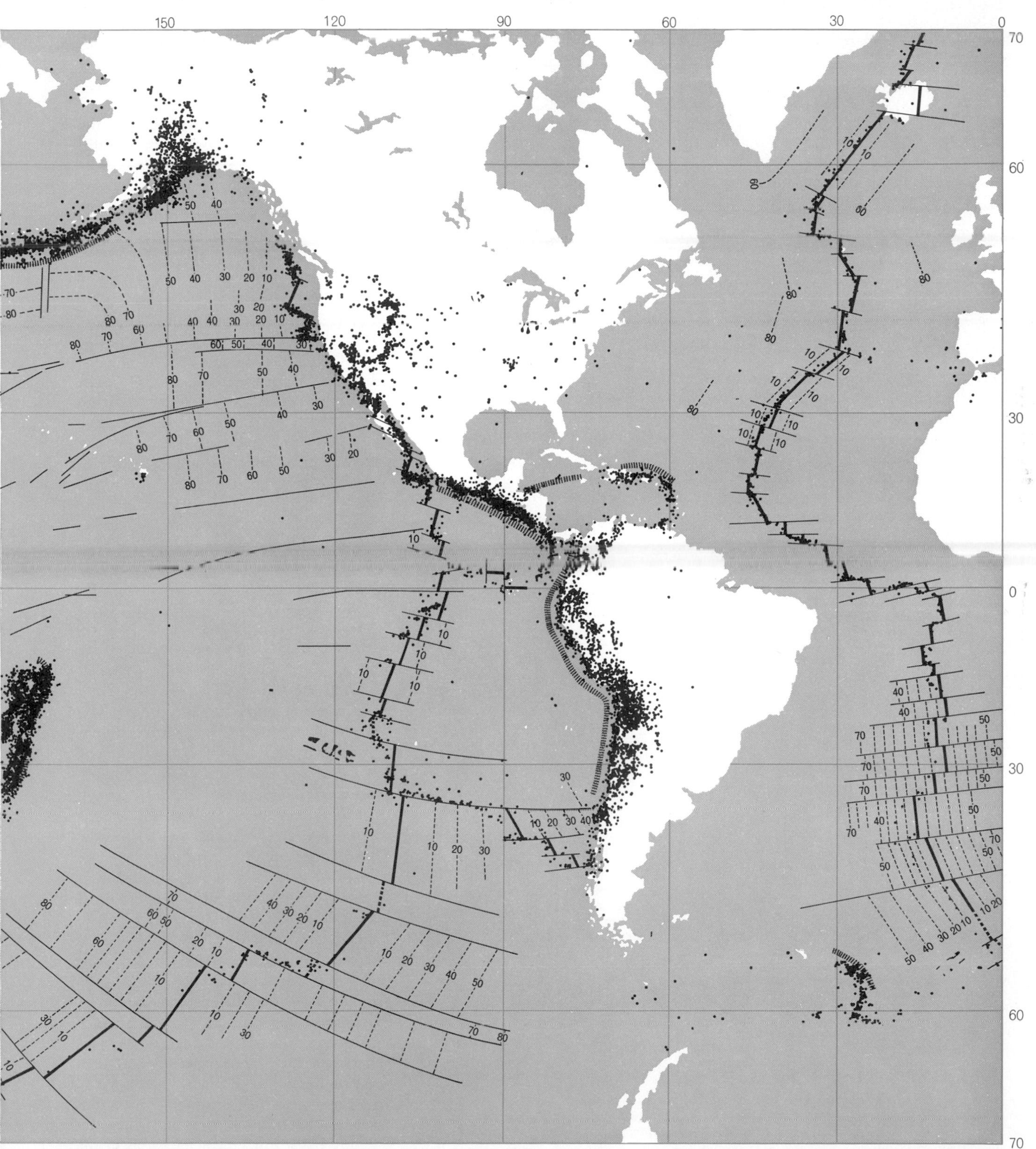

zones (*thin black lines*). Earthquakes (*black dots*) occur along the crests of ridges, on parts of the fracture zone and along deep trenches. These trenches, where the ocean floor dips steeply, are represented by hatched bands. At the maximum estimated rate of sea-floor spreading, about 16 centimeters a year, the entire floor of the Pacific Ocean could be created in perhaps 100 million years.

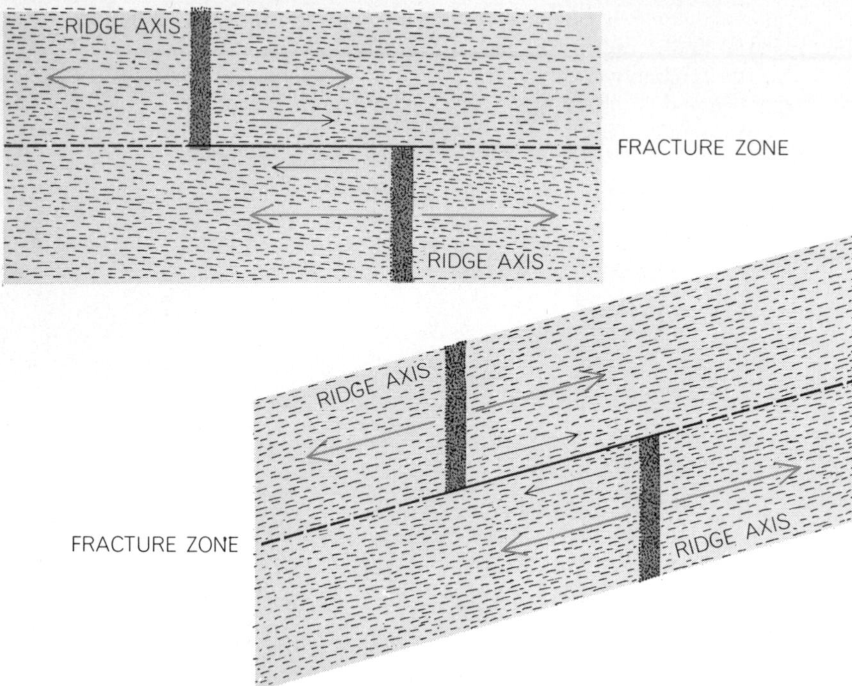

MOTION AT AXIS OF RIDGE consists of an opening of an axial crack (*vertical bands*) where two plates separate (*arrows*). Often the ridge is offset by a fracture zone, making a transform fault where one plate slips past another. The motion must be parallel to the fracture zone. It is usually at right angles to the ridge (*upper left*) but need not be (*lower right*).

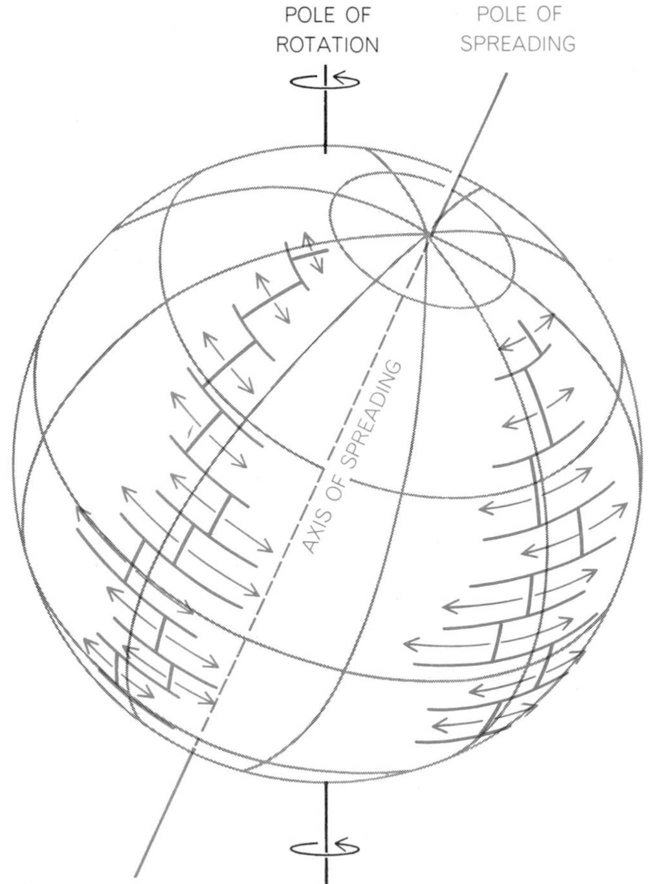

MOTION OF RIGID PLATES on a sphere requires that the plates rotate around a "pole of spreading" through which passes an "axis of spreading." Plates always move parallel to the fracture zones and along circles of latitude perpendicular to the axis of spreading. The rate of spreading is slowest near the pole of spreading and fastest 90 degrees away from it. The spreading pole can be quite remote from the sphere's pole of rotation.

PROBABLE ARRANGEMENT of continents before the formation of the Atlantic Ocean was determined by the author with the aid of a computer. The fit was made not at the present coastlines but at the true edge of each continent, the line where the continental shelf (*dark brown*) slopes down steeply to the sea floor. Overlapping land and shelf areas are reddish orange; gaps where the continental edges do not quite meet are dark blue. At present the entire western Atlantic is moving as one great plate carrying both North America and South America with it. At an earlier period the two continents must have moved independently.

tween the sea floor and the bordering continent. The continent can then be regarded as part of the same plate as the adjacent ocean floor; the rocks of the continental crust evidently ride on top of the plate and move with it. In other places there is another kind of coast, what Eduard Suess called "a Pacific coast." Such a coast is typified by the Pacific coast of South America. Here the oceanic plate dives under the continent and goes down at an angle of about 45 degrees. On the upper surface of this sloping plate there are numerous earthquakes—quite shallow ones near the coast and others as deep as 700 kilometers inland, under the continent. The evidence for the sinking plate has been beautifully confirmed by the discovery that seismic waves from shallow earthquakes and explosions, occurring near the place where the plate starts its dive, travel faster down the plate than they do in other directions. This is expected because the plate is relatively cold, whereas the upper mantle, into which the plate is sinking, is made of similar material but is hot.

Little is known of the detailed behavior of the plate; further study is vital for an understanding of the phenomena along the edges of continents. Near the point where the plate turns down there is an ocean deep, whose mode of formation is not precisely understood, but if a plate goes down, it is not difficult to imagine ways in which it could leave a depression in the sea floor. It is probable that, as the plate goes down, some of the sediment on its surface is scraped off and piled up in a jumbled mass on the landward side of the ocean deep. This sediment may later be incorporated in the mountain range that usually appears on the edge of the continent. The mountain range bordering the continent commonly has a row of volcanoes, as in the Andes. The lavas from the volcanoes are frequently composed of andesites, which are different from the lavas of the mid-

ocean ridges in that they contain more silica. It may reasonably be supposed that they are formed by the partial melting of the descending plate at a depth of about 150 kilometers. The first material to melt will contain more silica than the remaining material; it is also possible that the melted material is contaminated by granite as it rises to the surface through the continental rocks.

In many places the sinking plate goes down under a chain of islands and not under the continent itself. This happens in the Aleutians, to the south of Indonesia, off the islands of the Tonga group, in the Caribbean and in many other places. The volcanoes are then on the islands and the deep earthquakes occur under the almost enclosed sea behind the chain or arc of islands, as they do in the Sea of Japan, the Sea of Okhotsk and the Java Sea.

The destruction of oceanic crust explains one of the great paradoxes of geology. There have always been oceans, but the present oceans contain no sediment more than 150 million years old and very little sediment older than 80 million years. The explanation is that the older sediments have been carried away with the plates and are either piled up at the edge of a continent or are carried down with a sinking plate and lost in the mantle.

The picture is simple: the greater part of the earth's surface is divided into six plates [*see illustration on these two pages*]. These plates move as rigid bodies, new material for them being produced from the upper mantle by lava emerging from the crack along the crest of a mid-ocean ridge. Plates are destroyed at the oceanic trenches by plunging into the mantle, where ultimately they are mixed again with the material whence they came. The scheme is not yet established in all its details. Perhaps the greatest uncertainty is in the section of the ridge running south of South Africa; it is not clear how much of this is truly a ridge and a source of new crust and how much is a series of transform faults with only tangential motion. It is also uncertain whether the American and Eurasian plates meet in Alaska or in Siberia. It appears certain, however, that they do not meet along the Bering Strait.

A close look at the system of ridges, fracture zones, trenches and earthquakes reveals many other features of great interest, which can only be mentioned here. The Red Sea and the Gulf of Aden appear to be embryo oceans [*see illustration on page 46*]. Their floors are truly oceanic, with no continental rocks; along their axes one can find offset lengths of crack joined by fracture zones, and magnetic surveys show the

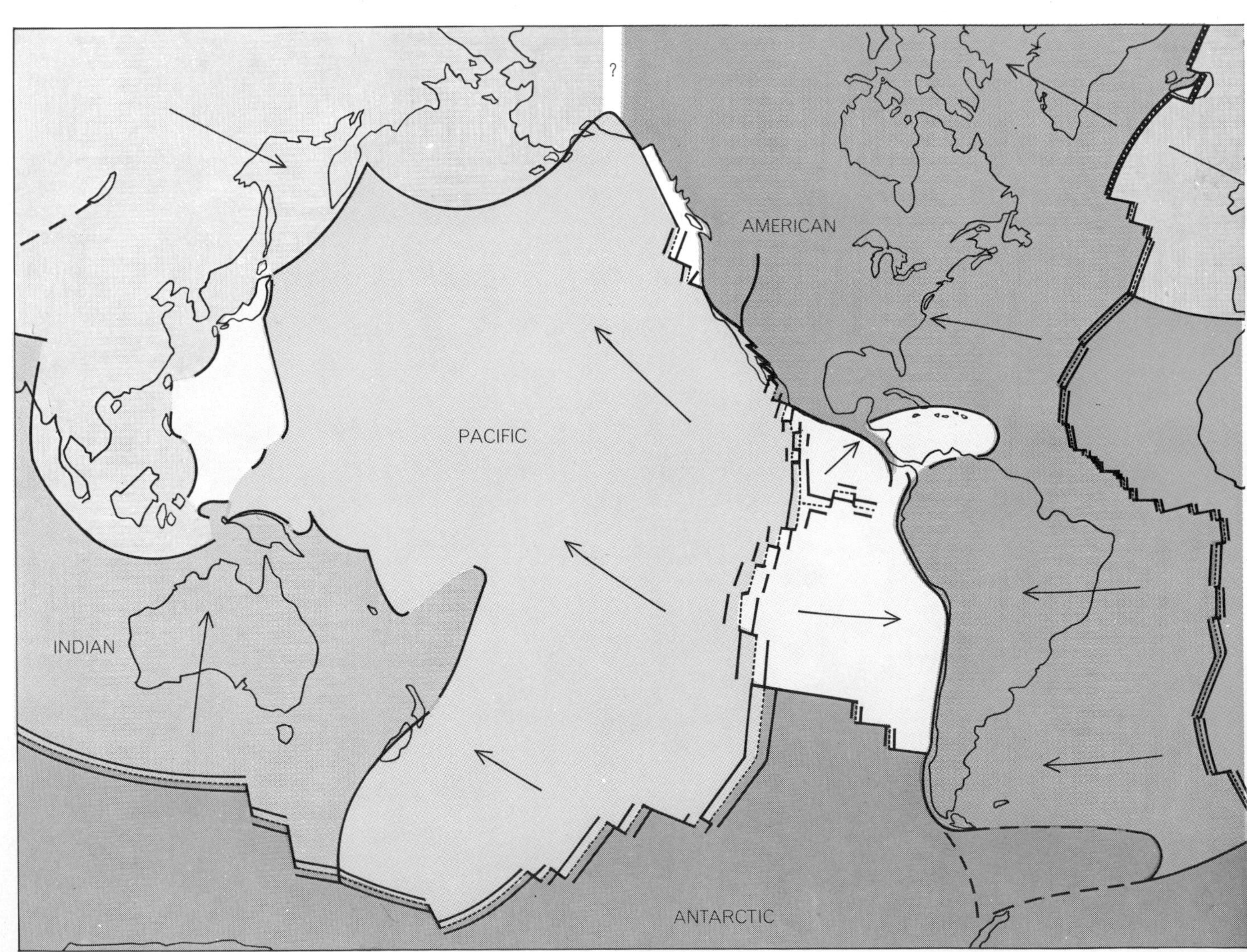

SIX MAJOR PLATES are sufficient to account for the pattern of continental drift inferred to be taking place today. In this model the African plate is assumed to be stationary. Arrows show the direction of motion of the five other large plates, which are generally bounded by ridges or trenches. Several smaller plates, unnamed, also appear. In certain areas, particularly at the junction of

worldwide magnetic pattern but only the most recent parts of it. These seas are being formed by the movement of Africa and Arabia away from each other. A detailed study of the geology, the topography and the present motion suggests that the separation started 20 million years ago in the Miocene period and that it is still continuing. If this is so, there must have been a sliding movement along the Jordan rift valley, with the area to the east having moved about 100 kilometers northward with respect to the western portion. There must also have been an opening of the East African rift valley by 65 kilometers or so.

The first of these displacements is well established by geological comparisons between the two sides of the valley, and it should be possible to verify the second. The reassembly of the pieces requires that the southwest corner of Arabia overlap the "Afar triangle" in northern Ethiopia [*see bottom illustration on page 47*]. This area should therefore be part of the embryo ocean. The fact that it is dry land presented a substantial puzzle, but recently it has been shown that the oceanic magnetic pattern extends over the area; it is the only land area in the world where this is known to happen. It seems likely that the Afar triangle is in some sense oceanic. The results of gravity surveys, seismic measurements and drilling will be awaited with interest. On this picture Arabia and the area to the north comprise a small plate separate from the African and Asian plates. The northern boundary of this small plate may be in the mountains of Iran and Turkey, where motion is proceeding today.

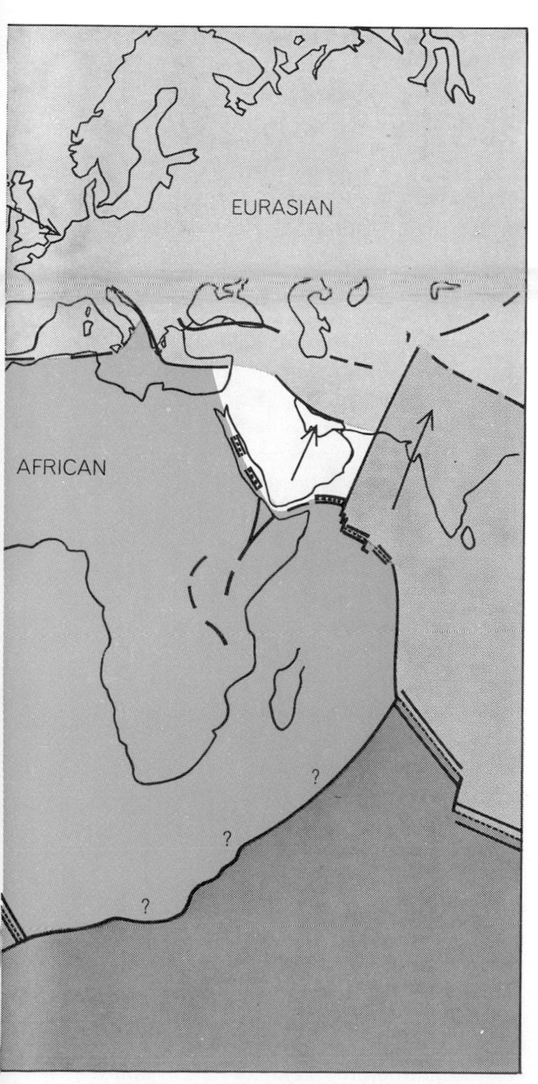

the American and Eurasian plates and in the region south of Africa, it is hard to say just where the boundaries of the plates lie.

A number of other small plates are known. There is one between the Pacific coast of Canada and the ridge off Vancouver Island; it is probable that this is being crumpled at the coast rather than diving under the continent. Farther south the plate and the ridge from which it spread may have been overrun by the westward motion of North America. The ridge appears again in the Gulf of California, which is similar in many ways to the Red Sea and the Gulf of Aden. From the mouth of the Gulf of California the ridge runs southward and is joined by an east-west ridge running through the Galápagos Islands. The sea floor bounded by the two ridges and the trench off Central America seems to constitute a separate small plate.

For the past four million years we can date the lavas on land with enough accuracy to give a timetable of magnetic reversals that can be correlated with the magnetic pattern on the sea bottom. For this period the rates of spreading from the ridges have remained constant. Further back we have a long series of reversals recorded in the ocean floor, but we cannot date them by comparison with lavas on land because the accuracy of the dates is insufficient to put the lavas in order. A rough guess can be made of the time since the oldest part of the magnetic pattern was formed by assuming that the rates have always been what they are today. This yields about 70 million years in the eastern Pacific and the South Atlantic. In fact the spacings of the older magnetic lineations are not in a constant proportion in the different oceans. The rates of spreading must therefore vary with time when long periods are considered. Directions of motion have also changed during this period, as can be seen from the departure of the older parts of some of the Pacific fracture zones from circles of latitude around the present pole of spreading. A change of direction is also shown by the accurate geometrical and geochronological fit that can be made between South America and Africa [*see illustration on page 51*]. A rotation around the present pole of spreading will not bring the continents together; it is therefore likely that in the early stages of the separation motion was around a point farther to the south.

The ideas of the development of the earth's surface by plate formation, plate motion and plate destruction can be checked with some rigor by drilling. If they are correct, drilling at any point should show sediments of all ages from the present to the time at which this part of the plate was in the central valley of the ridge. Under these sediments there should be lavas of about the same age as the lowest sediments. From preliminary reports of the drilling by the JOIDES project (a joint enterprise of five American universities) it seems that this expectation has been brilliantly verified and that the rate of spreading has been roughly constant for 70 million years in the South Atlantic. Such studies are of great importance because they will give firm dates for the entire magnetic pattern and provide a detailed chronology for all parts of the ocean floor.

The process of consumption of oceanic crust at the edge of a continent may proceed for tens of millions of years, but if the plate that is being consumed carries a continental fragment, then the consumption must stop when the fragment reaches the trench and collides with the continent beyond it. Because the fragment consists of relatively light rocks it cannot be forced under a continent. The clearest example is the collision of India with what was once the southern margin of Asia. Paleomagnetic work shows that India has been moving northward for the past 100 million years. If it is attached to the plate that is spreading northward and eastward from the Carlsberg Ridge (which runs down the Indian Ocean halfway between Africa and India), then the motion is continuing today. This motion may be the cause of the earthquakes of the Himalayas, and it may also be connected with the formation of the mountains and of the deep sediment-filled trough to the south of them. The exact place where the joint occurs is far from clear and needs study by those with a detailed geological knowledge of northern India.

It seems unlikely that all the continents were collected in a single block for

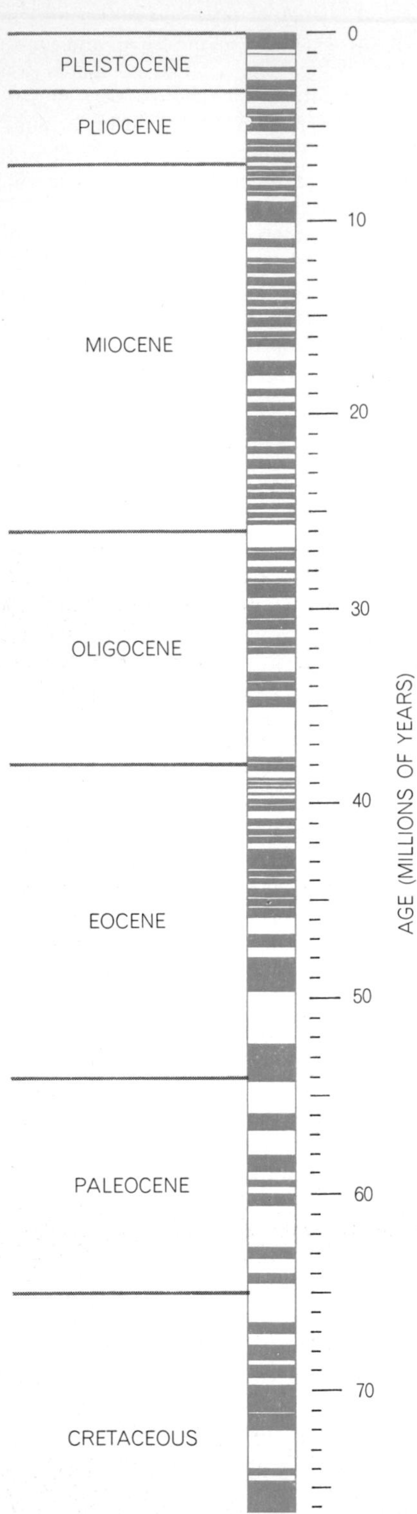

REVERSALS of the earth's magnetic field can be traced back more than 70 million years using magnetic patterns observed on the sea floor. The timetable of reversals for the most recent four million years was obtained by dating reversals in lava flows on land. Extrapolations beyond that assume that the sea floor spread at a constant rate. Colored bars show periods when the direction of the magnetic field was as it is now.

4,000 million years and then broke apart and started their wanderings during the past 100 million years. It is more likely that the processes we see today have always been in action and that all through geologic time there have been moving plates carrying continents. We must expect continents to have split many times and formed new oceans and sometimes to have collided and been welded together. We are only at the beginning of the study of pre-Tertiary events; anything that can be said is speculation and is to be taken only as an indication of where to look.

It is virtually certain that the Atlantic did not exist 150 million years ago. Long before that, in the Lower Paleozoic, 650 to 400 million years ago, there was an older ocean in which the sediments now in the Caledonian-Hercynian-Appalachian mountains of Europe and North America were laid down. Perhaps this ocean was closed long before the present Atlantic opened and separated the Appalachian Mountains of eastern North America from their continuation in northwestern Europe.

The Urals, if they are not unique among mountain ranges, are at least exceptional in being situated in the middle of a continent. There is some paleomagnetic evidence that Siberia is a mosaic of fragments that were not originally contiguous; perhaps the Urals were once near the borders of an ocean that divided Siberia from western Russia. Similarly, it is desirable to ask where the ocean was when the Rockies were being formed. A large part of California is moving rapidly northward, and the entire continent has overrun an ocean ridge; clearly the early Tertiary geography must have been very different from that of the present. Such questions are for the future and require that the ideas of moving plates be applied by those with a detailed knowledge of the various areas.

A history of the oceans does not necessarily require an account of the mechanism behind the observed phenomena. Indeed, no very satisfactory account can be given. The traditional view, put forward by Arthur Holmes and Felix A. Vening-Meinesz, supposes that the upper mantle behaves as a liquid when it is subjected to small forces for long periods and that differences in temperature under oceans and continents are sufficient to produce convection cells in the mantle—with rising currents under the mid-ocean ridges and sinking ones under the continents. These hypothetical cells would carry the plates along as on a conveyor belt and would provide the forces needed to produce the split along the ridge. This view may be correct; it has the advantage that the currents are driven by temperature differences that themselves depend on the position of the continents. Such a back-coupling can produce complicated and varying motions.

On the other hand, the theory is implausible in that convection does not normally happen along lines. It certainly does not happen along lines broken by frequent offsets, as the ridge is. Also it is difficult to see how the theory applies to the plate between the Mid-Atlantic Ridge and the ridge in the Indian Ocean. This plate is growing on both sides, and since there is no intermediate trench the two ridges must be moving apart. It would be odd if the rising convection currents kept exact pace with them. An alternative theory is that the sinking part of the plate, which is denser than the hotter surrounding mantle, pulls the rest of the plate after it. Again it is difficult to see how this applies to the ridge in the South Atlantic, where neither the African nor the American plate has a sinking part.

Another possibility is that the sinking plate cools the neighboring mantle and produces convection currents that move the plates. This last theory is attractive because it gives some hope of explaining the almost enclosed seas, such as the Sea of Japan. These seas have a typical oceanic floor except that the floor is overlain by several kilometers of sediment. Their floors have probably been sinking for long periods. It seems possible that a sinking current of cooled mantle material on the upper side of the plate might be the cause of such deep basins. The enclosed seas are an important feature of the earth's surface and urgently require explanation; in addition to the seas that are developing at present behind island arcs there are a number of older ones of possibly similar origin, such as the Gulf of Mexico, the Black Sea and perhaps the North Sea.

The ideas set out in this attempt at a history of the ocean have developed in the past 10 years. What we have is a sketch of the outlines of a history; a mass of detail needs to be filled in and many major features are quite uncertain. Nonetheless, there is a stage in the development of a theory when it is most attractive to study and easiest to explain, that is while it is still simple and successful and before too many details and difficulties have been uncovered. This is the interesting stage at which plate theory now stands.

The Deep-Ocean Floor 5

by H. W. Menard
September 1969

The discovery that it is growing outward from the mid-ocean ridges has suggested that it is formed in huge plates that act as units in the dynamic processes of the earth's crust

Oceanic geology is in the midst of a revolution. All the data gathered over the past 30 years—the soundings of the deep ocean, the samples and photographs of the bottom, the measurements of heat flow and magnetism—are being reinterpreted according to the concept of continental drift and two new concepts: sea-floor spreading and plate tectonics (the notion that the earth's crust consists of plates that are created at one edge and destroyed at the other). Discoveries are made and interpretations developed so often that the scientific literature cannot keep up with them; they are reported by preprint and wandering minstrel. At such a time any broad synthesis is likely to be short-lived, yet so many diverse observations can now be fitted into a coherent picture that it seems worthwhile to present it.

Before continental drift, sea-floor spreading and plate tectonics captured the imagination of geologists, most of them conceived the earth's crust as being a fairly stable layer enveloping the earth's fluid mantle and core. The only kind of motion normally perceived in this picture was isostasy: the tendency of crustal blocks to float on a plastic mantle. The horizontal displacement of any geologic feature by as much as 100 kilometers was considered startling. This view is no longer consistent with the geological evidence. Instead each new discovery seems to favor sea-floor spreading, continental drift and plate tectonics. These concepts are described elsewhere in this book [see the article "The Origin of the Oceans," by Sir Edward Bullard, beginning on page 45]. Here I shall recapitulate them briefly to show how they are related to the actual features of the deep-ocean floor.

According to plate tectonics the earth's crust is divided into huge segments afloat on the mantle. When such a plate is in motion on the sphere of the earth, it describes a circle around a point termed the pole of rotation (not to be confused with the entire earth's pole of rotation). This motion has profound geological effects. When two plates move apart, a fissure called a spreading center opens between them. Through this fissure rises the hot, plastic material of the mantle, which solidifies and joins the trailing edge of each plate. Meanwhile the edge of the plate farthest from the spreading center—the leading edge—pushes against another plate. Where that happens, the leading edge may be deflected downward so that it sinks into a region of soft material called the asthenosphere, 100 kilometers or more below the surface. This process destroys the plate material at the same rate at which it is being created at its trailing edge. Many of the fissures where plate material is being created are in the middle of the ocean floor, which therefore spreads continuously from a median line. Where the plates float apart, the continents, which are embedded in them, also drift away from one another.

The most obvious consequence of this process on the ocean floor is the symmetrical seascape on each side of a spreading center. As two crustal plates move apart (at a rate of one to 10 centimeters per year), the basaltic material that wells up through the spreading center between them splits down the middle. The upwelling in some way produces a ridge, flanked on each side by deep ocean basins and capped by long hills and mountains that run parallel to the crest. The flow of heat from the earth's interior is generally high along the crest because dikes of molten rock have been injected at the spreading center. A spreading center may also open under a continent. If it does, it produces a linear deep such as the Red Sea or the Gulf of California. If it continues to spread, or if the spreading center opens in an existing ocean basin, the same symmetrical seascape is ultimately formed.

This symmetry extends to less tangible features of the ocean floor such as the magnetic patterns in the basalt of the slopes on either side of the mid-ocean ridge. As the plastic material reaches the surface and hardens, it "freezes" into it the direction of the earth's magnetic field. The earth's magnetic field reverses from time to time, and as each band of new material moves outward across the ocean floor it retains a magnetic pattern shared by a corresponding band on the other side of the ridge. The result is a matching set of parallel bands on both sides. These patterns provide evidence of symmetrical flows and make it possible to date them, since they correspond to similar patterns on land that have been reliably dated by other means.

The steepness of the mid-ocean ridge is determined by a balance between the rate at which material moves outward from the spreading center and the rate at which it sinks as it ages after solidifying. The rate of sinking remains fairly constant throughout the ocean basin, and it seems to depend on the age of the

→

PILLOW LAVA (as pictured on the following page) assumes its rounded shape because it cools rapidly in ocean water. This flow lies on the western slope of the mid-ocean ridge in the South Atlantic at a depth of 2,650 meters. Flows like this erupt from the many volcanic vents and fissures that are created as the ocean floor spreads out from the mid-ocean ridges in the form of vast crustal plates. The photograph was made under the direction of Maurice Ewing of the Lamont-Doherty Geological Observatory.

crust. It can be calculated if the age of the oceanic crust (as indicated by the magnetic patterns) is divided into the depth at which a particular section lies. Such calculations show that the crust sinks about nine centimeters per 1,000 years for the first 10 million years after it forms, 3.3 centimeters per 1,000 years for the next 30 million years, and two centimeters per 1,000 years thereafter. Not all the crust sinks: on the southern Mid-Atlantic Ridge the sea floor has remained at the same level for as long as 20 million years.

The rate at which the sea floor spreads varies from one to 10 centimeters per year. Therefore fast spreading builds broad elevations and gentle slopes such as those of the East Pacific Rise. The steep, concave flanks of the Mid-Atlantic Ridge, on the other hand, were formed by slow spreading.

Whether the slopes are steep or gentle, the trailing edge of the plate at the spreading center is about three kilometers higher than the leading edge on the other side of the plate. The reason for this difference in elevation is not known. Heating causes some elevation and cooling some sinking, but the total relief appears much too great to be attributed to thermal expansion. Cooling might account for the relatively rapid sinking observed during the first 10 million years, but continued sinking remains a puzzle.

A decade ago scanty information suggested that the mid-ocean ridge in both the Atlantic and Pacific was continuous, with a few branches. More complete surveys have revealed that crustal plates have ragged edges. Instead of extending unbroken for thousands of kilometers, a mid-ocean ridge at the trailing edges of two crustal plates forms a zigzag line consisting of many short segments connected by fracture zones to other ridges, trenches, young mountain ranges or crustal sinks. The fracture zones connecting the ridge segments are associated with what are called "transform" faults. They provide important clues to the history of a plate. Because they form some of the edges of the plate, they delineate the circle around its pole of rotation, thereby indicating the direction in which it has been moving.

From what has been said so far it might appear that the spreading centers are fixed and stationary. The constantly repeated splitting of the new crust at the spreading center produces symmetrical continental margins, symmetrical magnetic patterns on the ocean floor, symmetrical ridge flanks and even

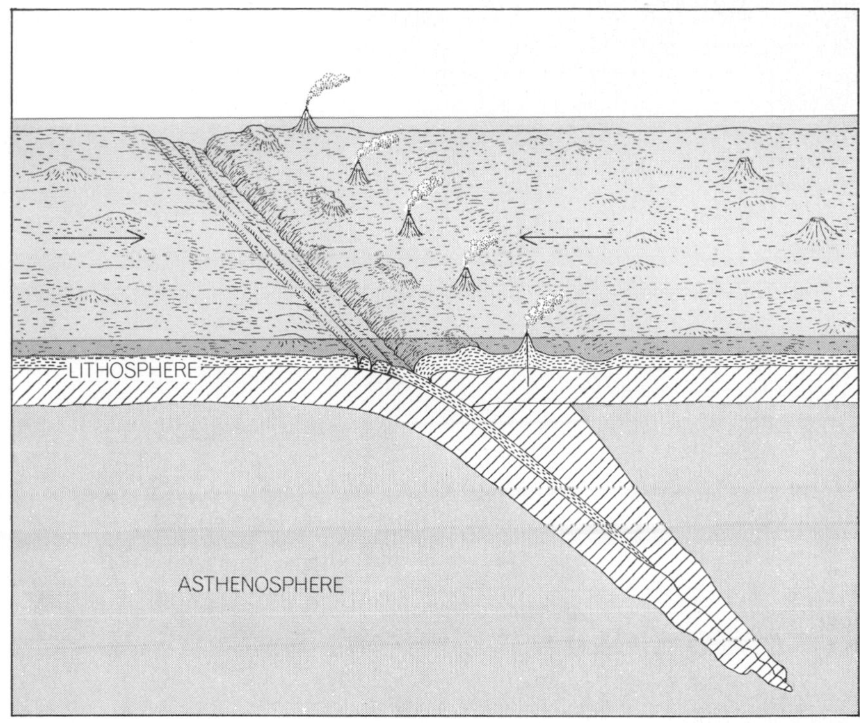

TRENCH IS CREATED where the leading edge of a plate that emerges from a fast spreading center collides with another plate. Because the combined speed of the two is more than six centimeters per year neither can absorb the impact by buckling. Instead one crustal plate (*in lithosphere*) plunges under the other to be destroyed in the asthenosphere, a hot, weak layer below. The impact produces volcanoes, islands and a deep, such as the Tonga Trench. Beside a trench are cracks that are produced by bending of the crust.

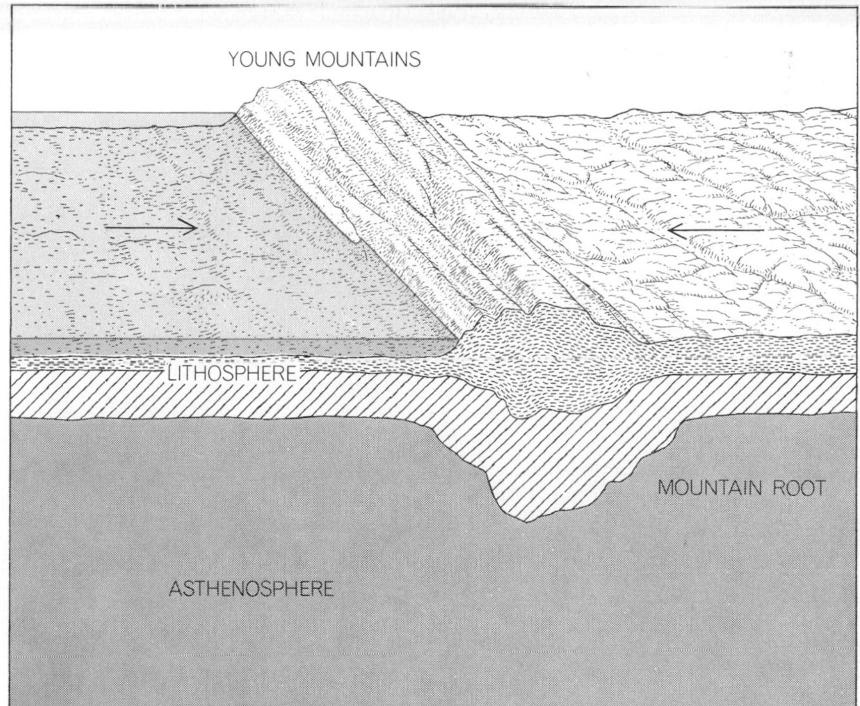

MOUNTAIN RANGE IS FORMED when the leading edges of two plates come together at less than six centimeters per year. Instead of colliding catastrophically, so that one plate slides under the other, both plates buckle, raising a young mountain range between them. The range consists of crustal material that folds upward under the compression exerted by the two plates (and also downward, forming the root of the mountain). Such ranges can be identified because they contain cherts and other material typical of the ocean bottom.

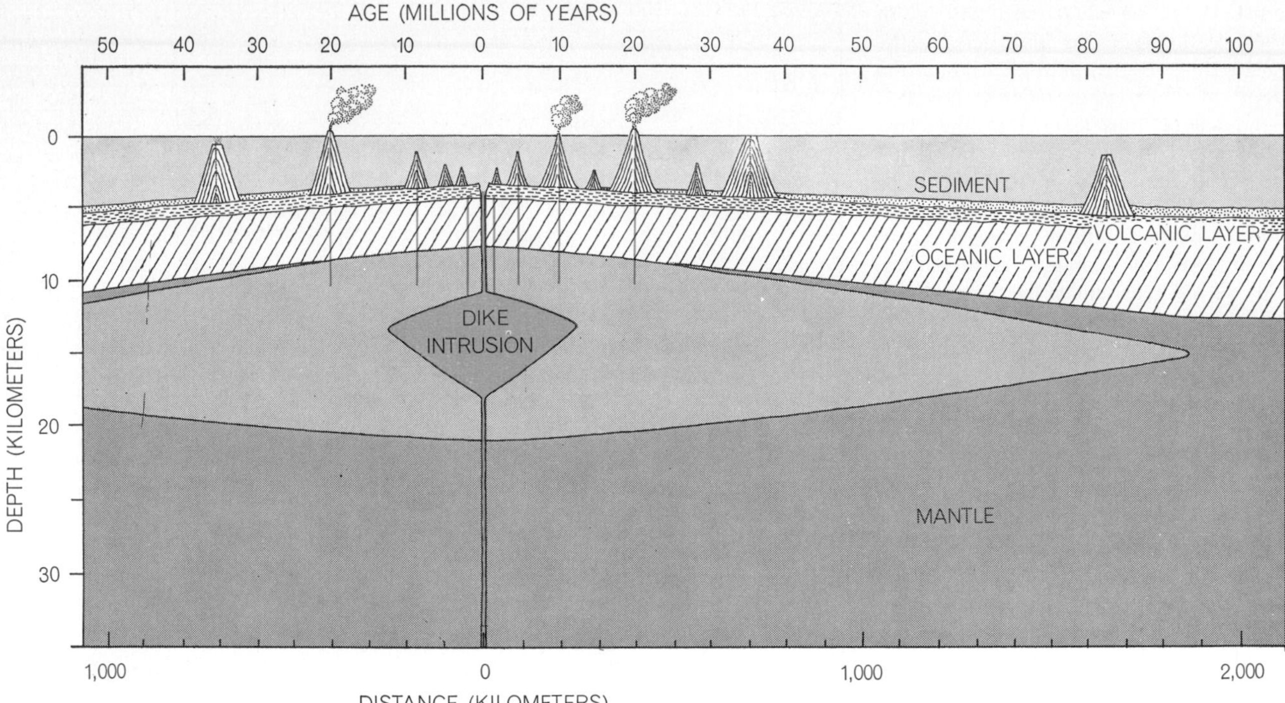

FAST SPREADING of the sea floor is revealed by gentle slopes. The sea floor is created at a spreading center that leaks molten rock from several dikes intruding from a pool in the low-density mantle (*light shading*). As the molten rock emerges it cools and adheres to the crust sliding away on each side of the fissure. If the crust moves at more than three centimeters per year, the slopes are gradual because spreading, which is horizontal, is rapid compared with the sinking of the crust. The balance between the two determines the steepness of the slope. Fast spreading also produces a thin volcanic layer because material moves so quickly from the fissure that it cannot accumulate. Islands built by eruptions are distant from the center because they grow on rapidly moving crust.

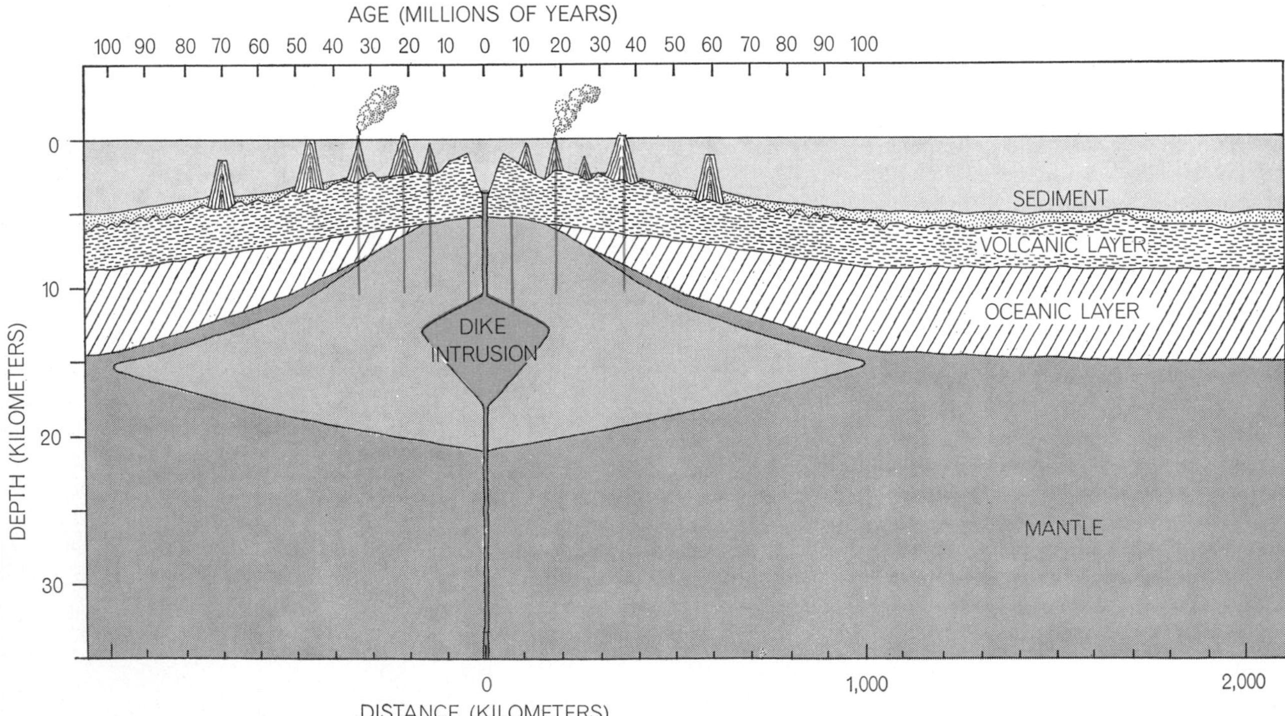

SLOW SPREADING produces steep slopes. Here the crust moves less than three centimeters per year; consequently sinking dominates the slope-forming process and produces steep escarpments. The volcanic layer is thicker at a slow center because material has time to accumulate. Mountains and volcanoes are high near the spreading center because the crust moves so slowly that the lava piles up. After 100 million years crust produced from a slow spreading center has strong similarity to crust from a fast center. Both kinds of crust have sunk to a depth of five kilometers. The oceanic layer is about five kilometers thick. Both fast- and slow-spreading crust are covered by the same kind of sediment. Slow spreading occurs mainly in the Atlantic, fast spreading in the Pacific.

symmetrical mountain ranges. More often than not, however, it has been found that the spreading center itself moves. Oddly enough such movement gives rise to the same symmetrical geology. All that is required in order to maintain the symmetry is that the spreading center move at exactly half the rate at which the plates are separating. If it moved faster or slower, the symmetry of the magnetic patterns would be destroyed.

Imagine, for instance, that the plate to the east of a spreading center remains stationary as the plate to the west moves. Since the material welling up through the fissure splits down the middle, half of it adheres to the stationary plate and the other half adheres to the moving plate. The next flow of material to well up through the split thus appears half the width of the spreading center away from the stationary plate. The flow after that appears a whole width of the spreading center away from the stationary plate, and so on. In effect the spreading center is migrating away from the stationary plate and following the moving one. If the speed of the spreading center exceeded half the speed of the migrating plate, however, a kind of geological Doppler effect would set in: the bands of the magnetic pattern would be condensed in the direction of the moving plate, and they would be stretched out in the direction of the stationary plate [*see illustration below*].

It might seem unlikely that the spreading center would maintain its even rate of speed and remain exactly between the two plates. W. Jason Morgan of Princeton University observes, however, that there is no impediment to such motion, provided only that the crust splits where it is weakest (which is where it split before, at the point where the hot dike was originally injected). As a result the spreading center is always exactly between two crustal plates whether it moves or not.

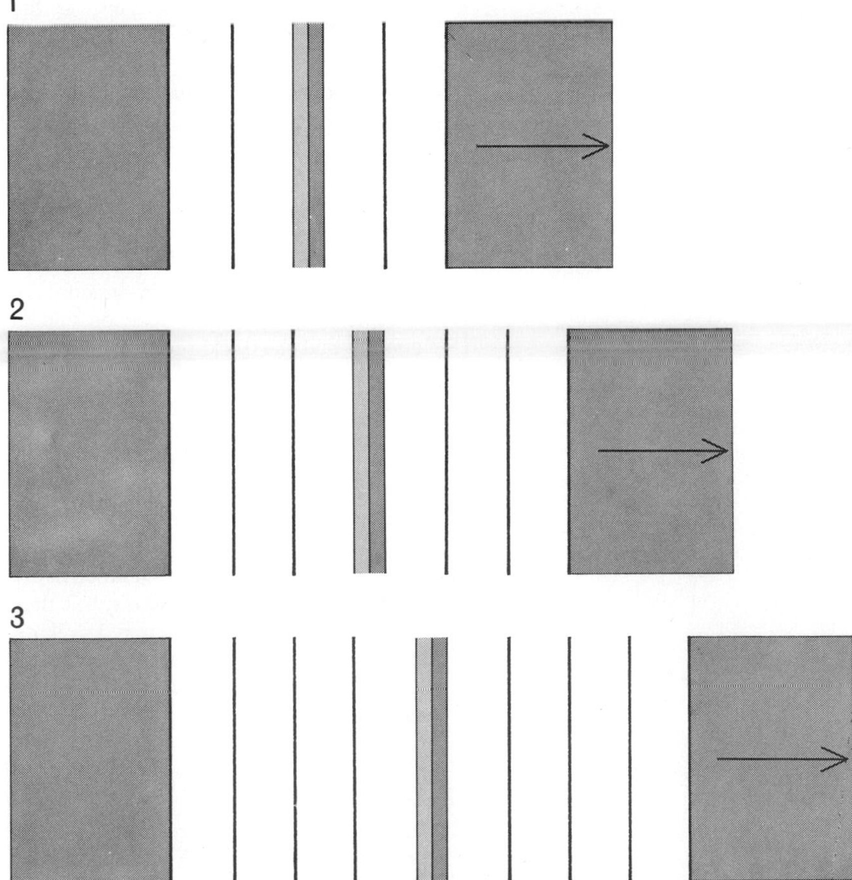

SPREADING CENTER MOVES, yet it can still leave a symmetrical pattern of magnetized rock. The molten material emerging from a spreading center becomes magnetized because as it cools it captures the prevailing direction of the earth's periodically changing magnetic field. In the instance illustrated here the right-hand plate moves out to the right while the left-hand plate remains stationary. In *1* hot material from a dike arrives at the surface, cools and splits down the middle. In *2* the next injection of material arrives in the crevice between the two halves of the preceding mass of rock. The new mass is therefore half the width of the preceding mass farther from the stationary plate than the preceding mass of material itself was. In *3* the new material has cooled and split in its turn and another mass has appeared that is a whole width farther from the left-hand plate. As long as the center moves at half the speed at which the right-hand plate moves away the magnetic bands remain symmetrical. If plate moved faster or more slowly, they would be jumbled.

Moving spreading centers account for some of the major features of the ocean floor. The Chile Rise off the coast of South America and the East Pacific Rise are adjacent spreading centers. Since there is no crustal sink between them, and new plate is constantly being added on the inside edge of each rise, at least one of the centers must be moving, otherwise the basin between them might fold and thrust upward into a mountain range or downward into a trench. Similarly, the Carlsberg Ridge in the Indian Ocean and the Mid-Atlantic Ridge are not separated by a crustal sink and hence one of them must be moving.

A moving center may have created the ancient Darwin Rise on the western edge of the Pacific basin and also the modern East Pacific Rise. As in the case of the Carlsberg Ridge and the Mid-Atlantic Ridge, the existence of two vast spreading centers on opposite sides of the ocean with no intervening crustal sink has puzzled geologists. If such centers can move, however, it is possible that the spreading center in the western Pacific merely migrated all the way across the basin, leaving behind the ridges of the Darwin Rise. In this way one rise could simply have become the other. Many other examples exist, and Manik Talwani of the Lamont-Doherty Geological Observatory proposes that all spreading centers move.

As a plate forms at a spreading center it consists of two layers of material, an upper "volcanic" layer and a lower "oceanic" one. Lava and feeder dikes from the mantle form the volcanic layer; its rocks are oceanic tholeiite (or a metamorphosed equivalent), which is rich in aluminum and poor in potassium. The oceanic layer is also some form of mantle material, but its precise composition, density and condition are not known. Farther down the slope of the ridge the plate acquires a third layer consisting of sediment.

The sediment comes from the continents and sifts down on all parts of the basin, accumulating to a considerable depth. It is mixed with a residue of the hard parts of microorganisms that is called calcareous ooze. Below a certain depth (which varies among regions) this

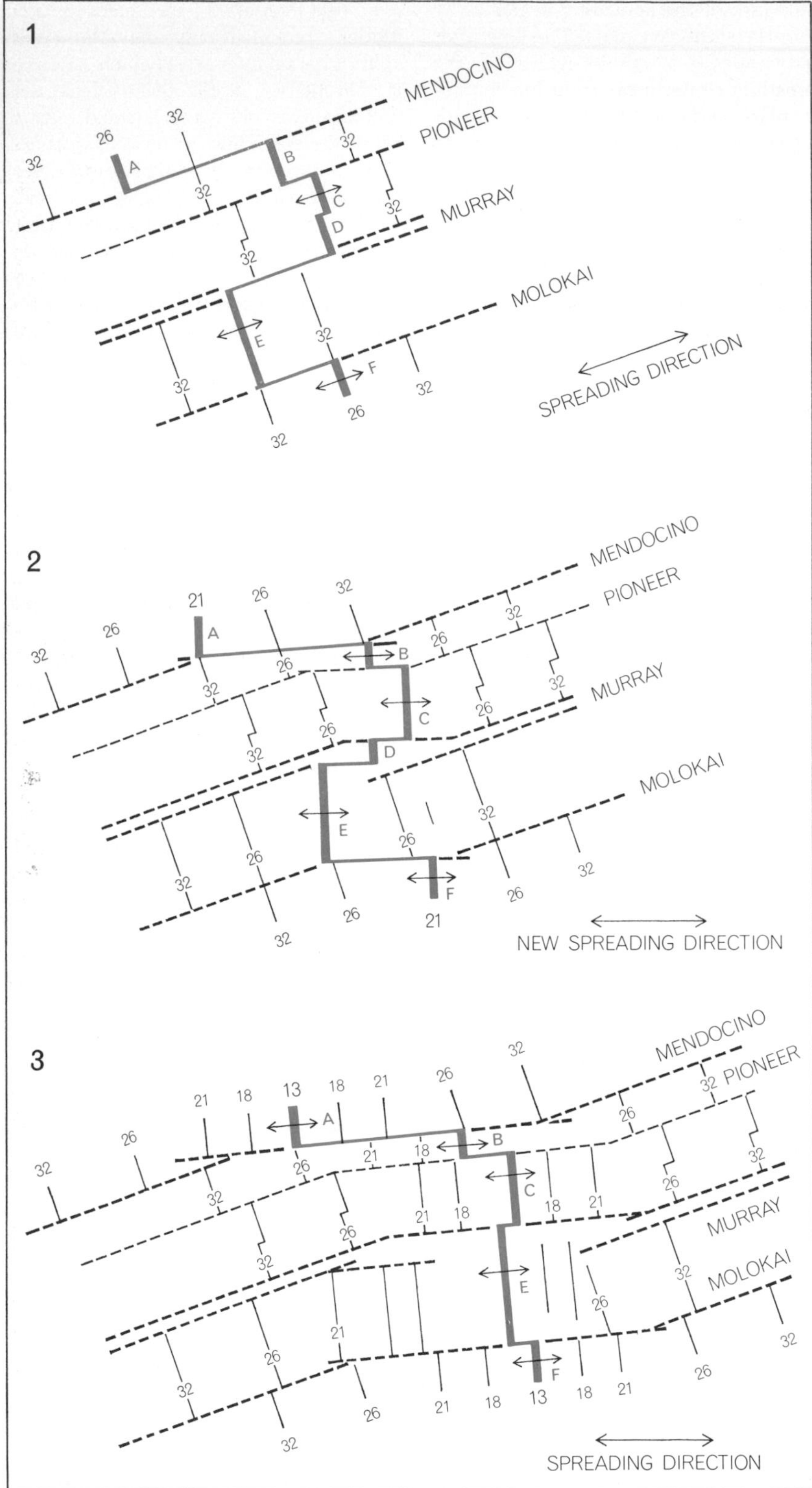

HOW A PLATE MOVES is revealed by the patterns of magnetic bands (time of formation is indicated by numbers) and by the relation between ridges and faults in the northeastern Pacific. At the time illustrated in *1* material from the Murray fault and other faults connected segments of the ridge (*indicated by letters*) offset from one another by plate motion. At the time shown in *2* the spreading direction changed. The readjustment of the plates has shortened the Pioneer-Mendocino ridge segment (*B*) while lengthening and reorienting the Pioneer, Mendocino and Molokai transform faults. In *3* the Mendocino fault remains the same length, but between most of the other ridges faults have been shortened or have almost disappeared as ridge segments tended to rejoin one another. Between Murray and Molokai faults ridge has jumped eastward, and one segment (*D*) has vanished.

material dissolves, and only red clay and other resistant components remain.

For reasons only partly known the sediment is not uniformly distributed. At the spreading center the newly created crust is of course bare of sediment, and within 100 kilometers of such a center the calcareous ooze is rarely thick enough to measure. The ooze accumulates at an average rate of 10 meters for each million years, during which time the plate moves horizontally from 10,000 to 100,000 meters and sinks 100 meters. Where the red clay appears, it accumulates at a rate of less than one meter per million years.

The puzzle deepens when one considers that sediment on oceanic crust older than 20 million years stops increasing in thickness after it sinks to the depth where the calcareous ooze dissolves. Indeed, in many places the age of the oldest sediment is about the same as the volcanic layer on which it lies. It would therefore seem that almost all the deep-ocean sediment accumulates in narrow zones on the flanks of the mid-ocean ridges. If this is correct, it has yet to be explained.

The volcanic layer forms mainly at the spreading center. Volcanoes and vents on the slopes of the mid-ocean ridge contribute a certain amount of oceanic tholeiite to it. It can be said in general that the thickness of the volcanic layer decreases as the spreading rate increases. If the crust spreads slowly, the material has time to accumulate. Fast spreading reduces this time and therefore the accumulation. The conclusion can be drawn that the rate at which the volcanic-layer material is discharged is nearly constant. These relations are based on only 10 observations, but they apply to spreading rates from 1.4 centimeters to 12 centimeters per year, and to thicknesses from .8 kilometer to 3.8 kilometers.

The total flow of volcanic material from all active spreading centers is about four cubic kilometers per year—four times the flow on the continents. Not the slightest sign of this volcanism on the ocean bottom can be detected at the surface of the ocean, with one possible exception: late in the 19th century a ship reported seeing smoke rising from the waters above the equatorial Mid-Atlantic Ridge. The British oceanographer Sir John Murray remarked that he hoped the smoke signified the emergence of an island, since the Royal Navy needed a coaling station at that point.

Like the volcanic layer, the oceanic

layer forms at the spreading center. Acoustical measurements of the thickness of this layer at spreading centers, on the flanks of mid-ocean ridges and on the deep-ocean floor show, however, that at least part of the oceanic layer evolves slowly from the mantle rather than solidifying quickly and completely at the spreading center. At the spreading center the thickness of the layer depends on how fast the ocean floor moves. In regions such as the South Atlantic, where the floor spreads at a rate of two centimeters per year, no oceanic layer forms within a few hundred kilometers of the spreading center. Farther away from the spreading center the oceanic layer accumulates rapidly, reaching a normal thickness of four to five kilometers on the flank of the mid-ocean ridge. A spreading rate of three centimeters per year is associated with an oceanic layer roughly two kilometers deep at the center that thickens by one kilometer in 13 million years. A plate with a spreading rate of eight centimeters per year is three kilometers deep at the center and thickens by one kilometer in 20 million years. Thus the thinner the initial crust is, the faster the thickness increases as the crust spreads.

As a plate flows continuously from the spreading center, faulting, volcanic eruptions and lava flows along the length of the mid-ocean ridge build its mountains and escarpments. This process can be most easily observed in Iceland, a part of the Mid-Atlantic Ridge that has grown so rapidly it has emerged from the ocean. A central rift, 45 kilometers wide at its northern end, cuts the island parallel to the ridge. The sides of the rift consist of active, steplike faults. There are other step faults on the rift floor, which is otherwise dominated by a large number of longitudinal fissures. Some of these fissures are open and filled with dikes. Fluid lava wells up from the fissures and either buries the surrounding mountains, valleys and faults or forms long, low "shield" volcanoes. Two hundred such young volcanoes, which have been erupting about once every five years over the past 1,000 years, dot the floor of the rift. Thirty of them are currently active.

Just as a balance between spreading and sinking shapes the slopes of the mid-ocean ridges, so does a balance between lava discharge and spreading build undersea mountains, hills and valleys. High mountains normally form at slow centers where spreading proceeds at two to 3.5 centimeters per year. In contrast, a spreading center that opens at a rate of five to 12 centimeters per year produces long hills less than 500 meters high. This relationship is a natural consequence of the long-term constancy of the lava discharge. Over a short period of time, however, the lava discharge may fluctuate or pulsate, a picture suggested by the fact that the thick volcanic layer associated with slow spreading can consist of volcanic mountains (which represent copious flow) separated by valleys covered by a thinner volcanic layer.

The volcanic activity and faulting that first appear near the spreading centers decrease rapidly as the plate ages and material moves toward its center, but volcanic activity in some form is never entirely absent. Small conical volcanoes are found on crust only a few hundred thousand years old near spreading centers, and active circular volcanoes such as those of Hawaii exist even in the middle of a plate. It would appear that the great cracks that serve as conduits for dikes and lava flows are soon sealed as a plate ages and spreads. Volcanic activity is then concentrated in a few central vents, created at different times and places.

Many of these vents remain open for tens of millions of years, judging by the size and distribution of the different classes of marine volcanoes. First, the biggest volcanoes are increasingly big at greater distances from a spreading center, which means they must continue to erupt and grow as the crust ages and sinks, even when the age of the crust exceeds 10 million years. In most places, in fact, a volcano needs at least 10 million years in order to grow large enough to become an island. Volcanoes that discharge lava at a rate lower than 100 cubic kilometers per million years never become islands because the sea floor sinks too fast for them to reach the sea surface.

Other volcanoes drifting with a spreading ocean floor may remain active or become active on crust that is 100 million years old (as the volcanoes of the Canary Islands have). Normally, however, volcanoes become inactive by the time the crust is 20 to 30 million years old. This is demonstrated by the existence of guyots, drowned ancient island volcanoes that were submerged by the gradual sinking of the aging crust. Guyots are found almost entirely on crust that is more than 30 million years old, such as the floor of the western equatorial Pacific.

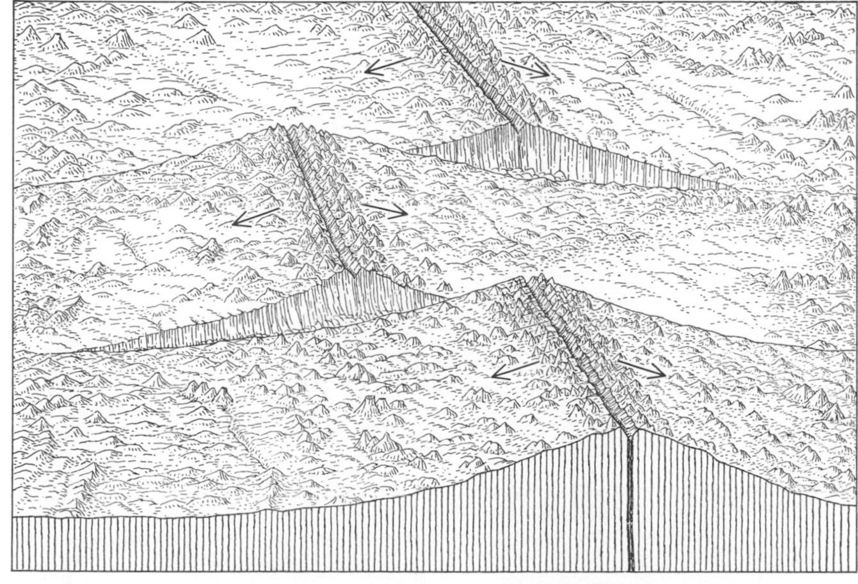

RIDGE-RIDGE TRANSFORM FAULT appears between two segments of ridge that are displaced from each other. Mountains are built, earthquakes shake the plate edges and volcanoes erupt in such an area because of the forces generated as plates, formed at the spreading centers under the ridges, slide past each other in opposite directions. On outer slopes of the mid-ocean ridges, however, this intense seismic activity appears to subside.

Traditionally it has been thought that marine volcanoes spew lava from a magma chamber located deep in the mantle. Some volcanoes have a top composed of alkali basalt, slopes with transitional basalt outcrops and a base of oceanic tholeiite, and it was therefore assumed that the lava in the magma chamber became differentiated into components that, rather like a pousse-café, separated into several layers of different kinds of material, each of which followed the layer above it up the spout. The emergence of plate tectonics and continental drift as respectable concepts have now brought this view of volcanic action into question.

It remains perfectly possible for a volcano to drift for tens of millions of years over hundreds of kilometers while tapping a single magma chamber embedded deep in the mantle. The motion of the plates, however, suggests another hypothesis. According to this view, the volcano and its conduit drift along with the crust as the conduit continually taps different parts of a relatively stationary magma that is ready to yield various kinds of lava whenever a conduit appears. In actuality the composition of the lava usually changes only slightly after the first 10 to 20 kilometers of drifting. Although the older hypothesis is still reasonable, the newer one must also be considered because it explains the facts equally well.

In addition to their characteristic volcanoes and mountains, spreading centers are marked by median valleys, which in places such as the North Atlantic or the northwestern Indian Ocean are deeper than the surrounding region. These rifts are commonly found in centers opening at a rate of two to five centimeters per year. The deepest rifts, which may go as deep as 1,000 to 1,300 meters below the surrounding floor, are associated with spreading at three to four centimeters per year. Only one valley is known to be associated with spreading at five to 12 centimeters per year. Although rifts are not found in all spreading centers, they usually do appear in conjunction with a slow center. Both of these features are also associated with volcanic activity.

The mid-ocean ridges, as we have noted, seldom run unbroken for more than a few hundred kilometers. They are interrupted by fracture zones, and the segments are shifted out of line with respect to one another. These fracture zones run at right angles to the ridge and connect the segments. Where they lie between the segments they are termed ridge-ridge transform faults, which are the site of intense geological activity. As the two edges of the fault slide past each other they rub and produce earthquakes. The slope of a transform fault drops steeply from the crest of one ridge segment to a point halfway between it and the adjacent segment and then climbs to the top of the adjacent segment, reflecting the fact that the crust is elevated at the spreading center and subsides at some distance from it [*see illustration on page 61*].

Like spreading centers, fracture zones have their own complex geology. In these

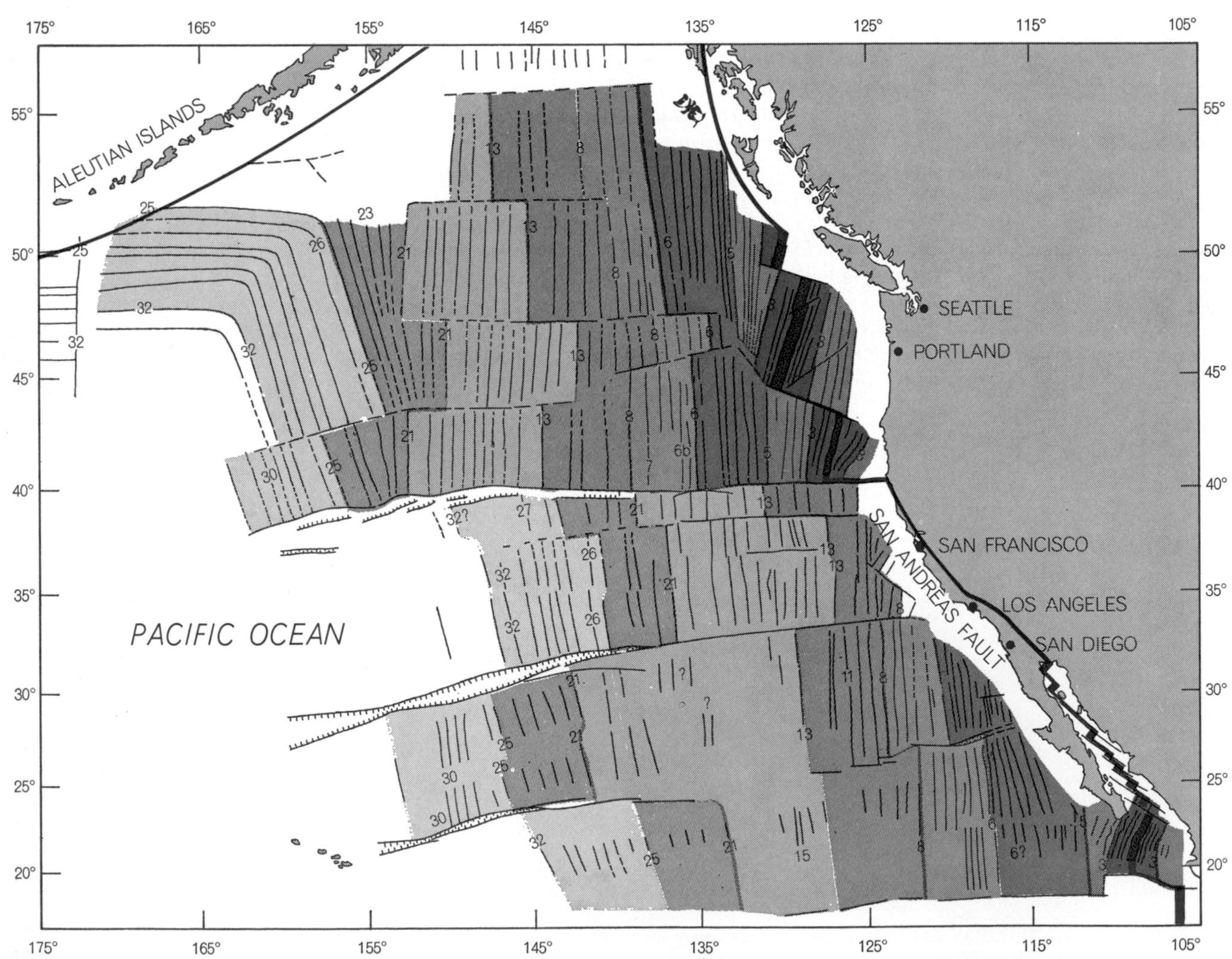

MAGNETIC PATTERNS reveal how the plate forming the floor of the northeastern Pacific has moved. Its active eastern edge now stretches from Alaska through California (where it forms the San Andreas fault) to the Gulf of California. In the gulf spreading centers break into short segments joined by active faults. Plate motion is opening the gulf and moving coastal California in the direction of the Aleutians. To the south lies the Great Magnetic Bight, formed by three plates that spread away from one another.

areas the ridges stand as much as several kilometers high, and the troughs are equally deep. It appears that the same volcanic forces that shape the main ridges produce the mountains and valleys of the faults. As fracture zones open they slowly leak lava from hot dikes. At the same time the crust sinks away from the fault line, and this balance produces high mountains.

Beyond the spreading centers the fracture zones become the inactive remains of earlier faulting. The different rates at which these outer flanks of the mid-ocean ridge sink do produce some vertical motion as the scarps of the fracture zone decay. This may account for the few earthquakes in these areas. I should emphasize that it is not known if horizontal motion is also absent from such dead fracture zones. It is not necessary, however, to postulate such motion in order to explain existing observations.

A fracture zone can become active again at any time, but if it does so, it becomes the side of a smaller new plate rather than part of the trailing edge of an old one. If the flank fracture zones are as quiescent as they appear to be, then the plates forming the earth's crust are large and long-lived. If these fracture zones were active, on the other hand, it could only be concluded that each one marked the flank of a small, elongated plate.

The direction the plate is moving can be deduced from the magnetic pattern that runs at right angles to the fractures in the fracture zone. When the direction of plate motion changes, the direction in which the fracture zone moves also changes. This change in direction can be most clearly seen in the northeastern Pacific, where our knowledge is most detailed. On this part of the ocean floor the changes of direction have taken place at the same time in many zones, indicating that the entire North Pacific plate has changed direction as a unit [*see illustration on page 62*].

On the bottom of the Pacific and the North Atlantic the magnetic patterns are sometimes garbled. Old transform faults may have vanished if short segments of spreading center have been united by reorientation. By the same token new transform faults may have formed if the change in plate motion has been too rapid to be accommodated by existing motions. Thus fracture zones may be discontinuous. They may start and stop abruptly, and the offset of the magnetic patterns may change from place to place along them without indicating any activity except at the former edges of plates.

Some patterns are even harder to interpret. Douglas J. Elvers and his colleagues in the U.S. Coast and Geodetic Survey discovered an abrupt boomerang-shaped bend in the magnetic pattern south of the Aleutians. The arms of this

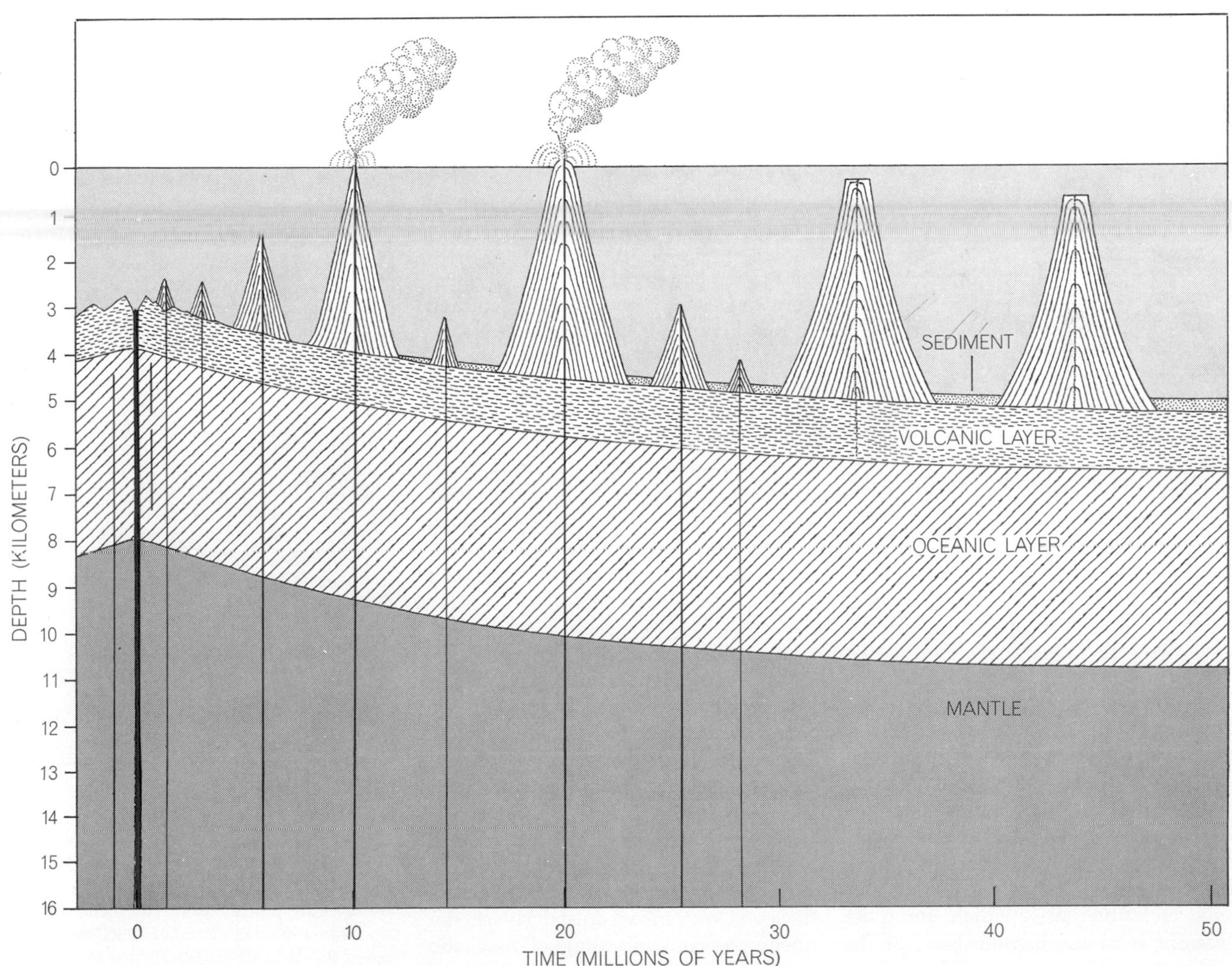

UNDERSEA VOLCANOES normally begin to rise near spreading centers. Then they ride along on the moving plate as they grow. If a volcano rises fast enough to surmount the original depth of the water and the sinking of the ocean floor, it emerges as an island such as St. Helena in the South Atlantic. To rise above the water an undersea volcano must grow to a height of about four kilometers in 10 million years. Island volcanoes sink after 20 to 30 million years and become the sediment-capped seamounts called guyots.

configuration, which Elvers calls the Great Magnetic Bight, are offset by fracture zones at right angles to the magnetic pattern. The Great Magnetic Bight seems to have required impossible forms of sea-floor spreading and plate movement. However, Walter C. Pitman III and Dennis E. Hayes of the Lamont-Doherty Geological Observatory, among others, have been able to show that the configuration is fully understandable if it is assumed that the trailing edges of three plates met and formed a Y. Similar complex patterns have now been found in the Atlantic and the Pacific. Indeed, if the transform faults are perpendicular to the spreading centers and two spreading rates are known, both the orientation and the spreading rate of the third spreading center can be calculated even before it is mapped. One can also calculate the orientation and spreading rate for a third center that has already vanished in a trench.

As a crustal plate grows, its leading edge is destroyed at an equal rate. Sometimes this edge slides under the oncoming edge of another plate and returns to the asthenosphere. When this happens, a deep trench such as the Mariana Trench and the Tonga Trench in the Pacific is formed. In other areas the movement of the crust creates young mountain ranges. Xavier Le Pichon of the Lamont-Doherty Geological Observatory concludes that the occurrence of one event or the other is a function of the rate at which plates are moving together. If the rate is less than five to six centimeters per year, the crust can absorb the compression and buckles up into large mountain ranges such as the Himalayas. In these ranges folding and overthrusting deform and shorten the crust. If the rate is higher, the plate breaks free and sinks into the mantle, creating an oceanic trench in which the topography and surface structure indicate tension.

Several crustal sinks are no longer active. Their past, however, can be deduced from their geology. Large-scale folding and thrust-faulting can be taken as evidence of the former presence of a crustal sink, although such deformation can also arise in other ways. Certain types of rock may also indicate the formation of a trench. The arcs of islands that lie parallel to trenches, for instance, are characterized by volcanoes that produce andesitic lavas, which are quite different from the basaltic lavas of the ocean floor. The trenches themselves, and the deep-sea floor in general, are featured by deposits of graywackes and cherts. These rocks are commonly found exposed on land in the thick prisms of sediment that lie in geosynclines: large depressed regions created by horizontal forces resembling those generated by a drifting crustal plate. Thus the presence of some or all of these types of rock may indicate the former existence of a crustal sink. This linking of marine geology at spreading centers with land geology at crustal sinks is becoming one of the most fruitful aspects of plate tectonics. Still, crustal sinks are by no means as informative about the history of the ocean floor as spreading centers are, because in such sinks much of the evidence of past events is destroyed. Even if the leading edge of a plate was once the side of a plate (or vice versa), there would be no way to tell them apart.

At the boundary between the land and the sea a puzzle presents itself. The sides of an oceanic trench move together at more than five centimeters per year, and it would seem that the sediment sliding into the bottom of the trench should be folded into pronounced ridges and valleys. Yet virtually undeformed sediments have been mapped in trenches by David William Scholl and his colleagues at the U.S. Naval Electronics Laboratory Center. Furthermore, the enormous quantity of deep-ocean sediment that has presumably been swept up to the margins of trenches cannot be detected on sub-bottom profiling records. There are many ingenious (but unpublished) explanations of the phenomenon in terms of plate tectonics. One of them may conceivably be correct. According to that hypothesis, the sediments are intricately folded in such a way that the slopes and walls of trenches cannot be detected by normal survey techniques, which look at the sediments from the ocean surface and along profiles perpendicular to the slopes. This kind of folding could be detected only by trawling a recording instrument across the trench much closer to the bottom or by crossing the slope at an acute angle.

The concepts of sea-floor spreading and plate tectonics allow a quantitative evaluation of the interaction of many important variables in marine geology. By combining empirical observation with theory it is possible not only to explain but also to predict the thickness and age of sediments in a given locality, the scale and orientation of topographic relief, the thickness of various crustal layers, the orientation and offsetting of magnetic patterns, the distribution and depth of drowned ancient islands, the occurrence of trenches and young mountain ranges, the characteristics of earthquakes, and many other previously unrelated and unpredictable phenomena. This revolution in marine geology may take some years to run its course. Ideas are changing, and new puzzles present themselves even as the old ones are solved. The only certainty is that the subject will never be the same again.

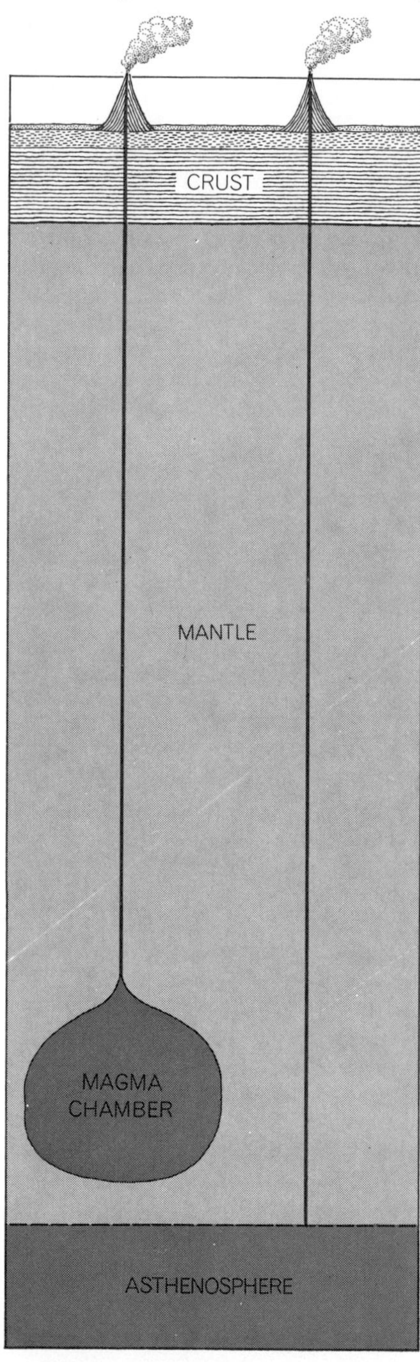

UNDERSEA ERUPTIONS can be explained in two ways, both consistent with observed facts. Since a volcano consists of different kinds of rock, it was originally thought that its conduit carried different forms of lava up from a magma chamber 50 kilometers or more down in the mantle. Now that crustal plates have been found to move, another theory must be considered. According to this idea, the conduit reaches through the mantle and taps several different kinds of magma at different places in the asthenosphere.

The Floor of the Mid-Atlantic Rift

6

by J. R. Heirtzler and W. B. Bryan
August 1975

Last summer U.S. and French submersibles explored a rugged area at some 8,400 feet where lava wells up and the ocean floor moves outward, bearing with it the continents to the east and the west

Some 20 years ago it became apparent that a continuous range of undersea mountains twists and branches for a total length of some 40,000 miles through all the world's oceans. This system of mid-ocean ridges is the site of frequent earthquakes with shallow foci. Geologists had no satisfactory explanation for these earthquakes, and they were even more mystified by the fact that samples of the ocean floor were geologically young. No samples seemed to be more than about 135 million years old, and the closer to the ridge the samples were taken, the younger they were. These findings, together with other discoveries (such as the regular alternation of magnetic polarity in broad bands of the ocean floor running parallel to the ridge), finally led to the hypothesis that the mid-ocean ridge is actually a system of parallel ridges centered on a continuous rift in the ocean floor, and that the floor itself is everywhere spreading outward from the rift. As the rift widens it is filled with lava that wells up from the earth's mantle. The ridges mark the edges of huge plates in the earth's crust that bear the continents. As the sea floor spreads, the continents at the margins of the oceans are either borne apart or compressed into new configurations.

Thus the mid-ocean-rift system reflects processes of the most fundamental importance in the evolution of the earth. Geologists now believe that the processes responsible for the rift system as we see it today have been creating and modifying the crustal plates throughout much of the four to five billion years of the earth's history. Although many details remain to be explained, the general concept of the creation of crust at the mid-ocean ridges has become widely accepted in less than a dozen years and forms the basis for much of present-day geological reasoning.

As has been the case with the spectacular growth of knowledge about the moon and the other planets, the expansion of our knowledge of ocean-bottom geology has been closely connected with technological advances that have made possible the increasingly rapid collection of data. Although much has been learned about the mid-ocean ridges by remote-sensing techniques, by drilling and by random sampling from surface ships, it finally became obvious that direct manned observation, by means of special submersible vessels, would ultimately be essential for any real understanding.

Starting late in 1971 a program was initiated by several interested groups to develop and harness specialized instruments and techniques for a detailed study of the central part of the Mid-Atlantic Ridge. A region of the ridge some 400 miles southwest of the Azores was selected for examination both because it is an area where one can expect good weather and because the port of Ponta Delgada in the Azores offered a convenient base of operations. The U.S. and France took leading roles in the effort because each had had experience with research submersibles of the type that would be needed for conducting manned observations in the later phases of the program. The cooperative venture was dubbed FAMOUS, for French-American Mid-Ocean Undersea Study.

Although the basic technology for deep-sea manned exploration was well developed, further refinements were needed. The French submersible *Archimède* had already gone much deeper than the 8,000-to-10,000-foot depths that would be encountered in the selected study area, but the *Archimède* was bulky, difficult to maneuver and provided only limited visibility. The U.S. submersible *Alvin* had already scored several technical successes in the Atlantic, including the recovery of the hydrogen bomb lost off the coast of Spain. Although the *Alvin* was small, maneuverable and afforded excellent visibility through its three portholes, it was limited to depths of less than 6,000 feet, too shallow for the depths of the Mid-Atlantic Ridge. Accordingly the French undertook to build a new submersible, the *Cyana,* similar in size and capability to the *Alvin,* and the U.S. fitted the *Alvin* with a new titanium pressure hull that would allow dives to at least 10,000 feet.

In further preparation for the venture plans were made to close an "observation gap" that would otherwise exist between the scale of detail that would be recorded from the submersibles, a scale measured in centimeters and meters, and the scale of observations that were then possible from surface vessels, a scale measured in hundreds of meters and kilometers. Strategies for closing the observation gap included echo sounding with a high-resolution narrow acoustic beam, photography with new techniques and the use of a variety of deep-towed geophysical instruments. All these methods would require the ability to position ships to an accuracy of a few tens of meters rather than the one kilometer that is usual in deep-sea navigation.

By 1971 there was no longer any doubt that our investigation would be confronted with detailed problems of volcanic geology. It was hard to realize that barely a dozen years earlier serious students of abyssal geology could still speculate that elevated portions of the mid-ocean-ridge system, such as the East Pacific Rise and the Azores Plateau, were remnants of sunken continental land areas. The deep central valley in the Mid-Atlantic Ridge and its transverse fracture zones, so well known today, were just beginning to be defined. The presence of outcrops of volcanic rock on the Mid-Atlantic Ridge was firmly established in 1961 by Earl Hayes and his co-workers on cruise No. 21 of the Woods Hole Oceanographic Institution vessel *Chain,* when they succeeded in photographing and sampling fresh, glassy volcanic rock at 28 degrees 53 minutes north latitude, due west of the Canary Islands. Over the next few years similar studies made by ships from various oceanographic institutions confirmed the presence of essentially similar volcanic rock on the crests of mid-ocean ridges in the South Atlantic, the Pacific and the Indian Ocean. Bizarre as the idea seemed at first, it was becoming evident that the mid-ocean-ridge system was nothing less than a vast unhealed volcanic wound.

Even more remarkable, it was learned that finely divided particles of iron oxide naturally present in the liquid volcanic rock become aligned with the earth's magnetic field as the molten rock freezes. Records of the fossilized magnetic polarity, obtained by magnetometers towed on the ocean surface, revealed symmetrical bands of periodic reversals of polarity on both sides of the mid-ocean ridges. Concurrent land-based studies of polarity reversals in precisely dated stratigraphic sections that included volcanic rocks provided a time scale for the global reversals in magnetic polarity [see "Sea-Floor Spreading," by J. R. Heirtzler; SCIENTIFIC AMERICAN Offprint 875], The conclusion seemed incontestable: the volcanic sea-floor basement was spreading away from the mid-ocean-ridge systems, growing by constant additions of fresh volcanic material at the

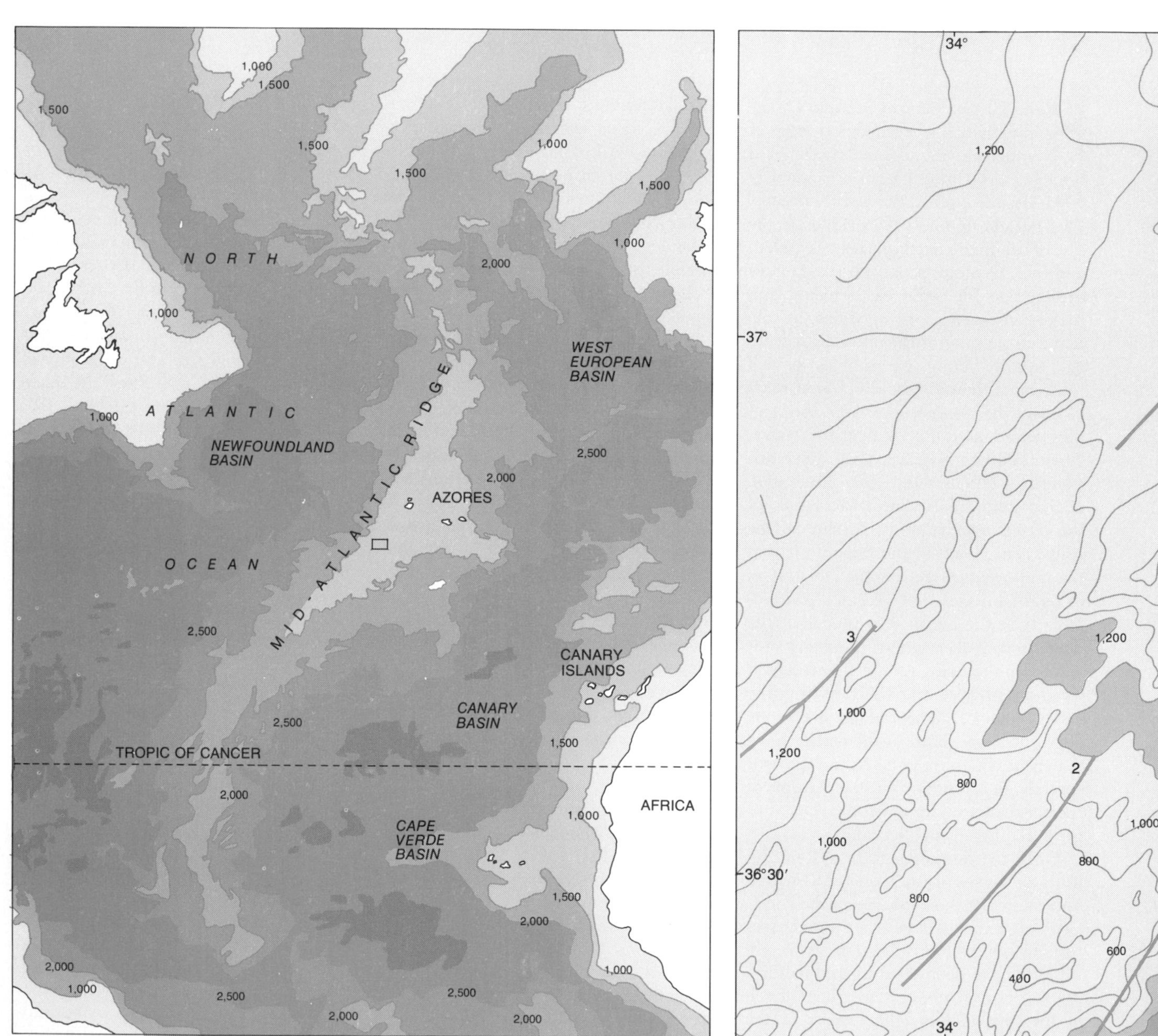

MID-ATLANTIC RIDGE *(map at left)* **is part of a continuous system of mid-ocean ridges some 40,000 miles long. Lava extrusions are centered on fissures along the crest of the ridges. These linear zones of extrusion mark the boundaries of huge lithospheric plates that are slowly moving apart, carrying the continents with them. As the lava solidifies, particles of iron oxide in the molten rock become aligned with the earth's magnetic field. Because of periodic reversals in the polarity of the earth's field the solidified and mag-**

central rift, and in the process was recording the polarity of the earth's magnetic field as a function of time, rather like a giant tape recorder.

Beginning in 1968, deep-sea drilling by the *Glomar Challenger* confirmed the hypothesis of sea-floor spreading by showing that sediments overlying the volcanic basement became progressively older with distance east or west from the Mid-Atlantic Ridge. In mid-1970 the oldest volcanic basement rock yet recovered in the Atlantic was brought up in a core drilled by the *Glomar Challenger* at the base of the continental slope east of Cape Hatteras. At least 150 million years old, the sample exhibits all the mineralogical and chemical characteristics of present-day volcanic extrusions on the Mid-Atlantic Ridge. It appears that the rock was extruded at the Mid-Atlantic Ridge shortly after North America broke away from Africa at the beginning of the current episode of sea-floor spreading.

Our primary purpose in Project FAMOUS would be to examine the details of the structure of the median valley in the Mid-Atlantic Ridge and to learn as much as possible about the extrusion and accretion of volcanic rock associated with the spreading process. We planned to begin with a series of broad surveys over the area of interest, gradually focusing more closely on the most promising area as both our knowledge and our technical capability improved. By the time the project was completed we had available for study the data collected by some 25 surface cruises, two aeromagnetic surveys and 47 coordinated submarine explorations of the rift valley.

In the fall of 1971 a regional aeromagnetic survey established the exis-

netized rock exhibits symmetrical bands of alternating polarity on both sides of the ridge. In the map at the right, which corresponds to the area within the rectangle in the map at the left, the numbers next to the colored lines that run parallel to the ridge axis represent the ages of the magnetized rocks in millions of years. (The contours are given in fathoms.) The region explored by Project FAMOUS lies within the rectangle in map at the right. A more detailed map of area explored by submersibles appears on next page.

tence of the typical pattern of magnetic anomalies on both sides of the axis of the mid-ocean ridge and indicated that the African and North American crustal plates were moving away from the axis at the rate of about 1.5 centimeters per year. Later in the year the first precisely navigated wide-beam acoustical survey revealed that the floor of the rift valley was fairly flat. Whereas the entire valley is about 30 kilometers wide, the inner floor is between one kilometer and two kilometers wide. This particular section of the Mid-Atlantic Ridge is offset by two fracture zones about 30 miles apart. The one to the north was designated Fracture Zone A, the one to the south Fracture Zone B [*see illustration on preceding two pages*].

Starting in the spring of 1972 the U.S. Naval Oceanographic Office began a narrow-beam echo-sounding survey of the area between the two fracture zones. The final product of the survey, completed early in 1973, was a set of bathymetric charts with a contour interval of five fathoms (about 10 meters). A different acoustical technique was simultaneously employed by British participants in the project. They used a side-scan instrument, embodied in a seven-ton submerged system, that directs its acoustic radiation at the sea bottom obliquely and thus can record echoes from many topographic features at different distances, all with one outgoing pulse. Since each pulse can irradiate large areas of the sea floor, it does not take long to assemble a regional echo map in which the major linear features of the area are clearly defined. The side-scan technique showed the steep west wall of the valley, the less steep east wall, the "corner hills" near Fracture Zone B and the hills down the center of the rift-valley floor, all in accurate relation to one another.

During the summer of 1972 the French research vessel *Charcot* carried out detailed bathymetry, a magnetic survey and a bottom-sampling program along the northern half of the rift valley south of Fracture Zone A. The survey showed the presence of a central hill on the valley floor (later to be known as Mount Venus) and confirmed that the magnetic-anomaly patterns were broader on the east than on the west, indicating that the spreading to the east is faster than that to the west. Rocks recovered

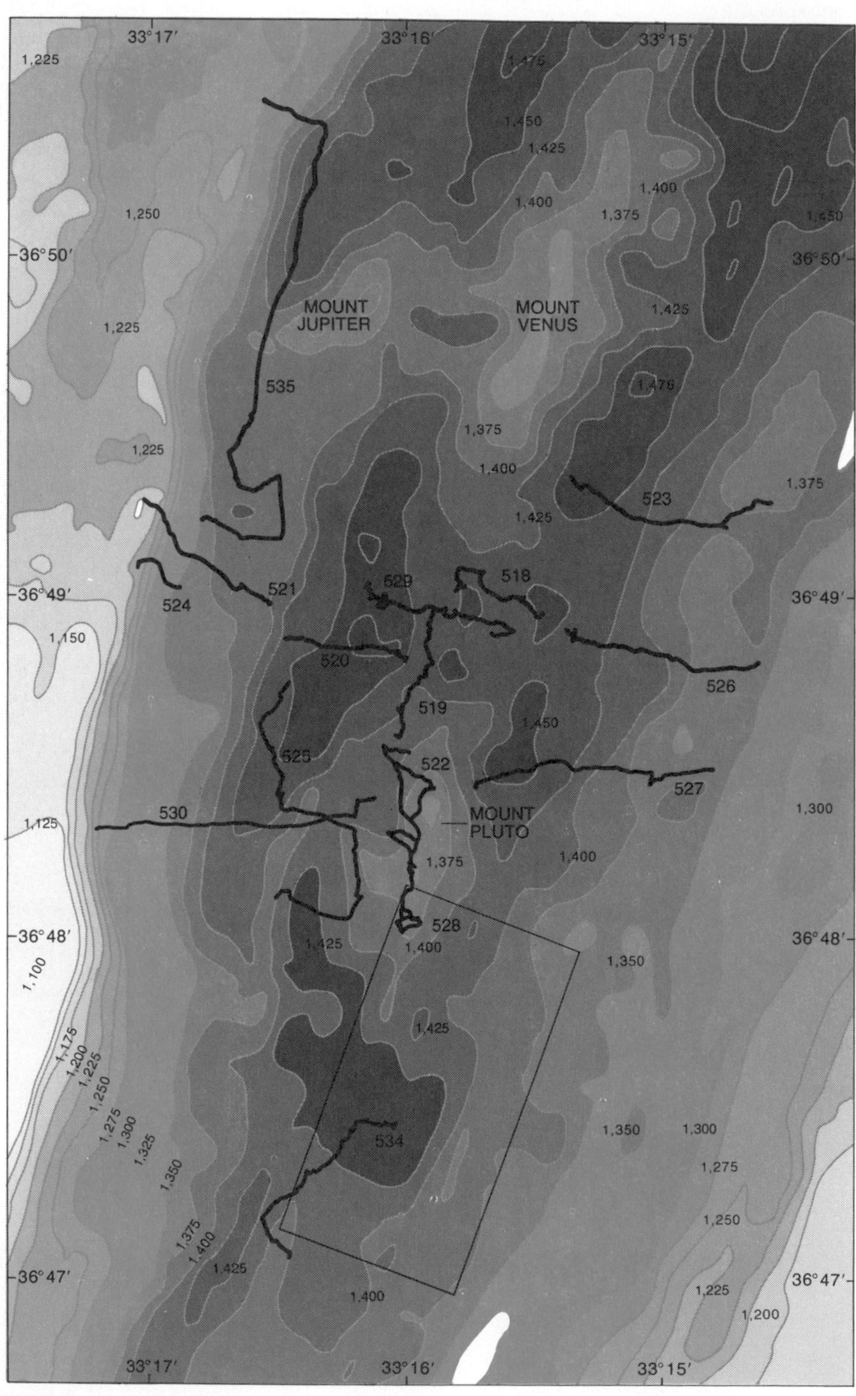

RIFT VALLEY explored in detail embraced an area approximately four kilometers by six kilometers. (Area inside rectangle appears in illustration on opposite page.) The rift valley is bounded on the west by a steep cliff, the West Wall, some 1,000 feet high, and on the east by a series of shallow terraces. The tracks of 15 exploratory dives made by the *Alvin* are depicted and numbered, beginning with dive No. 518. (The *Alvin* had previously made 517 dives elsewhere.) In addition the French submersibles *Cyana* and *Archimède* made dives in the rift-valley-and-fracture zone that lies to the north of Mount Venus in the Mid-Atlantic Rift. The *Cyana* made 12 dives in 1974; the *Archimède* made six in 1973 and 12 in 1974.

TENSION FRACTURE in the floor of the Mid-Atlantic Rift was photographed with an automatic camera mounted on the research submersible *Alvin*. Along the sides of the fissure are numerous "pillow lavas," which are formed when lava is extruded into water and rapidly cooled. At the left is part of the *Alvin*'s external gear; the round object at the bend of the pipe is a compass. Flowerlike objects at top of the fissure are crinoids, invertebrate animals attached to the bottom. The *Alvin* was one of three submersibles that explored the rift as part of Project FAMOUS (French-American Mid-Ocean Undersea Study).

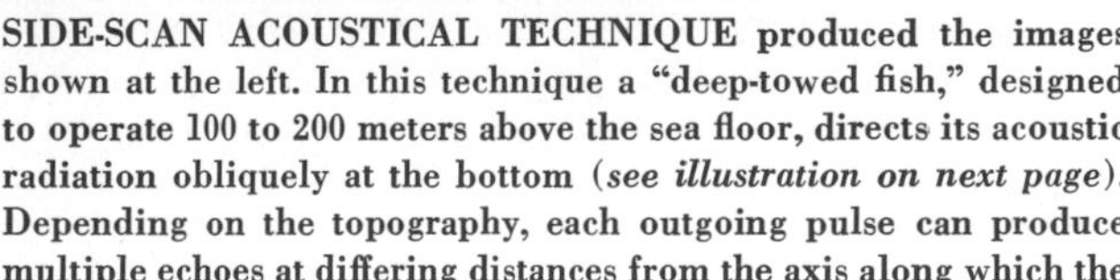

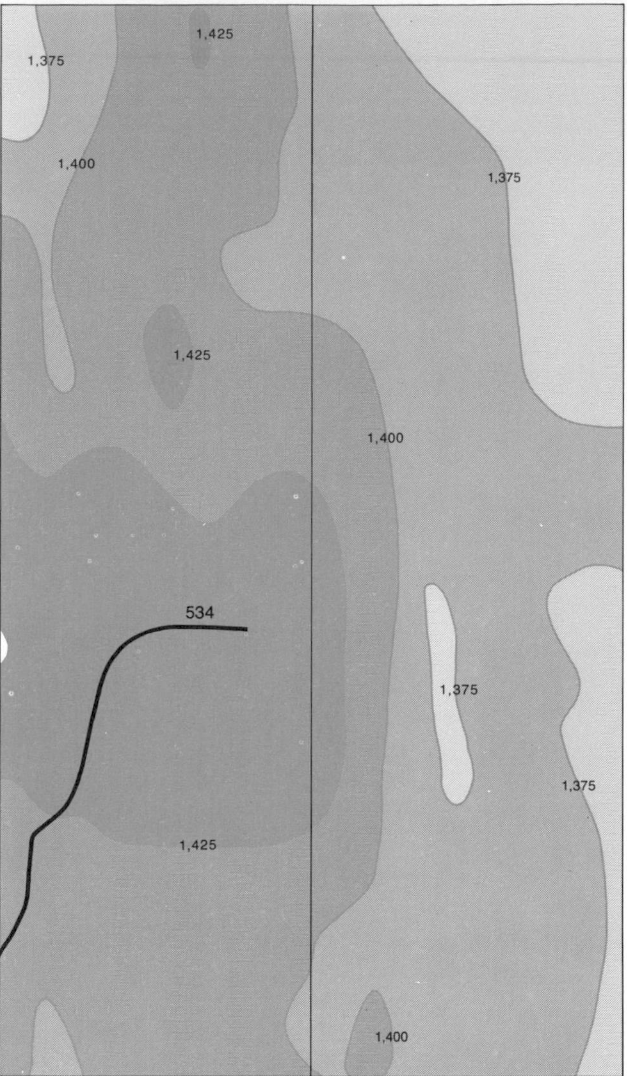

SIDE-SCAN ACOUSTICAL TECHNIQUE produced the images shown at the left. In this technique a "deep-towed fish," designed to operate 100 to 200 meters above the sea floor, directs its acoustic radiation obliquely at the bottom (*see illustration on next page*). Depending on the topography, each outgoing pulse can produce multiple echoes at differing distances from the axis along which the instrument is traveling. The two images that are reproduced here represent the left-hand and right-hand reflections from the sea floor depicted in the contour map of the same area at the right. Elevations in the floor produce reflections that appear bright in the acoustical record. It can be seen that in general the right (east) side of the record contains more linear features than the left (west) side.

from the rift valley popped and exploded on the deck of the ship, apparently as a result of the release of trapped gas. The rocks appeared to be very fresh, suggesting recent volcanic activity. A sled carrying cameras was towed along the bottom and obtained the first photographs of the bulbous and tubular lavas that were to become so familiar to the diving scientists.

The French survey was supplemented and extended south to Fracture Zone B in late 1972 by the *Atlantis II* from Woods Hole. Three radar transponder beacons were anchored on mountain peaks in the rift to serve as navigation reference marks for a series of closely spaced bathymetric and magnetic survey lines. Our data confirmed the asymmetry in the median valley, with the western wall rising very steeply and the eastern wall rising more gradually in a series of steps.

In order to survey and sample this portion of the valley in detail we moored two acoustical transponder beacons on a terrace near the top of the west wall. Since the beacons were supported on floats 100 meters above the sea floor, they would not be affected by storms, unlike the radar buoys, which could swing in a large circle around their mooring. In response to a signal transmitted from the ship, the beacons transmit a characteristic signal, each identifiable by its own wavelength located between 10 and 14 kilohertz. The two-way travel time provides a precise measure of the slant range between the ship and the transponder beacon. The depth of each transponder and the distance between transponders is established by running a survey pattern over them; a shipboard computer is then used to generate a best estimate of the depths and base-line length. With these parameters fixed, the slant ranges can be converted to horizontal distances from points directly above the transponders, thereby uniquely fixing the ship's position in relation to the base line. With a two-transponder base line it is necessary to know, of course, which side of the base line the ship is on, but this ambiguity is easily resolved from the known differences in bathymetry across the base line.

Our dredging and photographic surveys required the lowering of the dredge or the camera frame on a long cable behind the ship. The ship's position could now theoretically be plotted to an accuracy of a few tens of meters. (We now had to ask: For which part of the ship do we want a position?) The towed devices, however, would travel at some considerable but unknown depth and distance behind the ship. Thus a third transponder was attached to the cable close to the dredge or the camera. This transponder transmitted a signal that not only was received back at the ship but also triggered a second set of responses from the base-line transponders. The secondary base-line responses were received by the ship some seconds after the primary responses. Since the ship's position was continuously being fixed on the basis of the primary responses, the depth and position of the "relay" transponder on the cable could be computed. Any depth ambiguity in the solution could be resolved by entering into the calculation an estimated depth (based, for example, on the amount of cable paid out). Essentially the same system would be used later to navigate the manned submersibles, except that an automatic acoustic transmitter would be installed in the submarine to perform the same function as the towed relay transponder [*see top illustration on following page*].

In practice all the raw data (acoustic travel times) were stored on magnetic tape for later replay at home, to allow the filtering of spurious positions introduced by acoustic noise or "bottom bounce." Our real-time output on shipboard was a tabletop *X-Y* plotter that drew a small dash to represent the ship's position and a small cross for the position of the towed instrument. On later cruises real-time tracking was provided on a fluorescent screen coupled to a thermal printer. Both systems produced inked or printed records of the intricate maneuvers executed during camera or dredge stations, with the positions of the ship and the instrument being recorded every 20 to 40 seconds. By laying plots of portions of our bathymetric maps on the printer, or on transparent overlays on the screen, we could maneuver the ship and instrument package to photograph or sample any given bathymetric feature. Since the time of each acoustic fix was known, and since the time was recorded on each photograph, the position of any given feature photographed on the sea floor could be established.

We found that when the dredge touched bottom, vibration in the cable destroyed the acoustic signal; gaps in the track of the dredge indicated the positions of bottom contact and hence the location of any samples recovered. We also learned that we could reposition the dredge precisely in areas where rock had been revealed by photographs, and in these places we invariably recovered sizable quantities of rock fragments resembling those observed in the photographs. This procedure soon led to a system we called touchdown dredging; instead of dragging the dredge for half a mile or more across the bottom, as was the common traditional practice, we would maneuver it into position over a specific target, lower it to the bottom, hold it there for a few hundred yards of dragging and then bring it to the surface. In this way it was possible to relate recovered rock samples directly to specific bottom features.

Many of our photographs showed tubular or wormlike lava forms. Dredg-

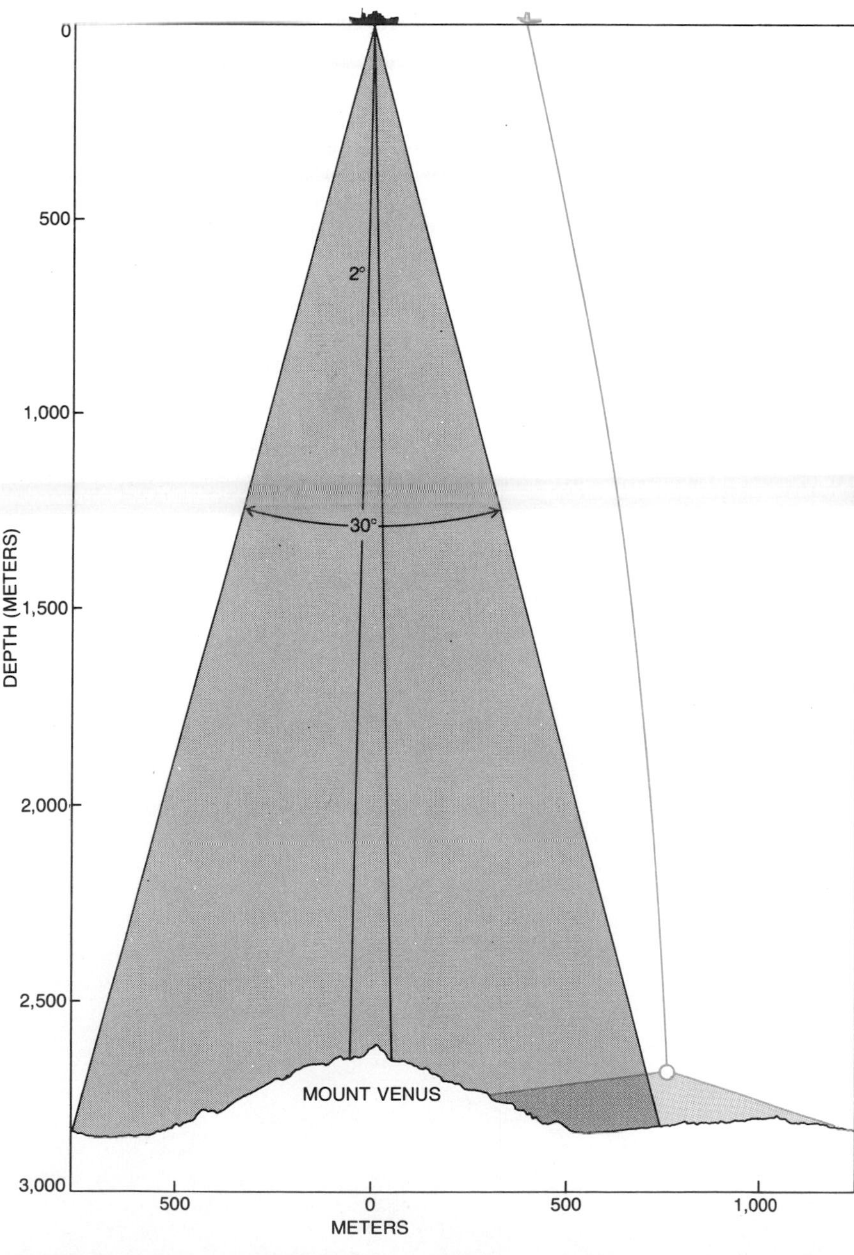

THREE ACOUSTICAL-MAPPING INSTRUMENTS supplement one another in recording the depth and the topography of the ocean floor. The conventional echo sounder, which has a beam angle of 30 degrees, provides quick, rough coverage of large areas. A newer surface instrument with a beam angle of only two degrees yields much finer detail. The deep-towed fish provides even finer detail of certain features, but its images require more interpretation.

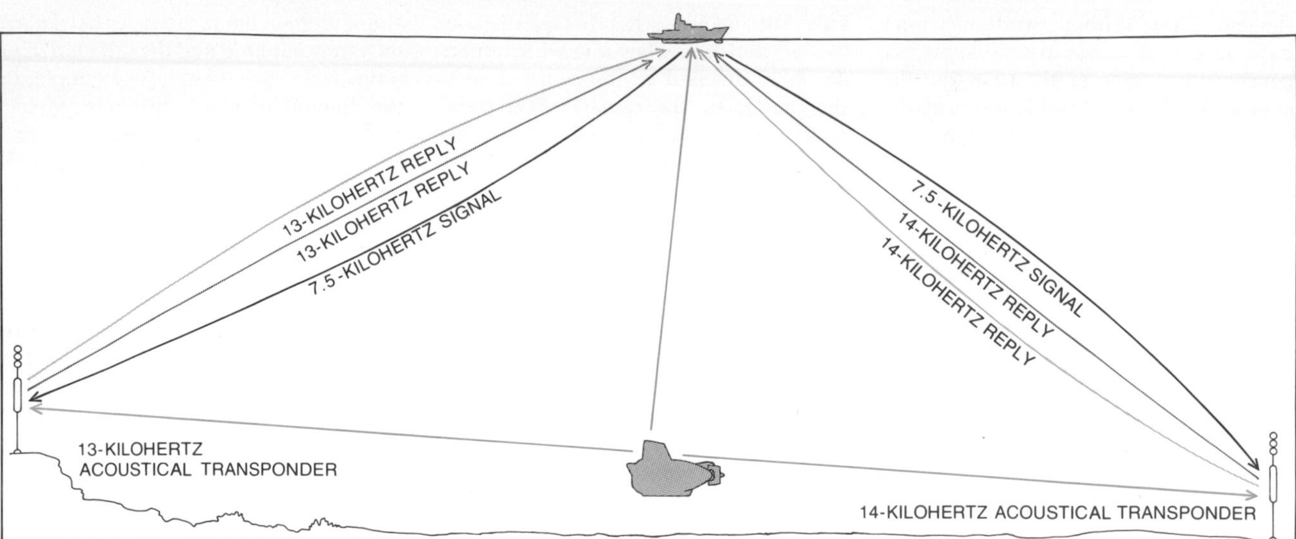

POSITION OF THE ALVIN and the other submersibles during the Project FAMOUS dives was established with the aid of acoustical transponders anchored to the ocean floor, which provided an accurately known 4.5-kilometer base line for computing the travel times of acoustic signals. Three transponders laid out in a shallow triangle were actually used; only two of them are shown here. The surface vessel guiding the submersible established its own position by sending out a 7.5-kilohertz pulse and timing the replies from the anchored transponders at two different frequencies (13 and 14 kilohertz). The submersible also emitted pulses at 7.5 kilohertz that were received directly by the surface tender and that also triggered separate responses from the anchored transponders. The difference in arrival times between the direct signal from the submersible and transponded signals were used to compute submersible's position.

ing in those areas, we recovered samples with a circular or wedge-shaped cross section. Many of the samples could be reassembled on board ship, much as an archaeologist reassembles broken pottery. These reconstructions reproduced exactly the tubular forms observed in our photographs and further confirmed the fact that most of the material had indeed been recovered from a small area, in most cases representing only one or two distinct outcrops. Photographs made in Fracture Zone B showed angular, irregular or slabby rock fragments; again our dredging recovered similar material. Many of these samples had been crushed, sheared and recemented, as one might expect in an active geological fracture zone.

In the summer of 1973 the *Atlantis II* returned to complete the survey of Frac-

LAVA EXTRUSIONS in rift take exotic forms. The large toothpastelike extrusion in this photograph taken from the *Alvin* is typical. A remote manipulator is sampling an adjacent blisterlike extrusion. Photograph was made on dive No. 522 on Mount Pluto.

ture Zone B and of the next rift-valley segment to the south. Data from this survey would help us to make the final decision on the dive site and would extend our regional coverage to help put the detailed dive operations into proper perspective. The acoustical navigation system had been refined on the basis of the previous year's experience. We would also attempt to make detailed studies of heat flow from the bottom in and near the median valley, and we would attempt to monitor and compute the position of earthquake shocks in the median valley and the fracture zones. The navigation system now was programmed to make use of a network consisting of three transponders, which both extended the range of the system and removed the base-line ambiguity.

By working with this system to position the ship over known pockets of sediment in the median valley, we were able to drop sensitive thermocouple probes into the sediment. Then, measuring the temperature differential between two or more points on the probe in the sediment, and knowing the thermal conductivity of the sediment, we were able to compute the rate of heat flow. Such measurements were obtained within a few kilometers of the valley axis and around the active transform section of Fracture Zone B. The results showed low to normal values in the median valley and above-average values on the flanks of the valley and in the fracture zone. In the fracture zone values 10 to 12 times normal were observed. These results, along with veining and cementation in the fracture-zone rock samples, seemed to provide evidence of escaping hydrothermal solutions in or near the fracture zone.

The earthquake survey also made use of a version of the acoustical navigation tracking program, modified to plot the position of three sonobuoys simultaneously. Each sonobuoy is a small floating instrument carrying a radio transmitter and a suspended hydrophone. Three of these sonobuoys were set out in a triangular array that was constantly tracked until the sonobuoys drifted out of range. Earthquake shocks picked up by the hydrophones were transmitted to the ship by radio and were recorded on magnetic tape; the signals were also amplified and broadcast in real time over loudspeakers in the laboratory. Since the positions of the three sonobuoys were known, differences in the arrival times of the earthquake shocks at the three hydrophones could be used to triangulate the position and approximate depth of the shock, just as is done with an array of three or more land-based seismic stations.

ELONGATED AND DRAPING LAVA PILLOWS, which are more characteristic of lava on land than they are of flows under water, appear in this photograph made on the *Alvin*'s dive No. 521 near base of West Wall of the rift. Canisters in foreground contain water samples.

What was to be one of our most exciting stations took place on a calm, starry night. With the sonobuoy array laid out, the *Atlantis II* moved slowly toward the steep west wall of the median valley as we lowered one of our new "Big Bertha" dredges. We planned to tow the dredge up the face of the wall. At this point, although we had begun to refer to the feature as a wall, we were aware of the way echo sounding tends to exaggerate such features, and we therefore assumed that the wall was actually no more than a steep slope, perhaps one of 30 to 40 degrees. As the dredge touched bottom at the base of the slope we began to receive ringing, banging sounds through the sonobuoy loudspeakers. When we slacked off on the cable, the noises stopped; as we took in the cable, they resumed. We watched the tension meter on the dredge cable expectantly as we approached the west wall. The echo sounder showed that the bottom was rising rapidly, when suddenly it disappeared entirely. We changed scales and tried our backup 12-kilohertz echo sounder. There was still no return. The bottom was so steep it could not reflect the signal!

Simultaneously the ringing of the sonobuoys was replaced by an ominous silence. The cable tension built up to 10,000, 12,000, then 15,000 pounds, more than twice its normal working range. Stopping the ship's engines, we drifted slowly backward, taking in cable carefully, trying to keep the tension below the 20,000 to 25,000 pounds that could break it. Finally we were directly back over the dredge, with bottom beginning to show again on the echo sounder. There was nothing more to do now but to take in the cable slowly, waiting for something to give. Suddenly there was a loud report from the loudspeakers, and the ship shuddered and reverberated as the tension meter swung wildly, then dropped back to the normal 5,000 to 6,000 pounds. We began reeling in cable cautiously, wondering when the frayed end would appear. A series of shocks and reverberations continued to roll in over the loudspeakers, although the dredge, if we still had it, should have been well clear of the bottom.

Finally the dredge appeared over the stern of the ship, with an immense freshly fractured hemisphere of pillow lava wedged in its jaws and several more large fragments caught in the chain bag. This was the largest single rock recovered during the project. It weighed about 400 pounds and had to be broken into three pieces for handling. The sonobuoys continued to record almost constant small shocks and vibrations for the rest of the night. We speculated that when the dredge broke loose, it triggered a series of rock slides on what must have been a very steep cliff. This speculation was given support the following year, when the submersible scientists observed piles of large talus blocks at the foot of the west wall. They also found many smaller piles of loose rock debris on terraces on the sides of the west wall, which turned out to be a spectacular cliff with a slope of nearly 80 degrees.

For a brief time the *Atlantis II* worked within sight of the French ships *Marcel le Bihan* and *Archimède,* as the first dives were being made near Mount Venus late in 1973. The initial dives showed that it was feasible to work with a submersible in this rugged terrain, to recover rock samples and to do geologic mapping. These preliminary submersible dives, before the major submersible effort planned for 1974, were of immense practical value in developing work and data-handling routines for the following year. For example, the *Archimède* detected bottom currents that were swifter than we had anticipated. A quickly instrumented program for metering ocean-bottom currents disclosed that the currents were nearly all tidal in nature and revealed no evidence of currents strong enough to be dangerous to submersibles.

A deep-towed instrument package, provided by the Scripps Institution of Oceanography, was brought into play at about the same time to study the microtopographic relief and the localized variations in the magnetic anomalies. This near-bottom survey confirmed the central magnetic and bathymetric asymmetry revealed by the surface-ship data, although it suggested that an asymmetry had existed in the opposite direction a few million years ago.

Following closely on the *Atlantis II* cruise the U.S. Naval Research Laboratory introduced a major new bottom photographic technique in the same area. By suspending a strong stroboscopic-flash lamp high above the camera, they were able to illuminate and to photograph an area nearly 100 times as large as the area covered by the usual ocean-bottom camera. In fact, the area photographed with each flash of the system was approximately the same as the area covered by each ping of the narrow-beam echo sounder. Mosaics made from these photographs were later laid out to scale on a gymnasium floor to enable the diving scientists to preview the terrain in which they would work.

Throughout 1972 and 1973 the French submersible *Cyana* and the U.S. submersible *Alvin* were undergoing their refitting for operation at rift-valley depths, and the diving scientists and pilots were engaged in a training program so that they would be able to work effectively over the volcanic terrain of the valley floor. In contrast to the practice of the Apollo lunar program, all the men selected to observe the rift valley from submersibles were professional earth scientists, and several had had previous experience with the submarine and over volcanic terrain. Nevertheless, we would rely heavily on our pilots' understanding of the features we would be searching for. In Hawaii the scientists and pilots spent five very profitable days, accompanied by several French colleagues. There they observed a variety of flow forms and fractured lava flows both above and under water, watched an active volcanic spatter cone and visited an area where divers had recently sampled and photographed flowing lava under water. The French made other trips to Iceland and Africa to observe areas generally supposed to be dry-land extensions of oceanic-rift systems.

In the summer of 1974 the French submersibles *Cyana* and *Archimède,* respectively accompanied by their mother ships *Le Noroit* and *Marcel le Bihan,* and the U.S. submersible *Alvin,* accompanied by its tender *Lulu* and the research vessel *Knorr,* all engaged in a coordinated diving and exploration program in the rift valley and the adjacent fracture zones. During the summer the *Cyana* completed 14 dives, the *Archimède* 13 dives and the *Alvin* 17 dives. These dives, and the preliminary seven made by the *Archimède* the year before, were man's first direct observation of the rift-valley floor.

By good fortune it was possible to schedule the *Glomar Challenger* to drill in the area. The deep-sea drilling vessel was able to penetrate some 600 meters into basement rock 18 miles west of the dive area. The rocks collected by the submersibles and by the *Glomar Challenger,* and dredged by surface ships in the area in between, comprise one of the most remarkable sets of samples ever obtained from the sea floor. The shipboard and submersible samples together will make it possible to examine regional variations in composition over an area measuring some 30 by 60 miles. In addition the drill core provides an opportunity to look at vertical variations in the crust.

The *Archimède* began its diving program on and near Mount Venus, gradually working north toward Fracture Zone A, while the *Alvin* began diving near the next central hill to the south, which we called Mount Pluto. A secondary dive site for the *Alvin* was designated in Fracture Zone B. The *Cyana* was scheduled to begin dives in Fracture Zone A, but rough weather at the start of the program damaged the *Cyana* slightly and delayed the start of the *Alvin* operations for several days. The *Alvin's* first few dives showed that visibility was excellent, that maneuvering was no problem and that sampling was as easy as we had hoped.

As the dives progressed the pilots found that it was possible to maintain almost continuous contact with the bottom. The scientists could recognize lava-flow fronts and conical structures named haystacks, which, because of their size and shape, were difficult to see in bottom photographs. The haystacks appeared to

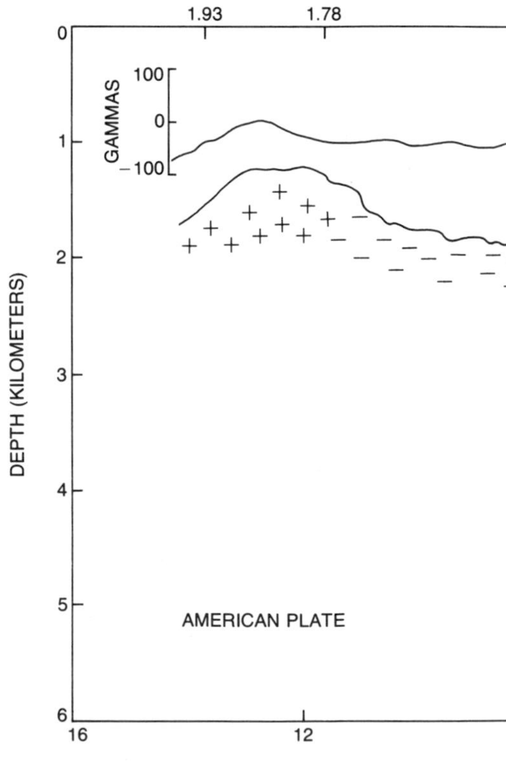

CROSS SECTION of rift is shown in relation to magnetic-field variations. Profile of the bottom is plotted from deep-towed

be small centers of lava extrusion. Major lava flows and vent structures were almost always confined to the central hills and were irregularly distributed along the central line of the valley, an observation strongly suggesting episodic volcanic activity. Minor lava flows and haystacks were observed on both sides of the central hills, particularly to the east. The submersibles collected precisely located and oriented rock samples, water samples and sediment cores. Almost continuous photographic coverage was obtained with semiautomatic cameras outside the hull of the vessels and both still and motion-picture cameras inside.

Small cracks in the sea floor near the axis of the mid-ocean ridge had been recognized first in photographs taken during the *Atlantis II* cruises. The extended-area bottom-camera system had clearly shown these features running for at least 100 meters and overlapping in an echelon fashion paralleling both the direction of the ridge and the major walls of the valley. The submersibles followed and crossed numerous fissures, measuring their heading, width, depth, longitudinal tilt, location and cross section. It was found that fissures exist everywhere from the valley's central line to its bounding walls on the east and the west, generally increasing in width with distance from the valley axis. The width varied from a few centimeters near the axis to tens of meters near the walls. Even the narrowest cracks were several meters deep. The wider ones were between 10 and 100 meters deep. In places there were small differences in elevation between one side of a fissure and the other. Across fissures up to a few meters in width it was possible to see matching halves of the same pillow lava on opposed walls.

Little sediment was found on the floor of the rift valley. Near the axis there was not even enough of it to half-bury pillow lavas a few tens of centimeters in diameter. Near the edges of the valley floor the sediment covered many such pillows and possibly reached a depth of a meter or more. Dives in the fracture zones encountered a great deal of semiconsolidated sediment fractured by horizontal shearing. In those zones no fresh volcanic rock was found, even though a high heat flow had been recorded along one narrow fracture line. In Fracture Zone A scientists aboard the *Cyana* observed two manganese-covered areas that appeared to have been created by hot water flowing up through the sea floor. Although sensitive thermocouple probes were carried on the *Alvin* and also were towed near the bottom from surface ships, no significant thermal anomalies were found in the water.

Toward the end of 1974 we began the task of assembling the great mass of data accumulated during the cruises of the surface ships. Although much still remains to be done, some interesting patterns have begun to emerge. For example, by plotting the location and composition of rocks collected by acoustically positioned dredges in and near Fracture Zone B we have found that the sheared and altered rocks, which are typical of fracture zones, are correlated with a band of subdued magnetic anomalies over the fracture zone. The band follows the trend of the fracture zone and is nearly 10 miles wide. The earthquake surveys show that the narrow active transform fault is located near the north side of this band and is connected almost exactly with the limits of the active volcanic rifts in the center of the rift valleys north and south of the fracture zone. Apparently the active fault has shifted over the width of this zone; the volcanic rifts

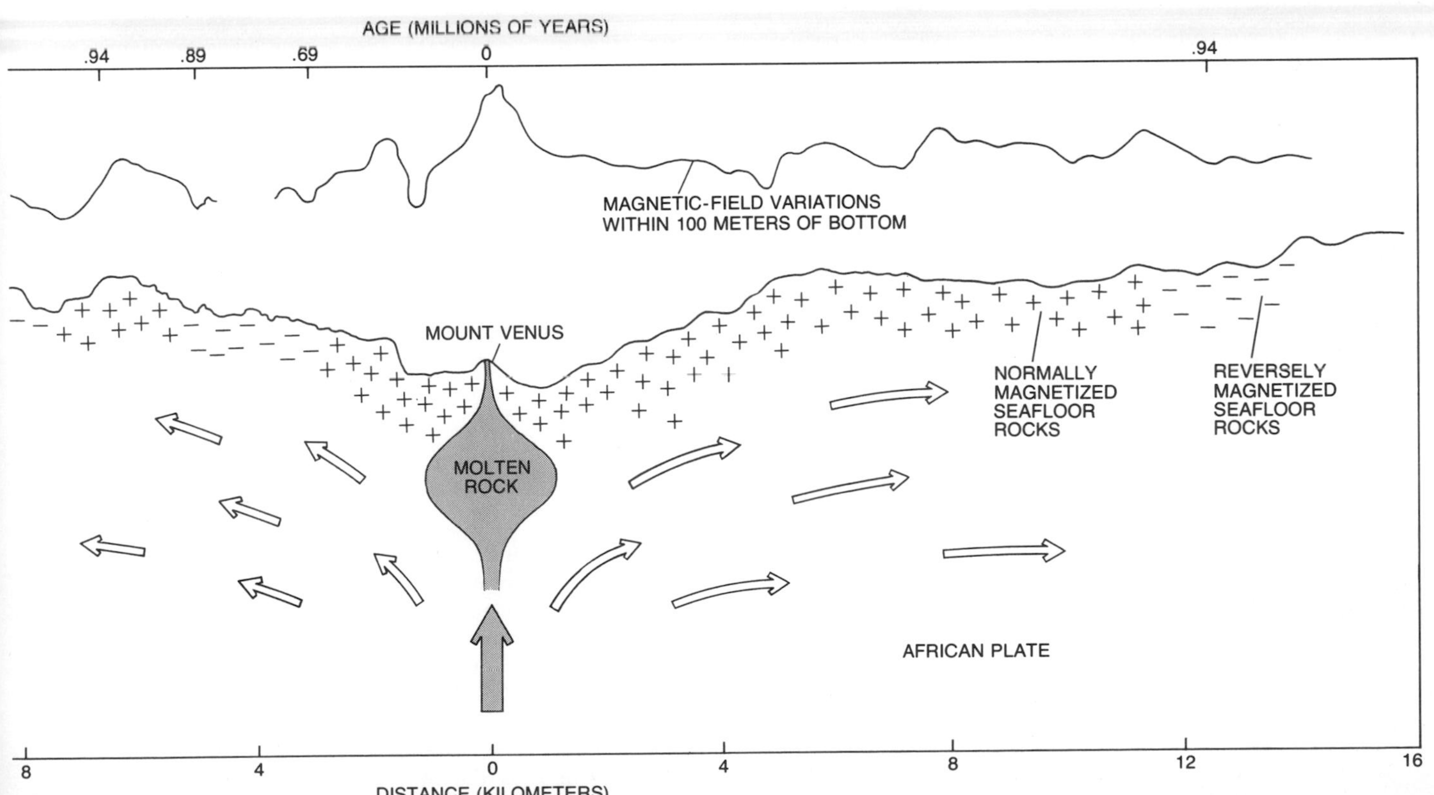

acoustical measurements. Rock of the floor is not symmetrically magnetized as measured from the center line, nor is it magnetized everywhere with equal strength. Ages of rock as identified from magnetic profile indicate that over the past 940,000 years average rate of spreading to the east has been about twice as fast as rate to the west: 1.3 centimeters per year compared with .74 centimeter.

BIZARRE SUBMARINE LANDSCAPE of lava extrusions was observed at the intersection of two lava-flow fronts in the Mid-Atlantic Rift. The drawing is based on a sketch of the scene made by one of the authors (Bryan), who served as one of the observers aboard the *Alvin* during its dives into the Mid-Atlantic Rift. The numbers identify some of the lava forms that were observed: (*1*) bulbous pillows with knobby budding, (*2*) a flattened pillow formed by the rapid drainage of lava while the skin was still plastic, (*3*) a hollow blister pillow formed by the drainage of lava after the skin had solidified, (*4*) a hollow layered lava tube formed by temporary halts in a falling lava level, (*5*) a bulbous pillow with a "trapdoor" and toothpaste budding, (*6*) an elongate pillow, typical of a lava extrusion on a steep slope, (*7*) a breccia cascade, formed on very steep slopes where the lower end of an elongate pillow has ruptured, releasing a cascade of fluid lava, and (*8*) an elongate pillow swelling into a bulbous form along a longitudinal spreading crack.

have probably moved north and south with the fault.

The bathymetric and magnetic asymmetry of the rift valley was confirmed in a striking way by the first analyses of the volcanic rocks. The analyses showed a regular variation in mineralogy and chemical composition with distance from the central volcanic hills toward the flanks of the valley. Moreover, the variations are more gradual to the east than they are to the west. In several dives the submersibles climbed the spectacular west wall, which rises some 300 meters in a series of closely spaced steps connected by nearly vertical fault scarps. No comparable feature was observed to the east; there, as the surface-ship data had suggested, the valley rises in a series of much wider steps, separated by short, steep slopes covered by broken fragments of lava.

Low areas flanking the central hills, which were originally thought to be grabens, or collapse depressions, were found to be simply low areas that had not quite been filled by lava flows converging on them from Mount Venus, Mount Pluto and the valley walls. True grabens were found at the base of the west wall and on the largest of the eastern fault scarps. These depressions suggest that the bases of the walls have pulled away from the valley floor, which has collapsed downward and inward toward the center of the valley. Indeed, one general impression is that the valley is near the end of a major period of structural extension and collapse and is just beginning to be inundated by new volcanic extrusions, represented by Mount Venus and Mount Pluto. Such episodes of alternating volcanic activity and quiescence may prove to be typical of slow centers of sea-floor spreading such as the Mid-Atlantic Ridge, where the rate of spreading may not be high enough to keep a rift constantly open to the underlying magma chamber that is the source of the lava flows.

No lava flows in actual progress were

observed by either the American or the French diving groups, although we suspect that the central volcanic hills may be very young; it is perhaps only a few hundred years since the lava last flowed out of them. Preliminary dating suggests that some of the lavas on the flanks of the valley may be much younger than their distance from the valley center would suggest, indicating that they must have emerged on the sides of the valley rather than in its center. The presence of these young flank lavas, together with extensive faulting and mechanical rotation of at least the upper 10 meters of the lava flows on the valley floor, presents difficult problems for the interpretation of the magnetic-anomaly pattern, which seems to persist in spite of such complications. It was formerly supposed that the magnetic patterns had to do with only the uppermost few tens of meters in the lava. That view may now have to be revised. The problem is further complicated by analyses of rock samples in cores taken by the *Glomar Challenger* at its drill site adjacent to the rift valley. The magnetic polarity measured in the rock samples does not seem to agree with the integrated polarity measured by surface-towed instruments. Preliminary results of magnetic-intensity measurements on samples collected by the *Alvin* also do not exhibit any simple relation to the magnetic intensity measured by deep-towed or surface-towed instruments. It now seems possible that the magnetic-anomaly patterns are related to intrusions more deep-seated than the lava flows. These deeper intrusions must be added to the diverging plates in a more regular and consistent way than the surface flows.

Many of the interpretations summarized here will necessarily be modified as the data are further refined and new facts come to light. Moreover, it may be that not all mid-ocean ridges, and probably not all parts of the Mid-Atlantic Ridge, have the same features as those of the area we have studied. Indeed, we expect that fast-spreading ridges, such as the ridge in the eastern Pacific, will turn out to be quite different. Perhaps most important, we have been able to demonstrate that the kind of fine-scale features that eluded detection in the unprecedented concentration of surface-ship studies pursued in the Project FAMOUS area can be examined routinely by manned submersibles. Further investigations of this kind are certain to add new dimensions to our understanding of processes operating on and below the deep-sea floor.

7 The Evolution of the Pacific

by Bruce C. Heezen and Ian D. MacGregor
November 1973

Deep-sea drilling shows that the bottom of the western Pacific basin is different from the bottom of the eastern basin. The slow movement of the crust underlying the basin seems to account for the difference

Advances in human knowledge often result from the coupling of two separate areas of human activity. An outstanding recent example of this phenomenon is the union of deep-water drilling technology and the scientific theories of continental drift, sea-floor spreading, plate tectonics and the youthfulness of the oceans. In 1968 the National Science Foundation inaugurated the Deep Sea Drilling Project, in which the ship *Glomar Challenger* began to systematically drill holes in the sediments that record the oceans' dynamic history. Among the oceans whose history has begun to emerge from this sampling program, coupled with the theoretical background, is the largest of all: the Pacific.

The *Glomar Challenger* is an unlikely hybrid: a ship with a hole in the middle and a 180-foot drilling rig amidships. With the ship's hull for a platform the men who operate her utilize the techniques of oil-well drilling to bring up the long cylindrical cores of deep-sea sediment. Holding a fixed position over a point on the sea bottom, compensating for the ship's vertical movements and guiding a fresh drill string into a hole only a foot in diameter several thousand feet below the ship's keel are only a few of the more formidable hurdles that the drillers have surmounted. The cores of sediment gathered by the *Glomar Challenger* from around the world constitute one volume after another, so to speak, in the growing library of oceanic history.

The deposition of sediments on the ocean bottom is a variable process. The continual rain of detritus, both organic and inorganic, that slowly sinks to the sea floor is not of the same composition everywhere, nor is it evenly distributed. Consider first the organic component. It is made up of fine fragments of the shells and skeletons and other parts of many different marine animal and plant species. The vast majority of these organisms are small or microscopic: planktonic protozoans and algae such as radiolarians, diatoms, foraminifera and coccoliths that inhabit the upper 400 meters of the ocean. The organisms are cosmopolitan in distribution, but they are found in the greatest numbers only in certain waters. For example, they are abundant in the nutrient-rich zones of deep oceanic upwelling along the western shores of continents. A similar but less well-known zone of abundance exists along the Equator, where the currents of the Northern and Southern hemispheres interact to produce another upwelling of nutrients. One planktonic group, the diatoms, is particularly numerous in two other zones of upwelling: the cold waters of the Arctic and the Antarctic.

Whether or not the calcareous and siliceous remains of these organisms ever actually reach the ocean floor depends on their solubility in seawater. Normally seawater is far from being saturated with calcium carbonate or silica. Because solubility increases with depth, almost all calcareous remains are completely dissolved before they can sink much deeper than 3,700 meters (a little more than two miles). This point is known as the carbonate compensation depth. The point where silica is completely dissolved is somewhat deeper. Compensation depths are, however, kinetic boundaries, and their location depends on both the rate of supply of organic detritus and its rate of dissolution. For example, in areas of high planktonic productivity such as the equatorial zone the carbonate compensation depth may be as much as 5,000 meters below the ocean surface (a little less than three miles). The accumulation of organic detritus on the ocean floor therefore varies from region to region, depending on the ocean depth and the planktonic productivity in the surface waters. As a result in a static model of the Pacific basin the accumulation of organic sediment on the ocean floor would be predictably greatest where the planktonic productivity was highest and the floor itself was shallower than the carbonate compensation level.

What about the sea-floor sediments formed from inorganic detritus? These substances move seaward along a variety of pathways before reaching the bottom. Essentially all inorganic detritus is derived either from continental processes of erosion or from volcanic activity along the margins of the continents. The sediments carried in suspension by streams and rivers make up one class of inorganic detritus. Some of the very smallest particles remain in suspension for a long time and become quite evenly distributed throughout an ocean basin before sinking to the bottom. The same is true of the fine dust that is carried far out to sea by the winds. These two deposition processes are responsible for a thin veneer of red, gray and green abyssal clays that covers much of the floor of all ocean basins. The clays are recognized as distinct sedimentary deposits, however, only where they have not been mingled with other sediments. For example, in the equatorial regions of the Pacific the detritus that elsewhere makes up the abyssal clays is only a trace dilutant, inseparable from the organic oozes that accumulate there in large quantities.

Coarser inorganic sediments reach the ocean in river runoff or as products of shore erosion. They do not usually travel far out to sea before sinking, and so they form a thick wedge of sand, silt and clay adjacent to their continental source. Some may be moved farther offshore by the action of turbidity currents (swift, self-propelled downslope flows of sediment-laden water) but their connection with the mainland remains obvious. Still another characteristic inorganic sediment of the Pacific basin is windblown ash from the active volcanic zones along the margins of the basin. The ash collects in the lee of the volcanoes, forming wedge-shaped deposits that usually thin out to the point of being no longer identifiable 1,000 kilometers or so from their source.

Each sediment, organic or inorganic, accumulates at a different rate. In regions of average planktonic activity the passing of a million years produces a layer of organic detritus 10 meters thick. In zones of greater activity, such as the equatorial region, the rate rises to 15 meters or more per million years. The abyssal clays are laid down at a much lower rate: about one meter per million years. For a few hundred kilometers downwind from an active volcano the average accumulation rate for volcanic ash is about 10 meters per million years. Sediments resulting from river runoff and shore erosion build up the most rapidly: from 50 to 500 meters per million years. (Even this process is not particularly fast; accumulation at the rate of 500 meters per million years means that

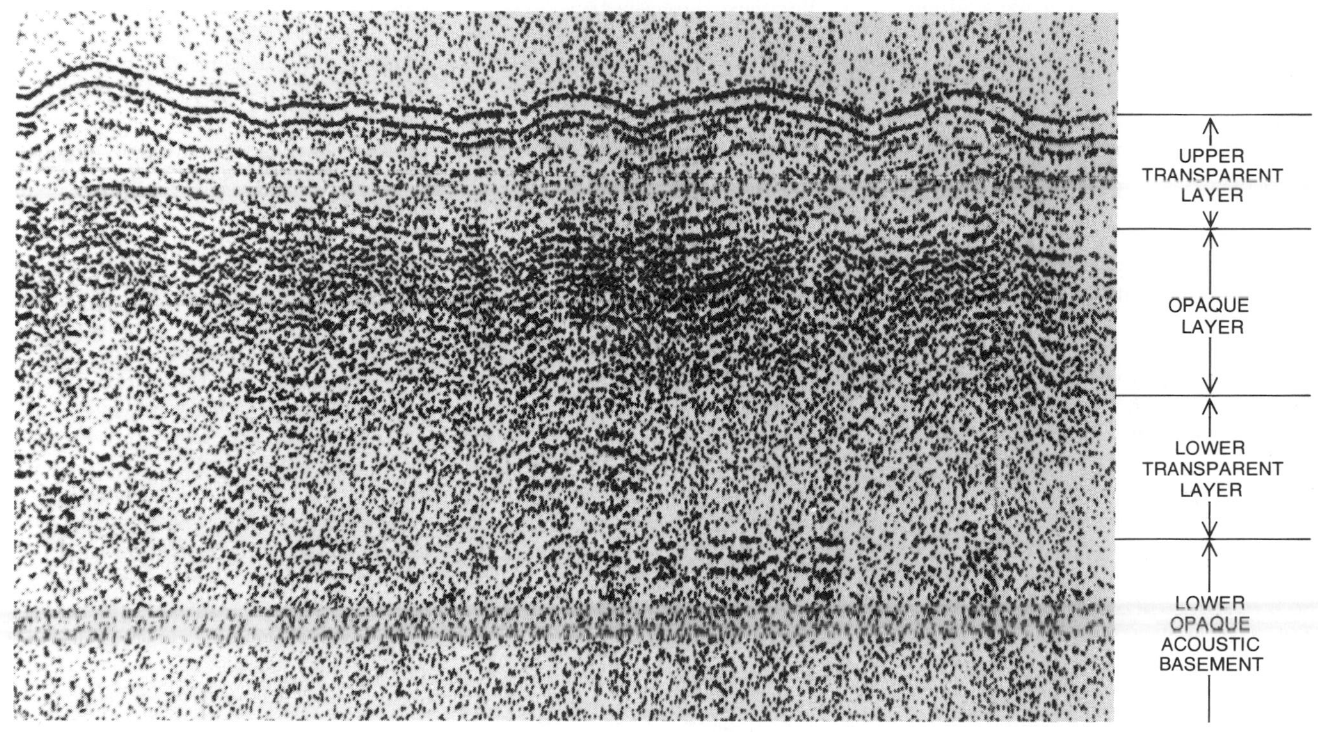

ACOUSTICAL PROFILE of the bottom of the Pacific in an area east of the Japan Trench shows four sedimentary strata, the number characteristic of the western Pacific. A "transparent" layer (*top*) overlies a darker "opaque" layer of sediments. Below the opaque layer lies a less well-defined transparent layer and below that is another opaque layer resting on the basalt oceanic crust.

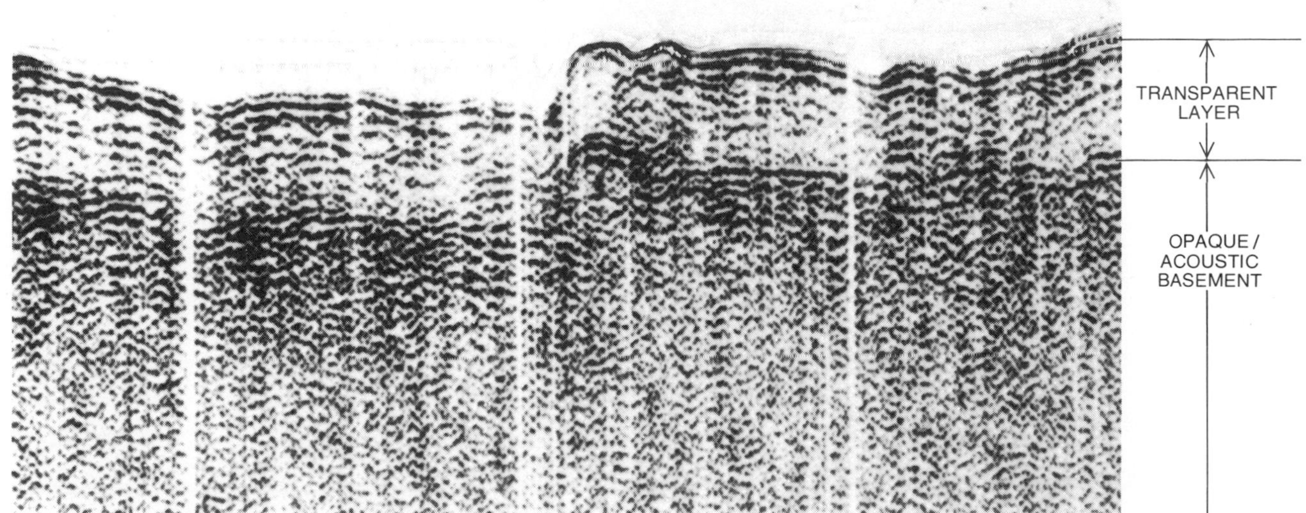

DIFFERENT PROFILE characterizes the eastern Pacific basin. Here only two layers of sediment cover the oceanic crust: a transparent top layer and an underlying opaque one. Deep-sea drill cores confirm the acoustical data. The two profiles are not reproduced at the same scale. In the top one the top three layers are 2,000 feet deep; in the bottom one the top layer is 200 feet deep.

only a twentieth of a centimeter is deposited each year.)

Sediments of these kinds are accumulating today on the floor of the Pacific basin in a clearly patterned fashion [*see illustrations on these two pages*]. In the northern waters and all along the margins of the basin volcanic ash and sediments from runoff and shore erosion build up, mixed in the subpolar region with the tiny skeletons of the diatoms that thrive there. In a broad zone along the Equator the steady rain of planktonic detritus gives rise to a blanket of organic oozes on the sea floor. Between the largely inorganic northern zone and the organic equatorial zone, below waters that are too deep to allow the buildup of organic debris, lies a vast expanse of abyssal clays.

If the present pattern of sedimentation has been constant over the past 100 million years or so, a static model of the Pacific basin would predict that any core of sea-floor sediment taken from one or another of these major zones should consist uniformly of, say, abyssal clays or organic oozes, or sands, silts and ash. In actuality no drill core is that uniform. We are left to conclude either that the pattern of sedimentation has not been constant or that a static ocean-basin model is wrong. Suppose we assume that any changes in the sedimentation pattern over the past 100 million years have been trivial. Can we, by changing the static model into a dynamic one, explain the observed nonuniform nature of the sedimentary deposits in the Pacific basin? We believe the answer is yes.

Current plate-tectonics theory states that the basement rock of the ocean floor is being formed continuously. In the Pacific basin the area of formation is the "rift valley" at the summit of the East Pacific Ridge. The new crust migrates essentially westward, sinking deeper as it creeps along. The generation of new crust at the ridge is presumably balanced by a comparable consumption of old crust as it plunges under the island arcs of the western Pacific. The horizontal and vertical motions of the crust are coupled; the rate of sinking decreases logarithmically with age.

Let us visualize a dynamic model of part of the basin, centered on the East Pacific Ridge. The movement of the newly formed crust will be westward and downward [*see top illustration on page 82*]. From the time that the crust is formed at the top of the ridge, and for a considerable period thereafter, it will be above the carbonate compensation depth. As a result it will acquire a blanket of organic oozes. The sediments will increase in thickness until the outward and downward motion of the crust carries it below the compensation depth. Thereafter, as the crust continues to creep westward into deeper water, a slowly accumulating blanket of inorganic abyssal clays will begin to cover the initial organic deposit. Because the carbonate compensation depth lies 1,000 meters below the ridge crest and because the subsidence rate of the new crust is about 50 meters per million years, organic detritus will accumulate on the new crust over a period of 20 million years. Thus the basal layer of organic sediments, accumulating at a rate of 10 meters per million years, should under

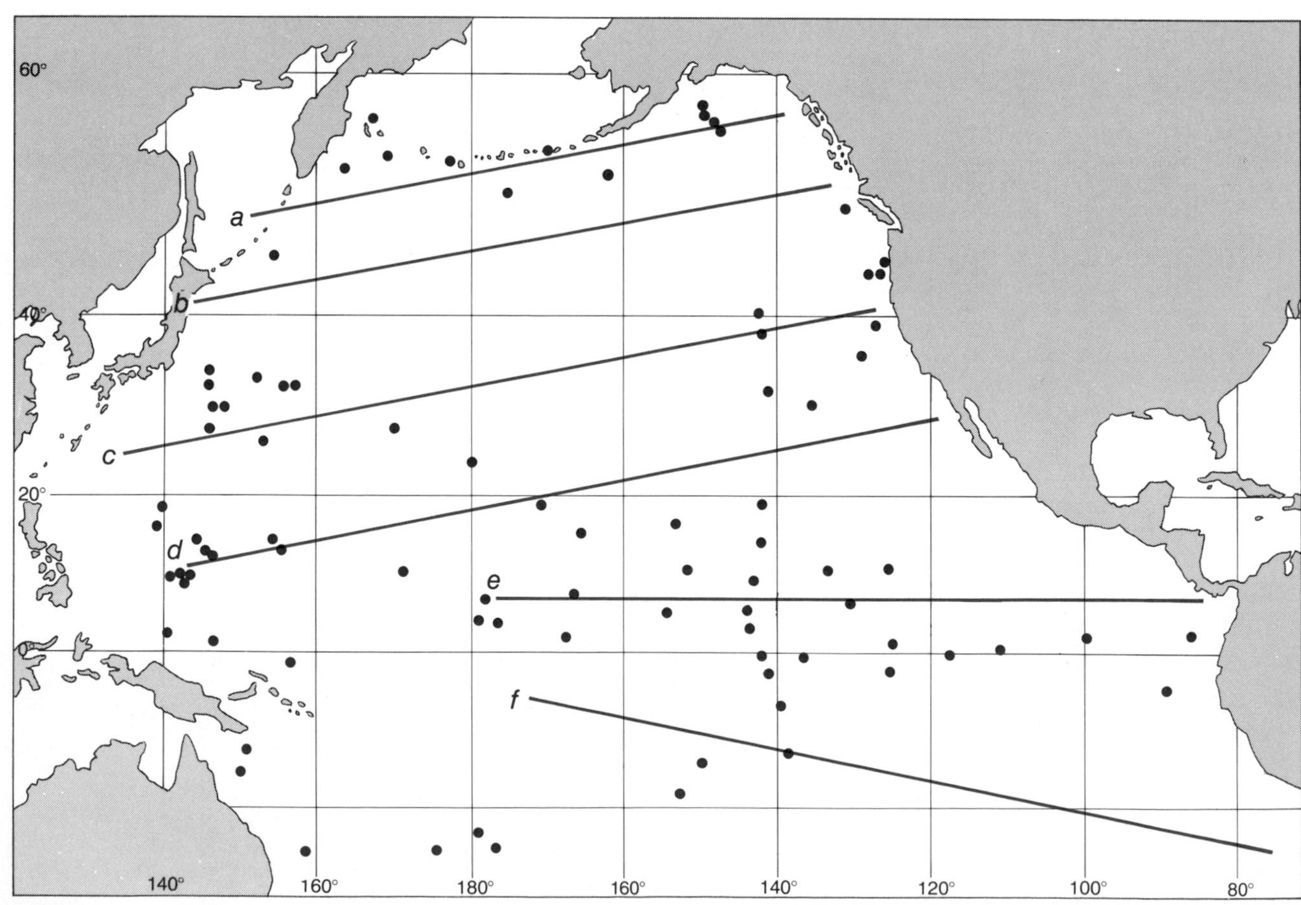

PACIFIC SEDIMENTS WERE SAMPLED by deep-sea cores taken at the sites marked with the black dots on this map. The colored lines labeled *a*, *b*, *c*, *d*, *e* and *f* relate the cores to the cross sections of the sediments shown on the opposite page. The cores were obtained by drilling from the *Glomar Challenger* in the Deep Sea Drilling Project sponsored by the National Science Foundation.

normal circumstances be no more than 200 meters thick.

Now, the sea floor in the western Pacific, the deepest and oldest part of the basin, is at least as old as early Cretaceous times, some 120 million years ago. If during the first 20 million years of its existence this part of the basin accumulated organic sediments at the rate proposed in our model, then 100 million more years were available for the deposition of abyssal clays on top of the organic oozes. At the rate of one meter of clay per million years, therefore, the western Pacific sea floor should have accumulated 100 meters of abyssal strata on top of the 200-meter organic layer. How do the sediments in the western Pacific compare with the predictions of the dynamic model? Before answering the question we must introduce another kind of information.

Drilling is the only way to collect samples of the layers of sediment below the ocean floor but it is by no means the only way to gather information about these strata. One method that has been used increasingly since World War II is acoustical profiling. In this method a strong acoustic signal is returned from the interfaces of layers of contrasty sea-floor sediment and is recorded in a continuous profile. Profiling thus makes it possible to trace the lateral distribution of different sedimentary layers for hundreds and even thousands of miles. In the Pacific profiling indicates that a layer of unconsolidated and relatively "transparent" sediments, less than 100 meters deep, covers the central part of the North Pacific floor [*see lower illustration on page 79*]. Such a region of shallow sediments would correspond to the area where our dynamic model predicts the buildup principally of abyssal clays. Along the northern and northwestern boundaries of the basin the profiles show a much thicker transparent layer; the greater thickness apparently corresponds to the rapid accumulation in these regions of continental debris and volcanic ash.

The acoustical profile along the Equator shows that the sediments there form a massive deposit, triangular in cross section and in places as much as 1,000 meters thick. Such a large accumulation corresponds to the high rate of sedimentation in a zone where upwelling nutrients produce an abundance of plankton. The acoustical profiles in the eastern Pacific are also what would be expected; the transparent layer of surface sediments, presumably abyssal clays, is present and is not unusually thick. In the western Pacific, however, not one or two but four separate sedimentary layers appear in the profiles. Just below the transparent layer on the sea-floor surface is an "opaque" layer. A second transparent layer underlies the opaque layer and a second opaque layer underlies the second transparent layer. This fourth layer rests directly on the oceanic crust [*see upper illustration on page 79*].

The acoustical data are in good agreement with the specific information about stratigraphic sequences provided by deep-sea drilling. The cores show that the bottom layer of sediments, resting directly on the oceanic crust both in the older, deeper western Pacific and in the younger, shallower eastern Pacific, is or-

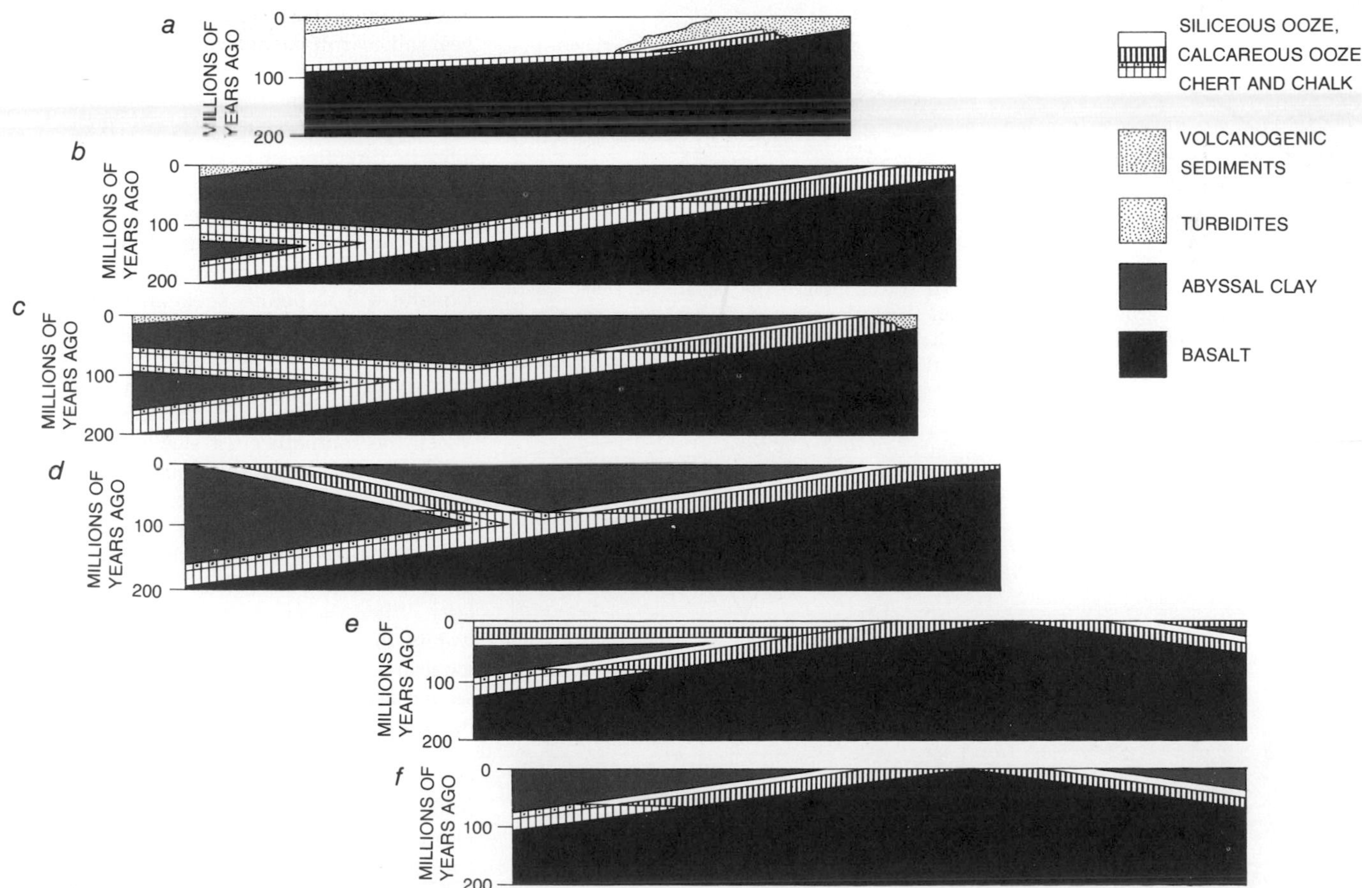

CROSS SECTIONS OF SEDIMENTS along the colored lines in the illustration on the opposite page are shown schematically. The wedge-shaped form of the cross sections results from the fact that, as the sediments were being deposited, the basalt crust below them was moving. The time at which the sediments were being deposited is given by the scale at the left side of each of the cross sections.

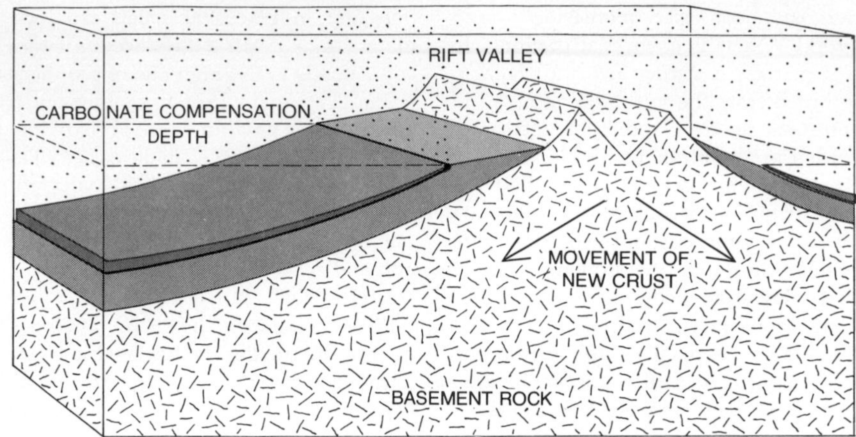

DYNAMIC MODEL of the Pacific basin in the vicinity of the East Pacific Ridge shows how different kinds of bottom sediment accumulate. Oceanic crust, newly generated in the rift valley at the crest of the ridge, slowly moves outward and downward. While the crust remains above the "carbonate compensation depth" most of the detritus that settles on it is organic (*black dots*) and a thick stratum of ooze is formed. When the crust enters deeper water, only inorganic detritus (*colored dots*) reaches it and forms a thin stratum of clay.

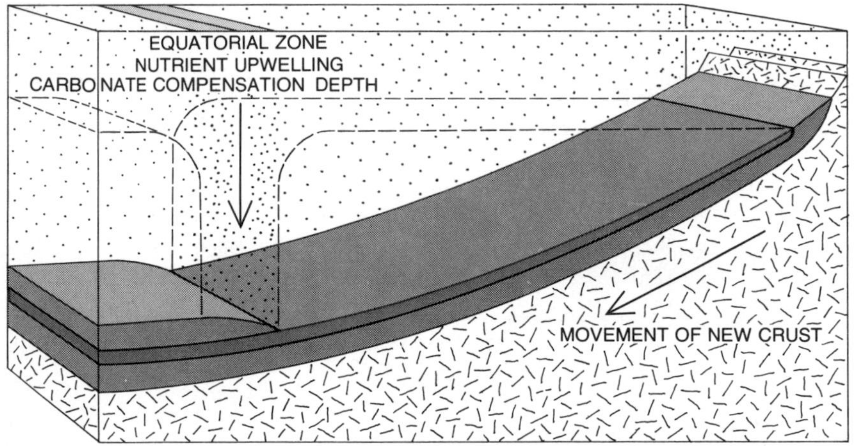

REVISED MODEL makes the oceanic crust travel through the equatorial zone, where abundant planktonic growth brings about an abrupt lowering of the carbonate compensation depth. The heavy "rain" of organic detritus (*black dots*) in this zone deposits a second layer of ooze on the abyssal clay. As a result three separate sedimentary strata overlie crust.

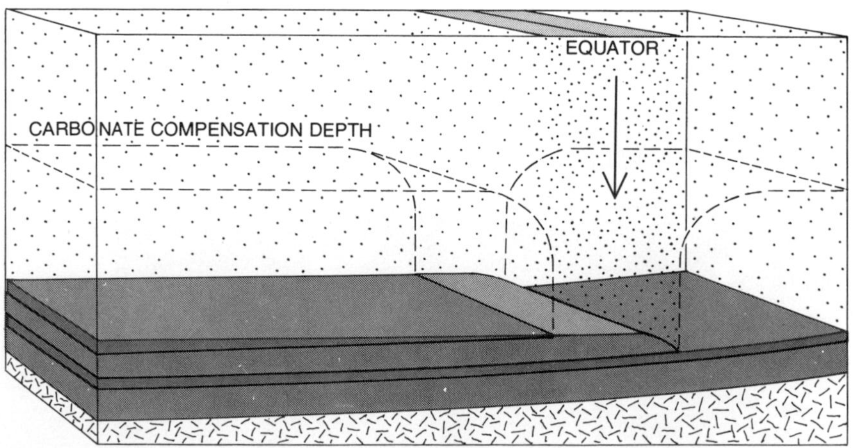

FOURTH LAYER OF SEDIMENTS accumulates above the three earlier layers after the oceanic crust passes beyond the equatorial zone. The crust is once again in water far below the carbonate compensation depth and only inorganic detritus (*colored dots*) reaches the floor of the ocean. As a result a thick second stratum of abyssal clays builds up, covering second stratum of organic ooze. This is the kind of layering characteristic of western Pacific.

ganic in origin. The only difference between west and east is that the older deposits have turned into chalks and cherts whereas the younger ones are still unlithified oozes. In both west and east, the cores show, the basal organic layer is overlain by a layer of abyssal clays.

The clays in the western Pacific are older than the clays in the eastern part of the basin. In addition the western clays are overlain by a second layer of chalks and cherts, and that organic layer is covered by a second layer of inorganic clays. Thus the cores show precisely the same four-layer sequence that is revealed by acoustical profiling, and it is obvious that serious discrepancies exist between the four-layer stratigraphy in the western Pacific and the two-layer prediction of our dynamic model. The model sequence—oceanic crust overlain first by organic oozes and then by inorganic clays—is correct as far as it goes, but it fails to describe the real sequence. There is another difficulty too. We find two depositions of organic sediments in the western Pacific rather than one, and the buildup of this double ration of sediment would have needed more time. The process may have taken 50 million years, or more than twice the 20 million years allotted in the model's timetable.

Can our defective dynamic model be repaired? The answer is yes, if we alter its direction of movement. Let us add a new component—northward drift —to the slow westward creep of the freshly formed oceanic crust. If we do that, the new crust formed north of the Equator will, as before, accumulate only two layers of sediment: first organic oozes and then abyssal clays. The same two layers will also build up on crust newly formed south of the Equator. When the southern crust slowly moves northwestward across the Equator, however, it will receive a further heavy rain of detritus from that nutrient-rich zone, thereby acquiring a second layer of organic oozes [*see middle illustration at left*]. When the crust once again enters waters that contain nothing but inorganic matter, it will start to acquire a second layer of abyssal clays on top of its second layer of organic oozes.

The predictions of our revised dynamic model compare favorably with the results of deep-sea drilling in the Pacific basin, although we should like to see cores taken at many more locations. In the cores from the western Pacific that are available now the oldest chalks and cherts are of lower Cretaceous age (and some are probably even older). This layer evidently represents the organic detritus accumulated on the new crust during its initial subsidence. The abyssal clays that overlie the basal organic layer similarly represent the deposition of inorganic detritus after the new crust sank below the normal carbonate compensation depth. The chalks and cherts in the layer above the lower abyssal clays are middle Cretaceous in age; they evidently record the passage of this part of the Pacific crustal plate northwestward across the Equator. Finally, the abyssal clays above the second organic layer are late Cretaceous or Cenozoic in age; they must represent the further accumulation of inorganic sediments after the equatorial zone was left behind.

Other kinds of evidence support the concept illustrated by our revised dynamic model. Provocative evidence on the probable directions of motion of the Pacific plate is provided by two long lines of submerged Pacific seamounts: the Hawaiian and the Emperor chains [*see illustration on this page*]. It has been proposed that the seamounts are the truncated remains of former volcanoes that were successively brought into existence by the action of a "hot spot" rooted to the deep mantle below the Pacific crustal plate. In principle the two long seamount chains combined represent a kind of fossil record of the slow movement of the Pacific plate across the hot spot over millions of years.

As a result of drilling and dredging the ages of such features can be estimated. The oldest seamount, at the northern end of the Emperor chain, was evidently submerged some 70 million years ago. Midway Island, near the western end of the Hawaiian chain, is only some 20 million years old. The volcano Kilauea, on the island of Hawaii at the eastern end of the chain, is still erupting. The ages of seamounts located at the "bend" where the two chains meet have not yet been determined, but on the basis of extrapolation they are thought to be from about 30 to 40 million years. Since the ages increase to the north and west it would appear that the Pacific plate has moved west-northwest in a direction parallel to the Hawaiian chain for the past 30 to 40 million years, and that for 30 to 40 million years before that time it had moved in the more northerly direction parallel to the Emperor chain.

It is possible to estimate on the basis of the deep-sea drilling data the average northward component of the plate's motion. The point in any core where the bottom layer of the youngest abyssal clays is in contact with the top layer of the underlying chalks and cherts can be interpreted as a temporal boundary; it indicates approximately when that part of the plate completed its transit of the equatorial zone. In cores taken at a variety of latitudes in the Pacific basin the age of this temporal boundary has been determined from the fossils preserved in the uppermost chalks and cherts. The results suggest that, for a period of some 80 million years beginning in the late Mesozoic and continuing until middle Cenozoic times, the northward movement of the Pacific plate proceeded at a rate of some 4.4 centimeters per year. Since then, for the past 30 million years or so, the northward component of motion has been smaller: some two centimeters per year [*see top illustration on next page*].

Still further evidence with respect to the motion of the Pacific plate comes from the observation that in oceanic crust generated along an east-west axis the bottom is "striped" with zones of alternating magnetic polarity, the result of past reversals of the earth's magnetic field. The magnetic striping is either nonexistent or poorly developed, however, in crust that has been generated in the equatorial zone; here the crust is magnetically quiet. Now surveys show that a huge area of the North Pacific is also magnetically quiet. The region starts near the present intersection of the Equator and the East Pacific Ridge, in the vicinity of 100 degrees west longitude, and continues generally north by west in a wide swath that terminates off the Kamchatka Peninsula in the northwest Pacific [*see bottom illustration on next page*].

Up to now the existence of this large quiet area has been explained as evidence that during a considerable part of the Cretaceous there were no reversals of the earth's magnetic field. Although there may have been such a period, our model predicts that a magnetically quiet zone should be present. A better explanation, in our opinion, is that the quiet region represents crust that was formed within the equatorial zone and that has

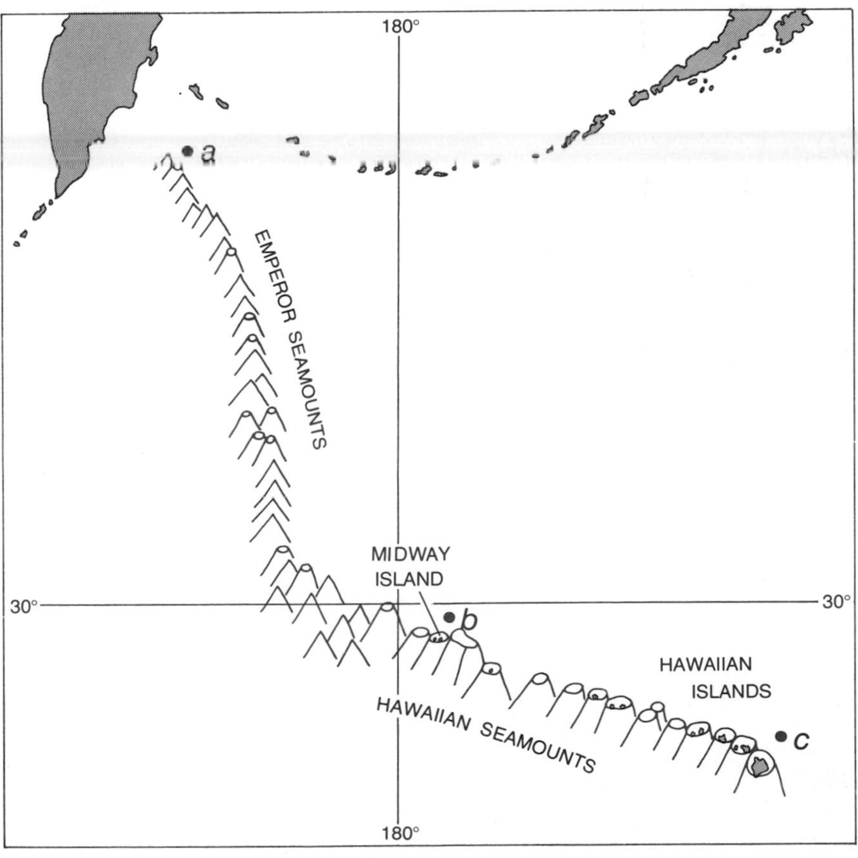

LINE OF SEAMOUNTS, the Emperor chain (*left*) and the Hawaiian chain (*right*), traces the past movement of the Pacific plate over a "hot spot" rooted in the mantle near Hawaii. As the plate moved northwestward over the hot spot, volcanoes appeared; the seamounts are their submerged remnants. A seamount (*a*) at the northern end of the Emperor chain is the oldest known: 70 million years. This contrasts with an age of 20 million years for Midway Island (*b*) and one of less than a million years for the volcano Kilauea on Hawaii (*c*).

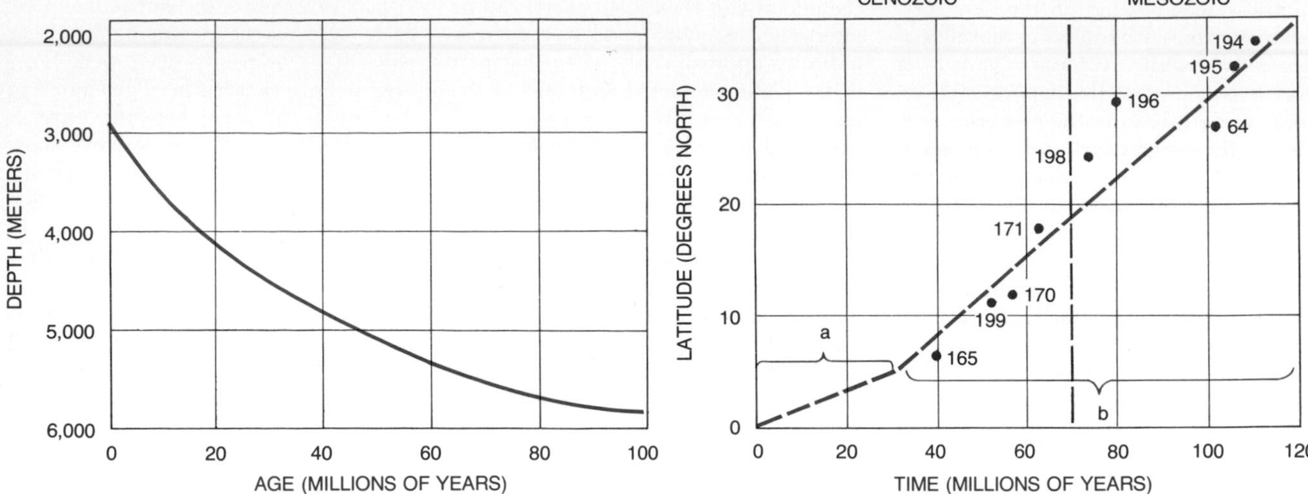

AGE OF FOSSILS in the uppermost organic strata of sea-floor sediments in the Pacific increases with depths (*left*) and distance north of the Equator (*right*). Two rates of plate movement become apparent (*broken line*) when the trend is averaged. Between 120 and 30 million years ago (*b*) the rate was some 4.4 centimeters per year. Since then (*a*) it has diminished to some two centimeters.

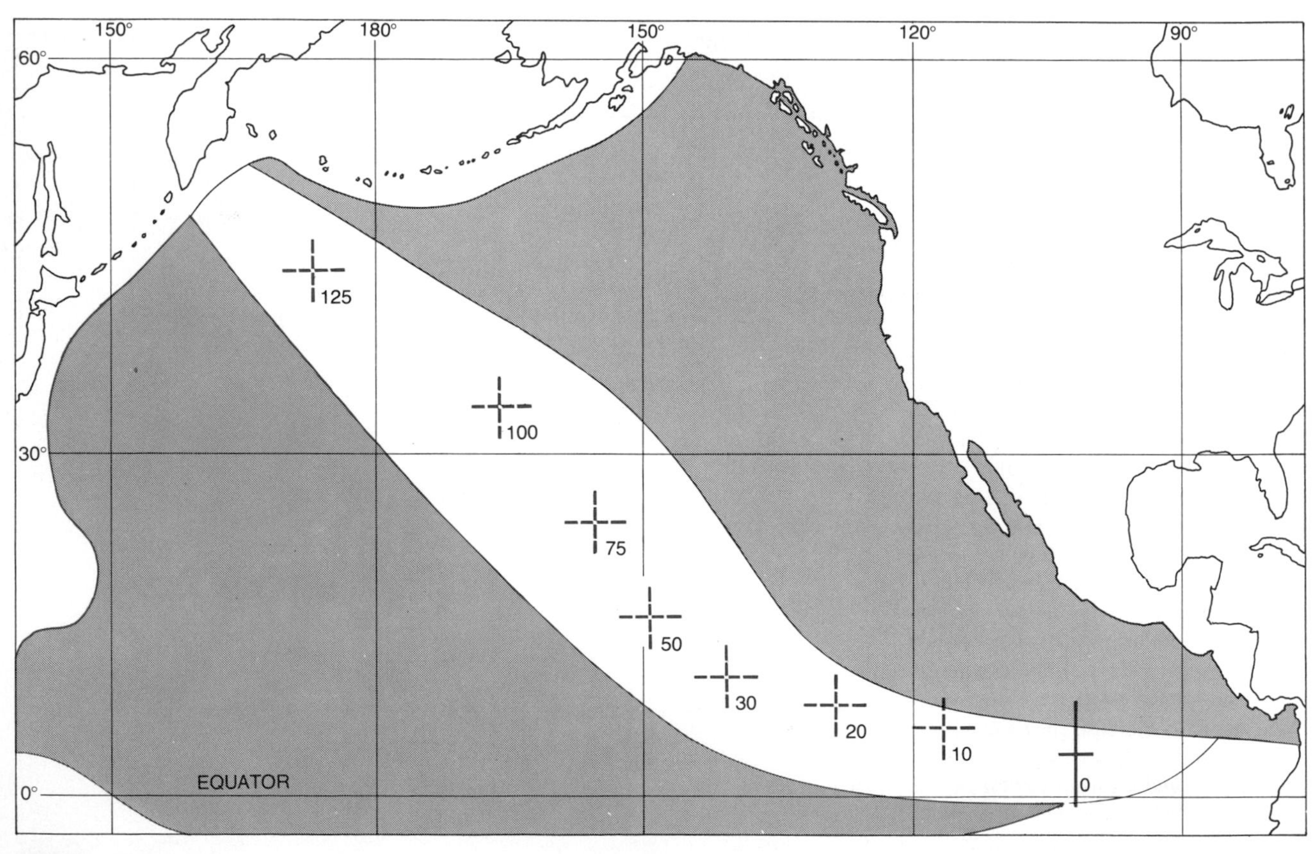

ZONE OF MAGNETIC QUIET, where evidence of reversals of the earth's magnetic field is weak or nonexistent in the crust of the ocean floor, extends from near South America on the Equator to the subpolar waters off the Kamchatka Peninsula. A solid cross (*right*) shows the present intersection of the Equator and the East Pacific Ridge, where new crust is formed continuously. Seven earlier intersections are marked by broken crosses and numbers that suggest how many millions of years ago these paleo-equators began to migrate northwestward; all seven lie within the magnetically quiet zone. The crust north of the quiet zone has never traveled across the Equator. By crossing the Equator the crust south of the zone has collected four layers of sediment rather than two layers collected elsewhere.

reached its present position as a result of the slow northwestward movement of the Pacific plate. An eastward extrapolation of paleo-equator positions determined from deep-sea drilling, together with a westward extrapolation of crustal age based on dated magnetic stripes, enables us to estimate the location and age of a series of points where the East Pacific Ridge and earlier "paleo-equators" once intersected.

In our interpretation the magnetically quiet region represents a boundary between the two principal components of the Pacific crust. North of the quiet region's northern boundary is crust that has never made an equatorial crossing. South of the southern boundary is crust that was formed before a crossing, so that during its subsequent equatorial transit it received a second ration of organic sediments. The quiet belt between these regions would of course be crust formed during the equatorial transit.

Our refined model applies to an oceanographic problem a particular kind of geological reasoning: the facies concept. The concept arises from the observation that sedimentary deposits reflect the circumstances and environment of their origin. In brief, it states that just because similar sets of strata are found in two separate sedimentary deposits this does not necessarily mean that the strata accumulated at the same time. It is just as likely that their similarity is due to the existence of similar environmental conditions at quite different times. For example, if the margin of a continent is subsiding, a shoreline environment will "migrate" inland; the sedimentary deposits of each successive shoreline will become a rock formation essentially identical with the one preceding it, but what may appear to be a uniform lithological unit will not have been deposited at the same time. In the same way our model of the Pacific basin assumes that the ocean's various sedimentary environments have remained constant while the plate comprising the floor of the basin has migrated. The strata that are similar are not contemporaneous; instead they represent the plate's passage through a series of constant environments.

It should be emphasized that the simplicity of our model necessarily gives it a number of shortcomings. For example, it is a depositional model that ignores episodes of erosion, changes in sea level and possible shifts in compensation depths. It scarcely considers such variations in plate motion as rotation or changes in direction and the rate of drift. Neither does it particularly concern

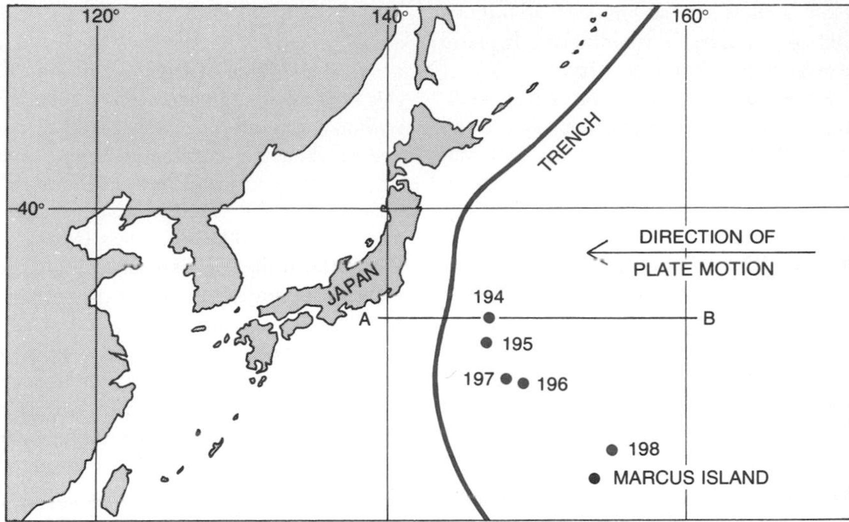

DEEP-SEA CORES from an area in the western Pacific between the Japan Trench and Marcus Island reveal the presence of a wedge of volcanic ash that overlies a part of the four-layer sequence of sediments in this area. In the first illustration below is a cross section.

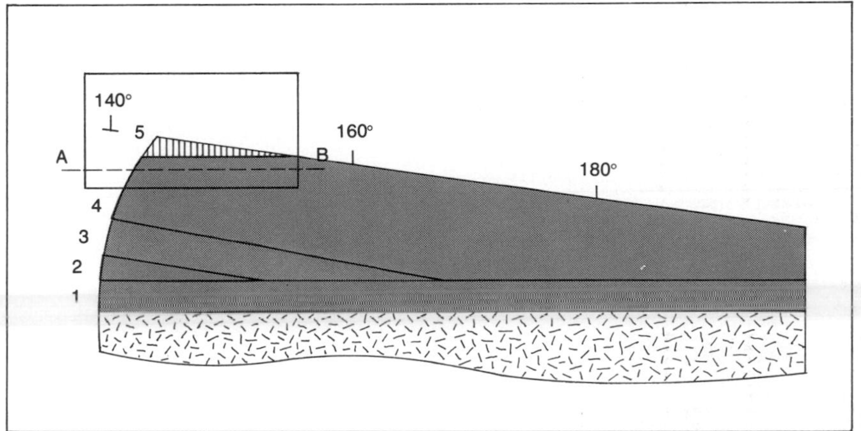

LAYER OF ASH is a fifth sedimentary deposit on the sea-floor crust. The lowest layer is the first deposit of organic detritus, accumulated after the crust was formed. The small wedge above it is the first deposit of inorganic abyssal oozes, which here pinches out. Above that are a second organic layer, which pinches out farther to the east, and a second, very thick inorganic layer that comprises the ocean bottom. A closer view of section is shown below.

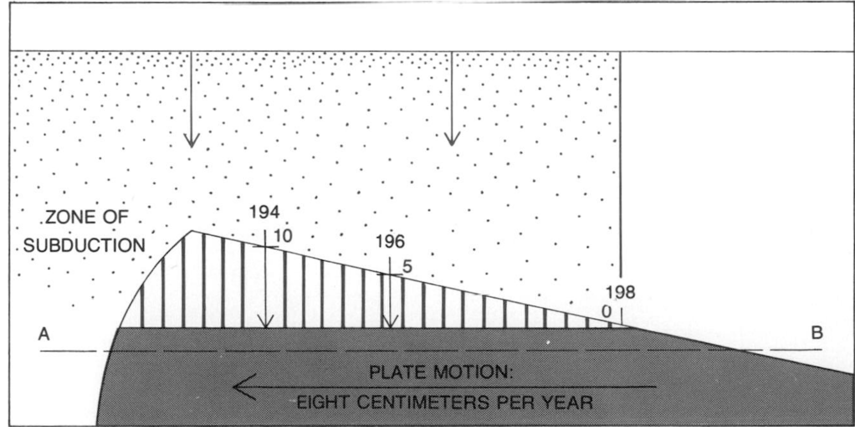

THICKNESS OF LAYER increases from east to west. Assuming a sedimentation rate of 15 meters per million years, the lowest ashes in drill hole No. 196 would be some five million years old and those in hole No. 194 would be some 10 million years old. Ash much older than that has been lost in the zone of subduction at western edge of moving Pacific plate.

itself with the great mass of sediments that comes from continental runoff, shore erosion and volcanic activity.

Concerning these continental sediments, however, one point should be made. The northward and westward motions of the Pacific plate and its subduction under the island arcs of the northern and western Pacific do affect the size and shape of the great sedimentary wedges that extend out from the continents. At the boundary of the subduction zone the dimensions of the wedge will be determined by the combined rates of sedimentation and subduction. A rapid rate of sedimentation will tend to advance the wedge seaward, whereas subduction will counteract this tendency by continuously moving the sediments back toward and under the continent.

Drill cores taken in the vicinity of the Japan Trench, northwest of Marcus Island, provide good examples of the layers of volcanic ash that began to build up on the surface of the abyssal mud in this area late in Miocene times, some 12 million years ago. The ash here has not yet reached the subduction zone, and the volcanic accumulation is some 150 meters thick [*see middle illustration on page 85*]. Some 800 kilometers farther to the southeast the wedge of ash thins to the vanishing point, leaving the uppermost layer of abyssal clay uncovered. Drillings in between show that the wedge is intermediate in thickness. If we assume that the sedimentation rate for volcanic ash has averaged about 15 meters per million years, then it appears that the Pacific plate is moving into the subduction zone at a rate of eight centimeters per year and that the volcanic ash deposited on the Pacific sea floor in middle Miocene and earlier times has already vanished under the continental plate.

In summary, the fact that our simple model gives a good first-order fit with the data so far available suggests that the evolution of the Pacific basin has been governed by a few equally simple and systematic factors. Among these are the carbonate and silica compensation depths, the gradual sinking of the oceanic crust as it creeps westward from the East Pacific Ridge and the simultaneous northward motion of the crust. Data already exist, of course, that deviate from the model, and future observations will surely produce more such data. Such deviations are welcome. Not only should they reveal local variations in the sedimentary history of the Pacific basin but also they may demonstrate irregularities in the motion of the Pacific plate itself.

When the Mediterranean Dried Up

by Kenneth J. Hsü
December 1972

Evidence acquired on a recent cruise by the deep-sea drilling vessel Glomar Challenger *has revealed that six million years ago the Mediterranean basin was a desert 10,000 feet deep*

Six million years ago a biological revolution swept across the Mediterranean Sea. The ancient marine fauna of the Mediterranean, descendants of mixed races from the Atlantic and Indian oceans, effected an unorganized mass exodus to find a refuge west of Gibraltar. Those that remained were soon to face annihilation, except for some hardy species that could tolerate the deteriorating environment. Thus ended the Miocene epoch, the less recent of the two dynasties that preceded our own Quaternary period. With the dawn of the Pliocene, or more recent, epoch the refugees returned, bringing with them new species from the Atlantic. They are the ancestors of the present marine fauna of the Mediterranean. This dramatic event, as recorded by fossils in certain sands and marls of Italy, did not escape the attention of Sir Charles Lyell, one of the founders of geology. The end of the revolution, signaled by the establishment of a new faunal dynasty, was chosen by Lyell in 1833 as the historical datum dividing the Miocene and Pliocene epochs. What was the cause of this revolution?

Near the end of the 19th century a deep gorge buried under the plain of Valence in southern France was discovered during a search for ground water. The gorge was cut into hard granite to a depth of hundreds of feet below sea level. Filling the gorge are Pliocene oceanic sediments, which in turn are covered by the sands and gravels of the Rhône river. When the gorge was first discovered, it was found to extend for some 15 miles between Lyons and Valence. Eventually the buried channel was traced for more than 100 miles downstream to La Camargue in the Rhône delta, where the valley was reached by drilling 3,000 feet below the surface. Obviously the modern Rhône is a lazy weakling compared with its ancestor, which sculptured a system of gorges almost comparable in size to the Grand Canyon of the Colorado. What caused the deep incision of the Rhône?

In 1961 the American oceanographic-research vessel *Chain* sailed to the Mediterranean with newly developed seismic equipment to explore the sea floor. The CSP, or continuous seismic-profiling, device aboard could be considered a super echo-sounder. Besides recording bottom reflections (the acoustic signals bounced back from the sea bottom), the instrument picked up signals transmitted by sound waves that were able to penetrate beyond the bottom and be reflected by hard layers hundreds of feet below. The new tool made possible a new discovery: it was found that the Mediterranean floor is underlain by an array of pillarlike structures, each a few miles in diameter and hundreds or thousands of feet high, protruding into the beds of sediments [*see top illustration on next page*]. Geophysicists were familiar with structures of this type; they looked very much like salt domes.

Salt domes are formed after rock salt from a deeply buried mother bed has forced its way upward into overlying sediments. Salt domes are common, for example, along the U.S. Gulf Coast, where many oil fields have been located around the domes. To find salt in coastal sediments is not unexpected, because geologists have long thought that the rock-salt formations were precipitated in coastal salinas, or lagoons. It was entirely unexpected, however, that salt domes would be discovered under the abyssal plains of the Mediterranean. Where could the salt have come from? Or are those structures indeed salt domes?

The exploration continued in the 1960's. William B. F. Ryan of the Lamont-Doherty Geological Observatory, a participant in four cruises to the Mediterranean, and others working in the area were soon impressed by the presence of a strong acoustic reflector everywhere in the Mediterranean [*see bottom illustration on next page*]. This reflector, the *M* reflector, is commonly found a few hundred feet below the sea bottom, and its geometry closely simulates the bottom topography.

A layer that could send back distinct echoes must be very hard. Yet ocean sediments are commonly soft oozes made up of minute skeletons of the small organisms called foraminifera and nannoplankton. What could this hard layer be? Why should such a hard rock be down there under the Mediterranean?

To solve these and many other puzzles a group of investigators constituting the Mediterranean Advisory Panel of the Joint Oceanographic Institutions for Deep Earth Sampling program (JOIDES) in 1969 recommended to the Deep Sea Drilling Project (which is funded by the National Science Foundation and administered by the Scripps Institution of Oceanography) that the deep-sea drilling vessel *Glomar Challenger* be sent to the Mediterranean. The proposal was approved, and a two-month cruise in the fall of 1970 yielded the surprising answer to our mysteries: The biological revolution of the Mediterranean, the deep incision of the Rhône and the oceanic salt are all silent witnesses to an event six million years ago when the Mediterranean was almost completely dry. The hard-rock layer serving as the strong acoustic reflector is composed of the inorganic residues left behind by the desiccated Mediterranean: minerals known as evaporites.

The *Glomar Challenger* left Lisbon on August 13, 1970, for the 13th cruise

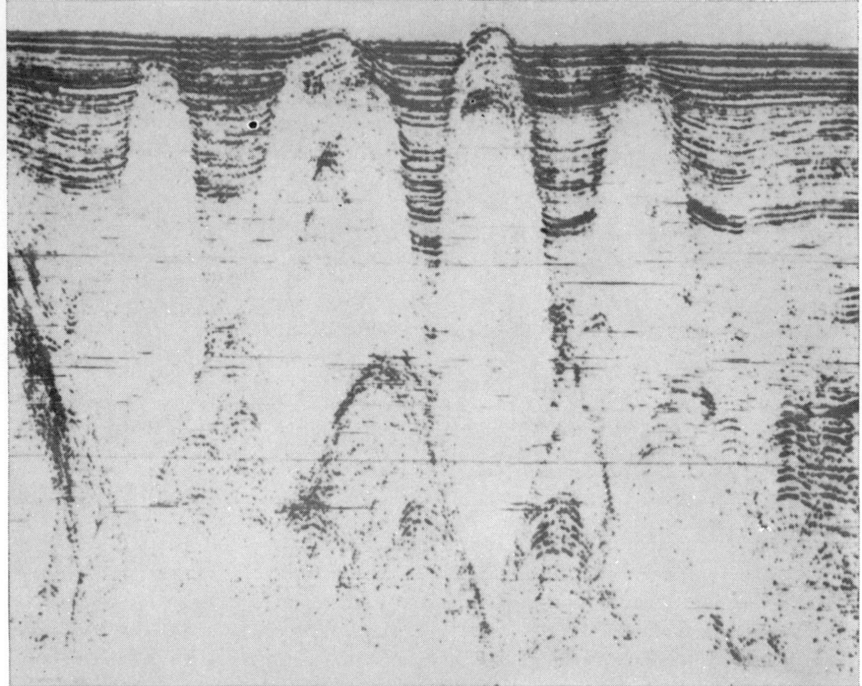

PILLAR-LIKE STRUCTURES, believed to be salt domes, are evident under the sea floor in this continuous seismic profile of a 10-mile-wide section of the Balearic abyssal plain in the western Mediterranean. The continuous-seismic-profiling device records not only reflections from the sea bottom but also signals transmitted by sound waves that penetrate beyond the bottom and are reflected by hard layers below. Some of the domes protrude as knolls above the sea floor; others are completely buried. The discovery of these pillar-like structures was the first hint that vast salt deposits are located under the Mediterranean floor.

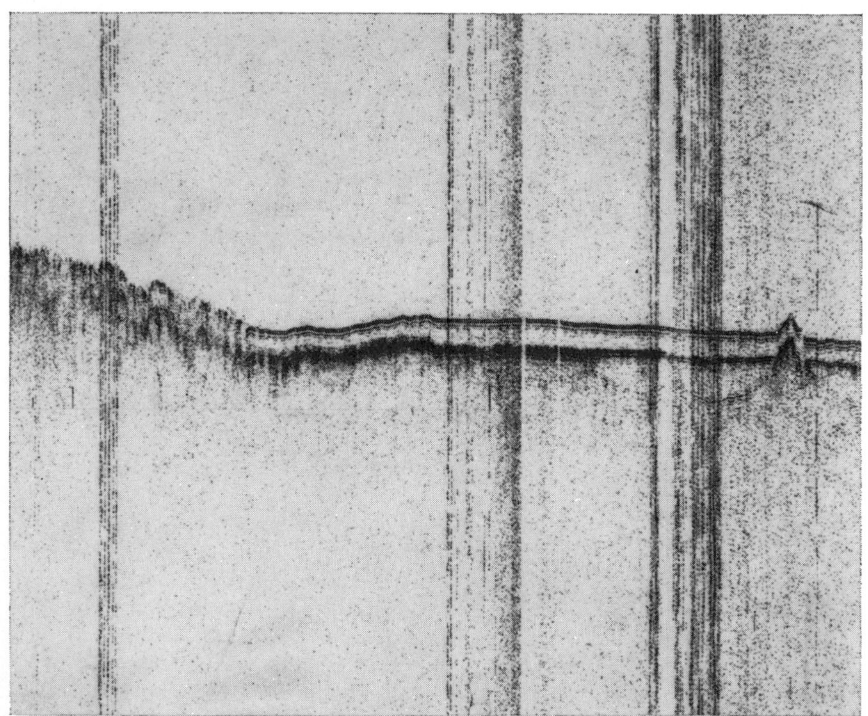

M REFLECTOR, a strong acoustic reflecting layer, underlies much of the Mediterranean floor. The relief of this layer (*lower dark contour*) closely simulates the bottom topography (*upper dark contour*). Drilling has shown that the *M* reflector corresponds to the top of an extensive underground evaporite formation consisting of inorganic residues precipitated from brines when the Mediterranean was isolated from the Atlantic Ocean some six million years ago. This continuous seismic profile was obtained by the *Glomar Challenger* as she worked her way northeastward (*right to left*) from a point southeast of Sicily toward Crete.

of the Deep Sea Drilling Project. The vessel is unique because of its ability to keep virtually stationary in a stormy sea. Guided by a system of radio beacons and computers, the four thrusters of the vessel can position her above the drill hole within a circle with a diameter of about 100 feet. Ryan and I were co-chief scientists on this cruise, heading an international team of 20 investigators and technicians. On the evening of August 23 we arrived at a spot some 100 miles southeast of Barcelona. After positioning our vessel above a buried submarine volcano we were ready to tackle the problem of identifying the *M* reflector.

The drill string was lowered to the bottom at a depth of 6,000 feet, and we bored some 600 feet into the sediments. According to our seismic record, we should have been hitting our goal and reaching the top of the hard layer. Since we had penetrated the oldest Pliocene sediments, the hard layer should date back to the late Miocene: some six million years ago. At the critical juncture, however, while we all waited eagerly for the next rock sample to come up, we ran into trouble: the core barrel got stuck inside the drill string. There was nothing to do but to haul all the drill pipes back on deck and pull the core barrel out of the pipe at the end of the string. Furthermore, we were advised to find another drill site; we could not expect to dig deeper at this spot.

Neither Ryan nor I slept that night as we supervised the drilling details. Now that the core barrel had come up, it was found to be buried in a tube full of sand. The sand was brought in a bucket to our shipboard laboratory. All morning long on the 24th Ryan and I busied ourselves sorting pea gravel out of the sand; we were too tired to work and too keyed up to sleep. We needed the menial labor to ease the tension. As the morning wore on, however, we became more and more amazed by what we saw.

Gravels are rare in the oceans. Submarine slumping can generate an underwater current, known as a turbidity current, that can transport coastal sands and gravels to deep abyssal plains. One would expect gravels of this type to be composed of many different kinds of rock, derived from erosion on land. As our pickings accumulated, however, we noticed that our gravel was made up of only three different types of rock: oceanic basalt, hardened oceanic ooze and gypsum. In addition there was an unusual dwarf fauna of small shells and snails. We found no quartz, no feldspars, no granites, no rhyolites, no gneisses, no

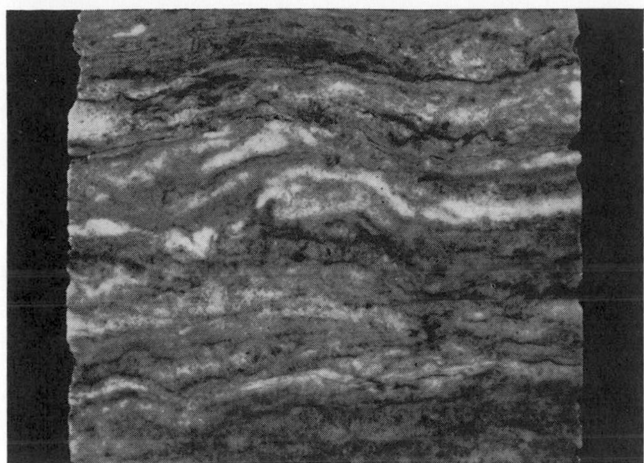

CONVEX-UPWARD LAMINATION of the sedimentary rock shown in this photograph of a Mediterranean *M*-reflector core obtained by the *Glomar Challenger* has what geologists refer to as a stromatolitic (literally "flat stone") structure. Stromatolites are typically layered carbonates that owe their origin to the fact that blue-green algae tend to grow in a succession of thin mats on certain coastal flats. Since the very existence of the algae depends on photosynthesis, the presence of a stromatolitic structure is considered evidence of shallow-water deposition, the deep-sea bottom being too dark to allow the survival of such photosynthetic plants.

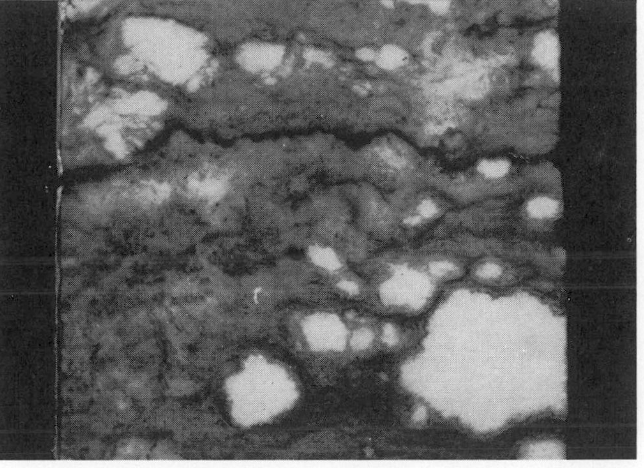

LIGHT-COLORED NODULES evident in this Mediterranean rock core are composed of anhydrite: a high-temperature form of calcium sulfate that can only be precipitated from brines at temperatures higher than 35 degrees Celsius. At lower temperatures the low-temperature form of the sulfate, gypsum, is precipitated. Anhydrite is formed today almost exclusively in the sedimentary layers of the hot and arid coastal deserts called sabkhas, where it is precipitated in the pore spaces of sediments near the ground-water table. The high-temperature form typically grows as irregular nodules, similar to the limy concretions one finds in arid soils.

schists, no quartzites, no sandstones. In fact, we found no trace of anything that could be identified as coming from the nearby continent. The gravels could not have been brought out here by turbidity currents from the Spanish coast. What was the meaning of this unusual gravel?

Toward the evening of the 24th the drill bit was brought in. Caught in the teeth of the bit, which had been stuck near the top of the hard layer, were fine aggregates of anhydrite, an anhydrous calcium sulfate. Anhydrite is a common mineral in evaporites, which represent the inorganic residues left behind by evaporated brines, and ancient evaporites are commonly lithified (converted into rock). Now we had the first solution to our mysteries: The *M* reflector, the hard layer under the Mediterranean, is a late Miocene evaporite.

This solution only deepened the mystery. Evaporites should be the deposits of coastal lagoons or of deserts. Why should we find an evaporite formation under the Mediterranean at depths of thousands of feet below sea level?

The gravel provided the clue. The fragments of gravel could not have come from land, only from a dried-up ocean. Was it possible, then, that the Mediterranean had been isolated from the Atlantic and had changed into a desert basin during the late Miocene?

One could imagine the gradual shrinkage of the Mediterranean and the increasing salinity of its waters, with the death of all normal marine animals except for some dwarf species of clams and snails tolerant of supersaline conditions. The inland sea would eventually be changed into a salt lake, like the Dead Sea, where the brine would be dense enough to precipitate gypsum. Continued evaporation would eventually have laid bare the Mediterranean bottom. The submarine volcano would be converted into a volcanic mountain, and the oceanic ooze on its flank would become lithified. Streams draining such a desiccated ocean bottom would produce an unusual gravel such as the one we had found.

It seemed preposterous to make up such a story on the flimsy bit of evidence we had. The Mediterranean abyssal plains are more than 10,000 feet deep and the basin holds almost a million cubic miles of water. It was unthinkable that this beautiful blue ocean should disappear and be replaced by a series of Dead Seas and Death Valleys.

Or was it so unthinkable? A few facts and figures showed that it would be rather easy to dry up the Mediterranean. The climate of the Mediterranean region is arid. The annual evaporation loss is approximately 1,000 cubic miles. Only a tenth of the loss is compensated by rainfall and by the influx of fresh water from rivers. The Mediterranean manages to maintain its normal marine salinity through an exchange of water masses with the Atlantic. If the Strait of Gibraltar were to be closed today, the annual evaporative loss could not be compensated, and the Mediterranean would be dried up in about 1,000 years.

Still, was it necessary to invoke such a drastic explanation? Many geologists had speculated that salts could be precipitated from a deep brine pool when the salt content in the brine exceeds the saturation concentration. The distribution of the evaporite layer indicated to us that this sedimentary formation was deposited in a deep Mediterranean basin. Is it not possible that the basin was filled to the brim with brines? True, a deep-water theory could not explain the genesis of our unusual gravel, but there should be more convincing evidence for desiccation.

The Arabic word sabkha is used in the Arabian Gulf countries to denote arid desert flats, particularly that part of a desert coastal plain situated slightly above the high-tide level. Sabkhas became an object of considerable interest to geologists soon after it was realized that certain types of ancient rock formation are practically identical with the sabkha sediments; both are characterized by the presence of nodular anhydrite and stromatolitic dolomite.

Anhydrite is a variety of calcium sulfate formed at high temperatures. At temperatures below 35 degrees Celsius in the presence of a brine that is saturated with sodium chloride (NaCl) anhydrite would be hydrated to form gypsum ($CaSO_4 \cdot 2H_2O$). (The hydration temperature would be higher if the brine were less salty). Since deep brine pools rarely exceed 35 degrees C. in bottom temperatures, anhydrite is formed today almost exclusively as a mineral in sabkha sediments. Since it is precipitated in the minute pore space of sediments near the ground-water table, it tends to grow as irregular nodules rather like the concretions one finds in arid soils.

Stromatolite (literally "flat stone") is a laminated carbonate that owes its genesis to the growth of algae [see "The Evolution of Reefs," by Norman D. Newell; SCIENTIFIC AMERICAN, June 1972]. A dense growth of blue-green algae forms a thin mat on coastal flats. After a severe storm the mat may be buried under a thin cover of sediments, but the algal growth persists and a new mat is constructed. This alternation ultimately results in the laminated rock called stromatolite. Stromatolite is common under the sabkhas of the Arabian peninsula, where nodular anhydrite also grows.

To return to our narrative, we sailed

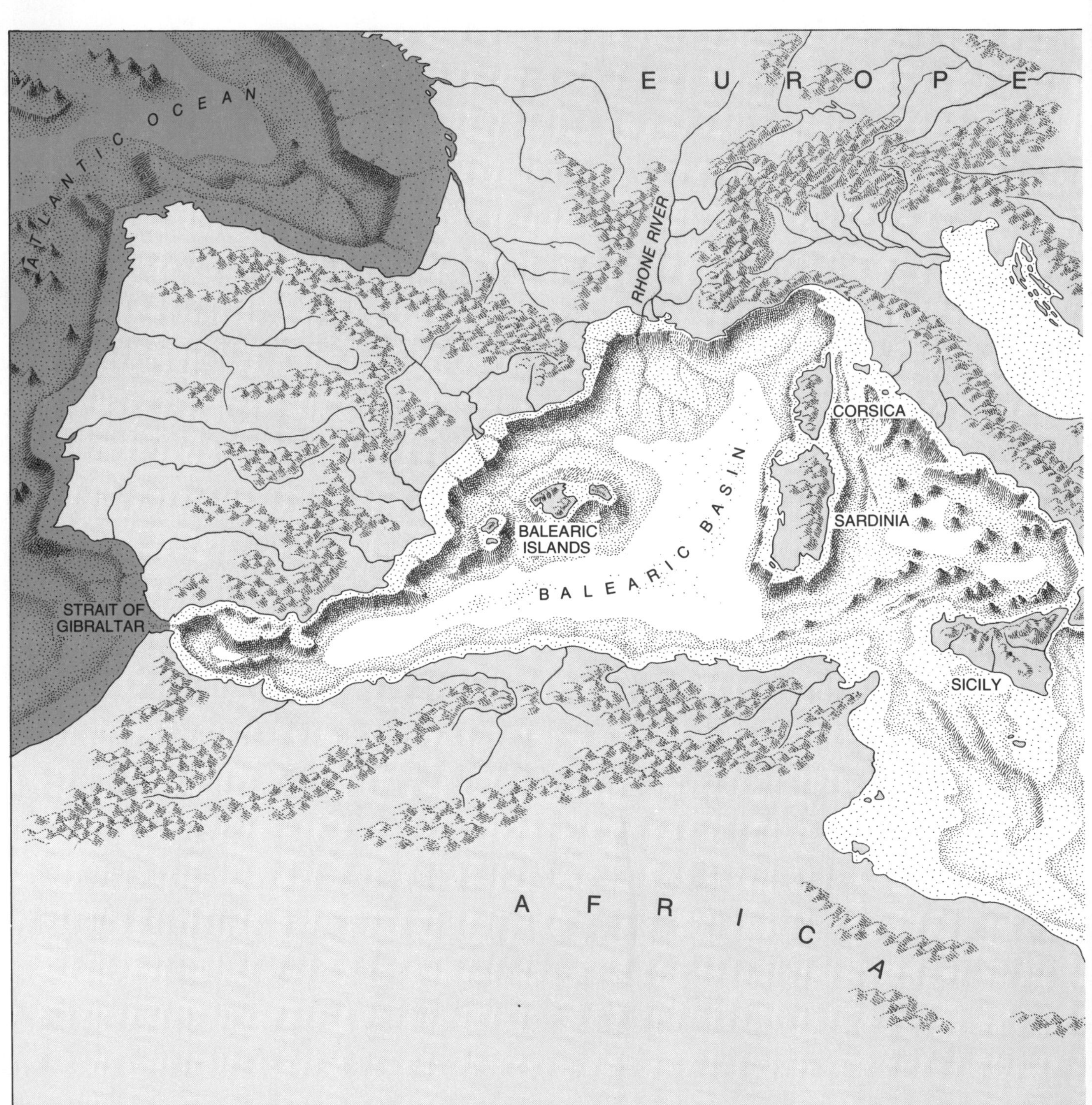

DESICCATED MEDITERRANEAN is represented by this panoramic drawing of the modern submarine topography of the Mediterranean basin. The Mediterranean must have looked something like this approximately six million years ago, when the basin was a great interior desert lying 10,000 feet below sea level. The Balearic abyssal plain was then a salt lake where evaporite minerals, including rock salt, were precipitated. Meanwhile nodular anhydrite grew in the soils on the lake shore. Gravels and variegated silts were

our vessel on August 27 to the Balearic Sea south of Majorca. Our drill site was positioned slightly north of the abyssal plain of the Balearic basin; our precision depth recorder registered 1,417 fathoms. The drill penetrated 1,000 feet of soft oozes before reaching the top of the hard layer. The core came up, and it was a late Miocene evaporite, as we expected. What was surprising, at least to those who advocated deep-water salt deposition, was the sampling of nodular anhydrite and stromatolitic dolomite [*see illustrations on page 89*].

Stromatolite cannot form in deep water because the growth of algae requires sunlight. Moreover, one could not expect the bottom of a deep-water Mediterranean to ever get as hot as the 35 degrees C. (95 degrees Fahrenheit) needed to precipitate anhydrite. Later a detailed petrological investigation by G. M. Friedman of Rensselaer Polytechnic Institute confirmed our prognosis that the Mediterranean evaporites were deposited on a desert flat; such sediments could not possibly have been deposited in several thousand feet of water.

More sophisticated methods of research led to the same conclusion. For

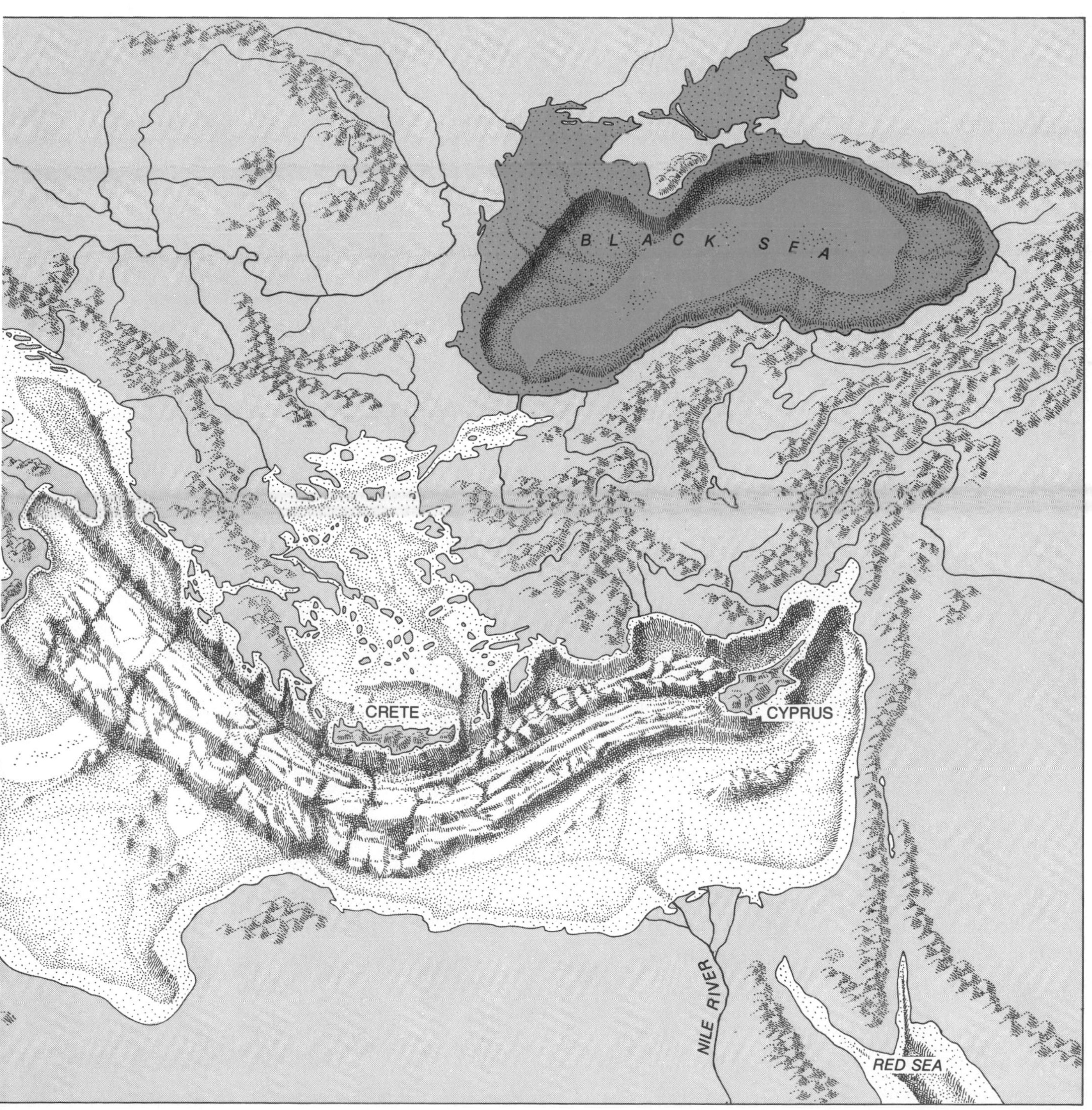

deposited around the edge of the basin at the foot of the steep slope, which is now the continental slope. At the end of the Miocene epoch, some 5.5 million years ago, an opening was breached at the Strait of Gibraltar. The inrushing water of the Atlantic constituted a great waterfall, which probably had a discharge rate 100 times greater than Victoria Falls. Within a few thousand years the desiccated Mediterranean would be filled to the brim, and deep marine sediments would again be deposited on the Balearic abyssal plain.

example, we know that the oxygen in sulfates and carbonates consists of two isotopes: oxygen 16, the common isotope, and oxygen 18, the heavy isotope. The ratio of oxygen 18 to oxygen 16 in a sample could reveal its genesis. Evaporites precipitated from evaporated seawater have a narrow range of isotopic compositions. In contrast, those deposited on playas, or desert lakes, have a wide range of values. Analyses of our samples by R. M. Lloyd of the Shell Oil Company laboratory in Houston showed a high variability of isotopic composition and thus provided additional confirmation for their playa origin.

Thus the anhydrite was deposited on a desert flat, but what a desert flat it must have been! The flat is now buried 9,000 feet below sea level. Do we have any unequivocal evidence that the floor of the Mediterranean basin was so very deep?

In fact, we do. Maria Cita of the University of Milan, one of our shipboard paleontologists, studied the microfossils in the marine sediments above and below the anhydrite. She told us that the sediments are normal deep-ocean ones. Clearly the basin was deep and submerged under marine waters when it was open to the Atlantic, but it turned into a deep hole when the floodgate was shut and the basin dried up. Because we discovered several oceanic sediments interbedded with the evaporites, we concluded that the floodgate swung open and shut repeatedly during an interval of about a million years.

After six weeks of drilling we were able to confirm that the *M* reflector is an evaporite everywhere. Nonetheless, we had sampled only evaporative carbonates and sulfates and not the rock salt. Some doubted that a piece of salt could ever be brought on deck before it had been pulverized and dissolved by the drilling. We were aware of the fact that we might be searching in the wrong places. The distribution of saline minerals in desert playas shows a bull's-eye pattern [*see illustration on page 95*]. The less soluble carbonates and sulfates, being the first to precipitate, are found at the edge of a salt pan, whereas the more soluble rock salt is usually deposited in the more central, deepest part of a basin, where the last bitter waters were concentrated. Our earlier drill sites had been on the peripheral, slightly elevated parts of the basin. To find rock salt we had to search under the abyssal plain. It was now October and we were heading home. A last borehole was spudded on the abyssal plain some 80 miles west of Sardinia. After drilling through 1,100 feet of soft oozes we hit pay dirt. The driller brought in a cylinder of shining, transparent crystals. Their bitter taste left us with no doubt that we had found rock salt 10,000 feet below sea level.

Interbedded in the salt layers are some windblown silts. This aeolian detritus includes land-formed quartz as well as the broken skeletons of forami-

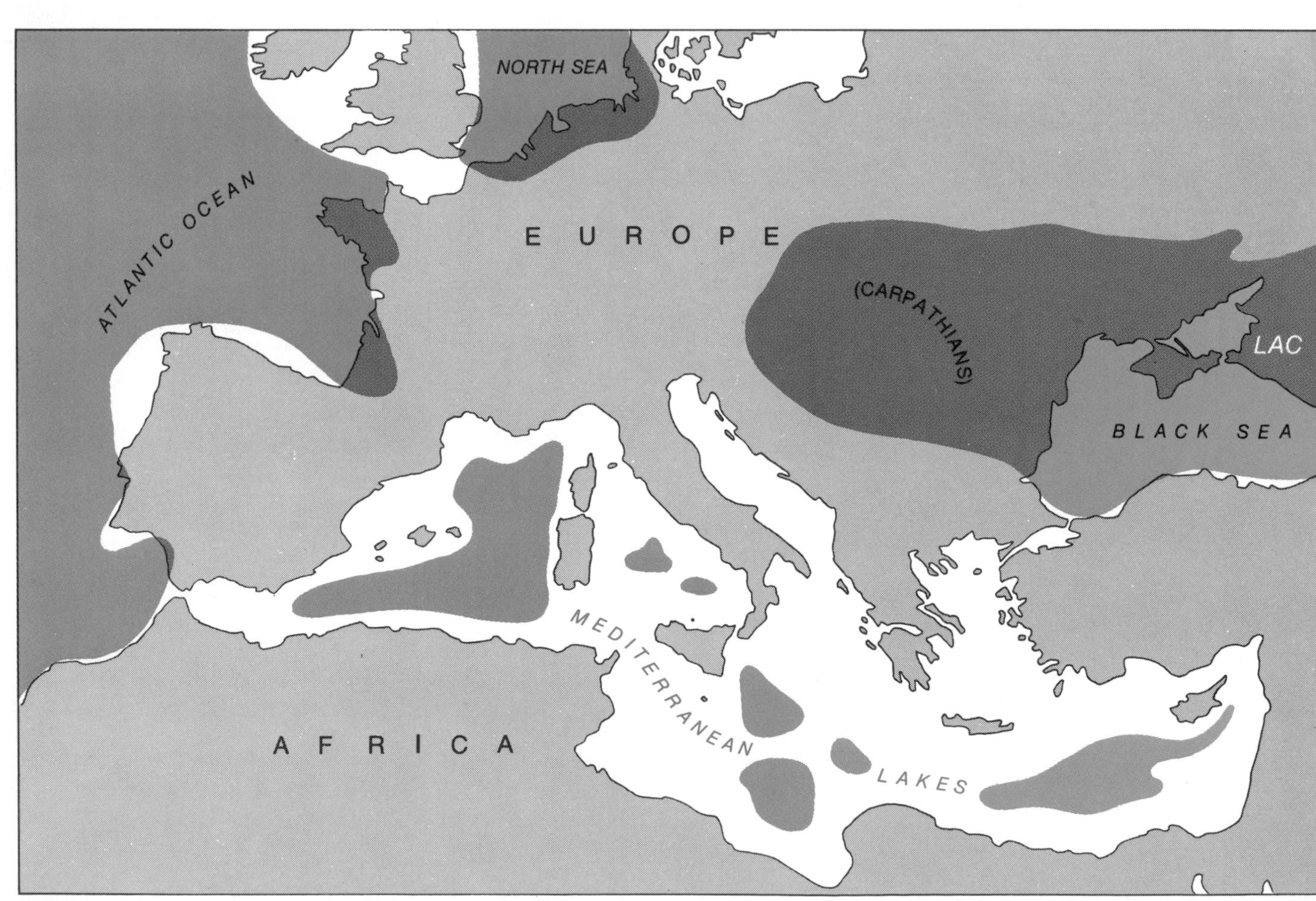

SEVEN MILLION YEARS AGO the geography of Europe was quite different from what it is today. Most of northeastern Europe was covered by a very large fresh-water-to-brackish-water lake that extended from the vicinity of Vienna eastward to beyond the Aral Sea. This great inland water body was named the Lac Mer by the French geologist Maurice Gignoux. At that time the Mediterranean was already separated from the Atlantic. Just before the uplift of the Carpathian Mountains, which took place roughly seven million years ago, the Lac Mer drained into the Mediterranean, supplying fresh and brackish waters to form a series of large inland lakes there. Laminated diatomites were deposited in such lakes. Eventually the Carpathians rose and formed a barrier, depriving the Medi-

nifera. Those tiny marine creatures flourished in the Mediterranean during the intervals when marine conditions prevailed. After the sea was isolated and desiccated the dead skeletons were blown across the desert flat by dust storms and laid down on the shore of a salt lake, to be buried eventually in rock salt. Bearing testimony to the alternate wetting and drying of the playa are nodular anhydrite and salt-filled mud cracks.

Examined under the microscope, the rock salt showed evidence of repeated solution and recrystallization, much like the salt in the modern coastal salinas of Lower California or in parts of Death Valley. The analogy to Death Valley can be carried a step further. We sampled red and green floodplain silts and well-rounded arroyo gravels from a nearby site. Those were carried to the base of an exposed continental slope by flash floods from the mountains of Sardinia and were deposited as alluvial fans fringing the salt pan. The similarity ends when one recalls that whereas Death Valley lies nearly at sea level, the floors of the Mediterranean desert basins were some 10,000 feet lower.

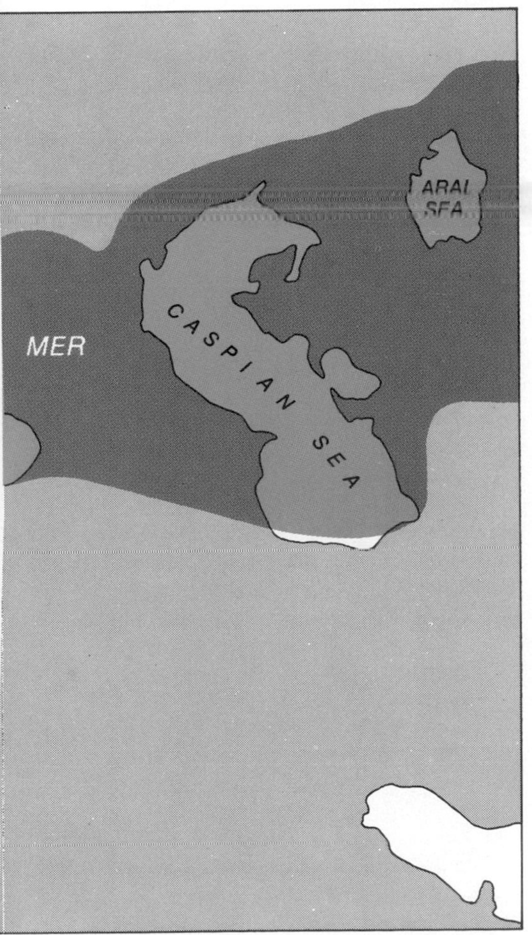

terranean of its water supply from the Lac Mer and turning the entire Mediterranean basin into a vast desert. The large fresh-water and brackish-water lakes were reduced to shallow lakes and playas, where salt and various other evaporites were precipitated.

Our drilling had barely scratched the top of a huge salt deposit. The Mediterranean salt should be 5,000 or 6,000 feet thick, according to geophysical surveys. This estimate is probably not too far off the mark, since we were told that late Miocene salt formations are present in Sicily and are several thousand feet thick. We now believe the Sicilian evaporite represents a segment of the Mediterranean sea bottom that was pushed up by mountain-building movements a few million years ago.

After the salt came the deluge. The evidence is unmistakable from our drill cores. We obtained the geological record from three drill sites (in the Balearic, Tyrrhenian and Ionian basins), showing that the separate parts of the Mediterranean were simultaneously flooded and submerged under deep marine waters at the end of the Miocene epoch some 5.5 million years ago. The first deposit is a dark gray marl five inches thick, deposited when the basin was being filled up, followed by a white ooze with local patches of red ooze.

One can picture the desiccated Mediterranean as a giant bathtub, with the Strait of Gibraltar as the faucet. Seawater roared in from the Atlantic through the strait in a gigantic waterfall. If the falls had delivered 1,000 cubic miles of seawater per year (equivalent to 30 million gallons per second, 10 times the discharge of Victoria Falls), the volume would not have been sufficient to replace the evaporative loss. In order to keep the infilling sea from getting too salty for even such a hardy microfauna as the one found in the dark gray marl the influx would have had to exceed evaporation by a factor of 10. Cascading at a rate of 10,000 cubic miles per year, the Gibraltar falls would have been 100 times bigger than Victoria Falls and 1,000 times more so than Niagara. Even with such an impressive influx, more than 100 years would have been required to fill the empty bathtub. What a spectacle it must have been for the African ape-men, if any were lured by the thunderous roar.

By the time the first Pliocene white ooze was deposited the Mediterranean must have been filled to the brim. Then the present system of exchange with the Atlantic would have been in operation. The white ooze is a typical oceanic sediment, made up almost entirely of the skeletons of microfossils and nannofossils. In addition to floating creatures bottom-dwelling organisms were found. William E. B. Benson of the National Science Foundation has examined the marine ostracods and Orville L. Bandy of the University of Southern California has studied the benthic foraminifera. They have both concluded that at that time the bottom of the Mediterranean was either colder or deeper than it is today. Those new immigrants from the Atlantic crawled through the deep gash at the western end of the Mediterranean. Eventually the cold-water bottom fauna died out when the Strait of Gibraltar was sufficiently shoaled to prevent the inflow of deep Atlantic waters, a condition that has persisted until today.

As we drilled through the evaporite formation we encountered an unusual laminated sediment. In addition to organic materials and minute crystals of dolomite the sediment contains fossil diatoms, whose skeletons consist of silica (SiO_2). Herbert Stradner of the University of Vienna, another of our shipboard paleontologists, recognized that some of the diatoms could only have lived in bodies of fresh or brackish water. This identification was later confirmed by Marta Hajós of the Hungarian Academy of Sciences; she found not only floating species but also bottom-dwelling ones. Later, at an eastern Mediterranean drilling site south of Crete, we sampled a late Miocene ostracod fauna. As Arredo Decima of the University of Palermo, an expert on ostracods, told us after he had examined our specimens, these tiny creatures also could have lived only on the bottom of brackish-water lakes.

We were puzzled by this discovery. Where could all that fresh water have come from? There was absolutely no reason to assume that the Mediterranean climate had been much more humid six or seven million years ago, nor could we find any evidence for sudden and drastic changes in precipitation or in evaporation that would have converted great lakes into Death Valleys. The mystery was resolved when we talked to our Austrian and Balkan colleagues, who told us about the Lac Mer. Apparently a large part of eastern Europe was covered by fresh or brackish waters during the late Miocene and the Pliocene. At one time a giant lake, called by the French Lac Mer, extended from Vienna to the Urals and the Aral Sea; its last descendants are the Caspian Sea and Black Sea of today. This body of water, collecting all the excess precipitation from the then wet and cold northeastern Europe, was draining into the Mediter-

NIGHT VIEW FROM THE BRIDGE of the *Glomar Challenger* shows, straddling the middle of the ship, the lower legs of the 142-foot drilling derrick, the vessel's most conspicuous structural feature. Beyond the derrick, on the foredeck, is the automatic pipe-racker, where 24,000 feet of drill pipe can be stored. The length of pipe suspended vertically above the derrick floor is the top end of a 15,000-foot drill string, which at the time the photograph was made was in the process of boring through the floor of the Mediterranean Sea 80 miles west of Sardinia. As the photograph suggests, such drilling and coring operations are carried out 24 hours a day. At present the *Glomar Challenger* is the only drilling ship capable of operating in the open ocean. It uses a dynamic positioning system to keep virtually stationary over the borehole, even in a stormy sea. The ship, which is owned by Global Marine Inc. and is operated by the Scripps Institution of Oceanography under a contract with the National Science Foundation, is named after the world's first full-time oceanographic research vessel, H.M.S. *Challenger*, which was launched in December, 1872, exactly 100 years ago.

ranean during the earlier part of the late Miocene, some seven or eight million years ago. Shortly thereafter tremendous earth movements led to the uplift of the Carpathian Mountains and a radical reorganization of the drainage system. The Lac Mer now found an outlet to the north. The eastern faucet to the Mediterranean was turned off. Cut off from its major supply of fresh water, the arid Mediterranean suffered the fate of desiccation.

From the eastern Mediterranean boreholes we obtained a series of middle Miocene cores. Aided by the outcrop section on Sicily, we were able to reconstruct the history of the Mediterranean during the past 15 or 20 million years. The Mediterranean was once a broad seaway linking the Indian and Atlantic oceans. With the collision of the African and Asiatic continents and the advent of mountain-building in the Middle East, the connection to the Indian Ocean was severed. Meanwhile Africa was also advancing toward Europe, and communication to the Atlantic was maintained by way of two narrow straits, the Betic in southern Spain and the Riphian in North Africa. We saw evidence in our cores of the gradual deterioration of the Mediterranean environment, the advancing stagnation of its waters, the inevitable extinction of its bottom-dwellers, the struggle for existence by its swimming and floating population and the evolution of a hardy race that could survive widely changing salinities. We saw a change from an inland sea to a series of great lakes, and we saw their desiccation and the complete extermination of the fauna and flora at the bottom of the Miocene Death Valleys, 10,000 feet below sea level. We saw also the deluge, the establishment of a new faunal dynasty and the changing marine population leading up to the population of the present.

The discovery of the Mediterranean desert was made possible by deep-sea drilling. The first people acquainted with the discovery were the shipboard investigators who were selected for qualifications other than their mastery of local geology. Hence the full impact of the discovery was only appreciated after the *Glomar Challenger* returned to Lisbon and the drill results were communicated to the public by press conferences in Paris and New York.

If the Mediterranean had indeed been emptied, the coastal plains of the sur-

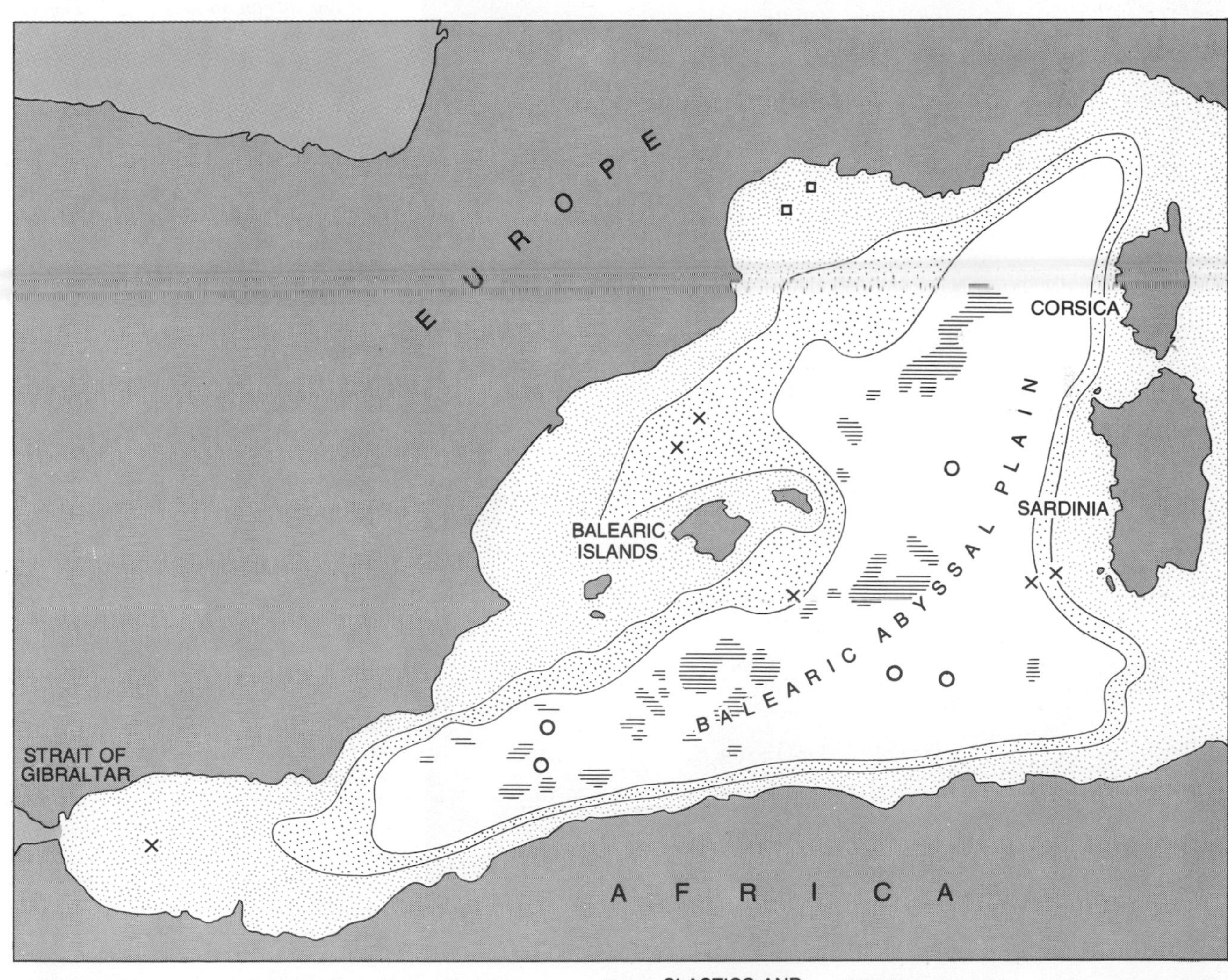

DISTRIBUTION OF EVAPORITES under the Balearic Sea shows the characteristic bull's-eye pattern one would expect in a completely enclosed basin. The less soluble carbonates and sulfates, being the first to precipitate, are found around the periphery of the basin; anhydrite and gypsum are found in a narrow ring just inside this outermost region, whereas rock salt, being the most soluble, is present only under the central, deepest part of the abyssal plain, where the last bitter waters must have been concentrated.

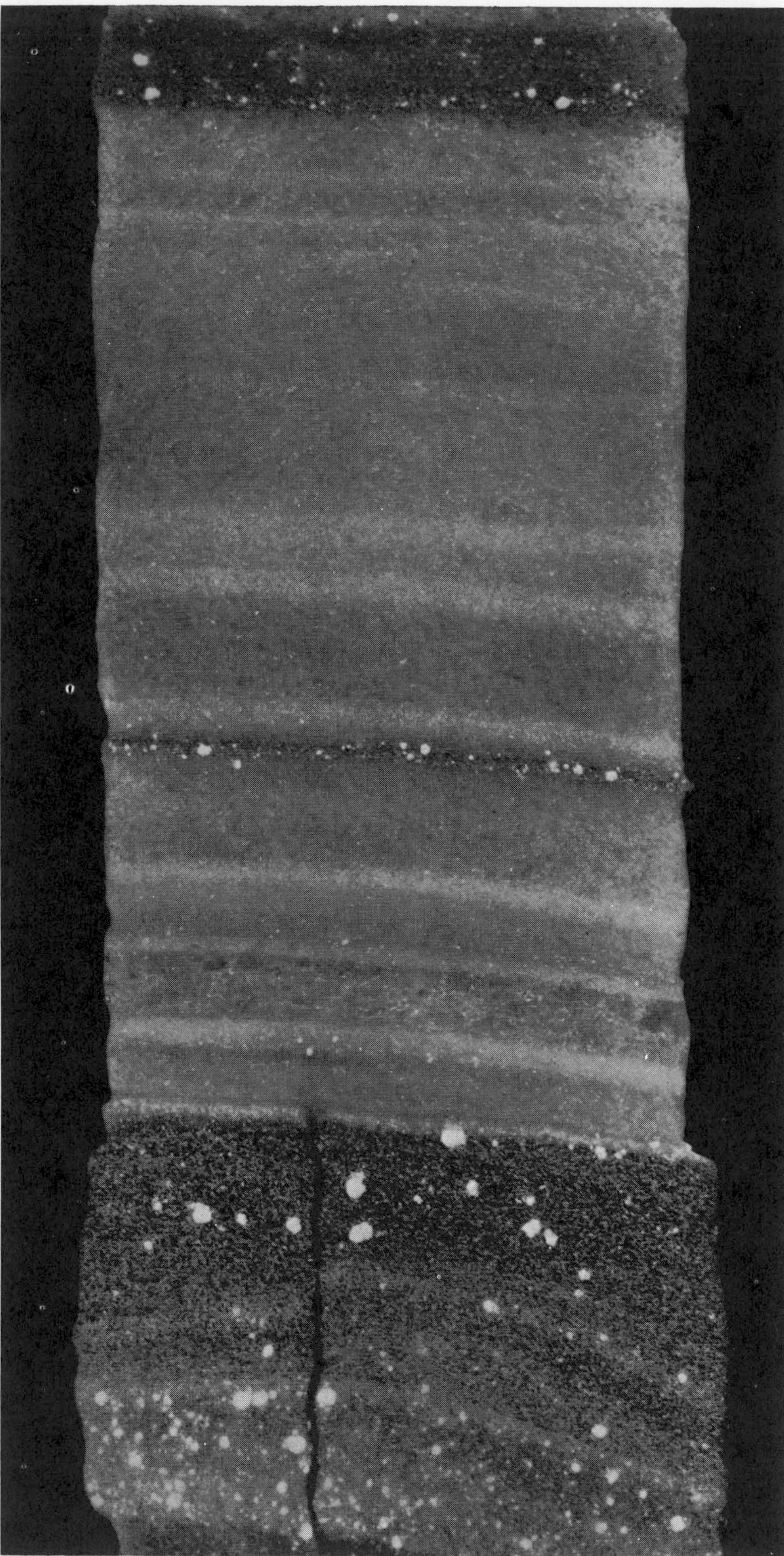

DEEP-SEA SALT CORE, the first of its kind ever obtained, was retrieved from a borehole drilled by the *Glomar Challenger* through some 1,100 feet of soft oozes underlying the Balearic abyssal plain about 80 miles west of Sardinia. The sea bottom in this region is about 10,000 feet deep. The vertical crack in the lower part of the core is believed to be a desiccation crack, further evidence that the Mediterranean was dry down to 10,000-foot level.

rounding lands would have become high plateaus, and islands would have been lofty peaks. The first response to a lowering of the water level would be rejuvenation of streams: a marked increase in their down-cutting power. The buried Rhône gorge, first discovered 80 years ago, should thus be one of the many surrounding the Mediterranean. Where are the other buried gorges?

Soon after we returned to port Ryan received a letter from a Russian geologist, I. S. Chumakov, who had learned of our findings through an article in *The New York Times.* Chumakov was one of the specialists sent by the U.S.S.R. to Aswan in Egypt to help build the famous high dam. In an effort to find hard rock for the dam's foundation 15 boreholes were drilled. To the Russians' amazement they discovered a narrow, deep gorge under the Nile valley, cut 700 feet below the sea level into hard granite [*see illustration on next page*]. The valley was drowned some 5.5 million years ago and filled with Pliocene marine muds, which are covered by the Nile alluvium. Aswan is 750 miles upstream from the Mediterranean coast. In the Nile delta boreholes more than 1,000 feet deep were not able to reach the bottom of the old Nile canyon. Chumakov estimated that the depth of the incision there might reach 5,000 feet, and he visualized a buried Grand Canyon under the sands and silts of the Nile delta.

Chumakov was not the only one who had been puzzled. Oil geologists exploring in Libya had also had their share of surprises. First, their seismograms would register anomalies; there were linear features underground transmitting seismic waves at abnormally high velocities. Drilling into the anomalies revealed that they are buried channels incised 1,300 feet below sea level. The geologic record tells the same story: vigorous downcutting by streams and sudden flooding by marine waters at the beginning of the Pliocene. Frank T. Barr and his coworkers of the Oasis Oil Company, based at Tripoli in Libya, concluded in a report that the Mediterranean must have been thousands of feet below its present level when the channels were cut. They could not get their manuscript published in a scientific journal, since no one would accept such an outrageous interpretation.

Still other buried gorges and channels have been found in Algeria, Israel, Syria and other Mediterranean countries. Rivers emptying into a desiccating Mediterranean not only would have in-

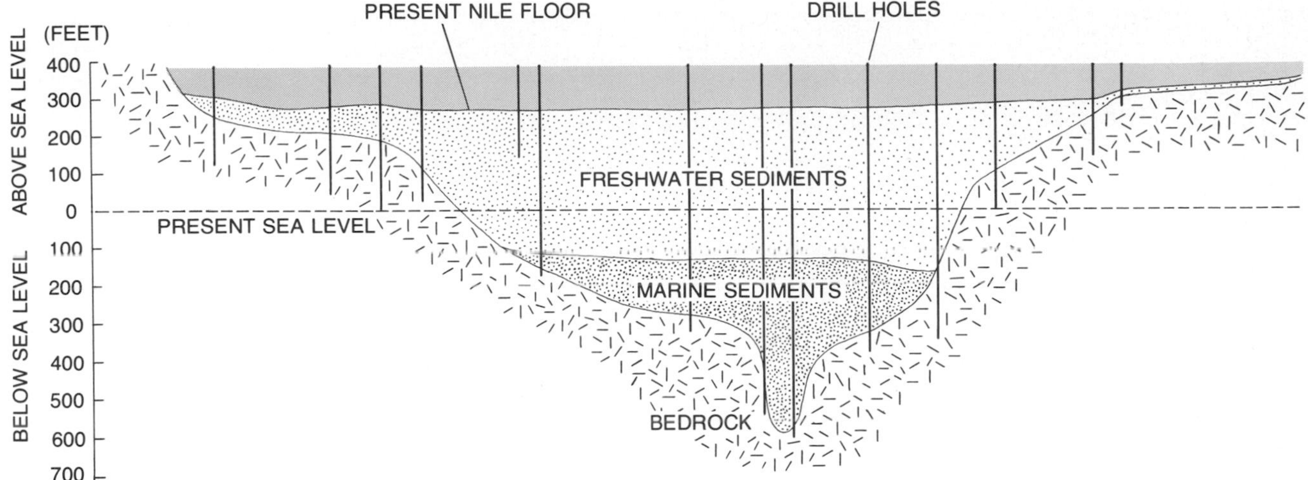

DEEP GORGE under the upper Nile valley near Aswan in Egypt was discovered by a team of Russian geologists while drilling test boreholes preparatory to the construction of the Aswan high dam. The narrow, deep central portion of the canyon cuts some 700 feet below the present sea level. The incision was apparently carved when the Mediterranean was dry. When the Mediterranean was flooded again at the beginning of the Pliocene epoch, the canyon was also drowned and marine sediments filled its lower section.

cised deeply on land but also would have continued on down across the exposed continental shelf and the continental slope to the flat bottom of the abyssal plain that was turning into a playa. In the course of a lecture at Yale University devoted to a discussion of the results of our expedition a student in the audience called my attention to such an eventuality and asked if such channels exist.

They do indeed. Extensive oceanographic surveys have been carried out by the French in the Balearic basin. The late Jacques Bourcart of the University of Paris reported in 1950 the discovery of numerous submarine canyons indenting the continental margins off the coast of France, Corsica, Sardinia and North Africa. The canyons are typically river-cut and are filled with alluvial gravels. Most of them can be related to a river on land and can be traced to a depth of 6,000 or 8,000 feet at the edge of the abyssal plain. They too were drowned by the early Pliocene deluge. Similar canyons have since been found in all parts of the Mediterranean. Their genesis had never been satisfactorily explained until it was realized that the Mediterranean was desiccated six million years ago.

The key opens the door to the solution of other mysteries. For example, one can now begin to understand the origin of the extensive caverns in the circum-Mediterranean lands and the peculiar karst topography of Yugoslavia, where sinkholes and pinnacles abound. One can also provide an answer to the long-standing question of why ground-water circulation once penetrated 10,000 feet below sea level in a mid-ocean island such as Malta. Not only the geomorphic changes but also the biological ones were catastrophic. Giuliano Ruggieri of the University of Palermo, an authority on the evolution of the shallow marine fauna of the Mediterranean, recently sent me a reprint of an article he wrote in 1955, in which he surmised that the Mediterranean must have been desiccated in order to explain the biological revolution six million years ago. Lyell's historical datum marked the return of marine waters to this inland basin.

To remove all the water from the Mediterranean and pile it elsewhere would raise the sea level of the world ocean by 35 feet, an event that would drown many of our coastal cities. The magnitude of the negative load on the desiccated Mediterranean was comparable to the weight of the Fennoscandian ice sheet on Europe during the last ice age. The resubmergence of the Mediterranean must have led to the subsidence of the basin and the uplift of surrounding lands. Oceanographers and geologists began to see such evidence in their records. The presence of a hot desert where the Mediterranean is now should have had a heavy climatic impact. Indeed, European paleobotanists have noticed a change toward warm aridity in central Europe during the late Miocene, when the Vienna woods were changed into steppes and when palms grew in Switzerland. With the return of marine waters to the Mediterranean in the Pliocene epoch the central European climate again became wet and cold and deteriorated gradually into the ice age. It is also interesting to note that the Arctic polar ice cap began to build up in the late Miocene. Was this a coincidence, or was the initiation of glaciation triggered by the drying up of the Mediterranean?

Was the disappearance of a large inland sea a unique event in the geological history? Probably not. The existence of large saline deposits indicates that there have been other desiccated oceans. The famous Zechstein salts of northern Europe may have been the residues of an inland sea that dried up 250 million years ago. The giant salt and potash deposit of Alberta and Saskatchewan, some 350 million years old, may have had a similar origin. In fact, the discovery that a small ocean basin can be converted into desert has led us to reexamine the entire problem of salt genesis. Geologists used to worry about the occurrence of oceanic salt deposits under the Gulf of Mexico, under the South Atlantic off the coast of the Congo and Angola and under the North Atlantic off the coast of Nova Scotia. We can now postulate that those salts were formed when the Gulf and the Atlantic were isolated inland seas undergoing desiccation.

It must of course seem somewhat farfetched to imagine the Mediterranean as a deep, dry, hot hell. We ourselves were reluctant to come to that conclusion until all other explanations had failed; the facts left us with no alternative. As Sherlock Holmes once remarked: "It is an old maxim of mine that when you have excluded the impossible, whatever remains, however improbable, must be the truth."

III

THE SEA AND ITS MOTIONS

III THE SEA AND ITS MOTIONS

INTRODUCTION

We now turn from the sea floor, the container of the sea, to the sea itself to consider such questions as why it is salt and why it moves, and to savor the delightful complexities of its saltiness and its motions and the interactions between them. Let us first address the most fundamental question about the ocean, namely, "Where did all the water come from?" This was answered in a series of famous papers published by W. W. Rubey in the early 1950s. There are only two possible sources for the water. Either it is the residue of a vast, dense primordial atmosphere and is almost as old as the earth, or else there was little or no ocean when the earth formed and it has since leaked out from the interior. Rubey showed that the geological record was incompatible with a primordial origin for the water, and that volatile substances, including water, pour out of volcanoes in amounts much greater than are necessary to form the ocean.

It was recognized that part of the water emerging from volcanoes was recycled groundwater, which had come down as rain rather than up from the mantle. However, the volume of volcanic outpourings was so great that this fact did not appear to present any impediment to acceptance of the theory. The development of isotopic geochemistry changed the situation. Tritium, an isotope of hydrogen with a half life of 12.5 years, is produced in the atmosphere and thus appears in rain. Harmon Craig and others measured the ratio of tritium to ordinary hydrogen in volcanic hot springs and found that, within the limits of accuracy of their measurements, all the water is recycled rain. This did not mean that Rubey's theory was wrong; indeed, there has been little question about his demonstration that the ocean water came from the interior. Geological time has been so long that only a minuscule amount of juvenile water from each volcanic eruption would be sufficient to have produced the ocean; in fact, the requirement per eruption is less than the range of error of the isotopic measurement; nonetheless, many earth scientists found it reassuring when it was discovered that the midocean rift was producing far more volcanic rock than anyone had realized. Thus, in his article "Why the Sea is Salt," written in 1970, Ferren MacIntyre refers to the opening crack as a major source of juvenile water. Since that time Craig has carried his research into new waters. So far, primordial neon, helium, and hydrogen have been discovered in the basalts of the sea floor, and from their ratio of hydrogen isotopes it appears that they also contain juvenile water, although even now this is not certain.

It seems that the water on the surface of the earth flowed out of the interior along with volcanic rocks, but when and at what rate? The simpler possibilities are that the flow was constant during geological time, or faster than average to begin with, or faster than average recently. All possibilities have been proposed. Rubey assumed a constant flow, essentially on the philosophical grounds that it is the simplest possibility. Roger Revelle, however, proposed that drowned ancient islands are at their present great depth mainly because of a eustatic rise in sea level. If so a third of the seawater has come out of the earth's interior during the last small percentage of geological time. Although it now appears that guyots are deep mainly because the sea floor sinks as it ages, Revelle's arguments show that it will be very difficult to determine the history of the water in the sea.

The history of the elements dissolved in seawater is probably even more complex than that of the water itself. Ferren MacIntyre's article discusses the chemistry of seawater and the involved cycles whereby various elements flow in and out of the sea. Physical, chemical, and biological processes interact to determine the local concentrations, residence times, and ultimate deposition of elements on the sea floor, whence they are recycled by subduction and vulcanism. The article also discusses the difficulties encountered by chemists in measuring the concentrations of elements in seawater and in comparing their data. These problems have now been largely overcome for many of the isotopes and elements of most interest to mankind. The Geochemical Oceans Sections Study (GEOSECS) program of the International Decade of Ocean Exploration (IDOE) is engaged in collecting water samples along cross sections of the ocean. The samples are being analyzed with precision in ways that make different measurements directly comparable. From this work will come a new understanding of the circulation of the ocean and the spread of pollution through it.

One element, gold, has been of interest not for its geochemistry but for its worth. It was discovered in seawater by E. Sonstadt in 1872; he reported a concentration of 65 milligrams of gold per metric ton of sea water. Some analyses in the following decades confirmed this concentration but others found only a tenth as much. It was early proposed that the gold might be extracted commercially, if not from water directly, then from brines concentrated in the manufacture of salt and bromine.

After World War I, the famous German scientist Fritz Haber proposed that gold from the sea might be used to pay his country's war debts. Some of the best oceanography of the period was directed toward determining the feasibility of the project. Unfortunately, as the chemical analyses improved, the gold content of seawater appeared to be less and less, and the scheme was abandoned. Subsequent studies suggest that the gold content of seawater is extremely variable, depending on the chemistry and biological activity in a water mass. Haber thought that the concentration was 0.004 milligrams per metric ton; at $35 per ounce of gold, that gave only 0.01 cents per ton. With the highest concentrations recently reported and the much higher current price of gold, the value would be enormously greater. Nonetheless, current thinking on mineral resources, as indicated in the article "Physical Resources of the Ocean" (in Section V), is based on Haber's analyses. It appears that we still have much to learn about the chemistry of the sea.

The motions of the ocean are extremely complex because they are generated in many ways, which produce superimposed effects. "Ocean Waves," by Willard Bascom, discusses all the waves, of all periods from seconds to hours, that various phenomena generate simultaneously in the sea. The familar short-period surface waves of the ocean are caused by wind stresses, but they need not be local. This was brought home to us in Southern California when Walter Munk and others placed an array of wave sensors offshore and accurately measured the orientation of waves before they were refracted near the coast. It turned out that our ordinary summer surf is generated by storms in the Southern Hemisphere. In those days before weather satellites it was possible to track storms, as they drifted from New Zealand to Chile, by the waves they sent to California.

Near the centers of storms, particularly hurricanes and typhoons, waves reach frightening dimensions in the open sea. I can attest to this myself from experiencing typhoons at sea that sank three destroyers, bent the bow of a cruiser, and draped the flight decks of aircraft carriers down around the hull, like the ears of a spaniel. Such damage is rare at sea; it occurs mainly where storms or storm waves reach shore. The awesome power of great storm waves in shallow water is described in fascinating detail by Bascom. To an oceanographer, his account of relatively modest storm damage to the Long Beach breakwater is even more intriguing, because it was caused by subtle focusing effects rather than by brute force.

We may note in passing the fulfillment of Bascom's prediction in 1959 that the population explosion along the shore meant that the greatest wave disaster was yet to come. He reported a maximum disaster of 200,000 dead in the Bay of Bengal in 1876. Even more hundreds of thousands were killed in the same place in 1971, mainly by drowning in a typhoon. The land is densely populated and, although it is dangerously low and flat, the people have nowhere else to go. Sad as the thought is, I see no reason not to repeat Bascom's prediction.

Tsunamis are giant waves, with periods of tens of minutes, that are caused by geological phenomena. A few of the most spectacular are generated by exploding or collapsing volcanoes, such as the famous eruption of Krakatoa. Recent studies have shown, however, that, like most things geological, they are related to plate tectonics rather directly. Where oceanic plates are being subducted into trenches they tend to be fractured by minor faults. The characteristics of most tsunamis demonstrate that they are the result of a sudden drop of a large area of the sea floor on the margin of a trench. This means that the first part of the wave to reach shore is a trough. This fact is well known in many Pacific cultures, where a sudden recession of the sea is taken as a danger sign. It is only ignorant foreigners who rush out to see the bared sea floor, and are overwhelmed by the surging crest.

Tsunamis have been sources of information about the sea floor for more than a century. They are "shallow" waves, in that they touch bottom at any depth, so by wave physics their speed was used to calculate the average depth of the Pacific when it was hardly sounded at all.

"The Atmosphere and the Ocean," by R. W. Stewart, describes the interaction of these enormous, thin, fluid sheets, which are influenced by everything from the shape of the boundary surface between them to the rotation of the earth. Interest in this subject has gradually spread from the academic to the practical world during the last decade. This is because of the increasing importance of predicting the weather and the dawning understanding that long-term weather changes depend on the sea.

In 1972 Jerome Namias published an important analysis of the influence of sea temperatures in the North Pacific upon the weather of North America. If the ocean is warm in the northeast and cold in the center, the jet stream in the atmosphere is deflected so that the Pacific states are warm and the Atlantic states are cold. If the sea temperature pattern is reversed, so is the weather. The importance of predicting whether New England, for example, will have a cold or warm winter is manifest now that natural gas is in short supply. Thus, it is not surprising that one program of the International Decade of Ocean Exploration, the Northern Pacific Experiment (NORPAX), is concerned with the coupling of the ocean and atmosphere over the North Pacific.

A related IDOE program is the Mid-Ocean Dynamics Experiment (MODE), a study of intermediate-scale eddies, which may contain more energy than the general circulation of the ocean. This is a vast coordinated international project. When one examines a map of the tracks of ships and the locations of arrays of buoys and other instruments employed by half a dozen nations in the North Atlantic, one cannot but wonder how the currents manage to get through the equipment. Nonetheless, individual eddies have been successfully tracked as they developed and dissipated. Enough has been learned so that theorists can construct numerical models of eddies, whose motions can thus be simulated, at considerable expense, by computers. However, computers get cheaper and ships get more expensive, so this is a promising development.

"The Circulation of the Abyss" was written by Henry Stommel in 1958, long before such projects as the IDOE. It has the notable virtue that it presents the views of a pioneer scientist just at the exciting time of a major breakthrough in understanding of the oceans. To appreciate what was happening, consider the circulation of the Gulf Stream as analyzed in the enormously influential book *The Oceans*, by H. U. Sverdrup, Martin Johnson, and Richard Fleming, published in 1942 and in its eleventh printing twenty years later. It points out that if the oxygen minimum is selected as a reference surface of no motion, the Gulf Stream would be confined to shallow depths and a reverse flow would exist below. This seemed unreasonable at the time, but by 1958 Stommel had demonstrated that it was, in fact, more reasonable than not, and John Swallow had just found and measured the deep countercurrent.

Swallow invented a new type of buoy that would float at intermediate depths and that could be tracked by sonar to follow deep currents. Similarly, it required new instrumentation to make the remarkable advances reported in "The Top Millimeter of the Ocean," by Ferren MacIntyre, and "The Microstructure of the Ocean," by Michael Gregg. Observations can now be made of phenomena on a scale of angstroms at the surface and of centimeters in shallow water. Just as MODE analyzes individual eddies in the Gulf Stream, so these studies are concerned with eddies and localized phenomena on the most minute scale. High-speed cameras follow the collapse of individual bubbles, and continuously recording instruments track tiny steps in the variation of salinity and temperature. On these miniscule scales there are astonishingly intense chemical and physical gradients that are of great interest for understanding the phenomena of mixing and stirring. Moreover, as the articles indicate, the study of these small phenomena promises to improve our ability to control pollution and increase the harvest of food from the ocean. These papers are harbingers of great advances to come as we learn to examine the detailed structure and motion of the ocean.

The last article in this section, "Beaches," by Willard Bascom, deals with another subject of great importance to society, namely, the utilization and preservation of beaches. The final section of the reader will consider the social problems that arise from conflicting uses of beaches and other physical marine resources. However, unless beaches are preserved, people will be unable to bicker about how to use them.

Although many interesting scientific questions about beaches remain unanswered, the broad outline of how they function has been known at least since the 1950s. A beach is a transient, usually migrating, deposit of sand whose form is determined by a balance between the forces that supply sand and those that remove it. The information exists, yet people build houses and highways on beaches and seem surprised when erosion drops them into the sea. People seem surprised when a beach disappears after they dam its sediment supply. They find it mysterious when a beach builds outward and fills in an anchorage after they construct a breakwater that prevents wave energy from reaching the beach. Society's interactions with beaches, at least in the United States, are as complex as the linkage between the atmosphere and ocean. They provide a splendid example of the difficulties that may arise in putting environmental knowledge to use.

9

Why the Sea is Salt

by Ferren MacIntyre
November 1970

The sea contains more than 70 elements in addition to sodium and chlorine. The global cycles that remove and replenish them involve rainfall, volcanoes and the spreading of the ocean floor

According to an old Norse folktale the sea is salt because somewhere at the bottom of the ocean a magic salt mill is steadily grinding away. The tale is perfectly true. Only the details need to be worked out. The "mill," as it is visualized in current geophysical theory, is the "mid-ocean" rift that meanders for 40,000 miles through all the major ocean basins. Fresh basalt flows up into the rift from the earth's plastic mantle in regions where the sea floor is spreading apart at the rate of several centimeters per year. Accompanying this mantle rock is "juvenile" water—water never before in the liquid phase—containing in solution many of the components of seawater, including chlorine, bromine, iodine, carbon, boron, nitrogen and various trace elements. Additional juvenile water, equally salty but of somewhat different composition, is released by volcanoes that rim certain continental margins, such as those bordering the Pacific, where the sea floor seems to be disappearing into deep trenches [*see illustration on these two pages*].

The elements most abundant in juvenile water are precisely those that cannot be accounted for if the solids dissolved in the sea were simply those provided by the weathering of rocks on the earth's surface. The "missing" elements, such as chlorine, bromine and iodine, were once called "excess volatiles" and were attributed solely to volcanic emanations. It is now recognized that juvenile water may have nearly the same chlorinity as seawater but is much more acid due to the presence of one hydrogen ion (H^+) for every chloride ion (Cl^-). In due course, as I shall explain later, the hydrogen ions are removed and replaced by sodium ions (Na^+), yielding the concentration of ordinary salt (NaCl) that constitutes 90-odd percent of all the "salt" in the sea.

The chemistry of the sea is largely the chemistry of obscure reactions at extreme dilution in a strong salt solution, where all the classical chemist's "distilled water" theories and procedures break down. The father of oceanographic chemistry was Robert Boyle, who demonstrated in the 1670's that fresh waters on the way to the sea carry small amounts of salt with them. He also made the first attempt to quantify saltiness by drying seawater and weighing the residue, but his results were erratic because some of the constituents of sea salt are volatile. Boyle found that a better method was simply to measure the specific gravity of seawater and from this estimate the amount of salt present. Since the distribution of density in the sea is important to oceanographers, the same calculation is routinely performed today in reverse: the salinity is deduced by measuring the electrical conductivity of a sample of seawater, and from this and the original temperature of the sample one can compute the density of the seawater at the point the sample was taken.

In 1715 Edmund Halley suggested that the age of the ocean and thus of the world might be estimated from the rate of salt transport by rivers. When this proposal was finally acted on by John Joly in 1899, it gave an age of some 90 million years. The quantity that Joly measured (total amount of x in ocean divided by annual river input of x) is now recognized as the "residence time" of the constituent x, which is an index of an element's relative chemical activity in the ocean. Joly's value is about right for the residence time of sodium; for a more reactive element (in the ocean environment) such as aluminum the residence time is as brief as 100 years.

Not quite 200 years ago Antoine Laurent Lavoisier conducted the first analysis of seawater by evaporating it slowly and obtaining a series of compounds by fractional crystallization. The first compound to settle out is calcium carbonate ($CaCO_3$), followed by gypsum ($CaSO_4 \cdot 2H_2O$), common salt (NaCl), Glauber's salt ($Na_2SO_4 \cdot 10H_2O$), Epsom salts ($MgSO_4 \cdot 7H_2O$) and finally the chlorides of calcium ($CaCl_2$) and mag-

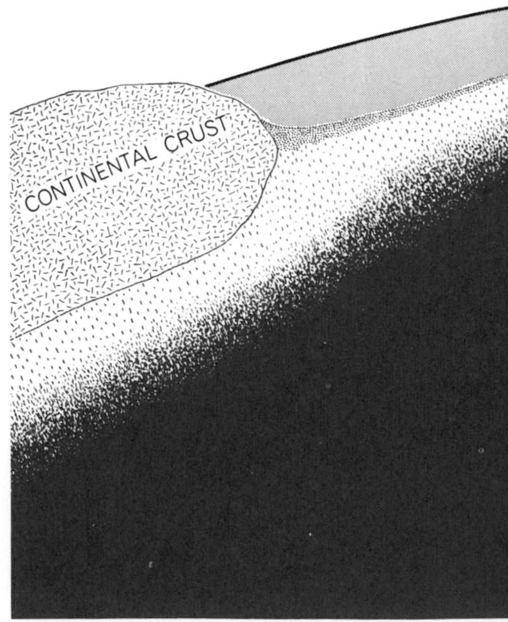

MAGIC SALT MILL at the bottom of the sea, imagined in the old Norse folktale, turns out to be not so fanciful after all. The modern explanation of why the sea is salt invokes the concept of the "mid-ocean" rift and sea-floor spreading, as depicted here in cross section. The rift is a weak point be-

nesium ($MgCl_2$). Lavoisier noted that slight changes in experimental conditions gave rise to large shifts in the relative amounts of the various salts crystallized. (In fact, some 54 salts, double salts and hydrated salts can be obtained by evaporating seawater.) To get reproducible results for even the total weight of salt one must remove all organic matter, convert bromides and iodides to chlorides, and carbonates to oxides, before evaporating. The resulting weight, in grams of salt per kilogram of seawater, is the salinity, S‰. (The symbol ‰ is read "per mil.")

In actual practice the total weight of salt in seawater is nowadays never determined. Instead the amount of chloride ion is carefully measured and a total for all other ions is computed by applying the "constancy of relative proportions." This concept dates back to the middle of the 19th century, when John Murray eliminated confusion about the multiplicity of salts by observing that individual ions are the important thing to talk about when analyzing seawater. Independently A. M. Marcet concluded from many measurements that various ions in the world ocean were present in nearly constant proportions, and that only the absolute amount of salt was variable. This constancy of relative proportions was confirmed by Johann Forchhammer and again more thoroughly by Wilhelm Dittmar's analysis of 77 samples of seawater collected by H.M.S. *Challenger* on the first worldwide oceanographic cruise. These 77 samples are probably the last ever analyzed for all the major constituents. Their average salinity was close to 35‰, with a normal variation of only ±2‰.

In the 86 years since Dittmar reported eight elements, 65 more elements have been detected in seawater. It was recognized more than a century ago that elements present in minute amounts in seawater might be concentrated by sea organisms and thereby raised to the threshold of detectability. Iodine, for example, was discovered in algae 14 years before it was found in seawater. Subsequently barium, cobalt, copper, lead, nickel, silver and zinc were all detected first in sea organisms. More recently the isotope silicon 32, apparently produced by the cosmic ray bombardment of argon, has been discovered in marine sponges.

There are also inorganic processes in the ocean that concentrate trace elements. Manganese nodules (of which more below) are able to concentrate elements such as thallium and platinum to detectable levels. The cosmic ray isotope beryllium 10 was recently discovered in a marine clay that concentrates beryllium. In all, 73 elements (including 13 of the rare-earth group) apart from hydrogen and oxygen have now been detected directly in seawater [*see illustration on page 107*].

It is only in the past 40 years that geochemists have become interested in the chemical processes of the sea for what they can tell us about the history of the earth. Conversely, only as geophysicists have pieced together a comprehensive picture of the earth's history has it been possible to bring order into marine chemistry.

The earth's present atmosphere and ocean are not primordial but have been liberated from chemical and mechanical entrapment in solid rock. Perhaps four billion years ago, or a little less, there was (according to many geophysicists) a "grand catastrophe" in which the earth's core, mantle, crust, ocean and atmo-

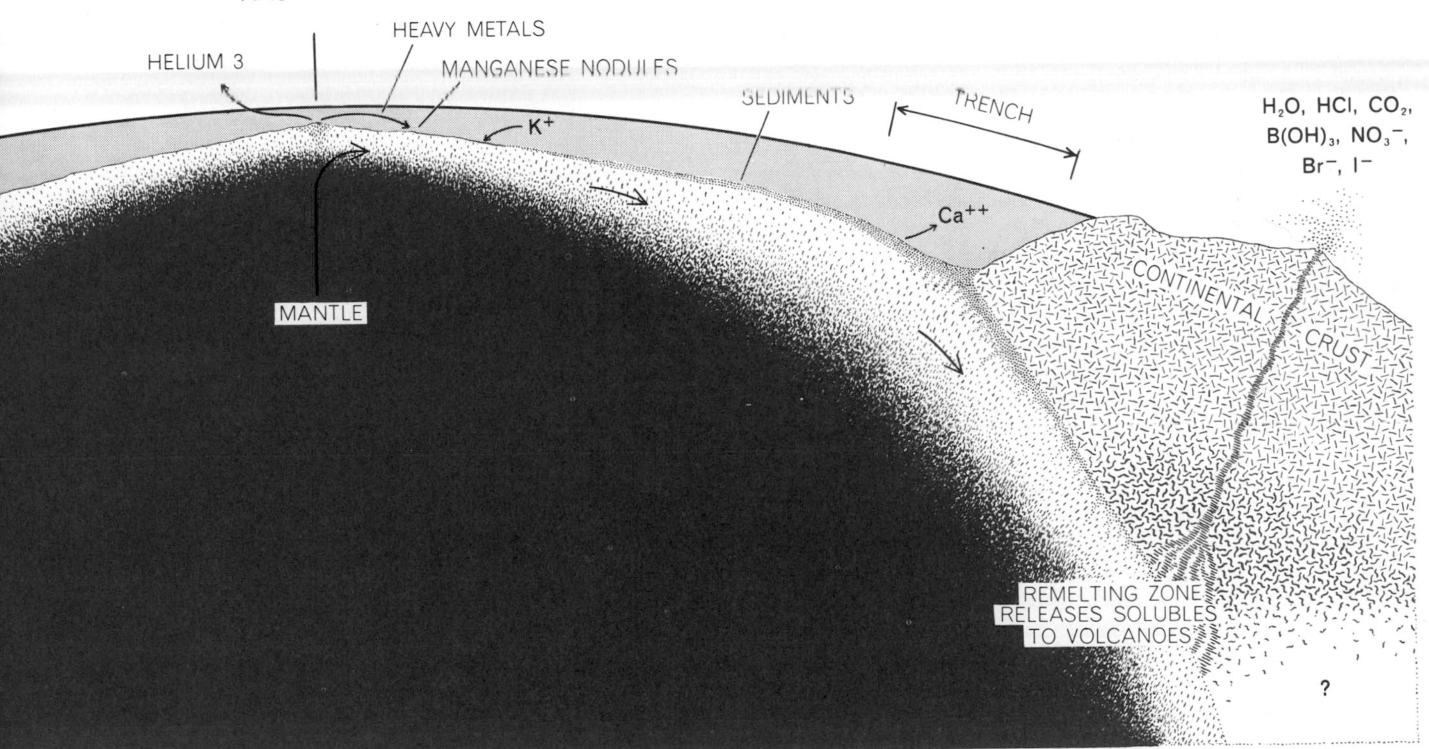

tween rigid plates, or segments, in the earth's crust. Although the driving mechanism is not yet understood, the plates move apart a few centimeters a year as fresh basalt from the plastic mantle flows up between them. The new basalt releases "juvenile" water (water never before in liquid form) and a variety of elements, including heavy metals that become incorporated in manganese nodules and the rare isotope helium 3, which escapes finally into space. At the continental margin (*right*) the lithospheric plate is subducted, forming a trench and carrying accumulated sediments with it. (The plate apparently thickens en route as plastic basalt "freezes" to its underside.) As it descends the plate remelts and releases soluble elements and ions that are ejected into the atmosphere by volcanoes. They maintain the saltiness of the sea and together with weathered crustal rock, such as granite, provide the stuff of sediments.

sphere were differentiated from an original homogeneous accumulation of material. Estimates of water released during the catastrophe range from a third to 90 percent of the present volume of the ocean. The catastrophe is not finished even yet, since differentiation of the mantle continues in regions of volcanic activity. Most of the exhalations of volcanoes and hot springs are simply recycled ground water, but if only half of 1 percent of the water released is juvenile, the present production rate is sufficient to have filled the entire ocean in four billion years.

There is evidence that the salinity of the ocean has not changed greatly since the ocean was formed; in any event the salinity has been nearly constant for the past 200 million years (5 percent of geologic time). The composition of ancient sediments suggests that the ratio of sodium to potassium in seawater has risen from about 1 : 1 to its present value of about 28 : 1. Over the same period the ratio of magnesium to calcium has risen from roughly 1 : 1 to 3 : 1 as organisms removed calcium by building shells of calcium carbonate. It is significant, however, that the total amount of each pair of ions varied much less than the relative amounts.

If we look at rain as it reaches the sea in rivers, we find a distinctly nonmarine mix to its ions. If we catch it even earlier as it tumbles down young mountains, the differences are even more pronounced. This continual input of water of nonmarine composition would eventually overwhelm the original composition of the ocean unless there were corrective reactions at work.

The overall geochemical cycle that keeps the marine ions closely in balance involves a complex interchange of material over decades, centuries and millenniums among the atmosphere, the ocean, the rivers, the crustal rocks, the oceanic sediments and ultimately the mantle [*see "a" in illustration on page 108*]. Because this overall picture is too general to be of much use, we abstract bits from it and call them thalassochemical models (*thalassa* is the Greek word for "sea"). One model involves simply the cyclic exchange of sea salt between the rivers and the sea; the cycle includes the transport of salt from the sea surface into the atmosphere, where salt particles act as condensation nuclei on which raindrops grow [*see "b" in illustration on page 108*]. This process accounts for more than 90 percent of the chloride and about 50 percent of the sodium carried to the sea by rivers.

Another useful abstraction is the "steady state" thalassochemical model. If the ocean composition does not change with time, it must be rigorously true that whatever is added by the rivers must be precipitated in marine sediments [*see "c" in illustration on page 108*]. Oceanic residence times computed from sedimentation rates, particularly for reactive trace metals, agree well with the input rates from rivers. Unfortunately residence times do not reveal the mechanism by which an element is removed from seawater. For residence times greater than a million years it is often helpful to invoke the "equilibrium" model, which deals only with the rate of exchange between the ocean and its sediments [*see "d" in illustration on page 108*].

To understand how the earth maintains its geochemical poise over a billion-year time scale we must return to the circle of arrows—the weathering and "unweathering" processes—of the geochemical cycle. This circle starts with primordial igneous rock, squeezed from the mantle. Ignoring relatively minor heavy metals such as iron, we can assume that the rock consists of aluminum, silicon and oxygen combined with the alkali metals: potassium, sodium and calcium. The resulting minerals are feldspars (for example $KAlSi_3O_8$). Rainwater picks up carbon dioxide from the air and falls on the feldspar. The reaction of water, carbon dioxide and feldspar typically yields a solution of alkali ions and bicarbonate ions (HCO_3^-) in which is suspended hydrated silica (SiO_2). The residual detrital aluminosilicate can be approximated by the clay kaolinite: $Al_2Si_2O_5(OH)_4$ [*see Step 1 in illustration on page 109*]. A mountain stream carries off the ions and the silica. The kaolinite fraction lags behind, first as a friable surface on weathering rock, then as soil material and finally as alluvial deposits in river valleys. If the stream evaporates in a closed basin, such as one finds in the western U.S., the result is a "soda lake" containing high concentrations of carbonates and amorphous silica.

In mature river systems the kaolinite fraction reaches the sea as suspended sediment. Encountering an ion-rich environment for the first time, the aluminosilicate must reorganize itself into new minerals. One such mineral, which seems to be forming in the ocean today, is the potassium-containing clay illite [*see Step 2 in illustration on page 109*]. These "clay cation" reactions may take decades or centuries. They are poorly understood because graduate students who study them invariably leave before the reactions are complete. The net effect of such reactions is to tie up and remove some of the potassium and bicarbonate ions, along with aluminum, silicon and oxygen.

A biologically important reaction, usually confined to shallow water, allows marine organisms to build shells of calcium carbonate, which precipitates when calcium (Ca^{++}) and bicarbonate ions react. If dilute hydrochloric acid is present (it is released by volcanoes), it reacts even more rapidly with bicarbonate, forming water and carbon dioxide and leaving free the chloride ion. When marine organisms die and sink to about 4,000 meters, they cross the "lysocline," below which calcium carbonate redissolves because of the high pressure. We have now traced the three metallic ions removed from igneous rock to three separate niches in the ocean. Sodium remains dissolved, potassium precipitates in clays on the deep-sea floor and calcium precipitates in shallow water as biogenic limestone: coral reefs and calcareous oozes.

Ages pass and the geochemical cycle rolls on, converting ocean-bottom clay into hard rock such as granite. When old sea floor finally reaches a region of high pressure and temperature under a continental block, it still contains some free ions that can react with the clay to reconstitute hard rock. A score of reaction

CURRENTLY RECOVERED FROM SEAWATER

ELEMENTS IN SHORT SUPPLY

RANGE OF BIOLOGICALLY CAUSED CHANGE

RANGE OF ANALYSES

METALS CONCENTRATED IN MANGANESE NODULES

COMPOSITION OF SEAWATER has been a challenge to chemists since Antoine Laurent Lavoisier made the first analyses. The logarithmic chart on the opposite page shows in moles per kilogram the concentration of 40 of the 73 elements that have been identified in seawater. A mole is equivalent to the element's atomic weight in grams; thus a mole of chlorine is 35 grams, a mole of uranium 238 grams. Only four elements are now recovered from the sea commercially: chlorine, sodium, magnesium and bromine. Recovery of other scarce elements is not promising unless biological concentrating techniques can be developed. Manganese nodules are a potential source of scarce metals but gathering them from the deep-sea floor may not be profitable in this century.

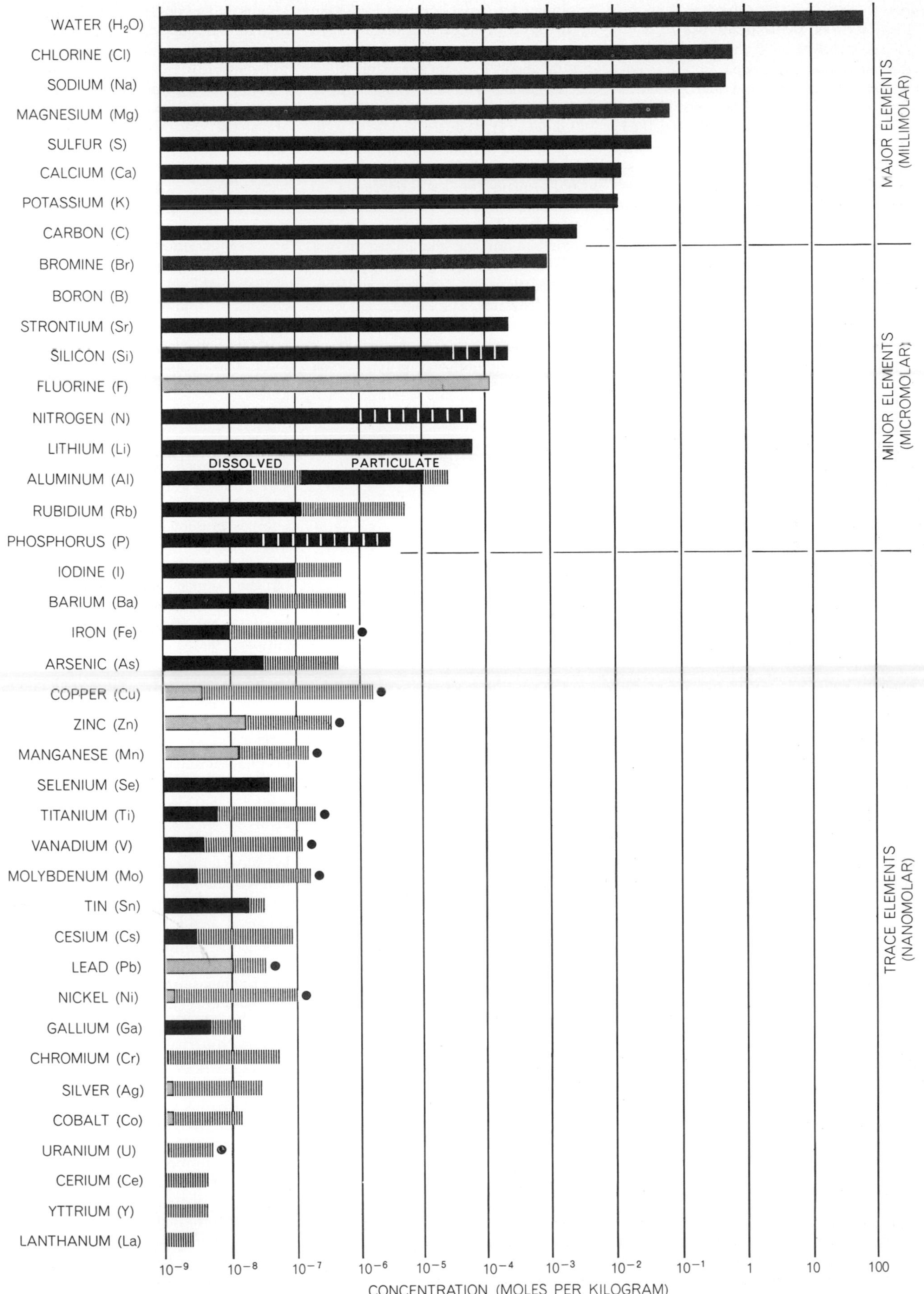

WATER (H_2O)
CHLORINE (Cl)
SODIUM (Na)
MAGNESIUM (Mg)
SULFUR (S)
CALCIUM (Ca)
POTASSIUM (K)
CARBON (C)
BROMINE (Br)
BORON (B)
STRONTIUM (Sr)
SILICON (Si)
FLUORINE (F)
NITROGEN (N)
LITHIUM (Li)
DISSOLVED
PARTICULATE
ALUMINUM (Al)
RUBIDIUM (Rb)
PHOSPHORUS (P)
IODINE (I)
BARIUM (Ba)
IRON (Fe)
ARSENIC (As)
COPPER (Cu)
ZINC (Zn)
MANGANESE (Mn)
SELENIUM (Se)
TITANIUM (Ti)
VANADIUM (V)
MOLYBDENUM (Mo)
TIN (Sn)
CESIUM (Cs)
LEAD (Pb)
NICKEL (Ni)
GALLIUM (Ga)
CHROMIUM (Cr)
SILVER (Ag)
COBALT (Co)
URANIUM (U)
CERIUM (Ce)
YTTRIUM (Y)
LANTHANUM (La)
MAJOR ELEMENTS (MILLIMOLAR)
MINOR ELEMENTS (MICROMOLAR)
TRACE ELEMENTS (NANOMOLAR)
10^{-9}
10^{-8}
10^{-7}
10^{-6}
10^{-5}
10^{-4}
10^{-3}
10^{-2}
10^{-1}
1
10
100
CONCENTRATION (MOLES PER KILOGRAM)

schemes are possible. In Step 3 in the illustration on the opposite page I have chosen to build a "granite" from equal parts of potassium feldspar, sodium feldspar, potassium mica and quartz. (Notice that calcium is missing because it has dissolved from the sediments during their descent into the deep-ocean trenches that carry the sediments under the continental blocks.) The reaction written in Step 3 uses up all the silica formed in Step 1.

The goal of this geochemical exercise has now been reached. First, we have shown that of all the substances that enter the ocean, only sodium and chlorine remain abundantly in solution. Of the other elements, the amount remaining in solution is less than a hundredth of the amount delivered to the ocean and precipitated from it. Second, we have made a start at explaining the observed sodium-potassium ratios: in basalt this ratio is about 1 : 1, in seawater 28 : 1 and in granite 1 : 1.2. If the weight of sodium tied up in granite were about 140 times as great as the weight of sodium dissolved in the sea, the slight excess of potassium over sodium in granite would explain the sea's deficiency in potassium.

We now have working models for thinking about the circulation of the major elements, but we have barely scratched the true complexity and subtlety of seawater. The sources and sinks of the minor elements are now being explored. In many cases we can only guess at what the natural marine form of an element is because our detection techniques either convert all forms to a common form for analysis or miss some forms completely. Moreover, certain ions seem to behave capriciously in the ocean. For example, at the *p*H (hydrogen-ion concentration) of seawater, vanadium should appear as $VO_2(OH)_3^{--}$, an ion with a double negative charge; instead it seems to exist in positively charged form, perhaps as VO_2^{+}.

Much of what is known about elements in the sea can be summarized in an oceanographer's periodic table [*see illustration on page 110*]. The usefulness of the usual kind of periodic table to the chemist is that it arranges chemically similar elements in vertical columns and presents behavioral trends in horizontal rows. The oceanographer's table shows how these regularities are disrupted in the ocean environment.

First of all, more than a dozen elements have never been detected in seawater, although two of them (palladium and iridium) exist in parts per billion in marine sediments and another (platinum) is present in manganese nodules. The second interesting feature of the oceanographer's table is the tendency for the "upper" and "outer" elements, those in the raised wings, so to speak, to be the most plentiful in the sea. The "upper" tendency simply reflects the greater cosmic abundance of light elements. (Lithium, beryllium and boron, however, are fairly scarce even cosmically.)

The "outer" trend can be explained in quantum-mechanical terms by the presence or absence of electrons in *d* orbitals, the electron shells principally involved in forming complexes. Elements in the first three columns at the left have no *d* orbitals; those in the last four columns at the right have full *d* orbitals. Both characteristics favor weak chemical bonds, with the result that these two groups of elements tend to ionize readily and remain in solution, either by themselves or in simple combination with oxygen and hydrogen. In contrast, the elements in the center of the table with partially filled *d* orbitals form strong chemical bonds and compounds that precipitate readily; thus they can exist only at low concentration in solution. For silver and the surrounding group of metals the most stable complexes are formed with the most abundant seawater ion: chloride. Most of the other elements that are hungry for *d* electrons form their complexes with oxygen, or oxygen plus some protons (hydrogen nuclei).

Ordinarily the oxidation state of metals

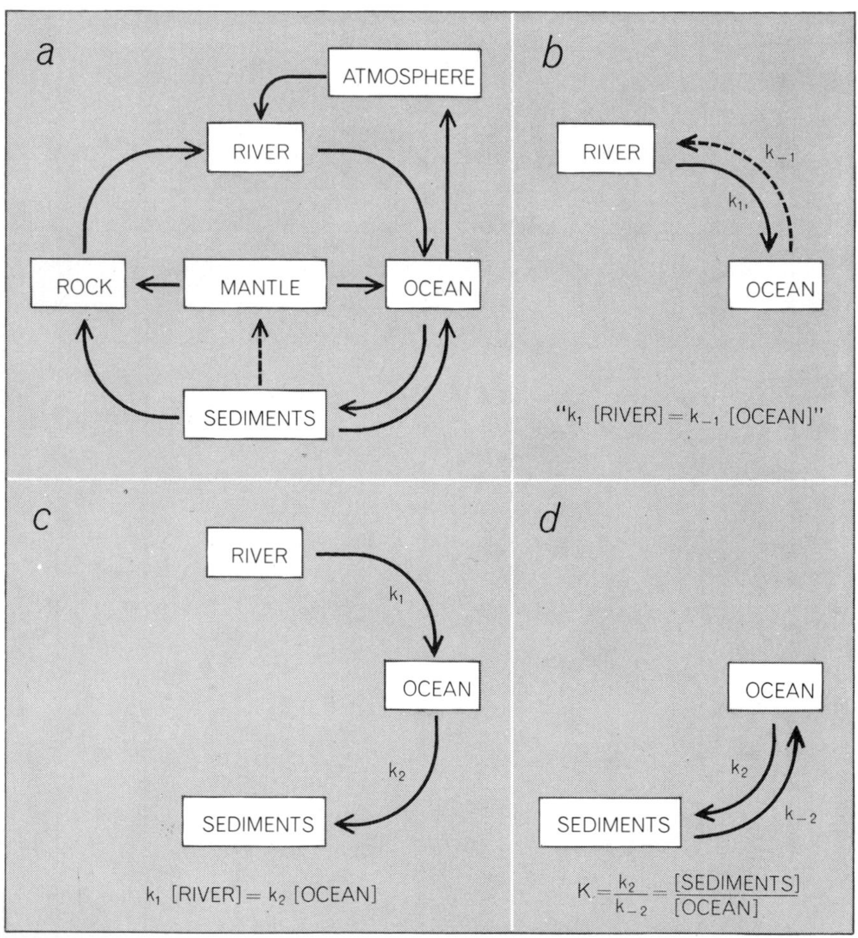

GRAND GEOCHEMICAL CYCLE (*a*) summarizes the global pathways taken sooner or later by the three-score elements that pass through the ocean and maintain its saltiness. The three "thalassochemical" models (*b, c, d*) abstracted from it are more helpful when trying to understand the rate laws governing the transport of specific elements. The rate constants, *k*, are expressed as a fraction: one over some number of years. The brackets enclose concentrations of the element being studied, specified according to its environment. The "cyclic" model (*b*) accounts for 90 percent of the chloride in river water. Its rate law is in quotation marks because extra factors, such as the area of the ocean, must be incorporated. The "steady state" model (*c*) works well for reactive trace metals; the reciprocal of k_2 is simply the residence time in the ocean. The "equilibrium" model (*d*) seems the most appropriate for the hydrogen ion (H^+) and the ions of the major metals, such as sodium.

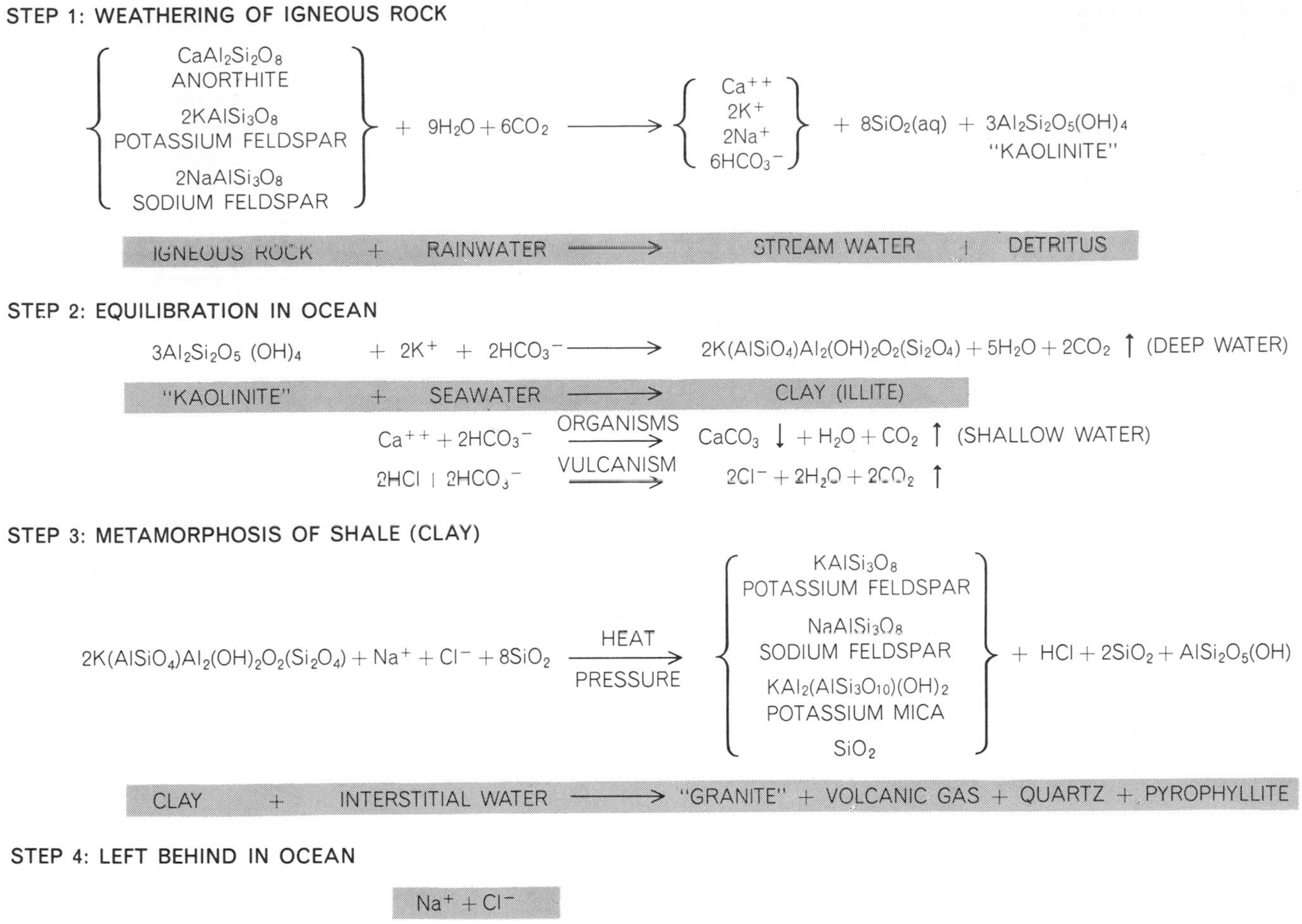

STEP 1: WEATHERING OF IGNEOUS ROCK

$$\left\{\begin{matrix} CaAl_2Si_2O_8 \\ \text{ANORTHITE} \\ 2KAlSi_3O_8 \\ \text{POTASSIUM FELDSPAR} \\ 2NaAlSi_3O_8 \\ \text{SODIUM FELDSPAR} \end{matrix}\right\} + 9H_2O + 6CO_2 \longrightarrow \left\{\begin{matrix} Ca^{++} \\ 2K^{+} \\ 2Na^{+} \\ 6HCO_3^{-} \end{matrix}\right\} + 8SiO_2(aq) + \underset{\text{"KAOLINITE"}}{3Al_2Si_2O_5(OH)_4}$$

IGNEOUS ROCK + RAINWATER ⟶ STREAM WATER + DETRITUS

STEP 2: EQUILIBRATION IN OCEAN

$$3Al_2Si_2O_5(OH)_4 + 2K^{+} + 2HCO_3^{-} \longrightarrow 2K(AlSiO_4)Al_2(OH)_2O_2(Si_2O_4) + 5H_2O + 2CO_2\uparrow \text{ (DEEP WATER)}$$

"KAOLINITE" + SEAWATER ⟶ CLAY (ILLITE)

$$Ca^{++} + 2HCO_3^{-} \xrightarrow{\text{ORGANISMS}} CaCO_3\downarrow + H_2O + CO_2\uparrow \text{ (SHALLOW WATER)}$$

$$2HCl + 2HCO_3^{-} \xrightarrow{\text{VULCANISM}} 2Cl^{-} + 2H_2O + 2CO_2\uparrow$$

STEP 3: METAMORPHOSIS OF SHALE (CLAY)

$$2K(AlSiO_4)Al_2(OH)_2O_2(Si_2O_4) + Na^{+} + Cl^{-} + 8SiO_2 \xrightarrow[\text{PRESSURE}]{\text{HEAT}} \left\{\begin{matrix} KAlSi_3O_8 \\ \text{POTASSIUM FELDSPAR} \\ NaAlSi_3O_8 \\ \text{SODIUM FELDSPAR} \\ KAl_2(AlSi_3O_{10})(OH)_2 \\ \text{POTASSIUM MICA} \\ SiO_2 \end{matrix}\right\} + HCl + 2SiO_2 + AlSi_2O_5(OH)$$

CLAY + INTERSTITIAL WATER ⟶ "GRANITE" + VOLCANIC GAS + QUARTZ + PYROPHYLLITE

STEP 4: LEFT BEHIND IN OCEAN

$Na^{+} + Cl^{-}$

ONLY SALT REMAINS after the ocean "laboratory" has finished processing the complex of chemicals removed from igneous rock by rainwater containing dissolved carbon dioxide. Step 1 yields a solution of alkali ions and bicarbonate (HCO_3^-) ions in which hydrated silica (SiO_2) and aluminosilicate detritus are suspended. In crystalline form the aluminosilicate would be kaolinite. In the ocean (*Step 2*) the "kaolinite" is complexed with potassium ions (K^+) to form illite clay. Marine organisms use the calcium ion (Ca^{++}) to make calcium carbonate shells, which form sediments in shallow water. Hydrochloric acid (HCl), injected by undersea volcanoes, reacts with bicarbonate ions, returning some carbon dioxide to the atmosphere. In Step 3 clay is metamorphosed into "granite." Sodium chloride (*Step 4*) remains. Although some of this sequence is hypothetical, something very similar seems to take place.

avid for *d* electrons would be determined by the oxidation potential of seawater, which is a measure of its ability to extract electrons from a substance just as its *p*H is a measure of its ability to extract protons. The oxidation potential of seawater has the high value of .75 volt, enabling it to extract the maximum possible number of electrons from nearly all elements except the noble metals (platinum group) and the halogens (fluorine family).

Surprisingly, however, the oxidation potential of seawater does not seem to control the oxidation states of many metals that have partially filled *d* shells. One reason is that most reactions proceed by a mechanism in which only a single electron is transferred at a time. Such transfers occur most readily when the reactants are adsorbed on surfaces where atomic geometry and electric-charge distribution are able to expedite the redistribution of electrons (hence the utility of catalysts, which provide such surfaces). But surfaces of any kind are few and far between in the ocean, and (with the exception of manganese nodules) those that do exist are poor catalysts. A second reason for the failure of the sea's oxidation potential to control valence states is that organisms sometimes excrete electron-rich substances, which then remain in that reduced state in spite of seawater's apparent capacity to oxidize them.

Manganese nodules are porous chunks of metallic oxides up to several centimeters in diameter, widely distributed over the ocean floor. They evidently exist because they are autocatalytic for the reaction that produces them. Because of their porous structure, nodules have a surface area of as much as 100 square meters per gram. The autocatalytic property seems to extend to an entire suite of metals that coprecipitate with manganese: iron, cobalt, nickel, copper, zinc, chromium, vanadium, tungsten and lead. Nodules found on the flanks of oceanic ridges contain significant concentrations of metals, such as nickel, that are scarce in seawater itself. This suggests that the nodules are collecting juvenile metals as the metals leak from the mantle at the fissure of the ridge. One would like to know why the nodule metals are present in oxide form rather than, as one would expect, in carbonate form.

The level of the discussion so far might best be called thalassopoetry. The discussion can be made more serious in two ways. One approach—the "geochemical balance"—has employed a computer to follow in detail as many as 60 elements as they move through the geochemical cycle, from igneous rock back

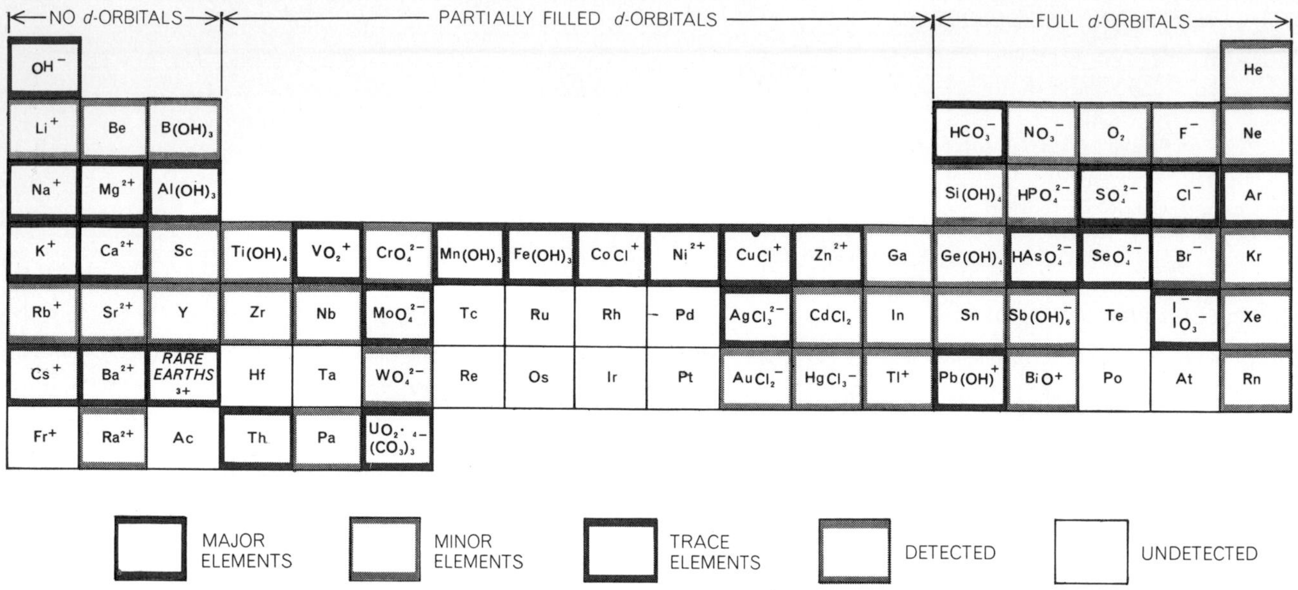

NO *d*-ORBITALS			PARTIALLY FILLED *d*-ORBITALS										FULL *d*-ORBITALS				
OH^-																	He
Li^+	Be	$B(OH)_3$											HCO_3^-	NO_3^-	O_2	F^-	Ne
Na^+	Mg^{2+}	$Al(OH)_3$											$Si(OH)_4$	HPO_4^{2-}	SO_4^{2-}	Cl^-	Ar
K^+	Ca^{2+}	Sc	$Ti(OH)_4$	VO_2^+	CrO_4^{2-}	$Mn(OH)_3$	$Fe(OH)_3$	$CoCl^+$	Ni^{2+}	$CuCl^+$	Zn^{2+}	Ga	$Ge(OH)_4$	$HAsO_4^{2-}$	SeO_4^{2-}	Br^-	Kr
Rb^+	Sr^{2+}	Y	Zr	Nb	MoO_4^{2-}	Tc	Ru	Rh	Pd	$AgCl_3^{2-}$	$CdCl_2$	In	Sn	$Sb(OH)_6^-$	Te	I^- IO_3^-	Xe
Cs^+	Ba^{2+}	RARE EARTHS 3+	Hf	Ta	WO_4^{2-}	Re	Os	Ir	Pt	$AuCl_2^-$	$HgCl_3^-$	Tl^+	$Pb(OH)^+$	BiO^+	Po	At	Rn
Fr^+	Ra^{2+}	Ac	Th	Pa	$UO_2 \cdot (CO_3)_3^{4-}$												

MAJOR ELEMENTS | MINOR ELEMENTS | TRACE ELEMENTS | DETECTED | UNDETECTED

PERIODIC TABLE, as prepared by the "thalassochemist," shows the form in which the detectable elements appear in seawater. In each box the element normally found in that place in the usual periodic table is shown in color; the elements associated with it are in black. Thus carbon appears predominantly as HCO_3^-, arsenic as $HAsO_4^{2-}$ and so on. The superscripts show the number of positive or negative charges carried by each ion. Iodine's two forms, I^- and IO_3^-, are about equally common. Except for the noble gases (*last column at right*), all the elements dissolved in the sea must be present as ions. When an element (other than a noble gas) is shown by itself, without a plus or minus charge, it means that its preferred ionic form in seawater is not yet established.

to metamorphosed sediments. In the second approach the actual chemistry of each element is followed by applying the thermodynamic methods of Josiah Willard Gibbs to systems regarded as being near equilibrium. This effort was launched by Lars Gunnar Sillén of Sweden and has been pursued by Robert M. Garrels of Northwestern University and by Heinrich D. Holland of Princeton University.

Of course no chemist in his right mind would talk seriously about equilibria in a system of variable temperature, pressure and composition that was poorly stirred, had variable inputs and contained living creatures. On the other hand, the observed uniformity of the ocean and the long periods available for reacting suggest that at least the major components are sufficiently close to equilibrium to make an investigation worthwhile. (We *know* the minor constituents are not in equilibrium.)

The equilibrium approach is based on Gibbs's phase rule, which states that the number of phases (P) possible in a system of C components at equilibrium is given by the equation $P = C + 2 - F$, where F is the number of "degrees of freedom," or quantities that may be independently varied without changing the number of phases or their composition (although F may change their relative proportions). The 2 enters the equation because only two variables, temperature and pressure, are important in most chemical reactions.

One of Sillén's most comprehensive ocean models has nine components: water, hydrochloric acid, silica, three hydroxides (aluminum, sodium and potassium), carbon dioxide and the oxides of magnesium and calcium. Observation of sea-floor sediments, aided by laboratory studies, suggests that a nine-phase ocean will result [*see illustrations on opposite page*]. If C and P both equal nine, the phase rule states that the number of degrees of freedom (F) must equal two. Logically these are temperature (which can vary over the oceanic range from −2 degrees Celsius to 30 degrees) and the chloride ion concentration (which can shift over the normal oceanic range without changing the composition of the stable phases).

A diagrammatic view of how the nine components sort themselves into phases is shown in the bottom illustration on the opposite page. Note that the liquid phase contains ions not listed either as components or phases (for example H^+ and OH^-). Thermodynamics need not consider them explicitly because they do not vary independently; their concentrations are fixed by the equilibrium constants that connect the observed phases. Thus $H_2O = H^+ + OH^-$. Moreover, one knows that the product of H^+ and OH^- is a thermodynamic constant, which equals 10^{-14} mole per liter. Similar relations tie the entire system into a comprehensible whole, so that when all the calculations are performed one has discovered the equilibrium concentrations of five cations (H^+, Na^+, K^+, Mg^{++} and Ca^{++}) and four anions (Cl^-, OH^-, HCO_3^- and CO_3^{--}).

It may seem peculiar to discuss an "atmosphere" containing only water vapor and carbon dioxide. One could easily add oxygen and nitrogen to the list of components. Since they would add no new phases, they would raise the number of degrees of freedom from two to four ($9 = 11 + 2 - 4$). The two new F's would be the total atmospheric pressure and the ratio of oxygen to nitrogen. In the study of the ocean, however, the partial pressure due to carbon dioxide is more significant than the total pressure of the atmosphere. Moreover, the presence of gaseous oxygen and nitrogen has little importance for the inorganic environment of the ocean, so that it is simpler to omit them and just as "real."

Suppose now we perturb the equilibrium of the model ocean by assuming that a submerged volcano has suddenly released enough hydrochloric acid (HCl) to double the amount of chloride ion (Cl^-). The dissociation of hydrochloric acid releases enough H^+ ions to raise the total number of hydrogen ions in the ocean from the former equilibrium value of 10^{-8} mole per liter to $10^{+.3}$. This excess of hydrogen ions almost immediately pushes all the available carbonate ions (CO_3^{--}) to bicarbonate ions (HCO_3^-) and the latter to carbonic acid (H_2CO_3). These shifts, however, only slightly depress the pH, which remains high until the slow circulation of the ocean brings the hydrogen ions in direct contact with the clay sediments on the sea floor.

The structure of clay is such that oxygen atoms at the free corners of polyhedrons carry unsatisfied negative charges, which attract positive ions [*see top illustration on next page*]. Because the ocean is so rich in sodium ions (Na^+), they occupy most of the corners of clay polyhedrons. When the excess hydrogen ions come in contact with the clay, they quickly replace the sodium ions and set them adrift. This fast reaction is limited in scope because the surface and interlayer ion-exchange capacity of clay is not very great. Much more capacity is provided when the structure of the clay is rearranged; for example, the conversion of montmorillonite to kaolinite also consumes hydrogen atoms and releases sodium. Given sufficient time—centuries—such rearrangements inexorably take place, and the pH of the ocean slowly drifts back to its equilibrium value. The charge on the excess chloride introduced by the volcano will then be balanced not by H^+ but by Na^+. This slow equilibration mechanism can be regarded as the ocean's "pH-stat" (in analogy with "thermostat"). This clay-cation model suggests that the pH of the ocean has been constant over the span of geologic time and that hence the carbon dioxide content of the atmosphere has been held within narrow limits.

If only the pH-stat were available for leveling surges in pH, the ocean might be subjected to violent local fluctuations. For fast response pH control is taken over by a carbonate buffer system [*see bottom illustration on next page*]. In fact, until recently oceanographers neglected the clay-cation reactions and assumed that the carbonate-buffer system almost completely determined the pH of the ocean.

One might think that if the carbon dioxide content of the atmosphere were to decrease, carbon dioxide would flow from the sea into the atmosphere, leading to a general depletion of all carbonate species in the ocean and eventually to the dissolution of some carbonate sediments. In actuality something quite different happens because the carbonate system is its own source of hydrogen ions. Removal of carbon dioxide from water reduces the concentration of carbonic acid (H_2CO_3), the hydrated form of carbon dioxide. Replacement of this acid from bicarbonate ions requires a hydrogen ion, which can only be obtained by converting another bicar-

COMPONENTS (C)	PHASES (P)	VARIABLES (F)
H_2O	1 GAS	TEMPERATURE
HCl	2 LIQUID	Cl^-
SiO_2	3 QUARTZ (SiO_2)	
$Al(OH)_3$	4 KAOLINITE (t-o CLAY)	
NaOH	5 MONTMORILLONITE (Na-t-o-t CLAY)	
KOH	6 ILLITE (K-t-o-t CLAY)	
MgO	7 CHLORITE (Mg-t-o-t CLAY)	
CO_2	8 CALCITE ($CaCO_3$)	
CaO	9 PHILLIPSITE (Na-K FELDSPAR)	

NINE MAJOR COMPONENTS IN SEA can, to a first approximation, be combined into nine distinctive phases to satisfy the "phase rule" that governs systems in equilibrium. The rule, formulated in the 19th century by Josiah Willard Gibbs, prescribes the number of phases P, components C and degrees of freedom F in such a system: $P = C + 2 - F$. When the number of phases and components are equal, the number of degrees of freedom, F, must be two, which allows both the temperature and the chloride-ion concentration to vary without altering the number of phases. In the clay-containing phases (*4, 5, 6, 7*) the letter "t" stands for a tetrahedral crystal structure; the letter "o" stands for an octahedral structure.

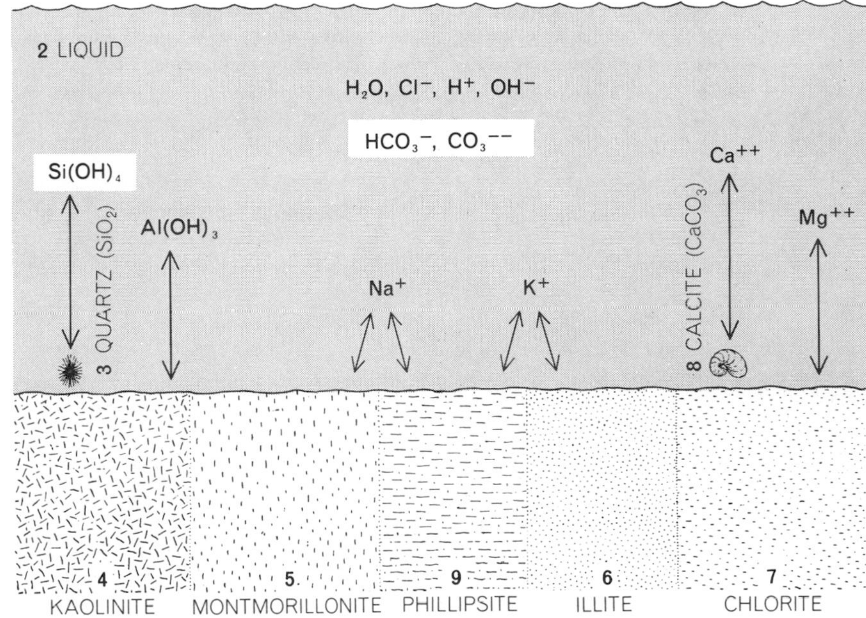

EQUILIBRIUM OCEAN MODEL, consisting of nine phases and nine components, shows how the principal constituents of the ocean distribute themselves among the atmosphere, the ocean and the sediments. Three of the constituents (HCO_3^-, CO_3^{--} and $Si(OH)_4$) are not included among nine listed components but appear as equilibrium products of those that are listed, as do seven ions ($H^+, K^+, Na^+, Ca^{++}, Mg^{++}, Cl^-, OH^-$). Two of the solids are shown as biological "precipitates": "quartz" (*3*) in the form of silicate structures built by radiolarians and "calcite" (*8*) in the form of calcium carbonate chambers built by foraminifera. The method of precipitation is unimportant as long as the product is stable. The equilibrium model goes far to explain why the ocean has the composition it does.

bonate ion to carbonate. The overall reaction is $2HCO_3^- \rightarrow H_2CO_3 + CO_3^{--}$. Thus instead of dissolving existing sediments, removing carbon dioxide from the sea may actually precipitate carbonate. This reaction can be seen in the "whitings" of the sea over the Bahama Banks, where cold deep water, rich in dissolved carbon dioxide and calcium, is forced to the surface and warmed. As carbon dioxide escapes into the air, the *p*H drops and aragonite ($CaCO_3$) precipitates, turning large areas of the ocean white with a myriad of small crystals.

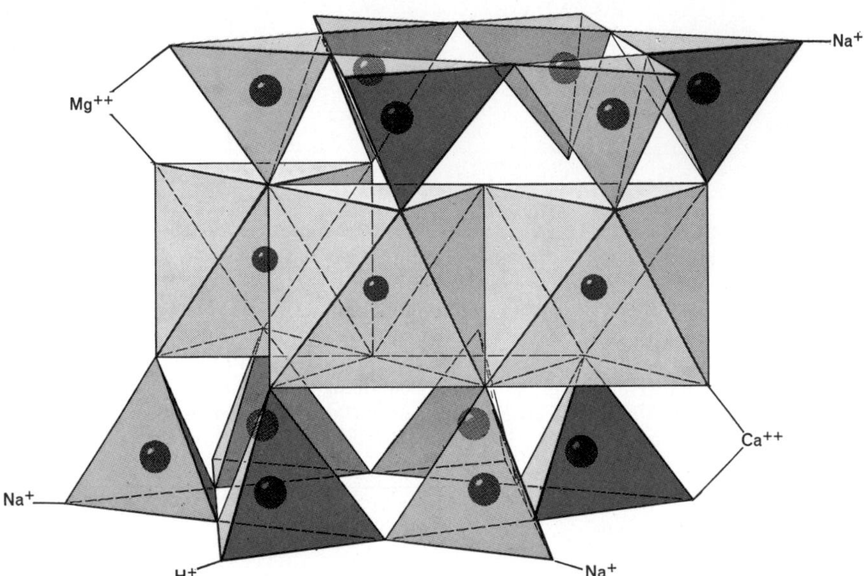

THREE-LAYER CLAY PARTICLE has a layer of octahedrons sandwiched between two layers of tetrahedrons. Each octahedron consists of an atom of aluminum surrounded by six closely packed atoms of oxygen. Each tetrahedron consists of a silicon atom surrounded by four atoms of oxygen. The polyhedrons are tied into layers at shared corners where a single oxygen atom is bonded to a silicon atom on one side and to an aluminum atom on the other. At the free corners the oxygen atoms bear unsatisfied negative charges that attract cations such as sodium (Na^+) and potassium (K^+). If the hydrogen-ion concentration should rise in the vicinity of clay, free hydrogen ions tend to be exchanged for sodium ions, which are released. In addition, many doubly charged metal ions can replace Si^{4+} at the centers of tetrahedrons and Si^{4+} can replace Al^{3+} in the octahedrons. Whenever this occurs, another cation is bound to the structure to conserve charge. Such reactions apparently exert considerable control over the ocean's composition and hydrogen-ion concentration.

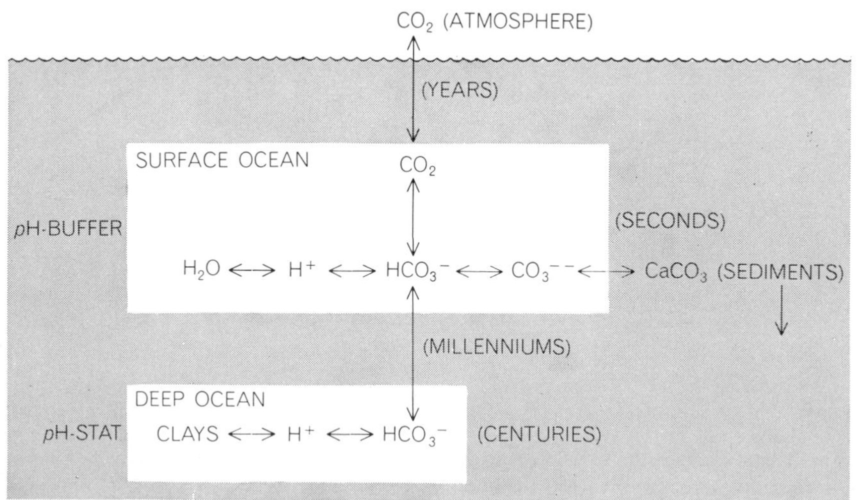

HYDROGEN-ION CONCENTRATION, or *p*H, of the ocean is controlled by two mechanisms, one that responds swiftly and one that takes centuries. The first, the "*p*H-buffer," operates near the surface and maintains equilibrium among carbon dioxide, bicarbonate ion (HCO_3^-), carbonate ion (CO_3^{--}) and sediments. The slower mechanism, the "*p*H-stat," seems to exert ultimate control over *p*H; it involves the interaction of bicarbonate ions and protons (H^+) with clays. Clay will accept protons in exchange for sodium ions (primarily).

The reaction above conserves charge, which means that the "alkalinity"—the traditional name for the concentration of sodium ion ("alkali") needed to balance this negative charge—is also conserved. The "carbonate alkalinity," defined as the bicarbonate concentration plus twice the carbonate concentration, is useful because it remains fixed even when the relative amounts of the two species vary.

The system can be visualized with the help of the illustration on the opposite page, which is the "Bjerrum plot" for carbonic acid at constant alkalinity. It takes its name from Niels Bjerrum, who introduced such plots in 1914; it shows the interrelations between the various compounds in the world carbonate system as a function of *p*H. Although the diagram ignores variations of pressure, temperature and salinity, it displays the essential features of the system.

The Bjerrum plot facilitates a semiquantitative discussion of the relation of atmospheric carbon dioxide to oceanic carbon dioxide. Over the next 20 years we shall burn enough fossil fuel to double the amount of carbon dioxide in the atmosphere from 320 parts per million to 640. On the plot this is indicated by shifting the line *A*, corresponding to 320 parts per million, to position *B*, 640 parts per million.

To produce this shift some 2.5×10^{18} grams of carbon dioxide must be added to the atmosphere. If the altered atmosphere were to come to equilibrium with the ocean, the *p*H of the ocean would drop from its present value of 8.15 to 7.89—still well within the range tolerated by marine organisms. This cannot happen, however, because the total mass of carbon dioxide in the ocean (Σ in the Bjerrum plot) plus the carbon dioxide in the atmosphere would have to increase from its present value, 128.9×10^{18} grams, to 138.3×10^{18} grams. The difference, 9.4×10^{18} grams, is nearly four times the amount added to the atmosphere.

The long-term equilibration process for such an atmospheric doubling can be broken down into two steps. First the *p*H-buffer system operates: 2.5×10^{18} grams, or 2 percent of the total mass, is added to the world system at constant alkalinity. The result of this step is the line *C* in the diagram, corresponding to a total mass of 131.4×10^{18} grams, an atmospheric carbon dioxide content of 390 parts per million and an oceanic *p*H of 8.08. Next, if the ocean has time to equilibrate with its sediments, the *p*H-stat will operate, returning the system to *p*H 8.15 at a constant total mass. The re-

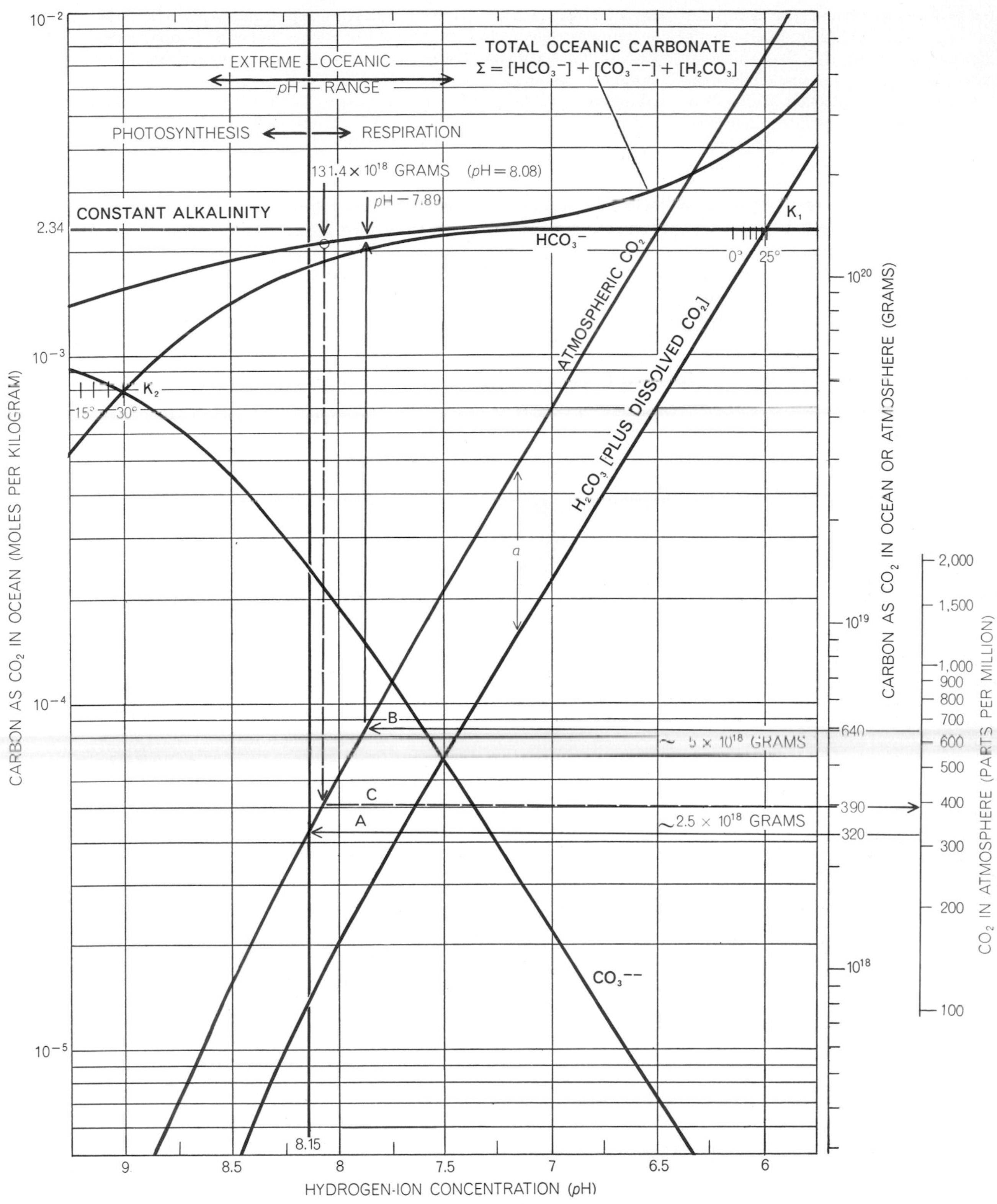

OCEANIC CARBONATE SYSTEM can be represented by a "Bjerrum diagram" that shows how carbonate in its several forms varies with the ocean's *p*H, or hydrogen-ion concentration. The diagram is plotted for a constant "carbonate alkalinity" of 2.34×10^{-3} moles of carbonate per kilogram of seawater (*scale at left*). "System point" K_1 shows where the concentrations of bicarbonate ion (HCO_3^-) and carbonic acid (H_2CO_3) are equal. At K_2 the concentrations of bicarbonate and carbonate (CO_3^{--}) are equal. The exact locations of K_1 and K_2 are shown for a range of temperatures (in degrees Celsius) at constant conditions of salinity and pressure. The top curve, Σ, is the sum of oceanic carbonate in all its forms. The normal *p*H of the ocean is 8.15. The two short arrows at top mark the normal biological limits: at 7.95 the available oxygen has been consumed by respiration; at 8.35 photosynthesis has removed so much carbon dioxide that absorption from the atmosphere rises sharply. The limits of oceanic *p*H lie between 7.45 and 8.6. The amount of carbon dioxide in the atmosphere (*colored curve and scale at far right*) is related to the amount of carbon dioxide dissolved in the ocean by alpha (α), the average worldwide solubility of carbon dioxide in seawater. The consequences of doubling the carbon dioxide in the atmosphere from 320 parts per million (*A*) to 640 parts (*B*) are discussed in the text of the article, as is line *C*.

sult of this step is that the alkalinity rises by 2 percent, which in terms of the Bjerrum plot means that the system will return to normal except that all the numbers on the concentration axes will be multiplied by 1.02. The long-range effect of a sudden doubling of the atmosphere's carbon dioxide, therefore, is to increase the ultimate value 2 percent, from 320 parts per million to 326, and some of that increase will ultimately find its way into vegetation and humus.

It is obvious that rates are crucial in the global distribution of carbon dioxide. The wind-stirred surface layer of the sea exchanges carbon dioxide rapidly with the atmosphere, requiring less than a decade for equilibration. Because this layer is only about 100 meters deep it contains only a tiny fraction of the ocean's total volume. Large-scale disposal of atmospheric carbon dioxide therefore requires that the gas be dissolved and transported to deep water.

Such vertical transport takes place almost exclusively in the Weddell Sea off the coast of Antarctica. Every winter, when the Weddell ice shelf freezes, the salt excluded from the newly formed ice increases the salinity and hence the density of the water below. This ice-cold water, capable of containing more dissolved gas than an equal volume of tropical water, cascades gently down the slope of Antarctica to begin a 5,000-year journey northward across the bottom of the ocean. The carbon dioxide in this "antarctic bottom water" has plenty of time to come to equilibrium with clay sediments.

Enough fossil fuel has been burned in the past century to have raised the carbon dioxide content of the atmosphere from about 290 parts per million to 350 parts. Since the actual level is now 320 parts per million, about half of the carbon dioxide put into the air has been removed. Although proof is lacking, a principal removal agent is undoubtedly antarctic bottom water. The process is so slow, however, that the carbon dioxide content of the atmosphere may reach 480 parts per million before the end of the century. By then it should be clear

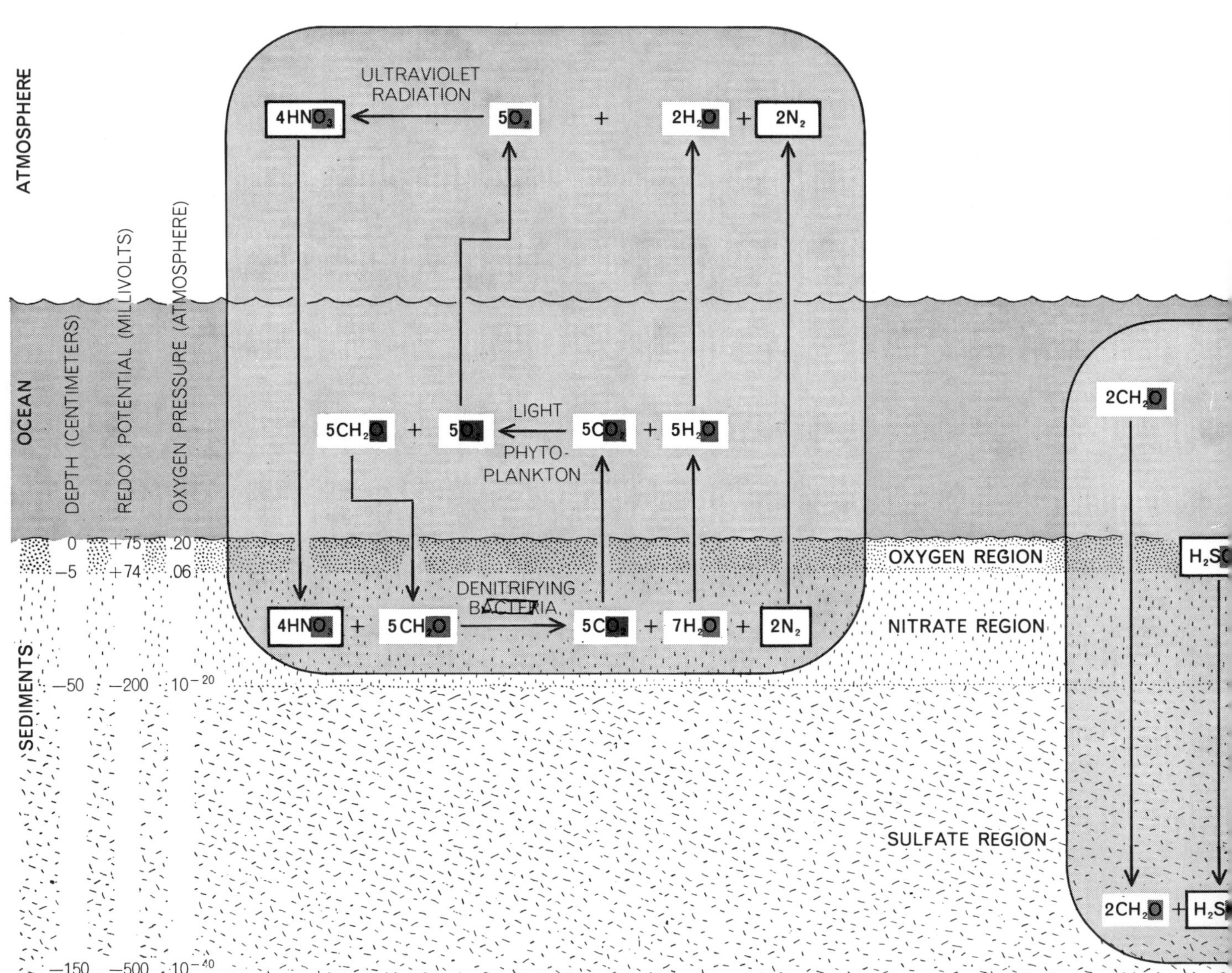

BACTERIA IN MARINE SEDIMENTS, although scarce by terrestrial soil standards, play a major role in replenishing the oxygen of the atmosphere and in limiting the accumulation of organic sediments. The bacteria concerned are buried in fine-grained sediments from several centimeters to several tens of centimeters below the ocean floor, with limited access to free oxygen for respiration. Thus deprived, they use the oxygen in nitrates and sulfates to oxidize organic compounds, represented by CH_2O. The actual reactions are far more complex than indicated here. The net result, however, is that denitrifying bacteria (*left*) release free nitrogen and convert carbon to a form (carbon dioxide) that can be reutilized by phytoplankton. These organisms, in turn, release free oxy-

if man's inadvertent global experiment (altering the atmosphere's carbon dioxide content) will have the predicted effect of changing the earth's climate. In principle an increase in atmospheric carbon dioxide should reduce the amount of long-wavelength radiation sent back into space by the earth and thus produce a greenhouse effect, slightly raising the average world temperature.

Having described an equilibrium model of the ocean that neglected the atmosphere's content of nitrogen and oxygen, I should not leave the reader with the impression that the continued presence of these two gases in the atmosphere is independent of the ocean. If the ocean were truly in equilibrium with the atmosphere, it would long since have captured all the atmospheric oxygen in the form of nitrates, both in solution and in sediments. This catastrophe has apparently been averted by the intervention of certain marine bacteria that have the happy faculty of releasing nitrogen gas from nitrate compounds and of converting the oxygen to a form that can later be liberated by phytoplankton.

The story is this. A variety of high-energy processes in the atmosphere continuously break the triple chemical bond that holds two nitrogen atoms together in a nitrogen molecule (N_2). The bonds can be broken by ultraviolet photons, by cosmic rays, by lightning and by the explosions in internal-combustion engines. Once dissociated, nitrogen atoms can react with oxygen to form various oxides, which are then carried to the ground by rainfall. In the soil these oxides are useful as fertilizer. Ultimately large amounts of them reach the sea. They do not, however, accumulate there and no one is really sure why.

The best guess is that denitrifying bacteria in oceanic sediments use the oxygen of nitrate to oxidize organic molecules when they run out of free oxygen [*see left half of illustration on these two pages*]. The nitrogen is released directly as a gas, which goes into solution but is available for return to the atmosphere. The oxygen emerges in molecules of water and carbon dioxide. The carbon dioxide is assimilated by phytoplankton, which build the carbon into organic compounds and release the oxygen as dissolved gas, also available for return to the atmosphere. Without these coupled biological processes the atmospheric fixation of nitrogen would probably exhaust the world's oxygen supply in less than 10 million years. Nevertheless, the amount of nitrogen returned to the atmosphere from the sediments is so small that we may never be able to measure it directly: the yearly return is less than one two-thousandth of the total nitrogen dissolved in the sea.

Another little-known epicycle in the global oxygen cycle probably has the effect of limiting the net accumulation of carbon in the form of oil-bearing shale, tar sands and petroleum. After denitrifying bacteria have consumed the nitrate in young sediments, sulfate bacteria begin oxidizing organic matter with the oxygen contained in sulfates [*see right half of illustration on these two pages*]. The product, in addition to water and carbon dioxide, is hydrogen sulfide, the foul-smelling compound that characterizes environments deficient in oxygen. In undisturbed mud the hydrogen sulfide never reaches the surface because it is inorganically reoxidized to sulfate as soon as it comes in contact with free oxygen. It seems likely that the bacterial turnover of oxygen in sulfate is so rapid that half of the world's oxygen passes through this epicycle in about 50,000 years.

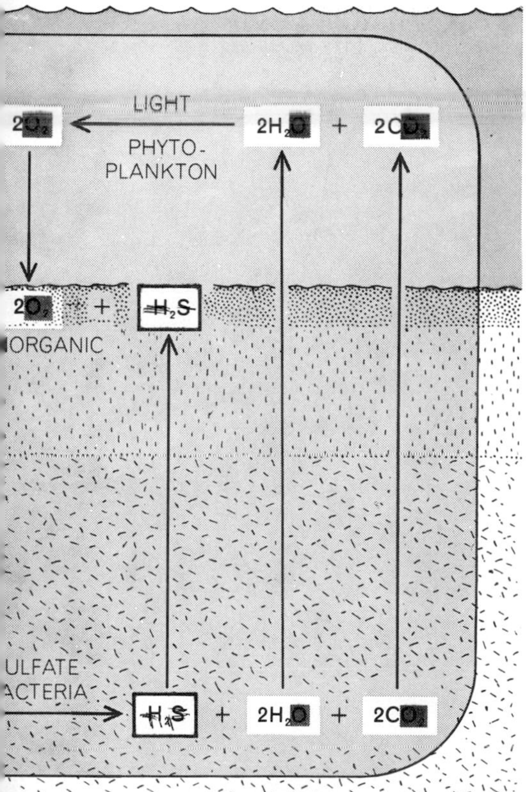

gen. Without the cooperative effort of these two groups of organisms the oxygen in the atmosphere might all be fixed by high-energy processes within some 10 million years. The sulfate bacteria (*right*) play a role in the recycling of sulfur and oxygen.

The global activities of man have now reached such a scale that they are beginning to have a profound effect on marine chemistry and biology. We are learning that even the ocean is not large enough to absorb all the waste products of industrial society. The experiment involving the release of carbon dioxide is now in progress. DDT, only 25 years on the scene, is now found in the tissues of animals from pole to pole and has pushed several species of birds close to extinction. The concentration of lead in plants, animals and man has increased tenfold since tetraethyl lead was first used as an antiknock agent in motor fuels. And high levels of mercury in fish have forced the abandonment of some commercial fisheries. (Lead and mercury are systemic enzyme poisons.) Of the total petroleum production some .2 percent gets slopped into the sea in half a dozen major accidents each year. (At least six of the rare gray whales died last year after migrating through the oil slick off Santa Barbara caused by the blowout of a well casing belonging to the Union Oil Company.) Conceivably a persistent oil film could change the surface reflectivity of the ocean enough to alter the world's energy balance. The rapid increase in the use of nitrogen fertilizers leaves a nitrate excess that runs into rivers, lakes and ultimately reaches the sea. The sea can probably tolerate the runoff indefinitely but along the way the nitrogen creates algal "blooms" that are hastening the dystrophication of lakes and estuaries.

It is fashionable today to view the ocean as the last global frontier, waiting only technological "development." Thermodynamically it is easier to extract fresh water from sewage than from seawater. Ecologically it is wiser to keep our concentrated nutrients on land than to dilute them beyond recall in the ocean. Sociologically, and probably economically, it makes more sense to process our junkyards for usable metals than to mine the deep-sea floor. The task is to persuade our engineers and business companies that working with sewage and junk is just as challenging as oceanography and thalassochemistry.

Ocean Waves **10**

by Willard Bascom
August 1959

Men have always been fascinated, and sometimes awed, by the rhythmic motions of the sea's surface. A century of observation and experiment has revealed much about how these waves are generated and propagated

Man is by nature a wave-watcher. On a ship he finds himself staring vacantly at the constant swell that flexes its muscles just under the sea's surface; on an island he will spend hours leaning against a palm tree absently watching the rhythmic breakers on the beach. He would like to learn the ways of the waves merely by watching them, but he cannot, because they set him dreaming. Try to count a hundred waves sometime and see.

Waves are not always so hypnotic. Sometimes they fill us with terror, for they can be among the most destructive forces in nature, rising up and overwhelming a ship at sea or destroying a town on the shore. Usually we think of waves as being caused by the wind, because these waves are by far the most common. But the most destructive waves are generated by earthquakes and undersea landslides. Other ocean waves, such as those caused by the gravitational attraction of the sun and the moon and by changes in barometric pressure, are much more subtle, often being imperceptible to the eye. Even such passive elements as the contour of the sea bottom, the slope of the beach and the curve of the shoreline play their parts in wave action. A wave becomes a breaker, for example, because as it advances into increasingly shallow water it rises higher and higher until the wave front grows too steep and topples forward into foam and turbulence. Although the causes of this beautiful spectacle are fairly well understood, we cannot say the same of many other aspects of wave activity. The questions asked by the wave-watcher are nonetheless being answered by intensive studies of the sea and by the examination of waves in large experimental tanks. The new knowledge has made it possible to measure the power and to forecast the actions of waves for the welfare of those who live and work on the sea and along its shores.

Toss a pebble into a pond and watch the even train of waves go out. Waves at sea do not look at all like this. They are confused and irregular, with rough diamond-shaped hillocks and crooked valleys. They are so hopelessly complex that 2,000 years of observation by seafarers produced no explanation beyond the obvious one that waves are somehow raised by the wind. The description of the sea surface remained in the province of the poet who found it "troubled, unsettled, restless. Purring with ripples under the caress of a breeze, flying into scattered billows before the torment of a storm and flung as raging surf against the land; heaving with tides breathed by a sleeping giant."

The motions of the oceans were too complex for intuitive understanding. The components had to be sorted out and dealt with one at a time. So the first theoreticians cautiously permitted a perfect train of waves, each exactly alike, to travel endlessly across an infinite ocean. This was an abstraction, but it could at least be dealt with mathematically.

Early observers noticed that passing waves move floating objects back and forth and up and down, but do not transport them horizontally for any great distance. From the motion of seaweeds the motion of the water particles could be deduced. But it was not until 1802 that Franz Gerstner of Germany constructed the first wave theory. He showed that water particles in a wave move in circular orbits. That is, water at the crest moves horizontally in the direction the wave is going, while in the trough it moves in the opposite direction. Thus each water particle at the surface traces a circular orbit, the diameter of which is equal to the height of the wave [*see illustration on next page*]. As each wave passes, the water returns almost to its original position. Gerstner observed that the surface trace of a wave is approximately a trochoid: the curve described by a point on a circle as it rolls along the underside of a line. His work was amplified by Sir George Airy later in the 19th century, by Horace Lamb of England in the present century, and by others.

The first wave experimentalists were Ernst and Wilhelm Weber of Germany, who in 1825 published a book on studies employing a wave tank they had invented. Their tank was five feet long, a foot deep and an inch wide, and it had glass sides. To make waves in the tank they sucked up some of the fluid through a tube at one end of it and allowed the fluid to drop back. Since the Weber brothers experimented not only with water and mercury but also with brandy, their persistence in the face of temptation has been an inspiration to all subsequent investigators. They discovered that waves are reflected without loss of energy, and they determined the shape of the wave surface by quickly plunging in and withdrawing a chalk-dusted slate. By watching particles suspended in the water they confirmed the theory that water particles move in a circular orbit, the size of which diminishes with depth. At the bottom, they observed, these orbits tend to be flattened.

As increasingly bolder workers contributed ideas in the 20th century, many of the complexities of natural waves found their way into equations. However, these gave only a crude, empirical answer to the question of how wind energy is transferred to waves. The necessity for the prediction of waves and surf for amphibious operations in World War II attracted the attention of Harald U.

Sverdrup and Walter Munk of the Scripps Institution of Oceanography. As a result of their wartime studies of the interaction of winds and waves they were the first investigators to give a reasonably complete quantitative description of how wind gets energy into the waves. With this description wave studies seemed to come of age, and a new era of research was launched.

Let us follow waves as they are generated at sea by the wind, travel for perhaps thousands of miles across the ocean and finally break against the shore. The effectiveness of the wind in making waves is due to three factors: its average velocity, the length of time it blows and the extent of the open water across which it blows (called the fetch).

Waves and the Wind

Waves start up when the frictional drag of a breeze on a calm sea creates ripples. As the wind continues to blow, the steep side of each ripple presents a surface against which the moving air can press directly. Because winds are by nature turbulent and gusty, wavelets of all sizes are at first created. The small, steep ones break, forming whitecaps, releasing some of their energy in turbulence and possibly contributing part of it to larger waves that overtake them. Thus as energy is added by the wind the smaller waves continually give way to larger ones which can store the energy better. But more small waves are continually formed, and in the zone where the wind moves faster than the waves there is a wide spectrum of wavelengths. This is the generating area, and in a large storm it may cover thousands of square miles. If storm winds apply more force than a wave can accept, the crest is merely steepened and blown off, forming a breaking wave at sea. This happens when the wave crest becomes a wedge of less than 120 degrees and the height of the wave is about a seventh of its length. Thus a long wave can accept more energy from the wind and rise much higher than a short wave passing under the same wind. When the wind produces waves of many lengths, the shortest ones reach maximum height quickly and then are destroyed, while the longer ones continue to grow.

A simple, regular wave-train can be described by its period (the time it takes two successive crests to pass a point), by its wavelength (the distance between crests) and by its height (the vertical distance between a trough and a succeeding crest). Usually, however, there are several trains of waves with different wavelengths and directions present at the same time, and their intersection creates a random or a short-crested diamond pattern. Under these conditions no meaningful dimensions can be assigned to wave period and length. Height, however, is important, at least to ships; several crests may coincide and add their heights to produce a very large wave. Fortunately crests are much more likely to coincide with troughs and be canceled out. There is no reason to believe that the seventh wave, or some other arbitrarily numbered wave, will be higher than the rest; that is a myth of the sea.

Since waves in a sea are so infinitely variable, statistical methods must be employed to analyze and describe them. A simple way to describe height, for example, is to speak of significant height—the average height of the highest third of the waves. Another method, devised in 1952 by Willard J. Pierson, Jr., of New York University, employs equations like those that describe random noise in information theory to predict the behavior of ocean waves. Pierson superposes the regular wave-trains of classical theory in such a way as to obtain a mathematically irregular pattern. The result is most conveniently described in terms of energy spectra. This scheme assigns a value for the square of the wave height to each

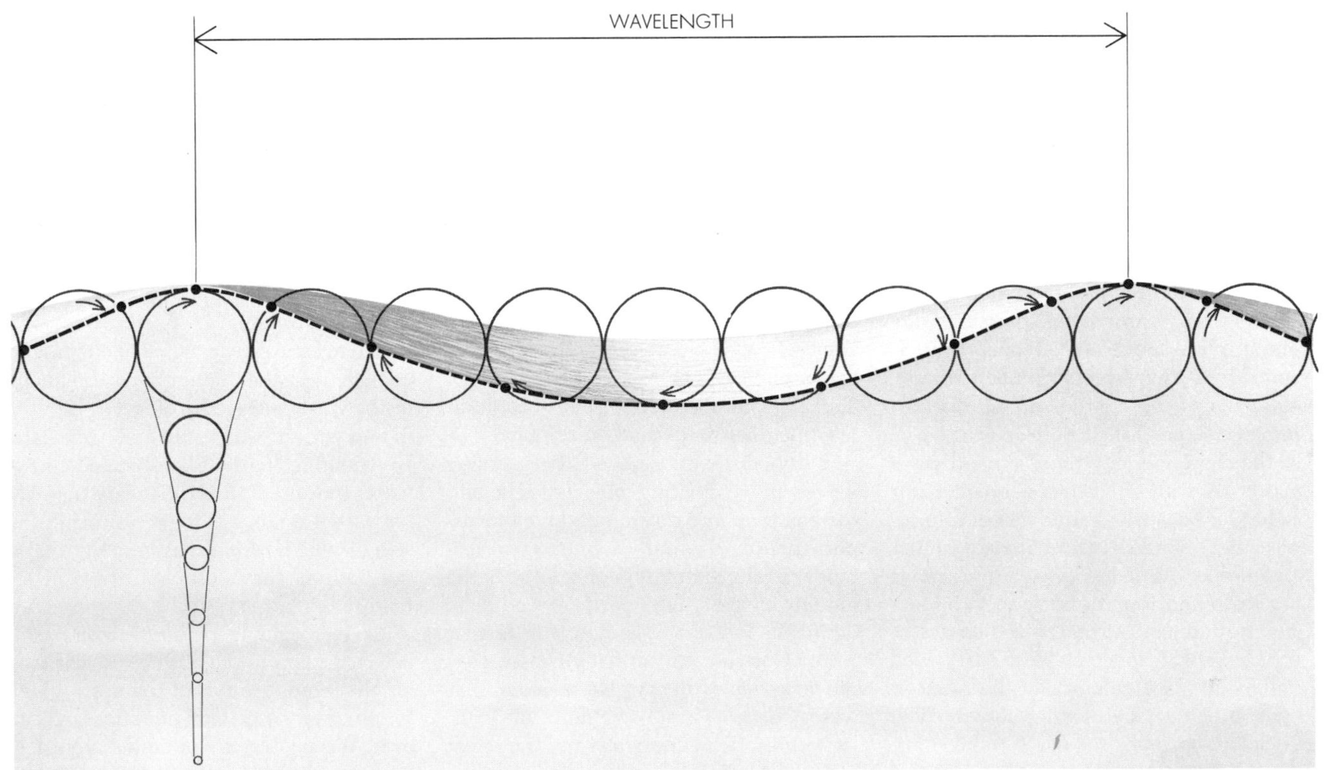

CROSS SECTION OF OCEAN WAVE traveling from left to right shows wavelength as distance between successive crests. The time it takes two crests to pass a point is the wave period. Circles are orbits of water particles in the wave. At the surface their diameter equals the wave height. At a depth of half the wavelength (*left*), orbital diameter is only 4 per cent of that at surface.

frequency and direction. Then, by determining the portion of the spectrum in which most of the energy is concentrated, the average periods and lengths can be obtained for use in wave forecasting.

Over a long fetch, and under a strong, steady wind, the longer waves predominate. It is in such areas of sea that the largest wind waves have been recorded. The height of the waves in a train does not, however, bear any simple relationship to their other two dimensions: the period and the wavelength. The mariner's rule of thumb relates wave height to wind velocity and says that the height ordinarily will not be greater than half the wind speed. This means that an 80-mile-per-hour hurricane would produce waves about 40 feet high.

The question of just how large individual waves at sea can actually be is still unsettled, because observations are difficult to make and substantiate from shipboard in the midst of a violent storm. Vaughan Cornish of England spent half a century collecting data on waves, and concluded that storm waves over 45 feet high are rather common. Much higher waves have been fairly well authenticated on at least two occasions.

In October, 1921, Captain Wilson of the 12,000-ton S.S. *Ascanius* reported an extended storm in which the recording barometer went off the low end of the scale. When the ship was in a trough on an even keel, his observation post on the ship was 60 feet above the water level, and he was certain that some of the waves that obscured the horizon were at least 10 feet higher than he was, accounting for a total height of 70 feet or more. Commodore Hayes of the S.S. *Majestic* reported in February, 1923, that his ship had experienced winds of hurricane force and waves of 80 feet in height. Cornish examined the ship, closely interrogated the officers and concluded that waves 60 to 90 feet high, with an average height of 75 feet, had indeed been witnessed.

A wave reported by Lieutenant Commander R. P. Whitemarsh in the *Proceedings of the U. S. Naval Institute* tops all others. On February 7, 1933, the U.S.S. *Ramapo*, a Navy tanker 478 feet long, was en route from Manila to San Diego when it encountered "a disturbance that was not localized like a typhoon . . . but permitted an unobstructed fetch of thousands of miles." The barometer fell to 29.29 inches and the wind gradually rose from 30 to 60 knots over several days. "We were running directly downwind and with the sea. It would have been disastrous to have steamed on any other course." From among a number of separately determined observations, that of the watch officer on the bridge was selected as the most accurate. He declared that he "saw seas astern at a level above the mainmast crow's-nest and at the moment of observation the horizon was hidden from view by the waves approaching the stern." On working out the geometry of the situation from the ship's plan, Whitemarsh found that this wave must have been at least 112 feet high [*see illustration at the bottom of the next two pages*]. The period of these waves was clocked at 14.8 seconds and their velocity at 55 knots.

As waves move out from under the winds that raise them, their character changes. The crests become lower and more rounded, the form more symmetrical, and they move in trains of similar period and height. They are now called swell, or sometimes ground swell, and in this form they can travel for thousands of miles to distant shores. Happily for mathematicians, swell coincides much more closely with classical theory than do the waves in a rough sea, and this renews their faith in the basic equations.

Curiously enough, although each wave moves forward with a velocity

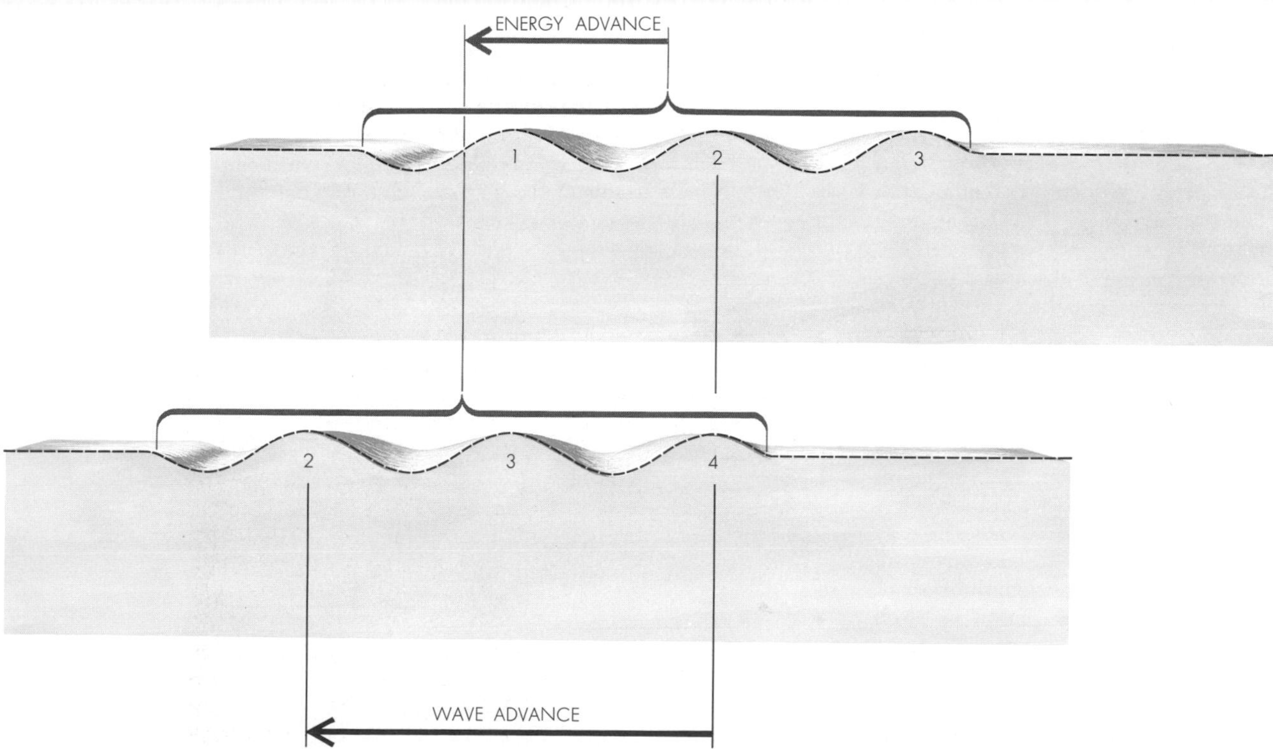

MOVING TRAIN OF WAVES advances at only half the speed of its individual waves. At top is a wave train in its first position. At bottom the train, and its energy, have moved only half as far as wave 2 has. Meanwhile wave 1 has died, but wave 4 has formed at the rear of the train to replace it. Waves arriving at shore are thus remote descendants of waves originally generated.

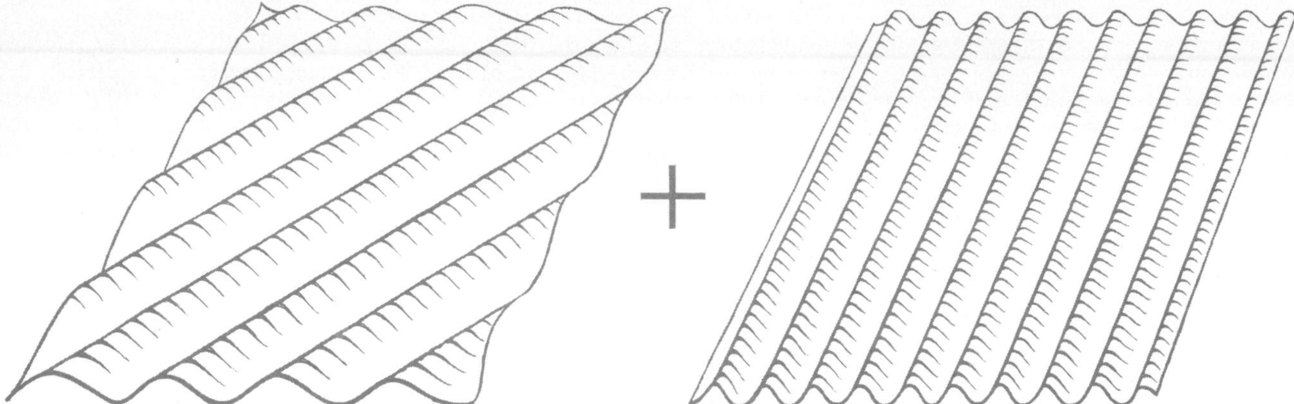

DIFFERENT TRAINS OF WAVES, caused by winds of different directions and strengths, make up the surface of a "sea." The various trains, three of which are represented diagrammatically here, have a wide spectrum of wavelengths, heights and directions. When

that corresponds to its length, the energy of the group moves with a velocity only half that of the individual waves. This is because the waves at the front of a group lose energy to those behind, and gradually disappear while new waves form at the rear of the group. Thus the composition of the group continually changes, and the swells at a distance are but remote descendants of the waves created in the storm [*see illustration on preceding page*]. One can measure the period at the shore and obtain from this a correct value for the wave velocity; however, the energy of the wave train traveled from the storm at only half that speed.

Waves in a swell in the open ocean are called surface waves, which are defined as those moving in water deeper than half the wavelength. Here the bottom has little or no effect on the waves because the water-particle orbits diminish so rapidly with depth that at a depth of half the wavelength the orbits are only 4 per cent as large as those at the surface. Surface waves move at a speed in miles per hour roughly equal to 3.5 times the period in seconds. Thus a wave with a period of 10 seconds will travel about 35 miles per hour. This is the average period of the swell reaching U. S. shores, the period being somewhat longer in the Pacific than the Atlantic. The simple relationship between period and wavelength (length$=5.12T^2$) makes it easy to calculate that a 10-second wave will have a deep-water wavelength of about 512 feet. The longest period of swell ever reported is 22.5 seconds, which corresponds to a wavelength of around 2,600 feet and a speed of 78 miles per hour.

Waves and the Shore

As the waves approach shore they reach water shallower than half their wavelength. Here their velocity is controlled by the depth of the water, and they are now called shallow-water waves. Wavelength decreases, height increases and speed is reduced; only the period is unchanged. The shallow bottom greatly modifies the waves. First, it refracts them, that is, it bends the wave fronts to approximate the shape of the underwater contours. Second, when the water becomes critically shallow, the waves break [*see illustration on page 126*].

Even the most casual observer soon notices the process of refraction. He sees that the larger waves always come in nearly parallel to the shoreline, even though a little way out at sea they seem to be approaching at an angle. This is the result of wave refraction, and it has considerable geological importance because its effect is to distribute wave energy in such a way as to straighten coastlines. Near a headland the part of the wave front that reaches shallow water first is slowed down, and the parts of it in relatively deep water continue to move rapidly. The wave thus bends to converge on the headland from all sides. As it does, the energy is concentrated in less length of crest; consequently the height of the crest is increased. This accounts for the old sailors' saying: "The points draw the waves."

Another segment of the same swell will enter an embayment and the wave front will become elongated so that the height of the waves at any point along the shore is correspondingly low. This is why bays make quiet anchorages and exposed promontories are subject to wave battering and erosion—all by the same waves. One can deal quantitatively with this characteristic of waves and can plot the advance of any wave across waters of known depths. Engineers planning shoreline structures such as jetties or piers customarily draw refraction diagrams to determine in advance the effect of waves of various periods and direction. These diagrams show successive

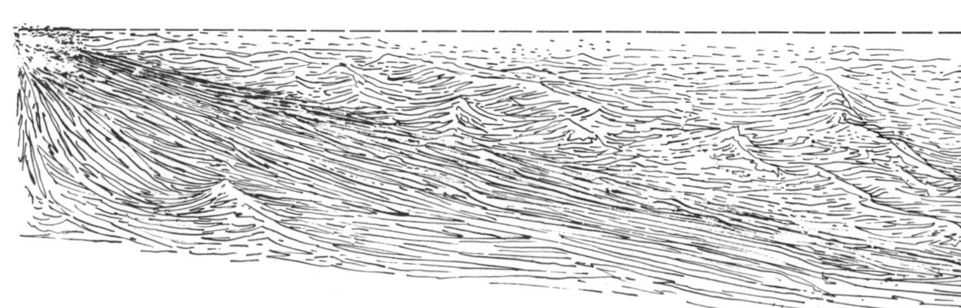

WAVE 112 FEET HIGH, possibly the largest ever measured in the open sea, was encountered in the Pacific in 1933 by the U.S.S. *Ramapo*, a Navy tanker. This diagram shows

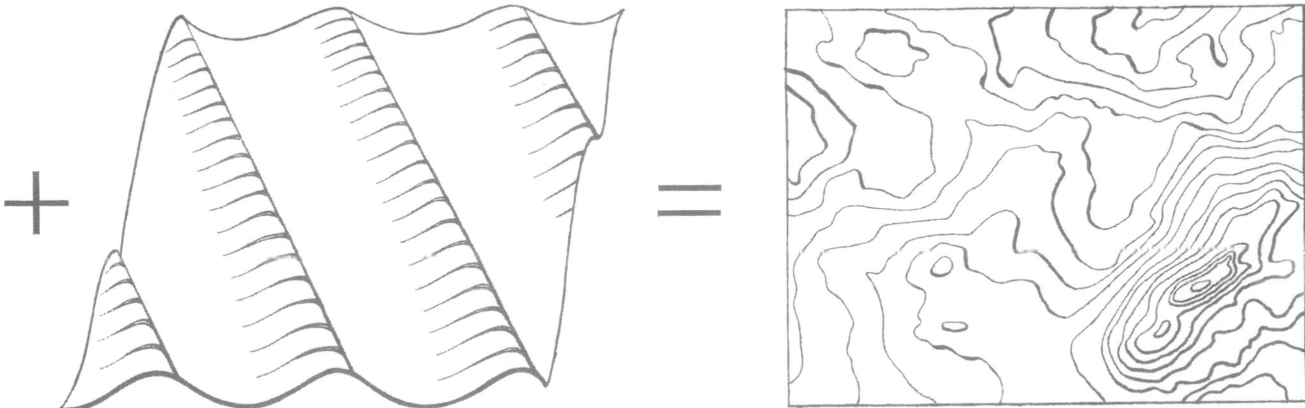

they meet, the result is apparent confusion, represented at far right by a topographic diagram drawn from actual photographs of the sea surface. The pattern becomes so complex that statistical methods must be used to analyze the waves and predict their height.

positions of the wave front, partitioned by orthogonals into zones representing equal wave energy [*see illustration on next page*]. The ratio of the distances between such zones out at sea and at the shore is the refraction coefficient, a convenient means of comparing energy relationships.

Refraction studies must take into account surprisingly small underwater irregularities. For example, after the Long Beach, Calif., breakwater had withstood wave attack for years, a short segment of it was suddenly wrecked by waves from a moderate storm in 1930. The breakwater was repaired, but in 1939 waves breached it again. A refraction study by Paul Horrer of the Scripps Institution of Oceanography revealed that long-period swell from exactly 165 degrees (south-southeast), which was present on only these two occasions, had been focused at the breach by a small hump on the bottom, 250 feet deep and more than seven miles out at sea. The hump had acted as a lens to increase the wave heights to 3.5 times average at the point of damage.

During World War II it was necessary to determine the depth of water off enemy-held beaches against which amphibious landings were planned. Our scientists reversed the normal procedure for refraction studies; by analyzing a carefully timed series of aerial photographs for the changes in length (or velocity) and direction of waves approaching a beach, they were able to map the underwater topography.

The final transformation of normal swell by shoal or shallow water into a breaker is an exciting step. The waves have been shortened and steepened in the final approach because the bottom has squeezed the circular orbital motion of the particles into a tilted ellipse; the particle velocity in the crest increases and the waves peak up as they rush landward. Finally the front of the crest is unsupported and it collapses into the trough. The wave has broken and the orbits exist no more. The result is surf.

If the water continues to get shallower, the broken wave becomes a foam line, a turbulent mass of aerated water. However, if the broken wave passes into deeper water, as it does after breaking on a bar, it can form again with a lesser height that represents the loss of energy in breaking. Then it too will break as it moves into a depth critical to its new height.

The depth of water beneath a breaker, measured down from the still-water level, is at the moment of breaking about 1.3 times the height of the breaker. To estimate the height of a breaker even though it is well offshore, one walks from the top of the beach down until the crest of the breaking wave is seen aligned with the horizon. The vertical distance between the eye and the lowest point to which the water retreats on the face of the beach is then equal to the height of the wave.

The steepness of the bottom influences

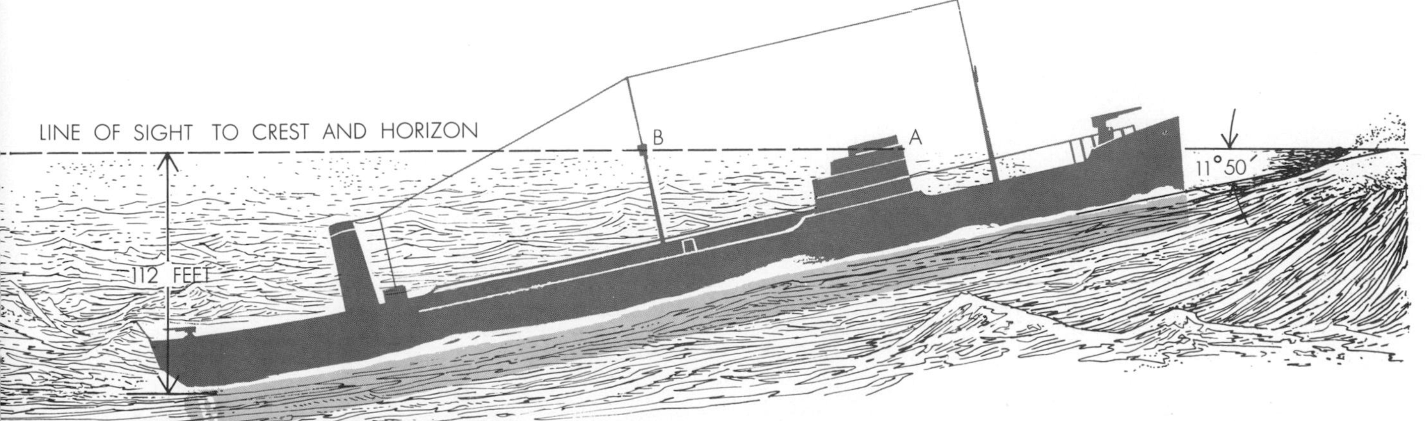

how the great wave was measured. An observer at A on the bridge was looking toward the stern and saw the crow's-nest at B in his line of sight to crest of wave, which had just come in line with horizon. From geometry of situation, wave height was calculated.

the character of the breakers. When a large swell is forced by an abrupt underwater slope to give up its energy rapidly, it forms plunging breakers—violent waves that curl far over, flinging the crest into the trough ahead. Sometimes, the air trapped by the collapsing wave is compressed and explodes with a great roar in a geyser of water [*see illustration on page 124*]. However, if the bottom slope is long and gentle, as at Waikiki in Hawaii, the crest forms a spilling breaker, a line of foam that tumbles down the front of the partly broken wave as it continues to move shoreward.

Since waves are a very effective mechanism for transporting energy against a coast, they are also effective in doing great damage. Captain D. D. Gaillard of the U. S. Army Corps of Engineers devoted his career to studying the forces of waves on engineering structures and in 1904 reported some remarkable examples of their destructive power. At Cherbourg, France, a breakwater was composed of large rocks and capped with a wall 20 feet high. Storm waves hurled 7,000-pound stones over the wall and moved 65-ton concrete blocks 60 feet. At Tillamook Rock Light off the Oregon coast, where severe storms are commonplace, a heavy steel grating now protects the lighthouse beacon, which is 139 feet above low water. This is necessary because rocks hurled up by the waves have broken the beacon several times. On one occasion a rock weighing 135 pounds was thrown well above the lighthouse-keeper's house, the floor of which is 91 feet above the water, and fell back through the roof to wreck the interior.

At Wick, Scotland, the end of the breakwater was capped by an 800-ton block of concrete that was secured to the foundation by iron rods 3.5 inches in diameter. In a great storm in 1872 the designer of the breakwater watched in amazement from a nearby cliff as both cap and foundation, weighing a total of 1,350 tons, were removed as a unit and deposited in the water that the wall was supposed to protect. He rebuilt the structure and added a larger cap weighing 2,600 tons, which was treated similarly by a storm a few years later. There is no record of whether he kept his job

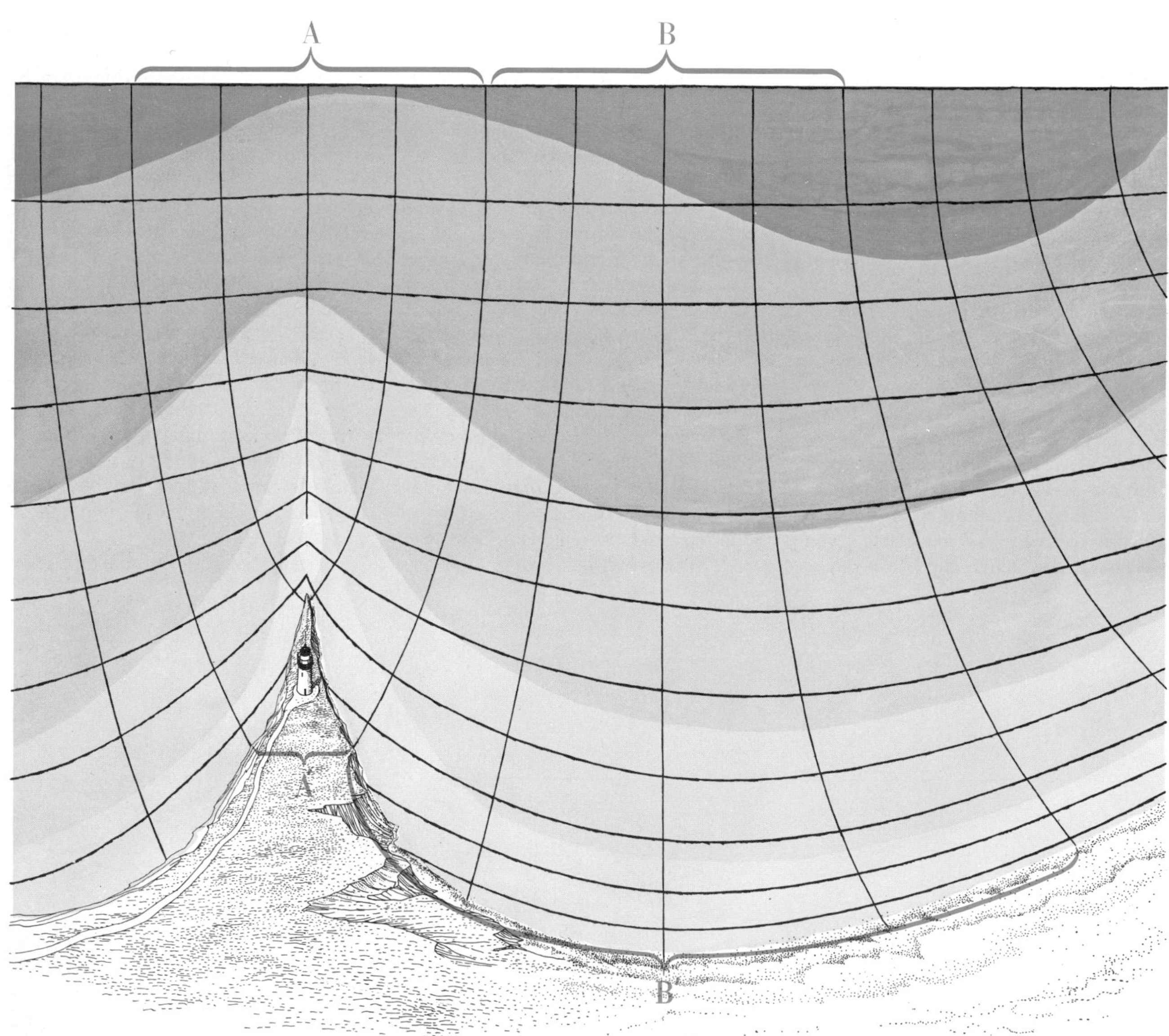

WAVE-REFRACTION DIAGRAM shows how energy of wave front at A is all concentrated by refraction at A′ around small headland area. Same energy at B enters a bay but is spread at beach over wide area B′. Horizontal lines are wave fronts; vertical lines divide energy into equal units for purposes of investigation. Such studies are vital preliminaries to design of shoreline structures.

DIFFRACTION OF OCEAN WAVES is clearly visible in this aerial photograph of Morro Bay, Calif. The waves are diffracted as they pass the end of the lower jetty. Variations in the way the waves break are caused by contours of the shore and the bottom.

and tried again. Gaillard's computations show that the wave forces must have been 6,340 pounds per square foot.

Tsunamis

Even more destructive than wind-generated waves are those generated by a sudden impulse such as an underwater earthquake, landslide or volcano. A man-made variation of the sudden impulse is the explosion of nuclear bombs at the surface of the sea, which in recent years have become large enough to be reckoned with as possible causes of destructive waves.

The public knows such waves as tidal waves, although they are in no way related to the tides and the implication has long irritated oceanographers. It was proposed that the difficulty could be resolved by adopting the Japanese word *tsunami.* Some time later it was discovered that Japanese oceanographers are equally irritated by this word; in literal translation tsunami means tidal wave! However, tsunami has become the favored usage for seismic sea waves.

Like the plunger in a wave channel, the rapid motion or subsidence of a part of the sea bottom can set a train of waves in motion. Once started, these waves travel great distances at high velocity with little loss of energy. Although their height in deep water is only a few feet, on entering shallow water they are able to rise to great heights to smash and inundate shore areas. Their height depends almost entirely on the configuration of the coastline and the nearby underwater contours. Tsunamis have periods of more than 15 minutes and wavelengths of several hundred miles. Since the depth of water is very much less than half the wavelength, they are regarded as long- or shallow-water waves, even in the 13,000-foot average depth of the open ocean, and their velocity is limited by the depth to something like 450 miles per hour.

These fast waves of great destructive potential give no warning except that the disturbance that causes them can be detected by a seismograph. The U. S. Coast Guard operates a tsunami warning network in the Pacific that tracks all earthquakes, and when triangulation indicates that a quake has occurred at sea, it issues alerts. The network also has devices to detect changes in wave period which may indicate that seismic waves are passing [see the article "Tsunamis," by Joseph Bernstein; Scientific American, August 1954]. Curiously the influence of the system may not be entirely beneficial. Once when an alert was broadcast at Honolulu, thousands of people there dashed down to the beach to see what luckily turned out to be a very small wave.

WAVE-CREATED "GEYSER" results when large breakers smash into a very steep beach. They curl over and collapse, trapping and compressing air. This compressed air then explodes as shown here, with spray from a 12-foot breaker leaping 50 feet into the air.

Certain coasts near zones of unrest in the earth's crust are particularly prone to such destructive waves, especially the shores of the Mediterranean, the Caribbean and the west coast of Asia. On the world-wide scale, they occur more frequently than is generally supposed: nearly once a year.

A well-known seismic sea wave, thoroughly documented by the Royal Society of London, originated with the eruption of the volcano Krakatoa in the East Indies on August 27, 1883. It is not certain whether the waves were caused by the submarine explosion, the violent movements of the sea bottom, the rush of water into the great cavity, or the dropping back into the water of nearly a cubic mile of rock, but the waves were monumental. Their period close to the disturbance was two hours, and at great distances about one hour. Waves at least 100 feet high swept away the town of Merak, 33 miles from the volcano; on the opposite shore the waves carried the man-of-war *Berow* 1.8 miles inland and left it 30 feet above the level of the sea. Some 36,380 people died by the waves in a few hours. Tide gauges in South Africa (4,690 miles from Krakatoa), Cape Horn (7,820 miles) and Panama (11,470 miles) clearly traced the progress of a train of about a dozen waves, and showed that their speed across the Indian Ocean had been between 350 and 450 miles per hour.

A tsunami on April 1, 1946, originating with a landslide in the Aleutian submarine trench, produced similar effects,

HUNDRED-FOOT "TIDAL WAVE," or tsunami, wrought impressive destruction at Scotch Cap, Alaska, in 1946. Reinforced concrete lighthouse that appears in top photograph was demolished, as shown in lower photograph, which was made from a higher angle. Atop the plateau a radio mast, its foundation 103 feet above sea, was also knocked down. Lighthouse debris was on plateau. Same tsunami, started by an Aleutian Island earthquake, hit Hawaiian Islands, South America and islands 4,000 miles away in Oceania.

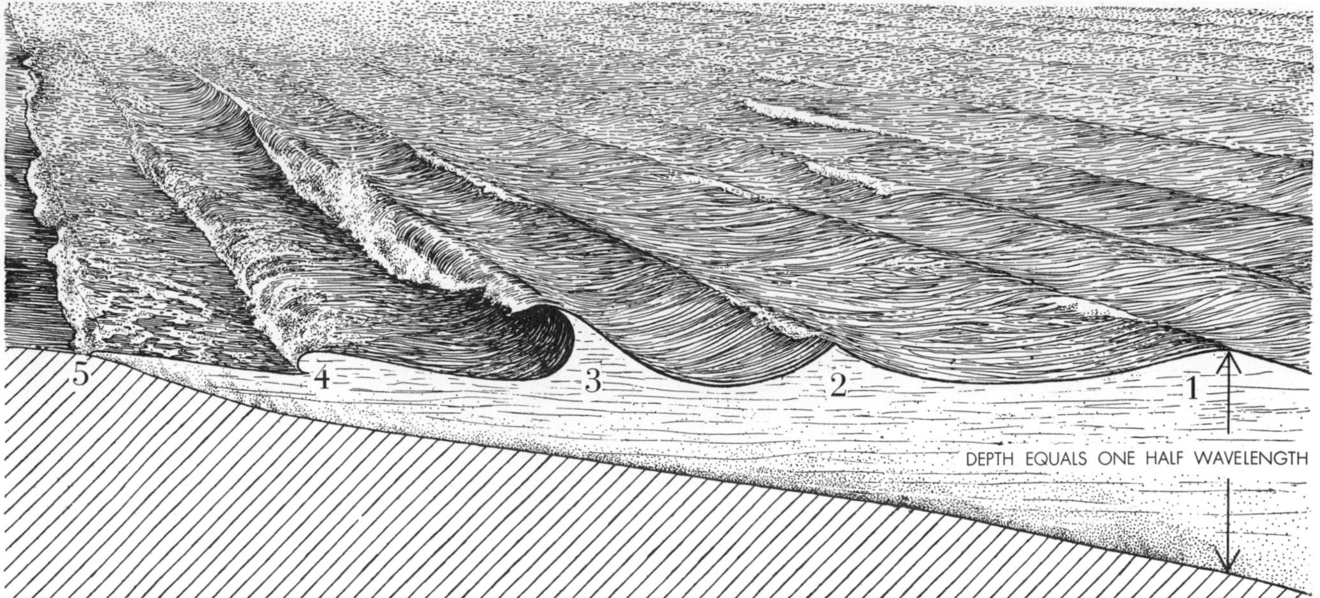

WAVE BREAKS UP at the beach when swell moves into water shallower than half the wavelength (1). The shallow bottom raises wave height and decreases length (2). At a water depth 1.3 times the wave height, water supply is reduced and the particles of water in the crest have no room to complete their cycles; the wave form breaks (3). A foam line forms and water particles, instead of just the wave form, move forward (4). The low remaining wave runs up the face of the beach as a gentle wash called the uprush (5).

fortunately on less-populated shores. It struck hard at the Hawaiian Islands, killing several hundred people and damaging property worth millions of dollars. At Hilo, Hawaii, the tsunami demonstrated that such waves are virtually invisible at sea. The captain of a ship standing off the port was astonished upon looking shoreward to see the harbor and much of the city being demolished by waves he had not noticed passing under his ship. The same waves caused considerable damage throughout the islands of Oceania, 4,000 miles from epicenter, and on the South American coast, but they were most spectacular at Scotch Cap in Alaska. There a two-story reinforced-concrete lighthouse marked a channel through the Aleutian Islands. The building, the base of which was 32 feet above sea level, and a radio mast 100 feet above the sea were reduced to bare foundations by a wave estimated to be more than 100 feet high [*see illustration on preceding page*].

Uncontrollable geologic disturbances will cause many more seismic sea waves in the future, and since the world's coastal population is continuously increasing, the greatest wave disaster is yet to come. Within the next century we can expect that somewhere a wave will at least equal the one that swept the shores of the Bay of Bengal in 1876, leaving 200,000 dead.

Tides and Other Waves

The rhythmic rise and fall of the sea level on a coast indicate the passage of a true wave we call a tide. This wave is driven, as almost everyone knows, by the gravitational influence of the sun and the moon. As these bodies change their relative positions the ocean waters are attracted into a bulge that tends to remain facing the moon as the earth turns under it; a similar bulge travels around the earth on the opposite side. The wave period therefore usually corresponds to half the lunar day.

When the sun and the moon are aligned with the earth, the tides are large (spring tides); when the two bodies are at right angles with respect to the earth, the tides are small (neap tides). By using astronomical data it is possible to predict the tides with considerable accuracy. However, the height and time of the tide at any place not on the open coast are primarily a function of the shape and size of the connection to the ocean.

Still another form of wave is a seiche, a special case of wave reflection. All enclosed bodies of water rock with characteristics related to the size of the basin. The motion is comparable to the sloshing of water in the bathtub when one gets out quickly. In an attempt to return to stability the water sways back and forth with the natural period of the tub (mine has a period of two seconds). Similarly a tsunami or a barometric pressure-change will often set the water in a bay rocking as it passes. In fact, the tsunami itself may reflect back and forth across the ocean as a sort of super-seiche.

In addition to seiches, tides, tsunamis and wind waves there are other waves in the sea. Some travel hundreds of feet beneath the surface along the thermocline, the interface between the cold deep water and the relatively warm surface layer. Of course these waves cannot be seen, but thermometers show that they are there, moving slowly along the boundary between the warm layer and the denser cold water. Their study awaits proper instrumentation. Certain very low waves, with periods of several minutes, issue from storms at sea. These long-period "forerunners" may be caused by the barometric pulsation of the entire storm against the ocean surface. Since they travel at hundreds of miles an hour, they could presumably be used as storm warnings or storm-center locators. Other waves, much longer than tides, with periods of days or weeks and heights of less than an inch, have been discovered by statistical methods and are now an object of study.

The great advances both in wave theory and in the actual measurement of waves at sea have not reduced the need for extensive laboratory studies. The solution of the many complex engineering problems that involve ships, harbors, beaches and shoreline structures requires that waves be simulated under ideal test conditions. Such model studies in advance of expensive construction permit much greater confidence in the designs.

Experimental Tanks

The traditional wave channel in which an endless train of identical small waves is created by an oscillating plunger is still in use, but some of the new wave tanks are much more sophisticated. In some the channel is covered, so that a high velocity draft of air may simulate the wind in making waves. In others, like the large tank at the Stevens Institute of Technology in Hoboken, New Jersey, artificial irregular waves approach the variability of those in the deep ocean. In such tanks proposed ship designs, like those of the America's Cup yacht *Columbia*, are tested at model size to see how they will behave at sea.

The ripple tank, now standard apparatus for teaching physics, has its place in shoreline engineering studies for conveniently modeling diffraction and refraction. Even the fast tsunamis and the very slow waves of the ocean can be modeled in the laboratory. The trick is to use layers of two liquids that do not mix, and create waves on the interface between them. The speeds of the waves can be controlled by adjusting the densities of the liquids.

To reduce the uncertainties in extrapolation from the model to prototype, some of the new wave tanks are very large. The tank of the Beach Erosion Board in Washington, D.C. (630 feet long and 20 feet deep, with a 500-horsepower generator), can subject quarter-scale models of ocean breakwaters to six-foot breakers. The new maneuvering tank now under construction at the David Taylor Model Basin in Carderock, Md., measures 360 by 240 feet, is 35 feet deep along one side and will have wave generators on two sides that can independently produce trains of variable waves. Thus man can almost bring the ocean indoors for study.

The future of wave research seems to lie in refinement of the tools for measuring, statistically examining and reproducing in laboratories the familiar wind waves and swell as well as the more recently discovered varieties. It lies in completing the solution of the problem of wave generation. It lies in the search for forms of ocean waves not yet discovered—some of which may exist only on rare occasions. Nothing less than the complete understanding of all forms of ocean waves must remain the objective of these studies.

The Atmosphere and the Ocean

11

by R. W. Stewart
September 1969

The two are inextricably linked. The ocean's circulation is driven by wind and by density differences that largely depend on the air. The atmospheric heat engine, in turn, is largely driven by the sea

The atmosphere drives the great ocean circulations and strongly affects the properties of seawater; to a large extent the atmosphere in turn owes its nature to and derives its energy from the ocean. Indeed, there are few phenomena of physical oceanography that are not somehow dominated by the atmosphere, and there are few atmospheric phenomena for which the ocean is unimportant. It is therefore hard to know where to start a discussion of the interactions of the atmosphere and the ocean, since in a way everything depends on everything else. One must break into this circle somewhere, and arbitrarily I shall begin by considering some of the effects of wind on ocean water.

When wind blows over water, it exerts a force on the surface in the direction of the wind. The mechanism by which it does so is rather complex and is far from being completely understood, but that it does it is beyond dispute. The ocean's response to this force is immensely complicated by a number of factors. The fact that the earth is rotating is of overriding importance. The presence of continental barriers across the natural directions of flow of the ocean complicates matters further. Finally there is the fact that water is a fluid, not a solid.

To simplify the picture somewhat, let us start by looking at what would happen to a slab of material resting on the surface of the earth. Let us further assume that the slab can move without friction. Consider the result of a sharp, brief impulse that sets the slab moving, say, due north [*see top illustration, page 132*]. Looked at by an observer on a rotating earth, any moving object is subject to a "Coriolis acceleration" directed exactly at a right angle to its motion. The magnitude of the acceleration increases with both the speed of the object's motion and the vertical component of the earth's rotation, and in the Northern Hemisphere it is directed to the right of the motion. An acceleration at right angles to the velocity is just what is required to cause motion in a circle, and in the illustration the center of the circle is due east of the original position of the slab. A circular motion of this kind is called an inertial oscillation, and something of this nature may sometimes happen in the ocean, since inertial oscillations are frequently found when careful observations are made with current meters.

An inertial oscillation requires exactly half a pendulum day for a full circle. (A pendulum day is the time required for a complete revolution of a Foucault pendulum. Like the Coriolis effect, it depends on the vertical component of the earth's rotation and therefore varies with latitude, being just under 24 hours at the poles and increasing to several days close to the Equator. To be precise, it is one sidereal—or star time—day divided by the sine of the latitude.) If there were a small amount of friction, the slab would gradually spiral to the center of the circle. Pushing it toward the north thus causes it to end up displaced to the east [*see bottom illustration on page 132*]. More generally, in the Northern Hemisphere a particle is moved to the right of the direction in which it is impelled, and in the Southern Hemisphere it is moved to the left.

Let us turn to what happens to our frictionless slab if, instead of giving it a short impulse, we give it a steady thrust. Again assume that the force is toward the north [*see upper illustration on page 133*]. Under the influence of this force the slab accelerates toward the north, but as soon as it starts to move it comes under the influence of the Coriolis effect and its motion is deflected (in the Northern Hemisphere) to the right—to the east. As long as the slab has at least some component of velocity toward the north the force will continue to add energy to it and its speed will continue to increase. After a quarter of a pendulum day, however, it will be moving due east. In this position the applied force (which is to the north) is pushing at a right angle to the velocity (east), opposing the influence of the Coriolis effect, which is now trying to turn the slab toward the south.

If there has been no loss of energy because of friction, the slab is moving fast enough so that the Coriolis effect dominates, and it turns toward the south. Now there is a component of velocity opposing the applied force, which acts as a brake and takes energy from the motion. At the end of half a pendulum day the process has gone far enough to bring the slab to a full stop, at which point it is directly east of its starting point. If the force continues, it will again accelerate toward the north and the entire process is repeated, so that the slab performs a series of these looping (cycloidal) motions, each loop taking half a pendulum day to execute. Overall, then, a steady force on a frictionless body resting on a rotating earth causes it to move at right angles to the direction of the force. What is happening is that the force is balanced—on the average—by the Coriolis effect.

Now let us look at the situation when there is a certain amount of friction between the slab and the underlying surface [*see lower illustration on page 133*]. Any frictional drag reduces the speed attained by the slab, reducing the Coriolis effect until it is no longer entirely able to overcome the driving force. As a result if the force is toward the north, the slab will move in a more or less north-

Wind force: 4 Wind speed: 5½ Wave period: 5 Wave height: 1

Wind force: 5 Wind speed: 11½ Wave period: 6 Wave height: 2

Wind force: 6 Wind speed: 13 Wave period: 7 Wave height: 3

Wind force: 8 Wind speed: 18 Wave period: 6 Wave height: 5

Wind force: 9 Wind speed: 21 Wave period: 9 Wave height: 8

Wind force: 10 Wind speed: 27 Wave period: 9 Wave height: 7

easterly direction—more northerly if the friction is large, more easterly if it is small.

A body of water acts much like a set of such slabs, one on top of the other [*see illustration on page 134*]. Each slab is able to move largely independently of the others except for the frictional forces among them. If the top slab is pushed by the wind, it will, in the Northern Hemisphere, move in a direction somewhat to the right of the wind. It will exert a frictional force on the second slab down, which will then be set in motion in a direction still farther to the right. At each successive stage the force is somewhat reduced, so that not only does the direction change but also the speed is a bit less. A succession of such effects produces velocities for which the direction spirals as the depth increases. It is known as the Ekman spiral, after the pioneering Swedish oceanographer V. Walfrid Ekman, who first discussed it soon after the beginning of the century. At a certain depth both the current and the frictional forces associated with it become negligibly small. The entire layer above that depth, in which friction is important, is termed the Ekman layer. Since there is negligible friction between the Ekman layer and the water lying under it, the Ekman layer as a whole behaves like the frictionless slab discussed above: its average velocity must be at a right angle to the wind.

The frictional mechanism, which involves turbulence, has proved to be extraordinarily difficult to study either theoretically or through observations, and surprisingly little is known about it. The surface flow does appear to be somewhat to the right of the wind. Primitive theoretical calculations predict that its direction should be 45 degees from the wind, but this theory is certainly inapplicable in detail. More complicated theoretical models have been attempted, but since almost nothing is known of the nature of turbulence in the presence of a free surface these models rest on weak ground. An educated guess, supported by rather flimsy observational evidence, suggests that the angle is much smaller, perhaps nearer to 10 degrees. All that seems fairly certain is that the average flow in the Ekman layer must be at a right angle to the wind and that there must be some kind of spiral in the current directions. We also believe the bottom of the Ekman layer lies 100 meters or so deep, within a factor of two or three. Of the details of the spiral, and of the turbulent mechanisms that determine its nature, we know very little indeed.

This Ekman-layer flow has some important fairly direct effects in several parts of the world. For example, along the coasts of California and Peru the presence of coastal mountains tends to deflect the low-level winds so that they blow parallel to the coast. Typically, in each case, they blow toward the Equator, and so the average Ekman flow—to the right off California and to the left off Peru—is offshore. As the surface water is swept away deeper water wells up to replace it. The upwelling water is significantly colder than the sun-warmed surface waters, somewhat to the discomfort of swimmers (and, since it is also well fertilized compared with the surface water, to the advantage of fishermen and birds).

The total amount of flow in the directly driven Ekman layer rarely exceeds a couple of tons per second across each meter of surface. That represents a substantial flow of water, but it is much less than the flow in major ocean currents. These are driven in a different way—also by the wind, but indirectly. To see how this works let us take a look at the North Atlantic [*see bottom illustration on page 135*]. The winds over this ocean, although they vary a good deal from time to time, have a most persistent characteristic: near 45 degrees north latitude or thereabouts the westerlies blow strongly from west to east, and at about 15 degrees the northeast trades blow, with a marked east-to-west component. The induced Ekman flow is to the right in each case, so that in both cases the water is pushed toward the region known as the Sargasso Sea, with its center at 30 degrees north. This "gathering together of the waters" leads not so much to a piling up (the surface level is only about a meter higher at the center than at the edges) as a pushing down.

(If it were not for the continental boundaries, the piling up would be much more important. Because water tends to seek a level, the piled-up water would push north above 30 degrees and south below; the pushing force, like any other force in the Northern Hemisphere, would cause a flow to its right, so that in the northern part of the ocean a strong eastward flow would develop and in the southern part a strong westward one. On the earth as it now exists, however, these east-west flows are blocked by the continents; only in the Southern Ocean, around the Antarctic Continent, is such a flow somewhat free. In the absence of the continents the oceans, like the atmosphere, would be dominated by east-west motion. As it is, only a residue of such motion is possible, and it is the pushing down rather than the piling up of water that is important.)

The downward thrust of the surface waters presses down on the layers of water underneath [*see illustration on page 136*]. For practical purposes water is incompressible, so that pushing it down from the top forces it out at the sides. It must be remembered that this body of underlying water is rotating with the earth. As it is squeezed out laterally its radius of gyration, and therefore its moment of inertia, increases, and so its rate of rotation must slow. If it slows, however, the rotation no longer "fits" the rotation of the underlying earth. There are two possible consequences: either the water can rotate with respect to the earth or it can move to a different latitude where its newly acquired rotation *will* fit. It usually does the latter. Hence a body of water whose rotation has been slowed by being squashed vertically will usually move toward the Equator, where the vertical component of the earth's rotation is smaller; on the other hand, a body whose rotation has been speeded by being bulged up to replace water that has been swept away from the surface will usually move toward the poles.

In the band of water a couple of thousand miles wide along latitude 30 degrees this indirectly wind-driven flow moves water toward the Equator. Of course the regions of the ocean closer to the poles do not become empty of water; somewhere there must be a return flow. The returning water must also attain a rotation that fits the rotation of the underlying earth. If it flows north, it must gain counterclockwise rotation (or lose clockwise rotation). It does this by running in a strong current on the westward side of the ocean, changing its rotation

EFFECT OF WIND on the surface of the sea is shown in a series of photographs made by the Meteorological Service of Canada. Much of the wind's momentum goes into generating waves rather than directly into making currents. The change in the surface as the wind increases is primarily a change in scale, except for the effect of surface tension: the waves break up more, making more whitecaps. For each photograph the wind force is given according to the Beaufort scale; the wind speed is given in meters per second, the wave period in seconds and the wave height in meters. (In the final photograph the waves are only about half as large as they might become if the force-10 wind, which had blown for less than nine hours, were to continue to blow.)

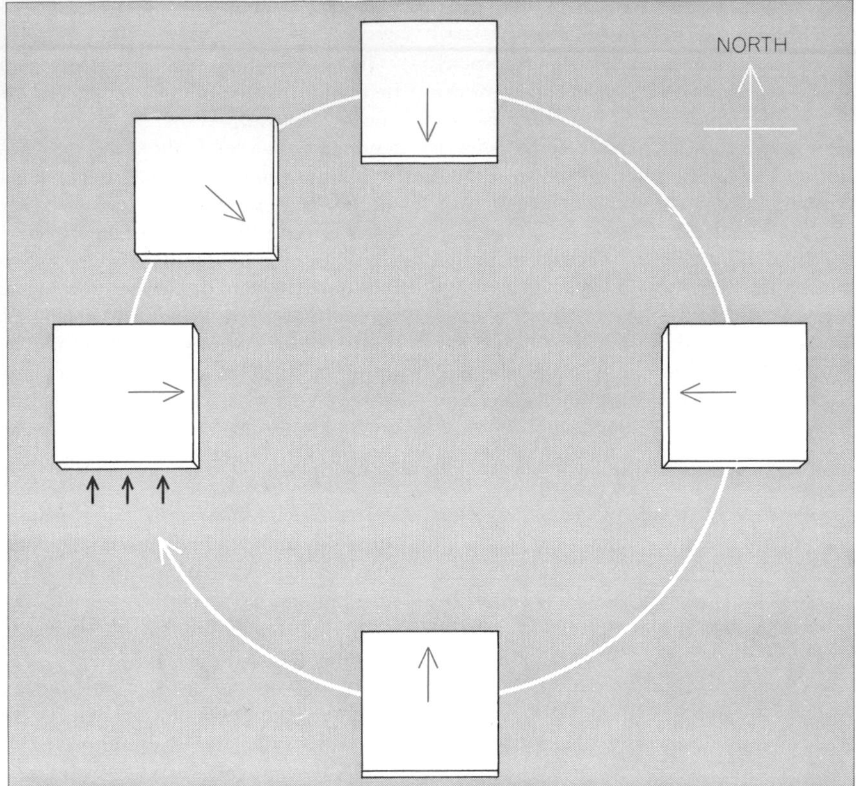

CORIOLIS ACCELERATION, caused by the earth's rotation, affects any object moving on the earth. It is directed at a right angle to the direction of motion (to the right in the Northern Hemisphere). If a frictionless slab is set in motion toward the north by a single impulse (*black arrows*), the Coriolis effect (*colored arrows*) moves the slab in a circle.

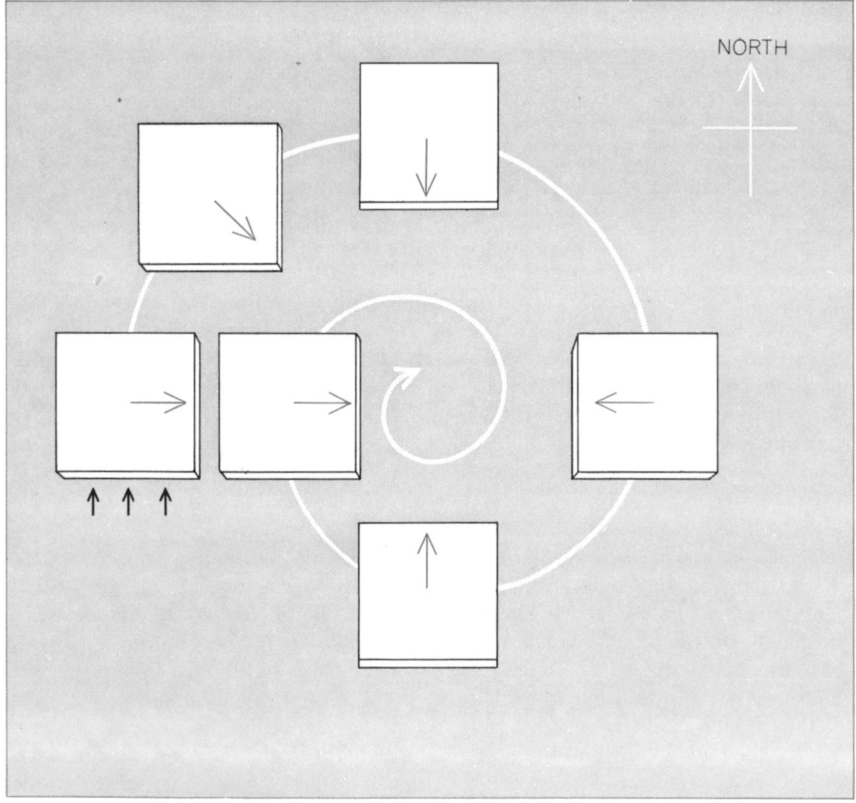

PRESENCE OF FRICTION causes the slab to slow down, spiraling in toward the center of the circle in the top illustration. A push to the north causes a spiral to the east.

by "rubbing its left shoulder" against the shore. The Gulf Stream is such a current; it is the return flow of water that was squeezed south by the wind-driven convergence of surface waters throughout the entire central North Atlantic. Most great ocean currents seem to be indirectly driven in this way.

It is worth noting that these return currents must be on the western side of the oceans (that is, off the eastern coasts of the land) in both hemispheres and regardless of whether the flow is northward or southward. The reason is that the earth's angular velocity of rotation is maximum counterclockwise to an observer looking down at the North Pole and maximum clockwise at the South Pole. Any south-flowing return current in either hemisphere must gain clockwise rotation (or lose counterclockwise rotation) if it is to fit when it arrives. It gains this rotation by friction on its right side, and so it must keep to the right—that is, to the west—of the ocean. On the other hand, a north-flowing return current must keep to the left—again the west!

This description of the general wind-driven circulation accords reasonably well with observations of the long-term characteristics of the ocean circulation. What happens on a shorter term, in response to changes of the atmospheric circulation and the wind-force pattern that results? The characteristic time constant of the Ekman layer is half a pendulum day, and there is every reason to believe this layer adjusts itself within a day or so to changes in the wind field. The indirectly driven flow is much harder to deal with. Its time constant is of the order of years, and we have no clear understanding of how it adjusts; the indirectly driven circulation may still be responding, in ways that are not clear, for years after an atmospheric change.

So far the discussion has been qualitative. To make it quantitative we need to know two things: the nature of the wind over the ocean at each time and place and the amount of force the wind exerts on the surface. Meteorologists are getting better at the first question, although there are some important gaps in our detailed information, notably in the Southern Ocean and in the South Pacific.

Investigation of the second problem, that of the quantitative relation between the wind flow and the force on the surface, is becoming a scientific discipline in its own right. Turbulent flow over a boundary is a complex phenomenon for which there is no really complete theory even in simple laboratory cases. Nevertheless, a great deal of experimental data has been collected on flows over solid surfaces, both in the laboratory and in nature, so that from an engineering point of view at least the situation is fairly well understood. The force exerted on a surface varies with the roughness of that surface and approximately with the square of the wind speed at some fixed height above it. A wind of 10 meters per second (about 20 knots, or 22 miles per hour) measured at a height of 10 meters will produce a force of some 30 tons per square kilometer on a field of mown grass or of about 70 tons per square kilometer on a ripe wheat field. On a really smooth surface such as glass the force is only about 10 tons per square kilometer.

When the wind blows over water, the whole thing is much more complicated. The roughness of the water is not a given characteristic of the surface but depends on the wind itself. Not only that, the elements that constitute the roughness—the waves—themselves move more or less in the direction of the wind. Recent evidence indicates that a large portion of the momentum transferred from the air into the water goes into waves rather than directly into making currents in the water; only as the waves break, or otherwise lose energy, does their momentum become available to generate currents or produce Ekman layers. Waves carry a substantial amount of both energy and momentum (typically about as much as is carried by the wind in a layer about one wavelength thick), and so the wave-generation process is far from negligible. So far we have no theory that accounts in detail for what we observe.

A violently wavy surface belies its appearance by acting, as far as the wind is concerned, as though it were very smooth. At 10 meters per second, recent measurements seem to agree, the force on the surface is quite a lot less than the force over mown grass and scarcely more than it is over glass; some observations in light winds of two or three meters per second indicate that the force on the wavy surface is less than it is on a surface as smooth as glass. In some way the motion of the waves seems to modify the airflow so that air slips over the surface even more freely than it would without the waves. This seems not to be the case at higher wind speeds, above about five meters per second, but the force remains strikingly low compared with that over other natural surfaces.

One serious deficiency is the fact that there are no direct observations at all in those important cases in which the wind speed is greater than about 12 meters per second and has had time and fetch (the distance over water) enough to raise substantial waves. (A wind of even 20 meters per second can raise waves eight or 10 meters high—as high as a three-story building. Making observations under such circumstances with the delicate instruments required is such a formidable task that it is little wonder none have been reported.) Some indirect studies have been made by measuring how water piles up against the shore when driven by the wind, but there are many difficulties and uncertainties in the interpretation of such measurements. Such as they are, they indicate that the apparent roughness of the surface increases somewhat under high-wind conditions, so that the force on the surface increases rather more rapidly than as the square of the wind speed.

Assuming that the force increases at least as the square of the wind speed, it is evident that high-wind conditions produce effects far more important than their frequency of occurrence would suggest. Five hours of 60-knot storm winds will put more momentum into the water than a week of 10-knot breezes. If it should be shown that for high winds the force on the surface increases appreciably more rapidly than as the square of the wind speed, then the transfer of momentum to the ocean will turn out to be dominated by what happens during

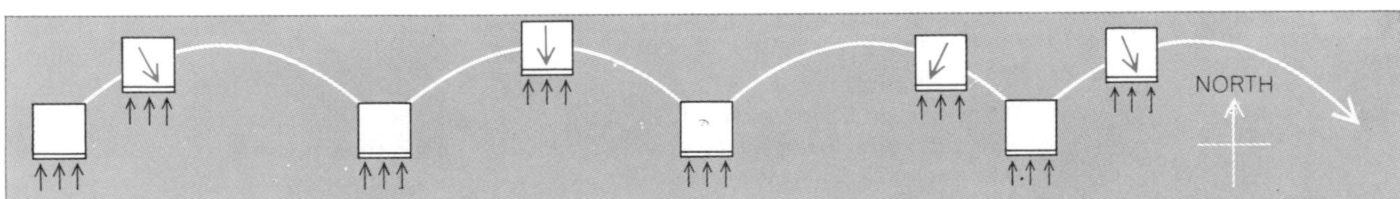

STEADY PUSH (*black arrows*), rather than a single impulse, is balanced, in the absence of friction, by the Coriolis effect (*colored arrows*), causing a series of loops. A steady force on a frictionless slab makes it move at a right angle to the direction of the force.

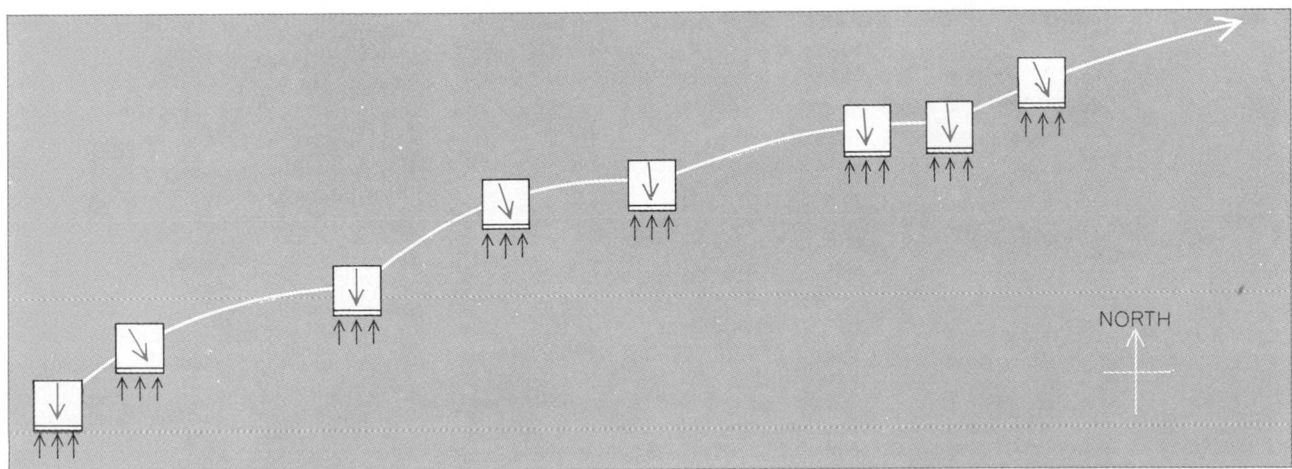

FRICTIONAL DRAG reduces the speed of the slab and thus of the Coriolis effect, which can no longer balance the driving force, and the amplitude of the loops is damped out gradually. A force toward the north therefore moves the slab toward the northeast.

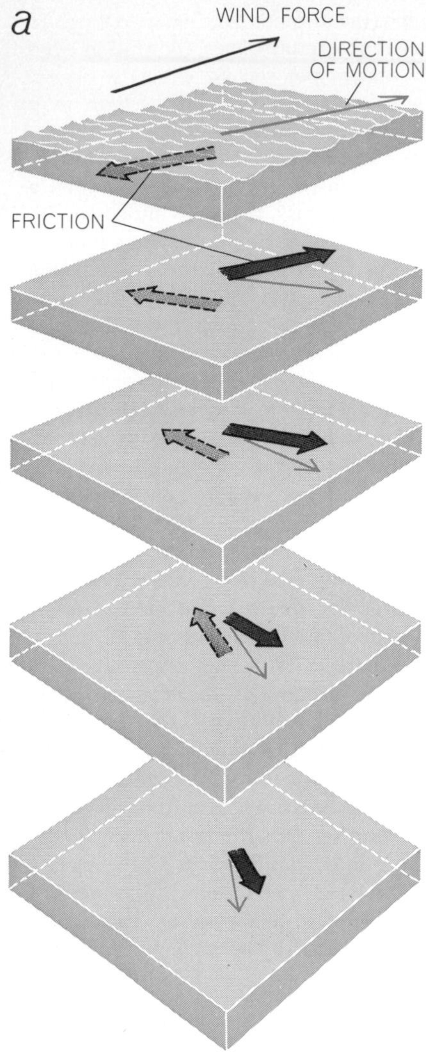

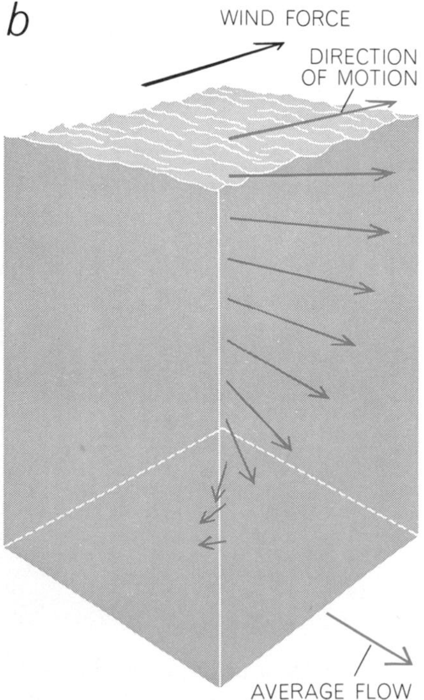

BODY OF WATER can be thought of as a set of slabs (*a*), the top one driven by the wind and each driving the one below it by friction. At each stage the speed of flow is reduced and (in the Northern Hemisphere) directed more to the right. This "Ekman spiral" persists until friction becomes negligible. The "Ekman layer" in which this takes place (*b*) behaves like the frictionless slabs in the preceding illustrations. Its average flow is at right angle to wind driving it.

the occasional storm rather than by the long-term average winds.

It is tempting to try to infer high-wind behavior from what we know about lower wind speeds. Certainly the shapes of wavy surfaces appear nearly the same notwithstanding the size of the waves—as long as one disregards waves less than about five centimeters long, which are strongly affected by surface tension. Yet, curious as it may seem, the only thing that makes one wind-driven wave field different in any fundamental way from another is surface tension, even though it directly affects only these very short waves. Indeed, surface tension is the basis of the entire Beaufort wind scale, which depends on the number and nature of whitecaps; only the fact that the surface tension is better able to hold the surface together at low wind speeds than at high speeds enables us to see a qualitative difference in the nature of the sea surface at different wind speeds [*see illustration on page 130*]. Otherwise the waves would look just the same except for a difference in scale. If we were sure we could ignore surface-tension effects, then we could calculate the force the wind would exert at high wind speeds on the basis of data obtained at lower speeds, but one should be extremely cautious about such calculations, at least until some confirming measurements are available.

Whereas the ocean seems primarily to be driven by surface forces, the atmosphere is a heat engine that makes use of heat received from the sun to develop the mechanical energy of its motion. Any heat engine functions by accepting thermal energy at a comparatively high temperature, discharging some of this thermal energy at a lower temperature and transforming the rest into mechanical energy. The atmosphere does this by absorbing energy at or near its base and radiating it away from much cooler high levels. A substantial proportion of the required heating from below comes from the ocean.

This energy comes in two forms. If cooler air blows over warmer water, there is a direct heat flow into the air. What is usually more important, though, is the evaporation of water from the surface into the air. Evaporation causes cooling, that is, it removes heat, in this case from the surface of the water. When the moisture-laden air is carried to a high altitude, where expansion under reduced atmospheric pressure causes it to cool, the water vapor may recondense into water droplets and the heat that was given up by the surface of the water is transferred to the air. If the cloud that is formed evaporates again, as it sometimes does, the atmosphere gains no net thermal energy. If the water falls to the surface as rain or snow, however, then there has been a net gain and it is available to drive the atmosphere. Typically the heat gained by the atmosphere through this evaporation-condensation process is considerably more than the heat gained by direct thermal transfer through the surface.

Virtually everywhere on the surface of the ocean, averaged over a year, the ocean is a net source of heat to the atmosphere. In some areas the effect is much more marked than in others. For example, some of the most important return currents, such as the Gulf Stream in the western Atlantic and the Kuroshio Current in the western Pacific off Japan, contain very warm water and move so rapidly that the water has not cooled even when it arrives far north of the tropical and subtropical regions where it gained its high temperature. At these northern latitudes the characteristic wind direction is from the west, off the continent. In winter, when the continents are cold, air blowing from them onto this abnormally warm water receives great quantities of heat, both by direct thermal transfer and in the form of water vapor.

The transfer of heat and water vapor depends on a disequilibrium at the interface of the water and the air. Within a millimeter or so of the water the air temperature is not much different from that of the surface water, and the air is nearly saturated with water vapor. The small differences are nevertheless crucial, and the lack of equilibrium is maintained by the mixing of air near the surface with air at higher levels, which is typically appreciably cooler and lower in water-vapor content. The mixing mechanism is a turbulent one, the turbulence gaining its energy from the wind. The higher the wind speed is, the more vigorous the turbulence is and therefore the higher the rates of heat and moisture transfer are. These rates tend to increase linearly with the wind speed, but even less is known about the details of this phenomenon than about the wind force on water. One source of complication is the fact that, as I mentioned above, the wind-to-water transfer of momentum is effected partly by wave-generation mechanisms. When the wind makes waves, it must transfer not only momentum but also important amounts of energy—energy that is not available to provide the turbulence needed to produce the mixing that would effect the transfer of heat and water vapor.

At fairly high wind speeds another phenomenon arises that may be of considerable importance. I mentioned that when surface tension is no longer able to hold the water surface together at high wind speeds, spray droplets blow off the top of the waves. Some of these drops fall back to the surface, but others evaporate and in doing so supply water vapor to the air. They have another important role: The tiny residues of salt that are left over when the droplets of seawater evaporate are small enough and light enough to be carried upward by the turbulent air. They act as nuclei on which condensation may take place, and so they play a role in returning to the atmosphere the heat that is lost in the evaporation process.

The ocean's great effect on climate is illustrated by a comparison of the temperature ranges in three Canadian cities, all at about the same latitude but with very different climates [*see top illustration at right*]. Victoria is a port on the southern tip of Vancouver Island, on the eastern shore of the Pacific Ocean; Winnipeg is in the middle of the North American land mass; St. John's is on the island of Newfoundland, jutting into the western Atlantic. The most striking climatic difference among the three is the enormous temperature range at Winnipeg compared with the two coastal cities. The range at St. John's, although much less, is still greater than at Victoria, probably because at St. John's the air usually blows from the direction of the continent and the effect of the water is somewhat less dominant than at Victoria, which typically receives its air directly from the ocean. St. John's is colder than Victoria because it is surrounded by cold water of the Labrador Current.

The influence of the ocean is associated with its enormous thermal capacity. Every day, on the average, the earth absorbs from the sun and reradiates into space enough heat to raise the temperature of the entire atmosphere nearly two degrees Celsius (three degrees Fahrenheit). Yet the thermal capacity of the atmosphere is equivalent to that of only the top three meters of the ocean, or only a few percent of the 100 meters or so of ocean water that is heated in summer and cooled in winter. (The great bulk of ocean water, more than 95 percent of it, is so deep that surface heating does not penetrate, and its temperature is independent of season.) If the ocean lost its entire heat supply for a day but continued to give up heat in a normal way, the temperature of the upper 100 meters would drop by only about a tenth of a degree.

Compared with the land, the ocean heats slowly in summer and cools slowly in winter, so that its temperature is much less variable. Moreover, because air has so much less thermal capacity, when it blows over water it tends to come to the water temperature rather than vice versa. For these reasons maritime climates are much more equable than continental ones.

Although the ocean affects the atmosphere's temperature more than the atmosphere affects the ocean's, the ocean is cooled when it gives up heat to the atmosphere. The density of ocean water is controlled by two factors, temperature and salinity, and evaporative cooling tends to make the water denser by affecting both factors: it lowers the temperature and, since evaporation removes water but comparatively little salt, it also increases the salinity. If surface water becomes denser than the water underlying it, vigorous vertical convective mixing sets in. In a few places in the ocean the cooling at the surface can be so intense that the water will sink and mix to great depths, sometimes right to the bottom. Such occurrences are rare both in space and in time, but once cold water has reached great depths it is heated from above very slowly, and so it tends to stay deep for a long time with little change in temperature; there is some evidence of water that has remained cold and deep in the ocean for more than

	VICTORIA	WINNIPEG	ST. JOHN'S
MEAN JULY MAXIMUM	68	80.1	68.9
MEAN JANUARY MINIMUM	35.6	−8.1	18.5

MODERATING EFFECT of the ocean on climate is illustrated by a comparison of the temperature range (in degrees Fahrenheit) at three Canadian cities. The range between minimums and maximums is much greater at Winnipeg than at coastal Victoria or St. John's.

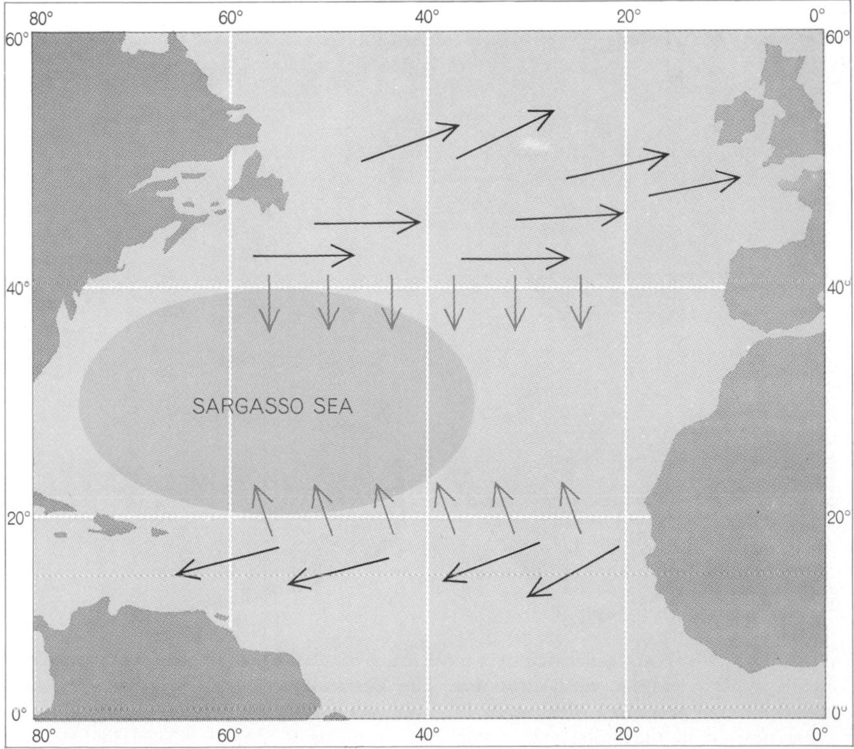

PREVAILING-WIND DIRECTIONS (*black arrows*) and the resulting Ekman-layer flows (*colored arrows*) in the North Atlantic drive water into the region of the Sargasso Sea.

1,000 years. With this length of residence not much of the heavy, cold water needs to be produced every year for it to constitute, as it does, the bulk of the ocean water.

The sinking of water cooled at the surface is one aspect of another important feature of the ocean: the flow induced by differences in density, which is to say the flow induced principally by temperature and salt content. This thermohaline circulation of the ocean is in addition to the wind-driven circulation discussed earlier.

In its thermohaline aspects the ocean itself acts as a heat engine, although it is far less efficient than the atmosphere. Roughly speaking, the ocean can be divided into two layers: a rather thin upper one whose density is comparatively low because it is warmed by the sun, and a thick lower one, a fraction of a percent denser and composed of water only a few degrees above the freezing point that has flowed in from those few areas where it is occasionally created. Somewhere—either distributed over the ocean or perhaps only locally near the shore and in other special places—there is mixing between these layers. The mixing is of such a nature that the cold deep water is mixed into the warm upper water rather than the other way around, that is, the cold water is added to the warm from the bottom [*see upper illustration on page 137*]. Once the water is in the upper layer its motion is largely governed by the wind-driven circulation, although density differences still play a role. In one way or another some of this surface water arrives at a location and time at which it is cooled sufficiently to sink again and thus complete the circulation.

This picture can be rounded out by consideration of the effects of the earth's rotation, which are in some ways quite surprising. The deep water that mixes into the upper layer must have a net upward motion. (The motion is far too

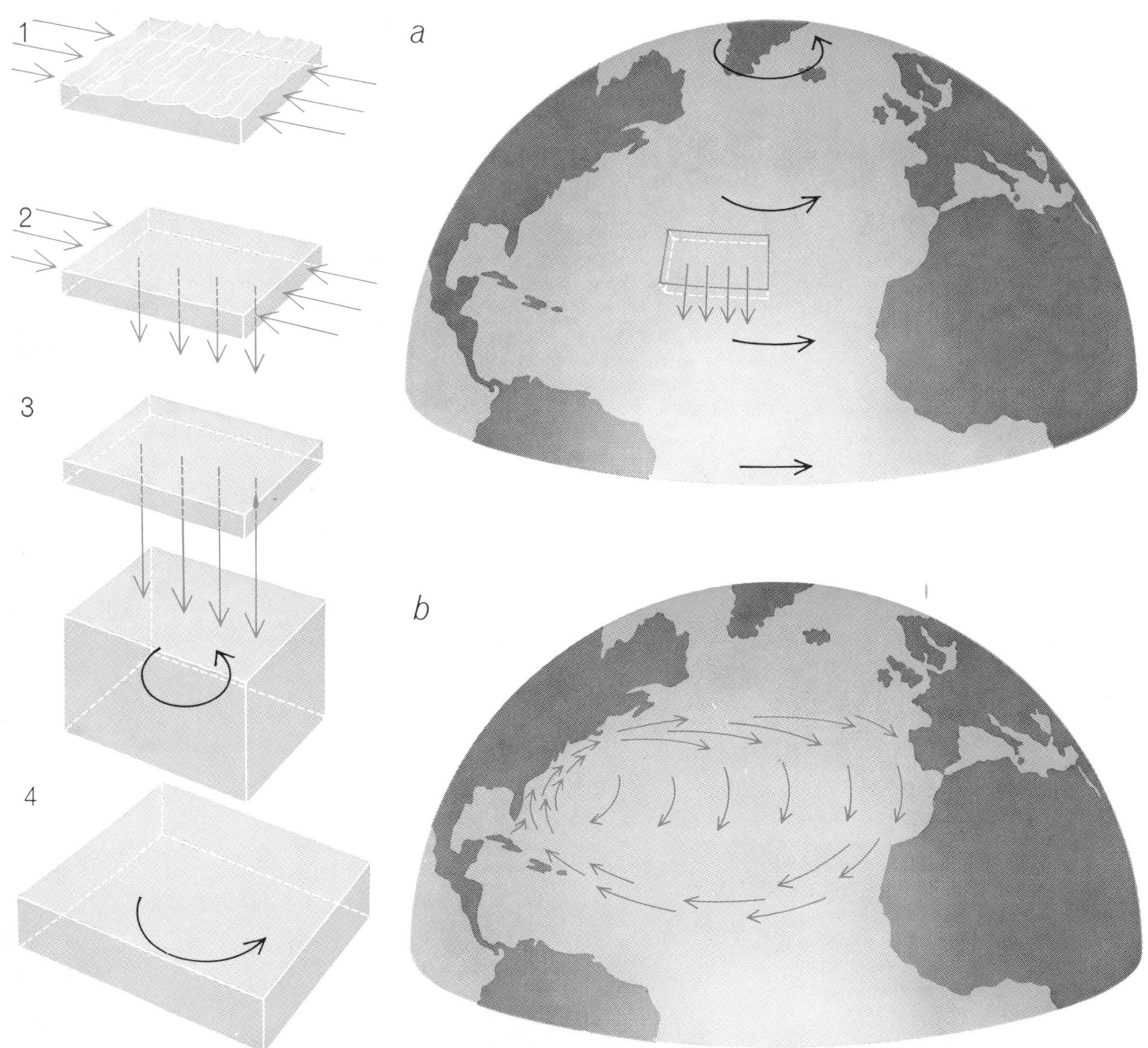

MAJOR CURRENTS are generated by a mechanism involving the Ekman-layer flow and the earth's rotation. The Ekman-layer inflow shown in the preceding illustration (*1*) produces a downflow (*2*) that presses on the underlying water (*3*), squeezing it outward (*4*) and thus reducing its rate of rotation (*curved black arrow*). There is a rate of rotation appropriate to each latitude, and when the rotation of a body of water is reduced, it must move (*colored arrows*) toward the Equator until its new rotation "fits" (*a*). For this reason there is a general movement of water from the mid-latitudes toward the Equator (*b*). That water must be replaced, and the water replacing it must have the proper rotation. This is accomplished by a return flow that runs along the western shore of the ocean, changing its rotation by "rubbing its shoulder" against the coast, as the Gulf Stream does in the Atlantic Ocean.

small to measure, only a few meters per year, but we infer its existence indirectly.) To make possible this upward flow there must be a compensating lateral inflow. Remember that on the rotating earth this lateral inflow results in an increase in speed of rotation, and so for it to continue to fit the rotation of the underlying earth the water must move toward the nearest pole; it must flow away from the equatorial regions. Yet the source of this cold deep water is at high latitudes! How does it get near the Equator to supply the demand?

The answer is similar to the one for the wind-driven circulation: The cold water must flow in a western boundary current, in order to gain the proper rotation as it moves [*see lower illustration on next page*]. There is some direct evidence of the inferred concentrated western boundary current in the North Atlantic, and there are hints of it in the South Pacific, but most of the rest is based on inference. There seems to be no source of cold deep water in the North Pacific, so that the deep water there must come from Antarctic regions.

We have seen that the atmosphere drives the ocean and that heat supplied from the ocean is largely instrumental in releasing energy for the atmo-

THERMOHALINE CIRCULATION, the flow induced by density rather than wind action, begins with the creation of dense, cold water that sinks to great depths. Under certain conditions this deep water mixes upward into the warm surface layer (*color*) as shown here. As it moves up, this water increases its rotation and so it must move generally from the Equator toward the two poles.

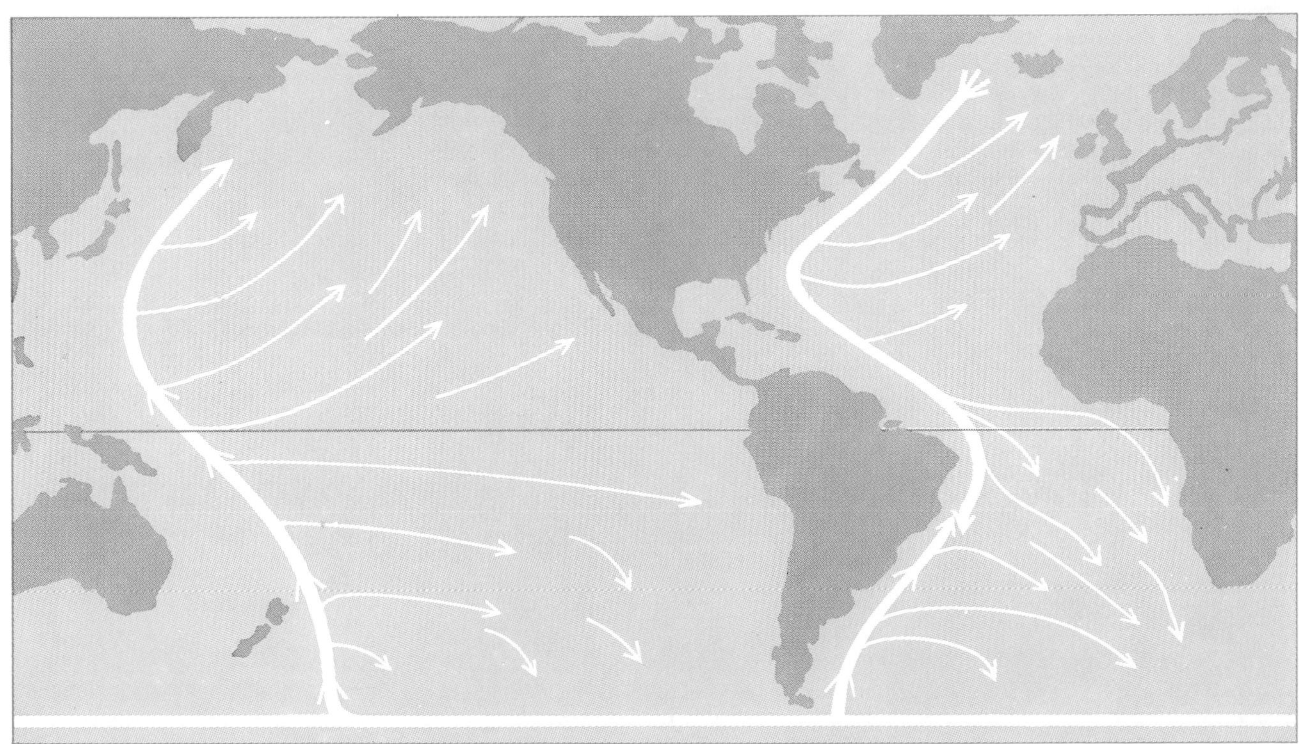

COLD DEEP WATER must flow in western boundary currents in order to arrive at the Equator and thus be able to move poleward as it mixes upward. Details of deep circulation are still almost unknown and the chart is intended only to suggest its approximate directions. There is some evidence of such boundary currents in the North Atlantic and there are some hints in the South Pacific.

INFRARED IMAGERY delineates the temperature structure of bodies of water and is used to study currents and wave patterns. This image of the shoulder of the Gulf Stream is from the Antisubmarine Warfare Environmental Prediction Services Project of the Naval Oceanographic Office. It was made by an airborne scanner at low altitude and shows several hundred yards of the boundary between the warm current and cooler water off Cape Hatteras. The range is from about 13 to 21 degrees Celsius, with the warm water darker.

sphere. There is a great deal of feedback between the two systems. The atmospheric patterns determine the oceanic flows, which in turn influence where—and how much—heat is released to the atmosphere. Further, the atmospheric flow systems determine how much cloud cover there will be over certain parts of the ocean and therefore how much—and where—the ocean will be heated. The system is not a particularly stable one. Every locality has its abnormally cold or mild winters and its abnormally wet or dry summers. The persistence of such anomalies over several months almost certainly involves the ocean, because the characteristic time constants of purely atmospheric phenomena are simply too short. Longer-term climatological variations such as the "little ice age" that lasted for about 40 years near the beginning of the 19th century are even more likely to have involved changes in the ocean's circulation. And then there are the more dramatic events of the great Pleistocene glaciations.

There are any number of theories for these events and, since experts disagree, it is incumbent on the rest of us to refrain from dogmatic statements. Nevertheless, it does not seem impossible that the ocean-atmosphere system has a number of more or less stable configurations. That is, there may be a number of different patterns in which the atmosphere can drive the ocean in such a way that the ocean releases heat to the atmosphere in the right quantity and at the right places to allow the pattern to continue. Of course the atmosphere is extremely turbulent, so that its equilibrium is constantly being disturbed. If the system is stable, then forces must come into play that tend to restore conditions after each such disturbance. If there are a number of different stable patterns, however, it is possible that a particularly large disturbance might tip the system from one stable condition to another.

One can imagine a gambler's die lying on the floor of a truck running over a rough road; the die is stable on any of its six faces, so that in spite of bouncing and vibration the same face usually remains up—until a particularly big bump jars it so that it lands with a different face up, whereupon it is stable in its new position. It seems not at all impossible that the ocean-atmosphere system behaves something like this. Perhaps in recent years we have been bouncing along with, say, a four showing. Perhaps 200 years ago the die flipped over to three for a moment, then flipped back to four. It could one day jounce over to a snake eye and bring a new ice age!

The Circulation of the Abyss

by Henry Stommel
July 1958

Unseen and still largely uncharted currents in the ocean depths are a major force in determining world climate. A new theory indicates that these currents have a surprising pattern of flow

There is a good deal of talk these days about controlling the world's climate. Optimistic promoters of the earth sciences hold out visions of turning tropic deserts or arctic wastes into temperate and fertile plains. Of course we could not hope to do this by brute force. To deflect major wind systems or ocean currents, or to heat the outdoors, would call for engineering works on a scale that man cannot even dream of. But, some people suggest, perhaps if we knew enough about the mechanics of the atmosphere and the ocean circulations we might be able to find some critical time and place where a relatively small man-made disturbance could set off a snowballing reaction which would produce a major alteration in weather patterns. Actually this prospect is quite remote. It does add some spice, however, to a study such as oceanography.

The general circulation of the world's oceans is a matter of great interest, not only from various practical points of view—climate, fishing, dumping of ra-

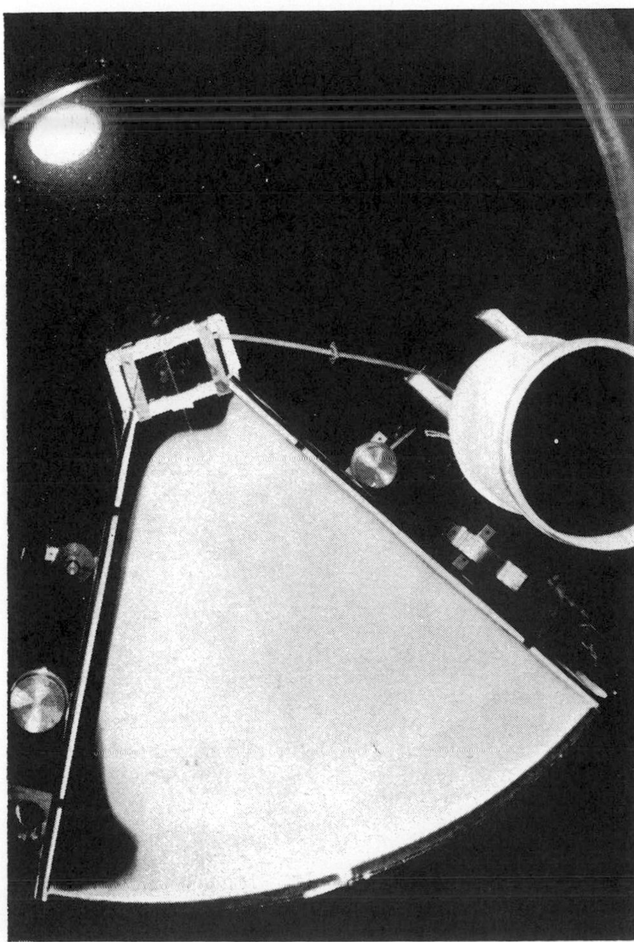

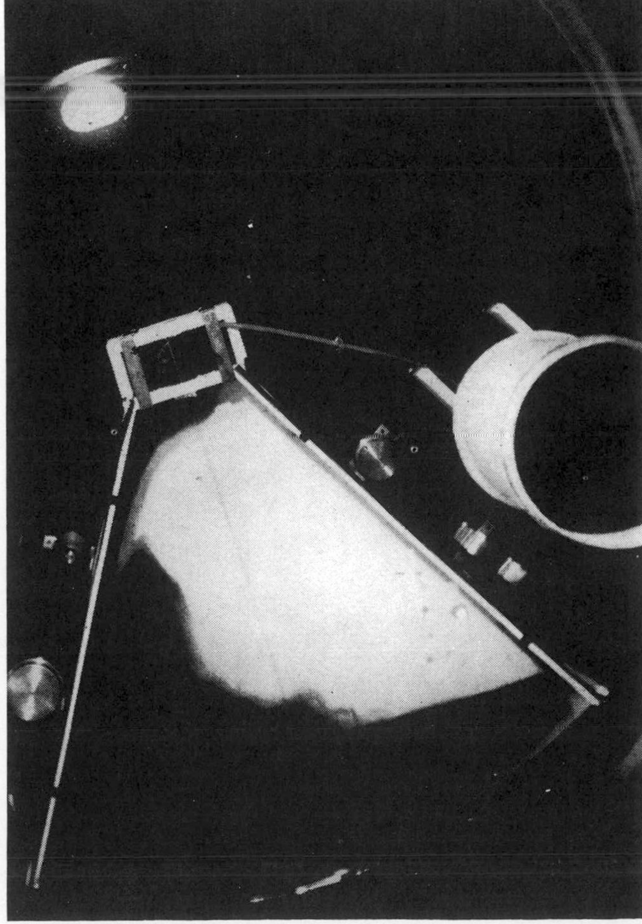

CIRCULATION PATTERNS in the deep ocean are simulated in a rotating, wedge-shaped basin. Ink poured in at the point of the wedge (the "pole") flows down along "western" edge (*left*), turns and flows over the rest of the bottom (*right*) toward the "pole".

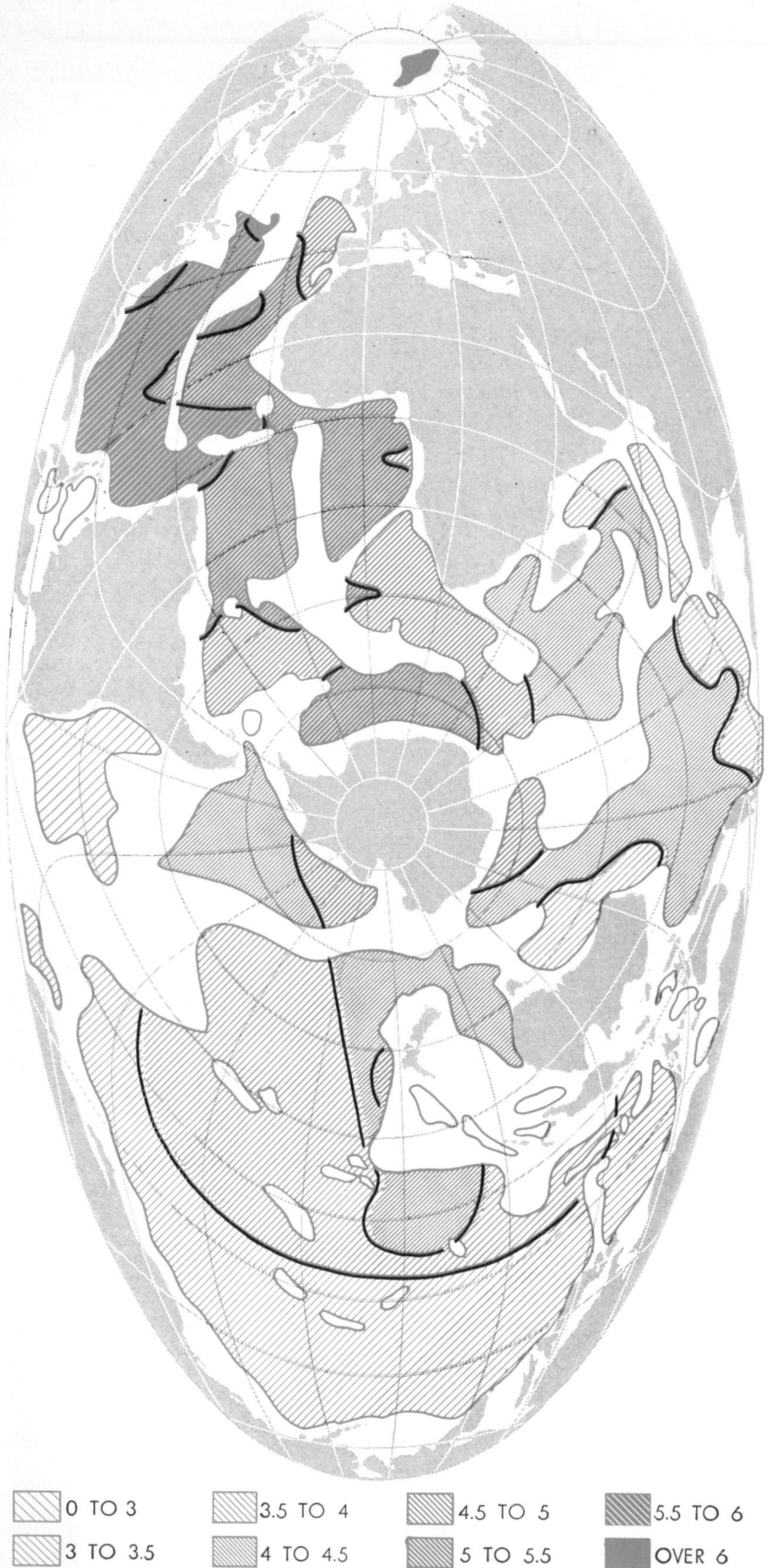

OXYGEN CONCENTRATION at a depth of 4,000 meters (about 13,000 feet) is indicated by colored areas on map. The numbers given in the key are milliliters of oxygen per liter of water. White areas are regions where the ocean bottom is less than 4,000 meters deep. This projection places the North Pole at top and the South Pole near the center of the map.

dioactive wastes and so forth—but primarily from the standpoint of understanding the dynamics and history of the planet on which we live. We know that the great surface currents (*e.g.*, the Gulf Stream) are important to us. Even more significant is the circulation of the oceanic water as a whole. And to get any sort of picture of this circulation we must find out about the movements of water in the ocean deeps. This is difficult to do: we have had no reliable means of getting direct, accurate measurements of the currents or the massive general flow at great depths. Our inferences have to be based mainly on indirect indications such as the comparative densities, salinities and pressures of the water at various places and various depths.

One item of evidence is the amount of dissolved oxygen in the waters of the abyss. Ocean water receives oxygen only at or near the surface—by direct contact with the atmosphere and by the action of photosynthesis in floating plants. After the water sinks to deeper levels it gradually loses some of its dissolved oxygen: the gas is consumed in chemical reactions with dead microorganisms. Therefore the amount of dissolved oxygen in a sample of deep water is a rough index to the "age" of the water since it sank from the surface.

Now when we measure the oxygen content of the deep water in various parts of the oceans, a distinctive pattern emerges [*see map at the left*]. The "youngest" deep waters (richest in oxygen) are found in the western North Atlantic and around the Antarctic Continent. The concentration in the other oceans diminishes with increasing distance from these sources: the water poorest in oxygen is that at the bottom of the vast and little explored basin off Peru. It thus appears that there are only two important regions where substantial amounts of water sink from the surface to the abyss: in the North Atlantic and around Antarctica. This water evidently spreads gradually into the Indian Ocean and finally northward into the Pacific.

The oxygen studies tell us, then, that there is a general sluggish circulation of water along the ocean bottoms from the North Atlantic and the Antarctic to the other oceans. We have another indication of this global flow. The deep water in all the oceans is very cold—only a few degrees above freezing. This is true even in the tropics; only the top 1,500 feet of water is warm. The warm surface water does not mix deeply with the colder water beneath it; this must mean that the

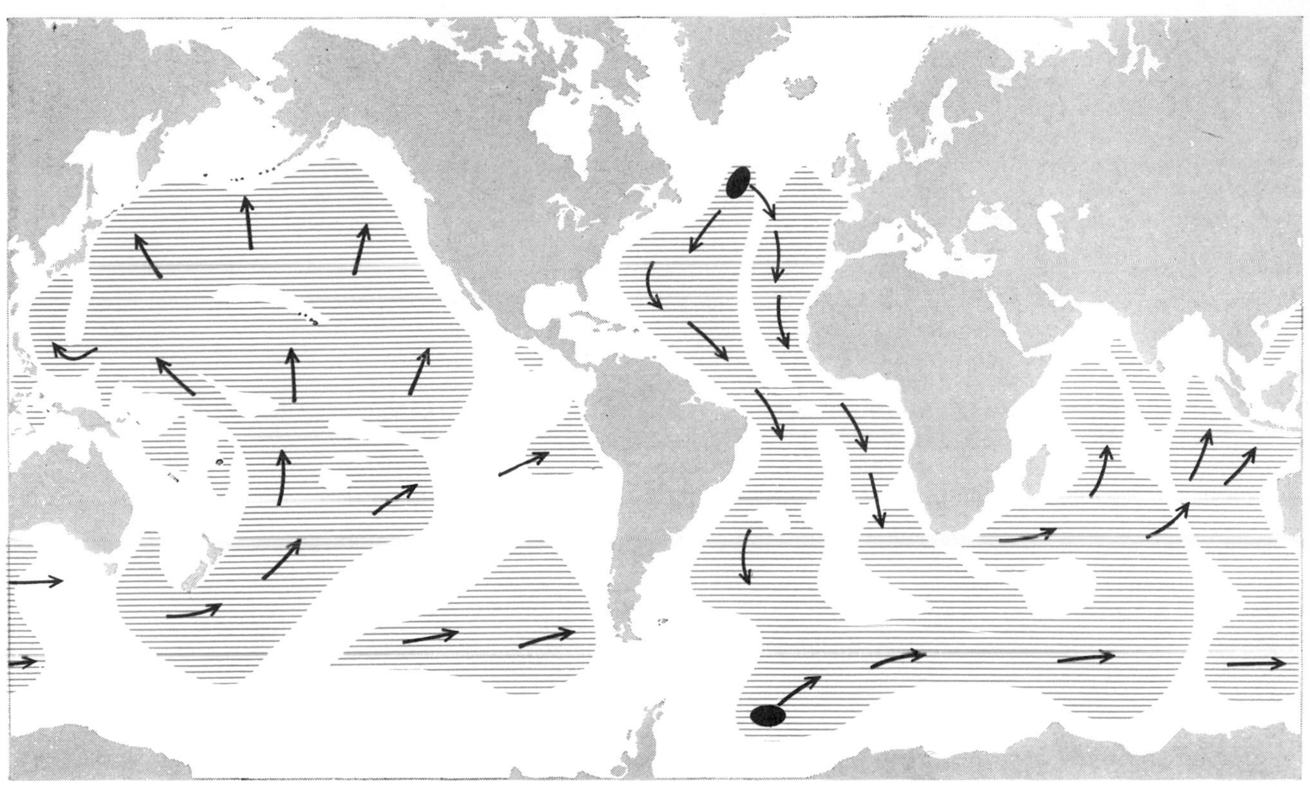

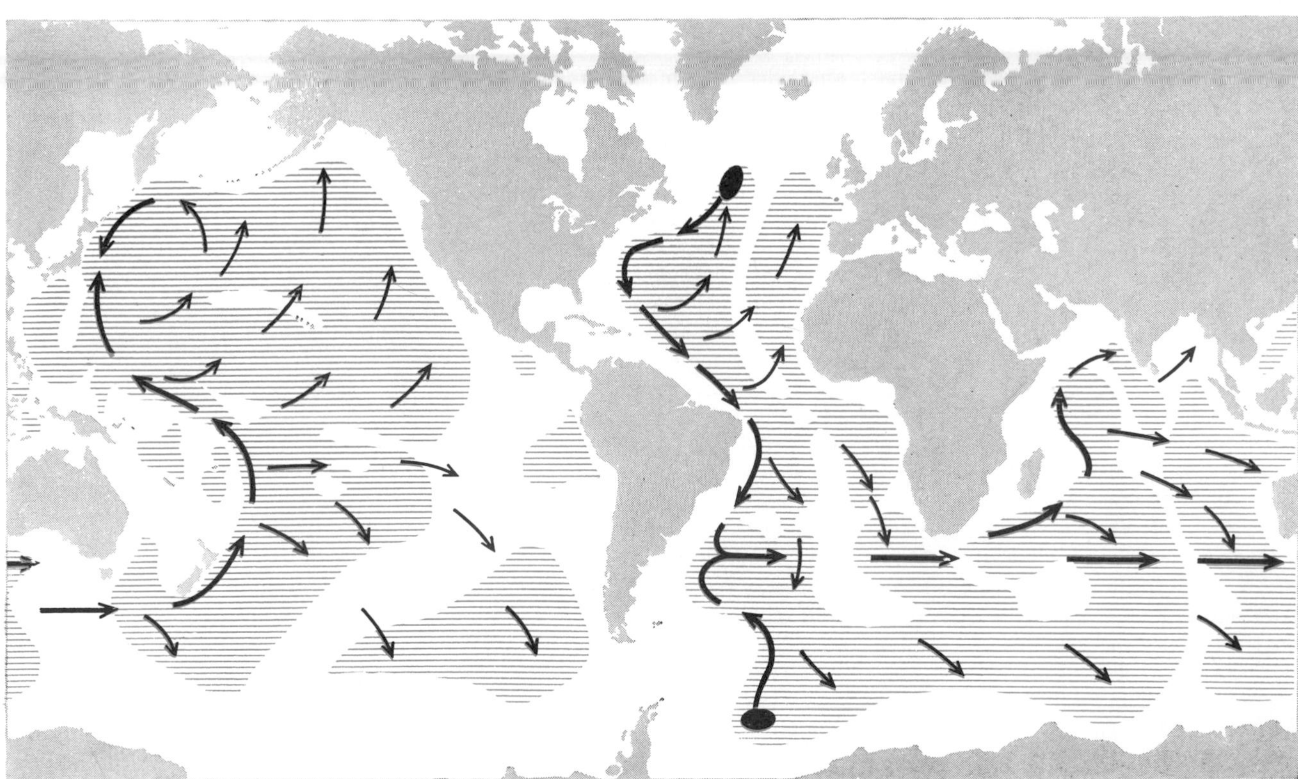

TWO THEORIES of the deep circulation are represented on these maps. On the older view (*top*) water flows in broad currents toward the Equator from its sources (*black ovals*) in the North Atlantic and the Antarctic, and then spreads through the Indian Ocean up into the Pacific. According to the author's theory (*bottom*) the flow away from the sources is confined to intense currents on the western edge of the oceans. These in turn feed a broad flow that carries water toward the poles in each of the ocean basins.

warm layer is held up by a continual upwelling of cold water from below. This slow rise of water (about an inch per day at mid-depths) represents a movement of water amounting to several thousand times the total daily discharge of the Mississippi River. Obviously all that water continually coming up from the depths must be replaced by an inflow into those depths—*i.e.*, it betokens a massive deep circulation.

The simplest circulation pattern that would fit the facts we have reviewed is a broad, sluggish, spreading current coming straight down the Atlantic and joining the current from the Antarctic to flow through the Indian Ocean and then up into the Pacific [*see upper map on the preceding page*]. Until recently most oceanographers accepted this general picture. But I believe that the rotation of the earth and considerations of fluid dynamics make another picture more likely. The general reasoning behind this new theory is as follows.

In the open ocean the flow of the bottom water must be different from that in a river. It should be more like the flow of air in the atmosphere. As every reader of weather maps knows, air does not travel directly from high-pressure to low-pressure areas. It circulates around the highs and lows, because of the so-called Coriolis force arising from the earth's rotation. That is, the air travels along the lines of equal pressure, or isobars. Now my own construction of the pattern of isobars in the deep ocean, based on dynamical considerations and on the requirements for the supply of upwelling water, shows a flow from the Equator toward the poles. The flow originating near the poles, it then appears, may be carried by strong currents along the western sides of the ocean basins, just as there are strong currents along these routes at the ocean surface [see the article "The Circulation of the Oceans," by Walter H. Munk; SCIENTIFIC AMERICAN Offprint 813]. We postulate a strong, deep current running down the western North Atlantic and up the western South Atlantic. When they come together, the currents merge and turn eastward, flowing up the eastern coasts of Africa and Asia [*see lower map on the preceding page*].

Two independent observations have partly confirmed this theoretical scheme. Last year a joint British-American oceanographic expedition found a strong deep current in the western Atlantic. It flows beneath the Gulf Stream but in the opposite direction, from north to south, just as the theory predicts. Detection of the current was made possible by a new device for investigating the motion of deep water, invented by the British oceanographer John C. Swallow. It is a float which is cleverly designed to seek a predetermined level under the surface and then to remain there, drifting with the current. The float has a small ultrasonic transmitter sending signals which locate its position. The expedition made the soundings off Charleston, S.C. They found that floats lowered to levels below 6,500 feet drifted southward at velocities between two and eight miles per day.

The other confirmation of the new circulation scheme was supplied by the German oceanographer Georg Wüst. Analyzing the pressure distributions in the South Atlantic, he was able to show that the flow of deep waters in that ocean is confined to relatively narrow streams along the continental slopes off Brazil and Argentina.

Thus we now can score two modest victories for the theory. I think it sporting to risk a few further predictions. We should find (1) a countercurrent underneath the Agulhas Current off the East Coast of Africa, (2) a narrow northward current in the western South Pacific, let us say along the slopes of the Tonga-Kermadec Trench, and (3) only a weak flow underneath the Japan current.

The theory has also been buttressed by some laboratory experiments at the Woods Hole Oceanographic Institution. Liquid in a flat-bottomed, rotating basin has the same type of circulation patterns as water on the spherical, rotating earth. Alan Faller at Woods Hole set up a wedge-shaped section in the basin to simulate a single ocean. The point of the wedge corresponds to a pole of the earth, the sides to the coastlines of the ocean and the outer perimeter to the Equator. When this sector was set into rotation and some dyed water was poured into the liquid at the "pole," the dye moved through the "ocean" just as the theory predicts. First it flowed in a narrow current along the "western" edge; at the "Equator" it turned and flowed slowly over the basin toward the center [*see photographs on page 139*].

With ingenuity you can reproduce this experiment yourself in a small tin dish on a phonograph turntable. The dish should be covered with a sheet of glass to prevent air currents from disturbing the water surface. Most turntables revolve in a direction opposite to that of the earth, so the boundary cur-

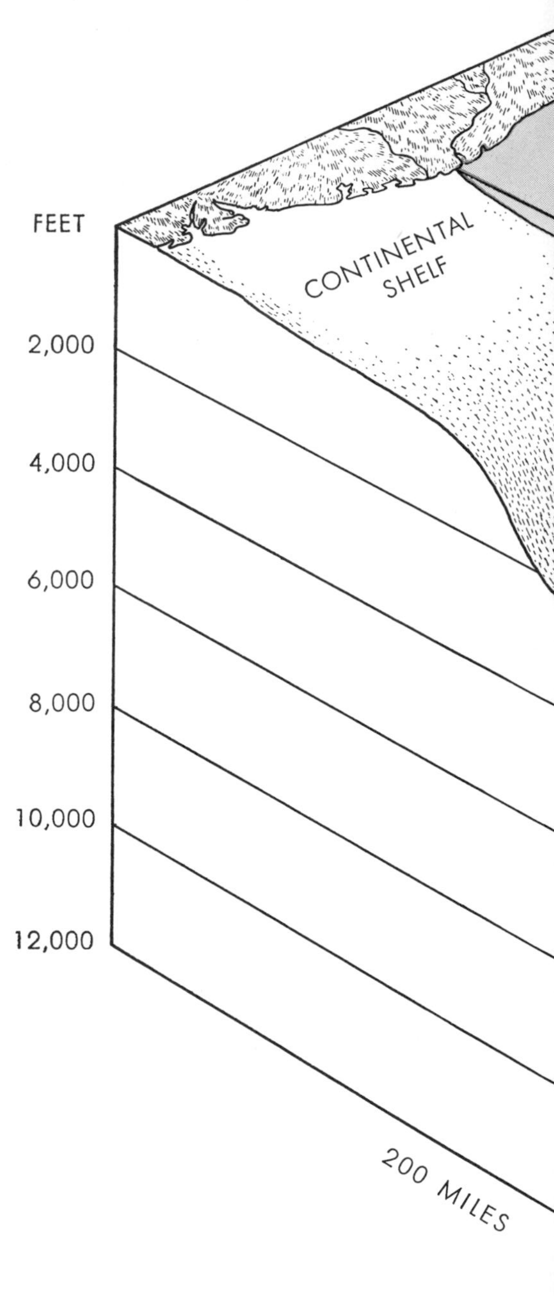

DEEP CURRENT flowing southward under the Gulf Stream off Charleston, S.C., was discovered last year by a joint British-

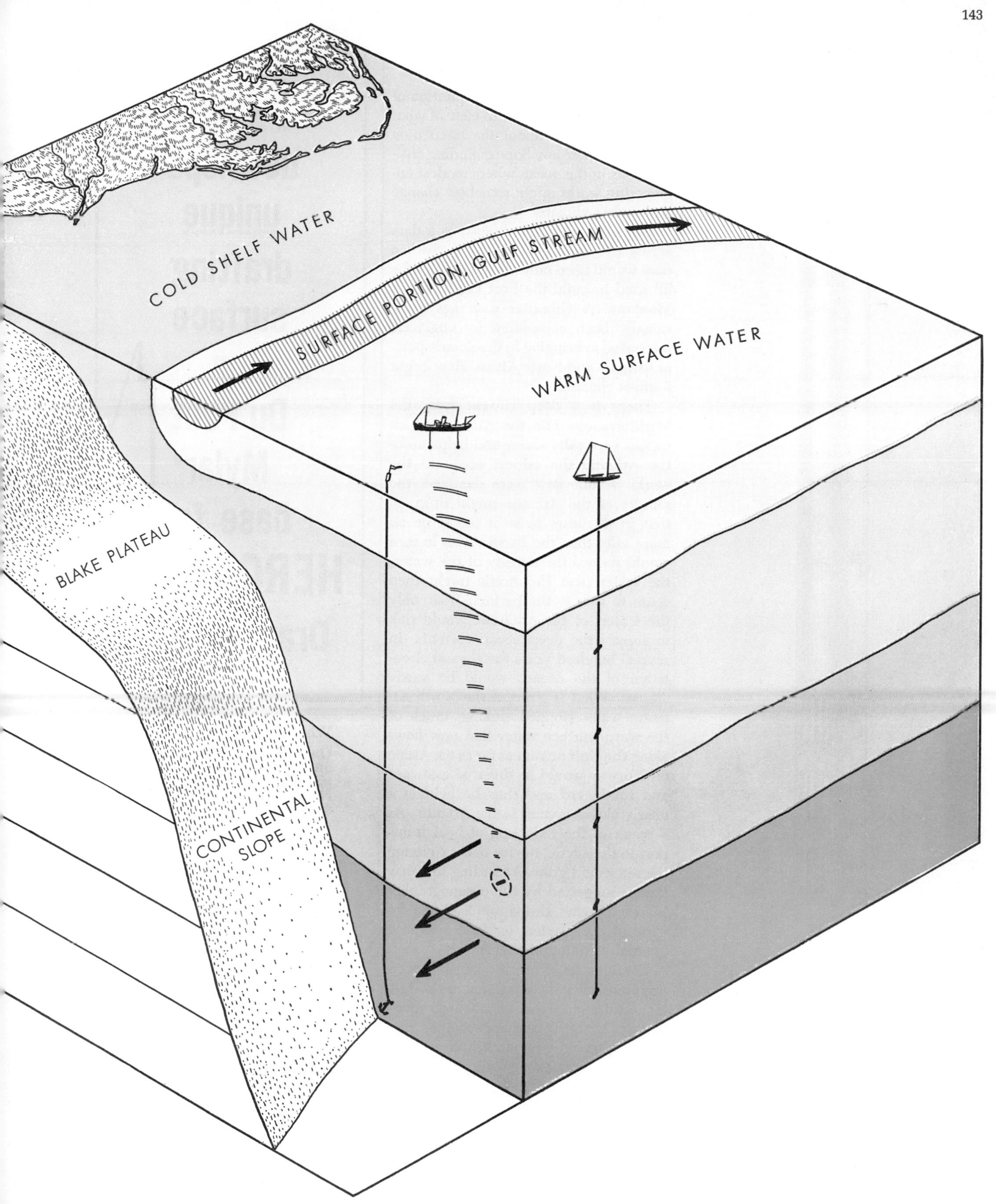

American oceanographic expedition. This cutaway view shows the British vessel *Discovery II* (*left*) receiving ultrasonic signals from a submerged float (*black bar*) which is drifting with the deep current. The American ship *Atlantis* (*right*) is recording other data such as temperature, salinity and density. The float is the first device with which undersea circulations can be measured directly.

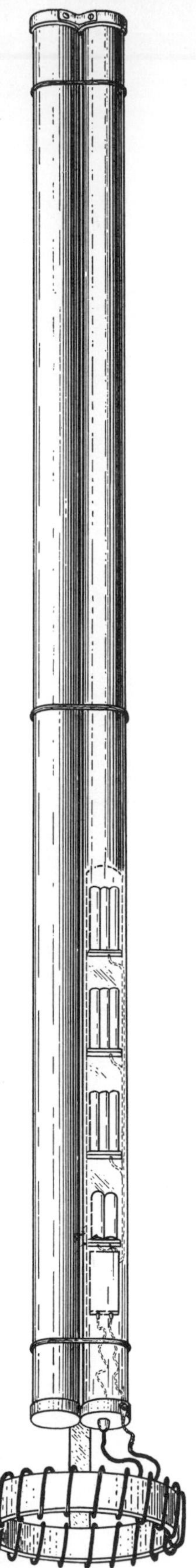

UNDERWATER FLOAT is slightly less compressible than sea water; thus it sinks until its density equals that of surrounding liquid. Oscillator in the tube at right sets up ultrasonic vibrations in the ring at bottom.

rent will probably show up on the "eastern" wall instead of the "western."

Now let us go back to the question of controlling climate. In view of what we have learned about the circulation pattern, is there any hope of finding critical areas in the ocean where modest engineering works might somehow change the world's climate?

The most attractive fantasy is a dam

dam would need only about 10 times the fill used to build the Fort Peck Dam in Montana. A Gibraltar dam has occasionally been considered by engineers interested in bringing hydroelectric power to Spain and North Africa. How might it affect climate?

There is a deep current from the Mediterranean into the Atlantic which carries very salty water and helps make the Atlantic the saltiest ocean in the world. If this flow were dammed, the salinity of the Atlantic might drop, so that in 30 years or so it might be no more salty than the Pacific. This in turn would reduce the density of the water; the water near the Arctic might then cease to sink to the bottom. If so, only the waters of the Antarctic would sink to supply the deep ocean currents. In several hundred years the abyssal circulation of the oceans would be vastly altered. What is more, if the North Atlantic water stopped sinking, much of the warm surface water that now flows along the Gulf Stream as far as the Arctic off Norway would be diverted eastward and southward and thus be held in a nearly closed system in the Atlantic. As a result of the reduction of heat transport to the Arctic, the ice packs covering the sea would grow. According to a new theory suggested by the oceanographer Maurice Ewing, this would lead to a decline of the glaciers on land and to a general warming of the earth.

Common sense rebels against such an argument. It is hard to imagine so fantastic an effect from so small an intervention by man. And indeed the argument is loaded with unproved assumptions and tenuous speculations. We could construct an equally plausible argument that the same stratagem might cool rather than warm the earth. I cite this entertaining fantasy only to show that we need a great deal more information before we can begin to talk knowledgeably about altering the climate. All such speculations merely illustrate how little actual knowledge we have and how valuable it would be to develop a better quantitative understanding of the ocean circulation.

The Top Millimeter of the Ocean

13

by Ferren MacIntyre
May 1974

Subtle events that take place in a thin film of liquid covering 70 percent of the earth's surface are decisive for the well-being of terrestrial life

Reduced to the scale of a tabletop globe, the oceans have the thickness of a sheet of onionskin paper. From a planetary point of view the ocean's single most important aspect is neither its depth nor its volume but its surface: the largest and most homogeneous environment on the planet. Through the ocean's 360 million square kilometers of surface pass 70 percent of the solar energy that the earth absorbs, most of its supply of fresh water, a large fraction of the annual production of carbon dioxide and oxygen, a huge tonnage of particulate matter and unmeasured volumes of man-made pollutants. Yet only in the past few years has a significant effort been made to study the ocean surface in detail or to understand how its properties modify the transport of matter or energy across it.

Traditionally oceanographers have defined "surface" as any sample they could catch in a bucket from shipboard. This view is changing as we learn more about the detailed structure of the surface. If one uses a logarithmic scale to plot a cross section of the ocean from the dimensions of a molecule of water lying on the surface to a maximum depth of some 10 kilometers, the top millimeter corresponds exactly to the top half of the ocean [*see illustration on following page*]. Thus it is convenient to regard the entire top half of the (logarithmic) ocean as the "surface." The use of a logarithmic depth scale is not simply a literary device; rather it provides a realistic and helpful way of thinking about the complexity of the ocean surface. It makes room for the multiple interacting events that crowd near the surface and helps to clarify their very different depth scales. In terms of events and processes the top millimeter of the ocean offers as rich a field of study as the lower "half" of the ocean.

The top millimeter of the ocean has come to be called the microlayer. As an example of the complexity of events in this layer, let us look at the transfer of carbon dioxide across it. The subject is of considerable importance because the burning of fossil fuel will release enough carbon dioxide over the next 20 years to more than double the amount of carbon dioxide in the atmosphere from 320 parts per million today to at least 650 parts per million. No one can predict with confidence how rapidly the excess gas will be absorbed by the ocean and what effect the remnant increase in atmospheric carbon dioxide will have on the world's climate.

A few years ago reported values of the rate of carbon dioxide transfer across the ocean surface varied by a factor of 330 (if both laboratory and field data were included). The uncertainty reflected our ignorance of the important variables in the process, but gradually things are being sorted out. For example, the rate of transfer across a still surface can be increased fourfold by even the smallest ripples, which have the effect of thinning the laminar boundary layer through which gas must pass by molecular diffusion. Large waves can provide another twofold increase. Paul F. Twitchell of the Office of Naval Research has recently shown that neuston, the tiny plants and animals that live in the microlayer, can increase evaporation (and presumably gas transfer as well) by a factor of three by stirring the laminar layer with their flagella. Of course, the neuston further complicate matters by consuming carbon dioxide as it is passing through the microlayer. These three identified factors still leave a fourteenfold range in transfer rates to be accounted for. One suspects that organic material on the surface (a factor of two?), whitecaps (a factor of three?) and turbulence in both air and sea (another factor of three?) are the most important remaining variables.

We know from estimates of the amount of carbon dioxide released into the atmosphere by the burning of fossil fuels over the past century that the gas disappears from the atmosphere faster than can be accounted for by the known escape routes. The wind-stirred layer of the ocean, the top 100 meters, is more or less in equilibrium with the atmosphere at all times; it can neither absorb more carbon dioxide nor pass it on to the deeper water because there is very little vertical mixing between layers in the ocean. The deep water, representing 90 percent of the ocean volume, has a residence time on the order of 1,000 years. It communicates directly with the atmosphere only in the polar regions, so that it absorbs carbon dioxide very slowly [see the article "The Circulation of the Abyss," by Henry Stommel, beginning on page 139].

A route into the deep water that may be important, but has not been included

CROSS SECTION OF OCEAN is plotted logarithmically on the following page, demonstrating that the top millimeter can be regarded as the upper half of the ocean. The scale starts at the top with the diameter of a water molecule: about 10^{-10} meter, or one angstrom. (The long dimension of a water molecule is actually about two angstroms.) The maximum depth of the ocean is about 10 kilometers, or 10^{4} meters. The midpoint of the 14 decades falls at 10^{-3} meter, or one millimeter. Surface chemists study the upper 30 angstroms. Hydrodynamicists study the region between 30 micrometers and one centimeter, where the shearing action of the wind creates boundary layers. The region between 30 angstroms and 30 micrometers is a *mare incognitum*, only now being investigated by microlayer oceanographers.

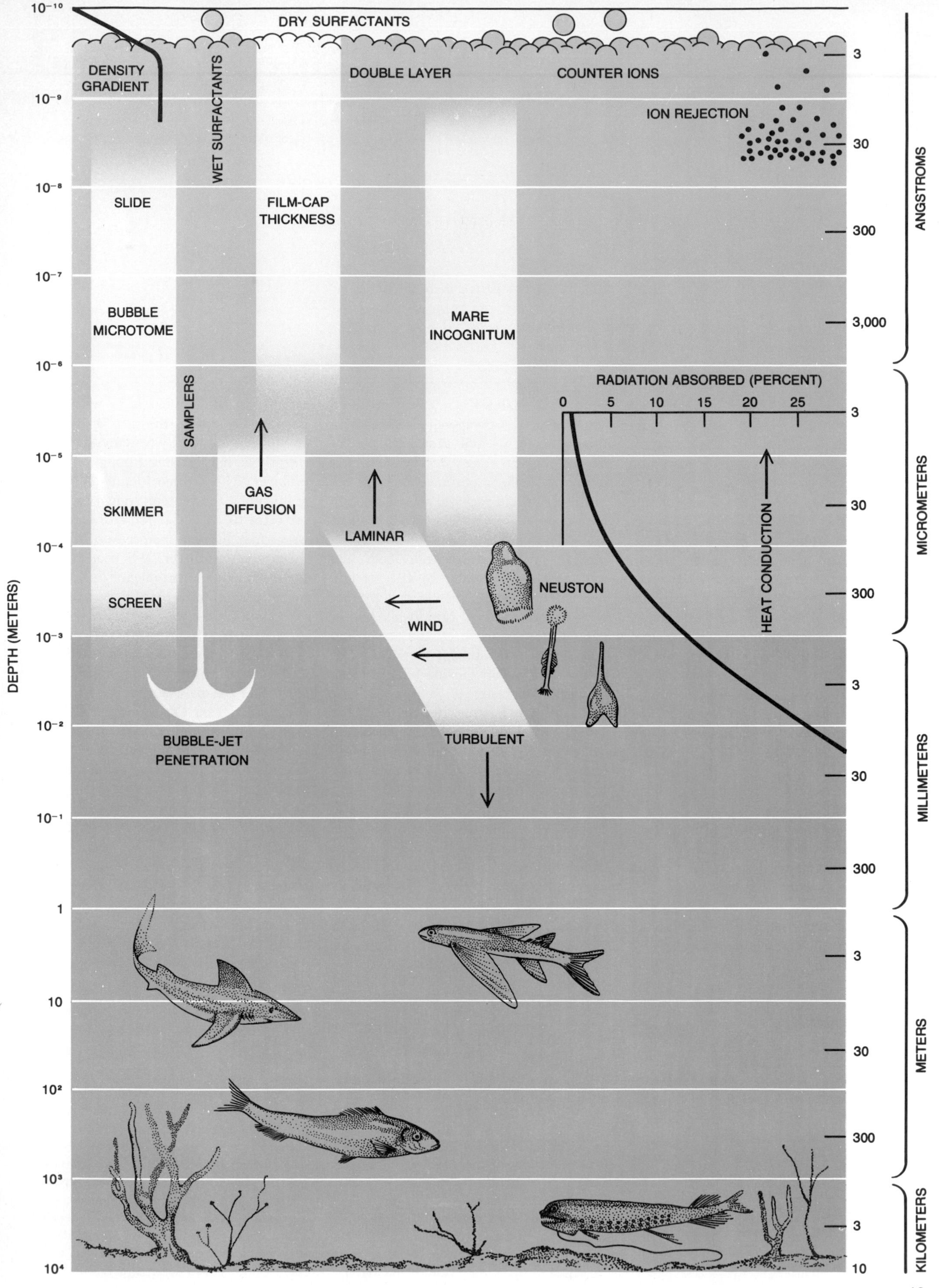

DRY SURFACTANTS
DENSITY GRADIENT
WET SURFACTANTS
DOUBLE LAYER
COUNTER IONS
ION REJECTION
SLIDE
FILM-CAP THICKNESS
BUBBLE MICROTOME
MARE INCOGNITUM
SAMPLERS
RADIATION ABSORBED (PERCENT)
0
5
10
15
20
25
GAS DIFFUSION
SKIMMER
LAMINAR
HEAT CONDUCTION
NEUSTON
SCREEN
WIND
BUBBLE-JET PENETRATION
TURBULENT
DEPTH (METERS)
10^{-10}
10^{-9}
10^{-8}
10^{-7}
10^{-6}
10^{-5}
10^{-4}
10^{-3}
10^{-2}
10^{-1}
1
10
10^{2}
10^{3}
10^{4}
3
30
300
3,000
ANGSTROMS
3
30
300
MICROMETERS
3
30
300
MILLIMETERS
3
30
300
METERS
3
10
KILOMETERS

in calculations, is the atmospheric carbon dioxide taken up by the neuston: carbon compounds of fecal pellets and dead organisms sink into the deep water. Once incorporated into a particle large enough to sink, the carbon need fall only 100 meters to short-circuit the normal storage time of carbon dioxide in the mixed layer.

Things That Coat the Surface

The transfer rates of gas, water vapor and momentum are known to be decreased by minute amounts of surfactants, or surface-active substances. These are long-chain molecules that ubiquitously coat the surface of liquids. The precise nature of the surfactants on the ocean, however, remains an open question. Early workers reported the presence of lipids: molecules that commonly contain 16 or 18 atoms in a hydrophobic tail sticking out of the water and an ionized or otherwise hydrophilic head sticking into the water. These substances, ordinary soap being a typical example, are known as "dry" surfactants because most of their structure sticks out of the water. Robert Morris of the National Institute of Oceanography in Britain reports finding on the ocean such lipid surfactants as hydrocarbons, sterols, esters, glycerides and phospholipids.

On the other hand, Robert E. Baier of the Calspan Corporation (formerly the Cornell Aeronautical Laboratory) finds that lipids are rare on the surface of the ocean except in regions of man-made pollution, whereas what we might call "wet" surfactants are common. They are long-chain proteinaceous substances that are essentially hydrophilic but that stick to the undersurface of the water by virtue of the fact that they have a few hydrophobic side chains [*see illustrations on this page*]. It should be possible to determine the relative amounts of dry and wet surfactants on the ocean surface and whether or not the proportion is changing by developing improved sampling techniques.

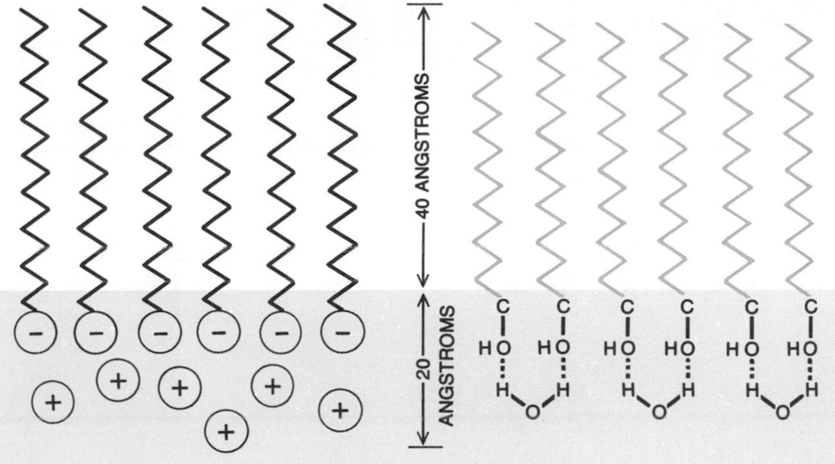

OILY OR FATTY SURFACE MOLECULES are described as "dry" surfactants because most of the molecule, consisting of a hydrocarbon chain, sticks up above the surface. A typical example is calcium stearate (*left*), which consists of an insoluble tail 17 carbon atoms long that projects upward and a hydrophilic head, a COO^- group, that projects below the surface. Hexadecanol, a 16-carbon alcohol (*right*), forms a liquid monolayer film that is held to the surface by hydrogen bonds linking the terminal OH group with water molecules.

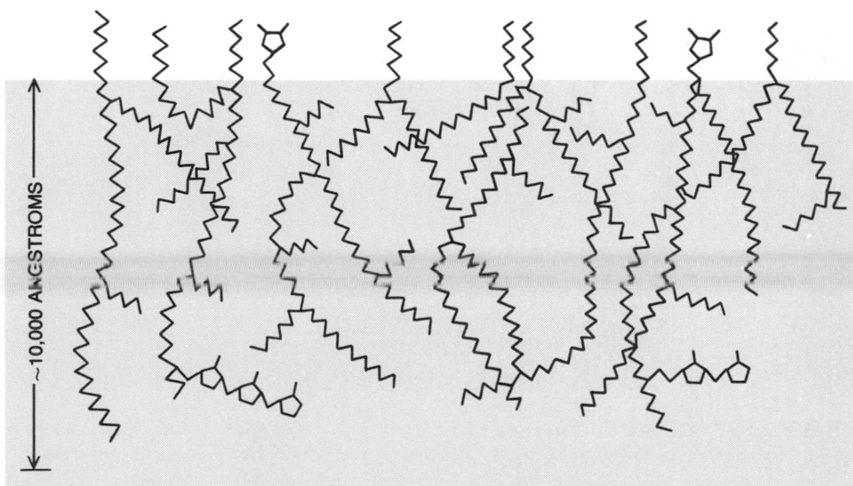

PROTEINACEOUS SURFACE MOLECULES form a broader and more complex class of substances that the author describes as wet surfactants because most of their structure is submerged. Typically proteins and glycoproteins, they are normally much more abundant on the ocean surface than the dry surfactants. Wet surfactants are attached to the surface by the occasional hydrophobic groups that seek to escape from their aqueous environment.

It is clearly not easy to invent a method that will efficiently pick up all the substances lying within, say, a micrometer of the ocean surface, or roughly the top half of the top half of the ocean. The earliest microlayer sampler, developed by John M. Sieburth and William Garrett, is a stainless-steel window screen that is immersed perpendicularly to the surface and withdrawn parallel to it. The screen gathers up virtually all surface molecules with a chain length of 16 carbon atoms or more, but its efficiency for collecting shorter molecules may be low. It also collects all the water and living matter lying within about 300 micrometers of the surface. The "skimmer," invented by George R. Harvey at the Scripps Institution of Oceanography, is a hydrophilic drum made of ceramic that rotates slowly on a horizontal axis, with its bottom just dipping into the water. The skimmer picks up a continuous film of water between 60 and 100 micrometers thick, which is scraped off by a blade resembling a windshield wiper.

Both of these traditional microlayer samplers collect neuston with high efficiency. Harvey's interest, in fact, lay with this specialized biota; samples taken with the skimmer are sometimes bright green with algae when "deep" water (10 centimeters below the surface) is perfectly clear. In order to extract dry surfactants from a skimmer sample one customarily shakes the sample with chloroform, which dissolves lipids. Unfortunately the chloroform also leaches lipids from the microscopic organisms that are present, confusing the results.

Baier has recently devised an elegant surface sampler that collects no neuston. It is an adaptation of the technique devised years ago by Irving Langmuir and Katharine B. Blodgett for collecting monolayer films from water surfaces. Baier's collector is a piece of germanium shaped like a microscope slide whose

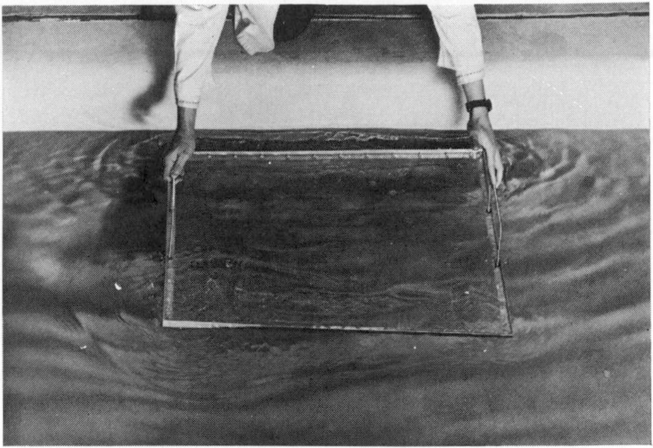

COMPOSITION OF OCEAN SURFACE has proved difficult to establish. The simplest device for sampling the top 300 micrometers of the ocean is a piece of stainless-steel mesh, which can be slipped into the water at an angle and withdrawn parallel to the surface, as is shown here. The device is effective in picking up certain kinds of oily film but leaves many typical surface molecules behind.

MICROLAYER "SKIMMER," devised by George R. Harvey at the Scripps Institution of Oceanography, is a drum coated with a hydrophilic substance. Rolled along the surface, it picks up a continuous film of water 60 to 100 micrometers thick. Both the skimmer and the steel mesh efficiently collect microscopic surface organisms known as the neuston in addition to various "surfactant" materials.

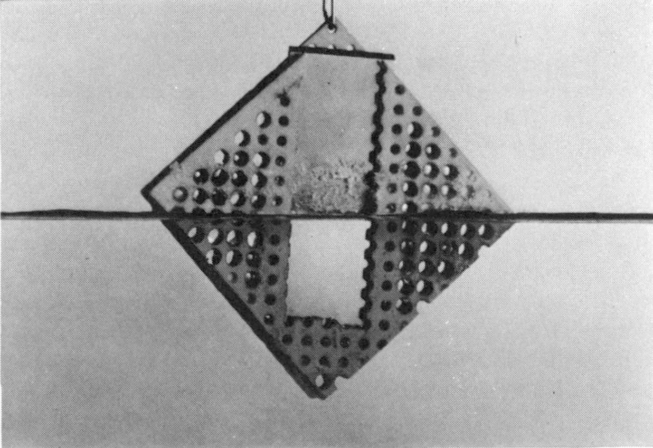

COHERENT LAYER OF SURFACE MOLECULES can be removed by a germanium slide resembling a microscope slide. The device, conceived by Robert E. Baier, adapts a method long used by chemists interested in surface films. At the left, the slide has been immersed. At the right, talc that had been floating on the surface adheres to the withdrawn slide. Baier's contribution was to make the slide of germanium and to bevel the ends so that the surface film adhering to the slide can be examined *in situ*. Since germanium is transparent to infrared radiation, an infrared beam directed into one beveled edge is reflected internally many times before it emerges from the opposite edge. The molecules adhering to the surfaces of the slide selectively absorb characteristic infrared wavelengths so that the emerging beam provides an absorption spectrogram of the adhering substances. All the photographs on this page were made in Baier's laboratory at Calspan Corporation, which was formerly Cornell Aeronautical Laboratory, in Buffalo, N.Y.

ends have been beveled [*see upper illustration below*]. When it is withdrawn perpendicularly from an aqueous solution, it picks up a coherent layer of any surfactants that may be present. The slide is made of germanium because the metal is transparent to infrared radiation. An infrared beam directed into one of the beveled edges of the slide is trapped by total internal reflection until it emerges from the other end. At each reflection the infrared radiation extends about one wavelength beyond the germanium into the adsorbed film of molecules, which absorb particular wavelengths exactly as they would in an infrared spectrophotometer. Thus the emerging beam carries an infrared spectrogram of the functional chemical groups in the adsorbed layer.

Still another microlayer collector coming into use is simply a large funnel, first employed as a sampler by Morris in his study of surfactants. The funnel is filled with surface water and allowed to drain slowly. The neuston float out with the water but the surface film adheres to the sides of the funnel. The film is then rinsed off with a solvent and analyzed by chromatography. Until someone undertakes to calibrate and compare the various samplers there will be no explanation of why the different techniques give such different results. A sampler that has been suggested by several people but not yet tried is simply toilet paper unrolled on the surface of the water. Recognized as a highly efficient way of cleaning a water surface in the laboratory, the toilet-paper method could be used from a dinghy even in moderate waves.

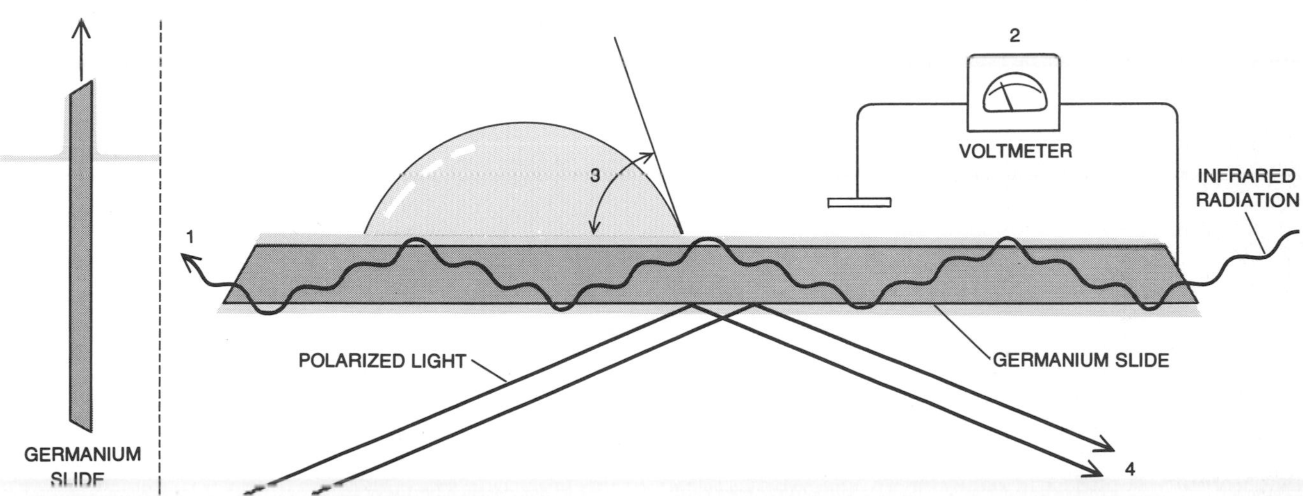

CHARACTER OF SURFACE MOLECULES can be studied in various ways using the sample obtained by Baier's germanium-slide technique. The slide collects a film of both dry and wet surfactants, leaving most organisms behind (*left*). The molecules clinging to the surface selectively absorb infrared radiation, providing clues to molecular structure (*1*). By measuring the surface electrical properties before and after sampling (*2*) one can tell where the molecular dipoles are located and what their density per unit area is. A drop of water placed on the surface assumes a geometry that is dictated by the critical surface tension and hence provides important clues to the free energy and the chemical constitution of the surface. Depending on the surface tension, the tangent angle formed where the drop meets the surface (*3*) can be either greater than or less than 90 degrees. Here it is less, indicating that the film is primarily composed of wet surfactants. The thickness and refractive index of the sample are obtained by reflected polarized light (*4*).

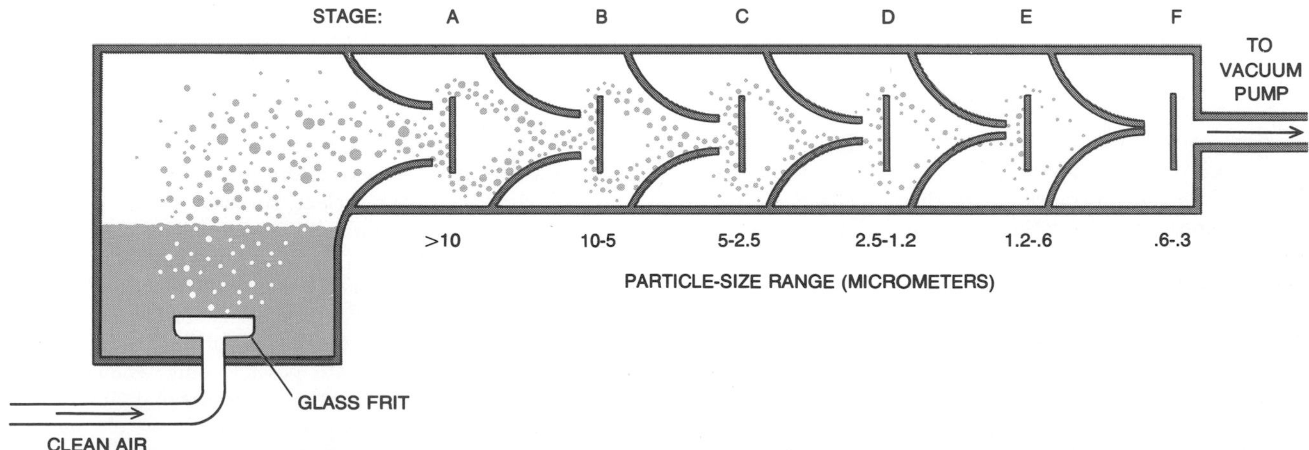

COMPOSITION OF DROPLETS produced when bubbles break the ocean surface is studied in the laboratory with the help of this apparatus. It is found not only that the droplets differ in composition from the bulk sample but also that the composition varies with droplet size. The droplets are sorted as they pass through the stages of an "impactor." Since the air velocity increases at each stage, successively smaller particles are unable to follow the airstream and are collected on impaction plates. In a typical experiment the bulk sample contains two elements whose ratio is carefully determined. The droplets collected on the impactor plates are then analyzed for the same elements. The ratio in the sample divided by the ratio in the original sample yields a value defined as "the fractionation."

The wet surfactants that Baier detects in his samples are not well-known molecules like the dry surfactants, which have been studied by surface chemists for decades. The most intensive work on wet surfactants has been medically oriented and has involved studies of how proteins are denatured (rendered inactive by unfolding) by contact with surfaces. Since the techniques for studying wet surfactants are primitive, almost nothing is known of their thermodynamic, viscous or electrokinetic properties. We have essentially no theoretical understanding of how they affect transport processes through the microlayer except that we can be sure they too retard all forms of transfer. It does appear, however, that the wet surfactants on the ocean, which are known to be glycoproteins and proteoglycans, are reasonably good carriers of phosphate, of various organic molecules, of the scarcer ions of seawater and of heavy metals. Thus they provide a mechanism for trapping and concentrating exotic substances at the ocean surface.

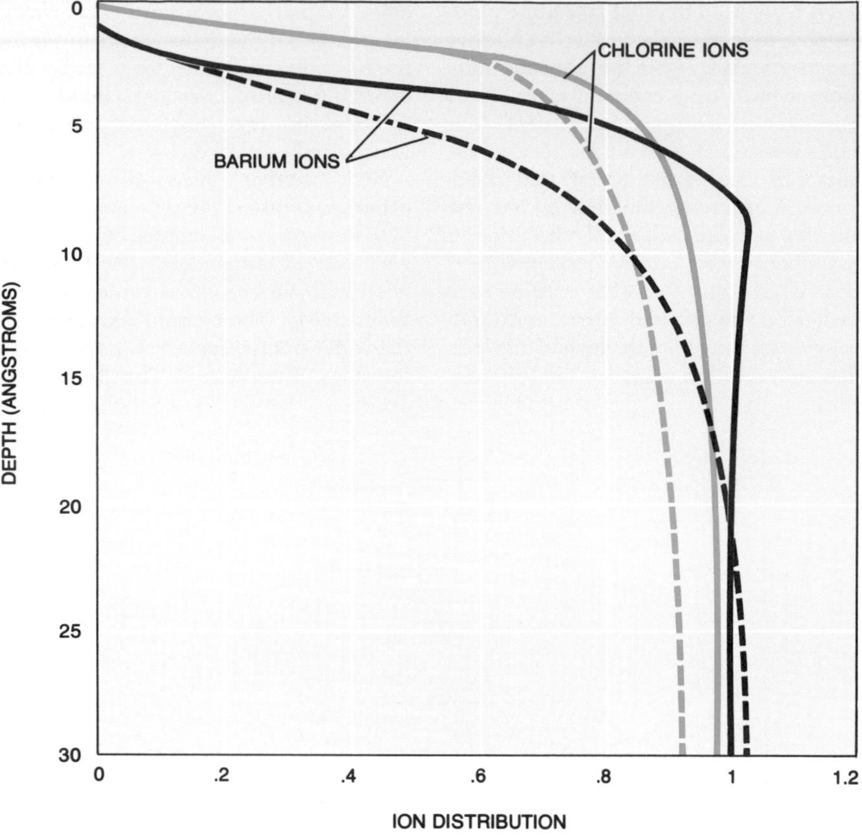

DISTRIBUTION OF IONS AT THE OCEAN SURFACE differs from the average distribution found in the bulk solution. Any substance that increases the surface tension of a liquid would increase the free energy of the surface and therefore must be rejected. The rejection thus operates against the ions of seawater. The theoretical rejection, however, is significant only in the top few angstroms, as these curves for barium and chlorine ions show. The rejection is less for a concentrated solution (*solid curves*) than for one 20 times more dilute (*broken curves*). The concentration of sodium and chlorine ions in seawater is actually five times greater than it is in the stronger of these two solutions. Hence seawater ions would be rejected even less than these minimum values. Thus thermodynamic rejection of ions from the ocean surface cannot explain the distorted composition of seawater particles injected into the atmosphere from the sea. Curves plotted here are based on calculations made by G. M. Bell and P. D. Rangecroft of Chelsea College of Science and Technology in London.

The Puzzle of Rainwater

The origins of microlayer oceanography curiously lie in meteorology. After John Dalton had demonstrated in 1822 that sea salt was present in rainwater a number of people suggested that the ocean was the source of other materials found in precipitation. By the early 1940's it was fairly clear that the mixture of marine substances found in rain and snow was not what one would obtain simply by evaporating a typical sample of seawater. In 1959 two meteorologists stated the issue prophetically. "The transport of salts from sea to air," said Carl-Gustaf Rossby, "is by no means a simple mechanical process." Erik Eriksson went on to suggest that "some fractionation process seems to take place at the sea surface, possibly involving material derived from surface films."

At about the same time the Japanese chemist Ken Sugawara reported that when he measured the ratios of iodine and sodium to chlorine in the spray of breaking bubbles, he found values vastly different from the normal ratios in bulk seawater. Sugawara's work was mostly ignored, perhaps because he reported only a single experiment and because the changes in ratios were incredibly large.

Since then a considerable amount of work on fractionation has shown that if Sugawara had conducted a second experiment, he would have got different numbers, and that his iodine data were surprisingly good. Fractionation experiments, whether they are done in the field or the laboratory, are plagued by nonreproducible results. Fractionation of the principal metallic ions in seawater (sodium, magnesium, calcium and potassium) appears to be slight; if it occurs at all, it alters the seawater ratios by less than 10 percent. It is still impossible to collect a suite of geochemical samples (rain, snow, dry fallout, aerosols and so on) and to estimate the contribution from continental dust with sufficient precision to see a 10 percent variability in the marine component. Gradually, however, from large-scale sampling programs (such as one run by Roger Chesselet in France) our understanding of regional variations in atmospheric constituents is growing, and eventually it may be possible to detect small changes.

Similarly, it is difficult to say anything about fractionation during ejections from the sea surface for substances (such as chlorine, bromine and iodine) that are chemically activated by sunlight or that are catalytically activated in the atmosphere by aerosols. In the laboratory few workers have appreciated how readily fractionation experiments can be confused by surface-active impurities and how extraordinarily difficult it is to prepare a clean surface. Just once have I been able to keep a surface clean for 24 hours. It took a year of preparation, and even then it was mostly good luck, since I never succeeded again. The experience taught me one useful fact: reproducible results are possible only when surface conditions are reproducible. Impurities of a few parts per billion can change the surface properties of water; at the part-per-million level impurities can alter fractionation results so completely as to mask any of the often sought physicochemical mechanisms.

Early searches for the fractionation

mechanisms at the sea surface ignored the unique properties of the top half of the ocean and relied instead on traditional physicochemical thinking. More than 100 years ago Josiah Willard Gibbs had shown that any substance that increases the surface tension of water will tend to be rejected as it diffuses upward from below. Since different salt solutions have different surface tensions, it follows that their ions are differentially rejected, so that the surface of a mixture such as seawater will differ in composition from the bulk. Although the difference is calculable, it extends into the liquid only a small fraction of a micrometer and seems far too small to be of any geochemical importance [*see illustration on preceding page*].

Nevertheless, some experiments designed to test the Gibbs equations reveal a startling disagreement with theory, both in the scale of fractionation and in the direction. For example, whereas theory predicts that a solution containing barium and cesium ions should be slightly depleted in barium ions immediately below the surface, amounting to a fractionation of about .005 percent, experiment shows a barium enrichment amounting to a fractionation of about 7 percent. This remarkable discrepancy is unexplained.

In other experiments M. R. Bloch of the Negev Institute for Arid Zone Research in Israel has found that under nonequilibrium conditions the fractionation of potassium and ammonium ions is strongly influenced by the humidity above the liquid surface [*see top illustration at left*]. Potassium in bubble spray is enriched as much as tenfold and ammonium as much as a hundredfold compared with their concentration in the bulk sample. The extraordinary values for ammonium ions (NH_4^+) are perhaps related to the hydrolytic formation of ammonia (NH_4OH), which, being uncharged, is not rejected from the surface by electrical effects. (It may be relevant that the dielectric constant of surface water has a value of only 9 whereas for water in bulk it is 81. This suggests that one might profitably study fractionation in solvents whose dielectric constant resembles that of surface water.) In any case it remains a mystery, at least to me, how humidity alone can produce the striking effects Bloch reports. There is also a question of whether even an extreme degree of fractionation within the range of electrical forces—say a few hundred angstroms from the surface—would be significantly reflected in the composition of film drops thrown up as an aerosol. Perhaps the flow of heat and water

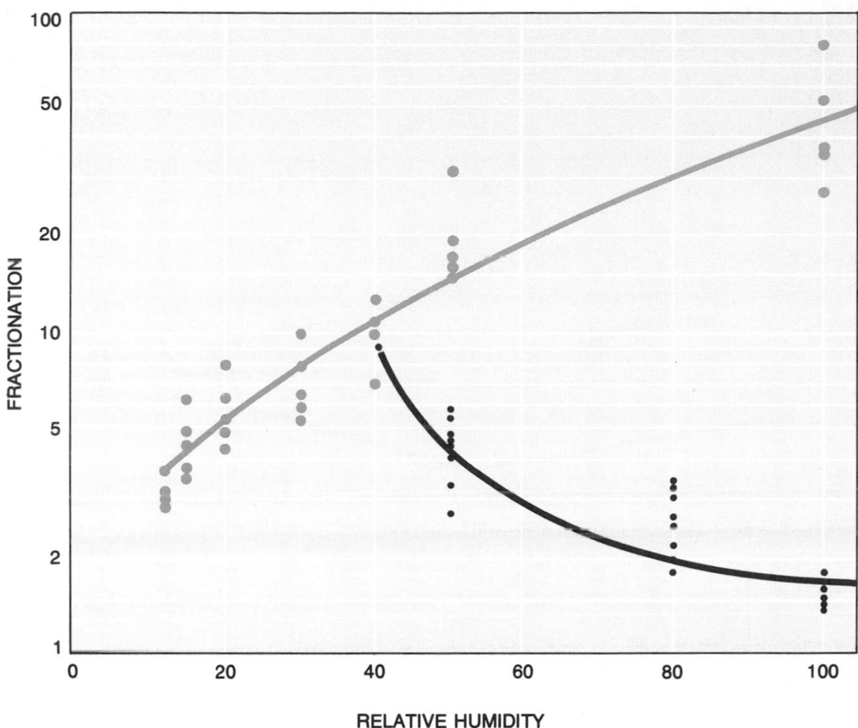

LARGE FRACTIONATION VALUES for potassium ions (*black*) and ammonium ions (*color*), far exceeding any possible fractionation based on ion rejection, have been obtained by M. R. Bloch of the Negev Institute for Arid Zone Research. The experiments were conducted using a version of the impactor illustrated at the bottom of page 149. The correlation between humidity and fractionation is very puzzling and is so far unexplained.

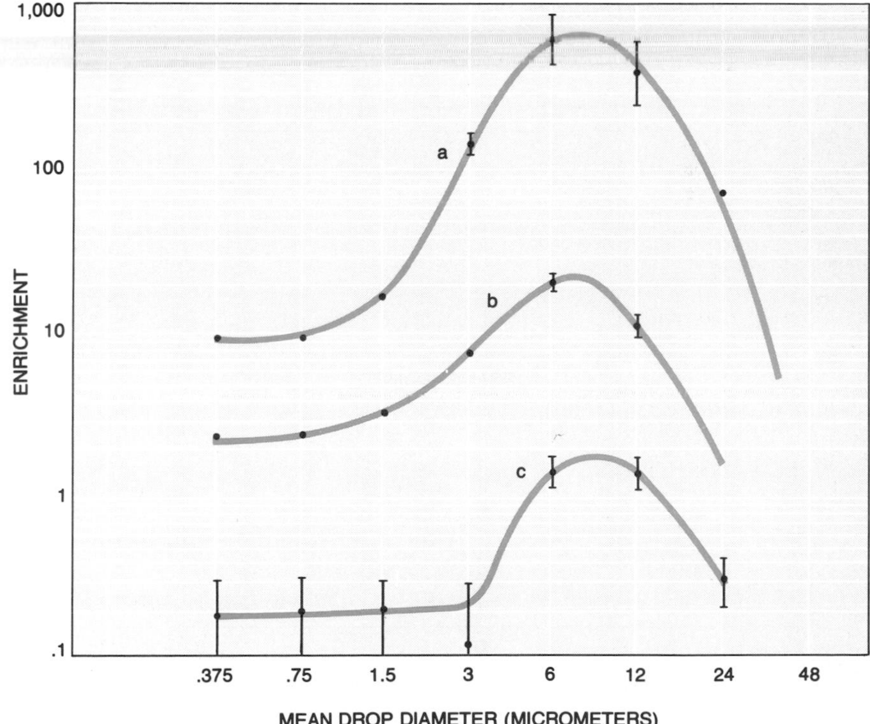

PHOSPHATE FRACTIONATION VALUES obtained by the author demonstrate the dramatic influence of trace amounts of surfactant molecules in experiments using the drop-impactor technique. The water used in the experiment that yielded curve *a* was assumed to be carefully distilled. Actually it seems to have contained something that strongly attracted phosphate ions to the surface. (The vertical coordinate, "enrichment," is simply "fractionation" minus 1.) Curve *b* was obtained after several hours of bubbling had removed the phosphate-attracting contaminant. Curve *c* shows how fractionation was depressed when a negatively charged surfactant was added to repel the negative phosphate ions.

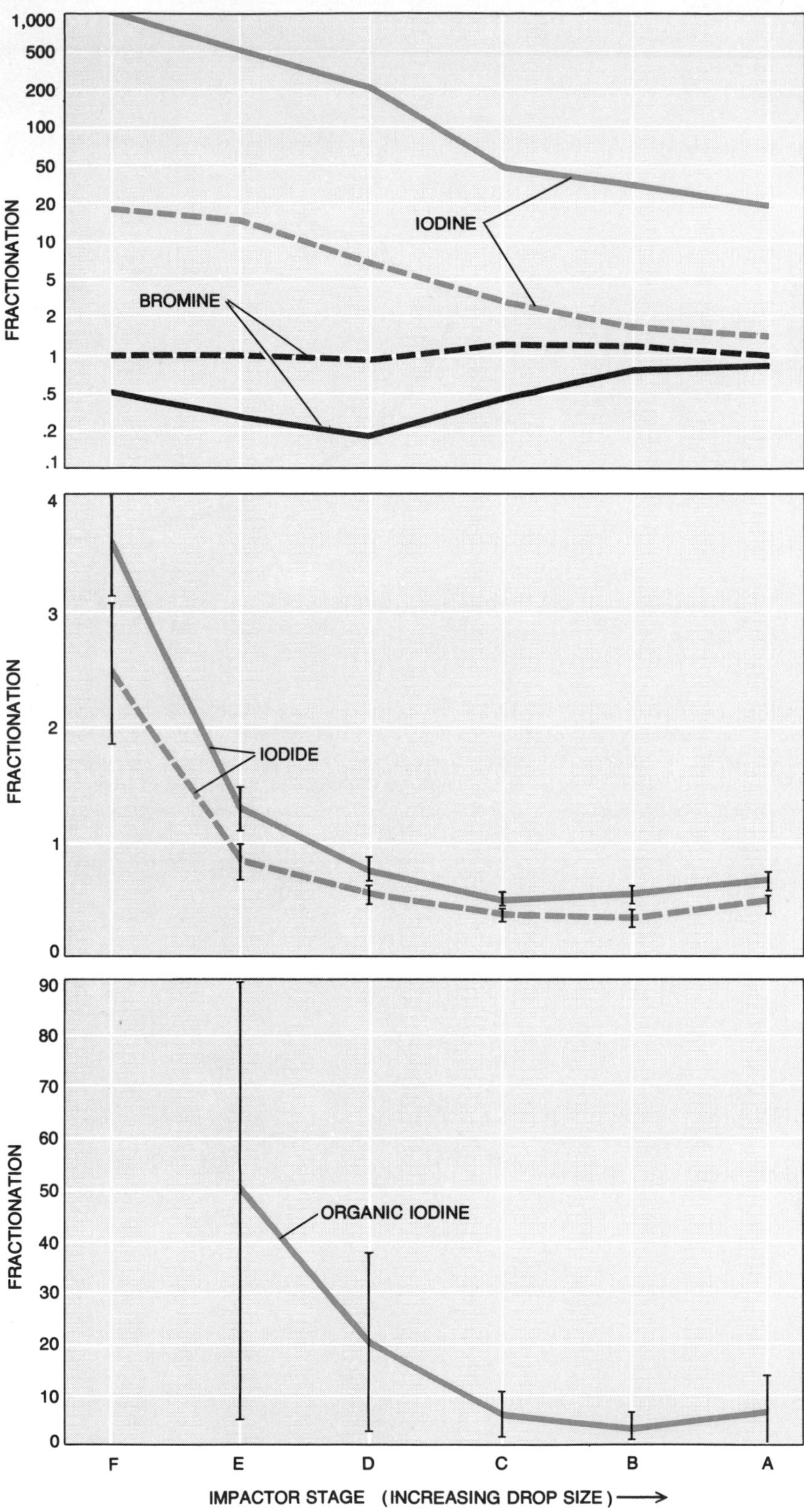

FRACTIONATION OF OCEANIC IODINE is geochemically important. The top set of curves shows that old oceanic aerosol from the trade winds (*solid lines*) contains more iodine and less bromine than aerosols freshly generated by ocean surf (*broken lines*). Bromine is unfractionated initially and is subsequently lost to the atmosphere. Iodine is not only fractionated initially but also gradually becomes concentrated in smaller drops. The two curves in the middle diagram, based on laboratory studies by F. Y. B. Seto of the University of Hawaii, demonstrate that iodide is rejected from large particles. The broken curve shows the additional effect of exposing droplets to 18 minutes of ultraviolet radiation, which volatilizes part of the iodine. The bottom curve shows that plankton grown with radioactive iodine yield organic material that collects preferentially on the small droplets.

vapor under nonequilibrium conditions are capable of altering the composition of the film cap as it is being formed, but this is a difficult question to approach either theoretically or experimentally.

Some of the most spectacular values of fractionation have been obtained in experiments with phosphate and iodine. In my own early work I correlated the fractionation of phosphate with drop size, using the aerosol-impactor technique shown in the lower illustration on page 66, and I found that the enrichment can be as high as 600 times and is rarely less than three [*see bottom illustration on preceding page*]. It was also clear that the degree of enrichment could be influenced by the presence of surfactants.

The fractionation of iodine and other halogens has been intensively studied by Robert A. Duce and his co-workers at the University of Hawaii and at the University of Rhode Island. Duce's colleague F. Y. B. Seto finds that iodine is rejected from large aerosol particles but that enrichment increases rapidly for particles that are smaller than about two micrometers. He also finds that if the particles are exposed to sunlight, a significant fraction of the iodine escapes as a gas [see *illustration at left*]. Seto's results are the only reproducible fractionation work of which I am aware. It is obvious that the large fractionation values observed for iodine cannot be caused by Gibbsian rejection of ions within a few angstroms of the surface.

The Recipe for Rainwater

The clue to the entire mystery of what happens at the surface was provided, in my opinion, by G. A. Dean of the New Zealand Medical Research Council, who was fascinated by the composition of rain. In 1963 he analyzed the rain at Taita, a small town on the New Zealand coast, and published a "recipe" for duplicating its content of the major ions found in seawater. The recipe called for .5 milliliter of seawater (to supply sodium, potassium, magnesium and calcium in their normal ratios), four milligrams of dried plankton and algae (to supply iodine, phosphorus, nitrogen and additional potassium), together with enough distilled water to make one liter. Dean's recipe provided a seminal insight into the relation between macroscopic meteorology and microlayer oceanography, by emphasizing the tremendous amount of biological material that accompanies the small particulate fraction (.05 percent) of the total water that moves from sea to air.

The biological material in Dean's

recipe, approaching the concentration found in a weak bouillabaisse, represents 2,000 times the amount of organics normally present in seawater. How does that much biological material find its way into the atmosphere? Duncan C. Blanchard, who was then working at the Massachusetts Institute of Technology, showed that drops from breaking bubbles could carry off an organic monolayer from a liquid surface, and I estimated that the thickness of sea surface so skimmed off could hardly be more than a tenth of a micrometer thick. At the time no one was able to see how so thin a slice could be ejected into the air.

Since then, by piecing together half a dozen arguments (no one of which is entirely convincing by itself), I am persuaded that a breaking bubble is a surface microtome of great finesse. It is

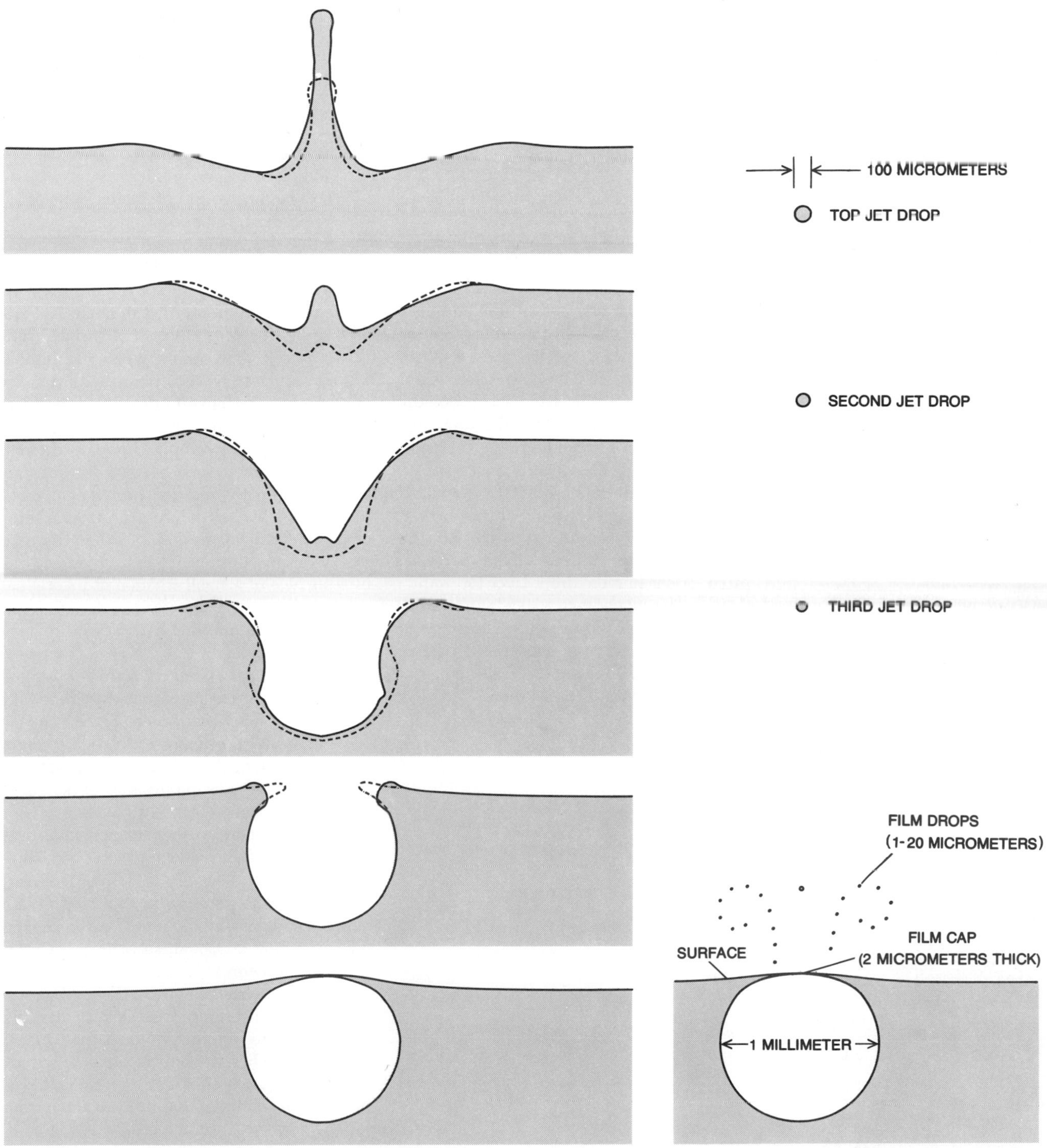

BREAKING BUBBLES, according to the author's hypothesis, provide the chief mechanism for injecting into the atmosphere a peculiar selection of the substances present in the top micrometer of the ocean. The six drawings at the left show how a collapsing bubble projects a high-velocity jet of liquid into the air. The sequence is based on high-speed photographs of a 1.7-millimeter bubble made by Duncan C. Blanchard, A. H. Woodcock and others at the Woods Hole Oceanographic Institution. The drawings depict the collapse at intervals of 1/3,000th second; profiles represented by broken lines show an intermediate stage 1/6,000th second earlier. The acceleration of surface material can reach 1,000 times gravity (1,000 *g*) for a 1.7-millimeter bubble and as much as a million *g* for a 10-micrometer bubble. The drawing at right shows schematically the variety of drops produced by a bursting bubble. Roughly a fourth of the available energy is carried off by the top drop of the jet, which then frequently evaporates into an airborne particle of salt.

capable of removing a sample of ocean surface as thin as a tenth of a micrometer, or about 1,000 times thinner than the best mechanical microtome yet devised for removing a slice of liquid.

High-speed photographs of breaking bubbles made at the Woods Hole Oceanographic Institution indicate that when a 1.7-millimeter bubble bursts, it can accelerate surface material to a value 1,000 times greater than the acceleration produced by the earth's gravity [*see illustration on preceding page*]. For bubbles of 10 micrometers the acceleration of surface material can reach a million g. Intuition is not much help in thinking about the consequences of such events. An analysis suggests that the average thickness of ocean surface ejected in the atmosphere by a bubble microtome is roughly .05 percent of the bubble diameter. A bubble one millimeter in diameter will remove a slice of ocean surface about .5 micrometer thick [*see illustrations on these two pages*].

The fastest-moving drop produced by the collapsing of a one-millimeter bubble is ejected upward at about 10 meters per second, carrying about a fourth of the energy released by the collapse. The drop reaches a height of between 10 and 15 centimeters and frequently evaporates into an airborne particle containing 30 nanograms of salt and perhaps .3 nanogram of dried plankton and algae. The organic material on the ejected drops is compressed, so that if the bubble interior is covered by a monolayer of organic molecules, the drop will be wrapped in perhaps 10 layers. The process delivers economically significant amounts of plant nutrients to coastal areas and, along with pollen and droplets of condensed volatiles from terrestrial plants, contributes enough food to high-altitude snow to support invertebrate life in the "aeolian" zone above the vegetation line. The organic material tossed up from the sea also makes the windshield of your car greasy if you park near the ocean and in times of "red tide" can include enough toxin to cause sore throats and eye irritation among shore dwellers.

The Ubiquity of Bubbles

Some 3 or 4 percent of the sea surface is covered with bubbles at any moment. Assuming that 10^{18} break per second, between one and 10 billion tons of salt are ejected into the atmosphere annually. Each drop also carries about 200 charges of positive electricity. The charge results from the peculiarities of flow at the moment the drop separates from the ascending jet. Air friction decelerates the surface of the jet by values that can reach 500,000 g while the interior is still moving rapidly upward; the resulting shear across the electrical double layer injects a few excess positive ions into the drop. Over the world ocean this amounts to a steady upward current of some 160 amperes. The excess of positive ions may account in part for the general feeling of well-being one experiences at the seashore.

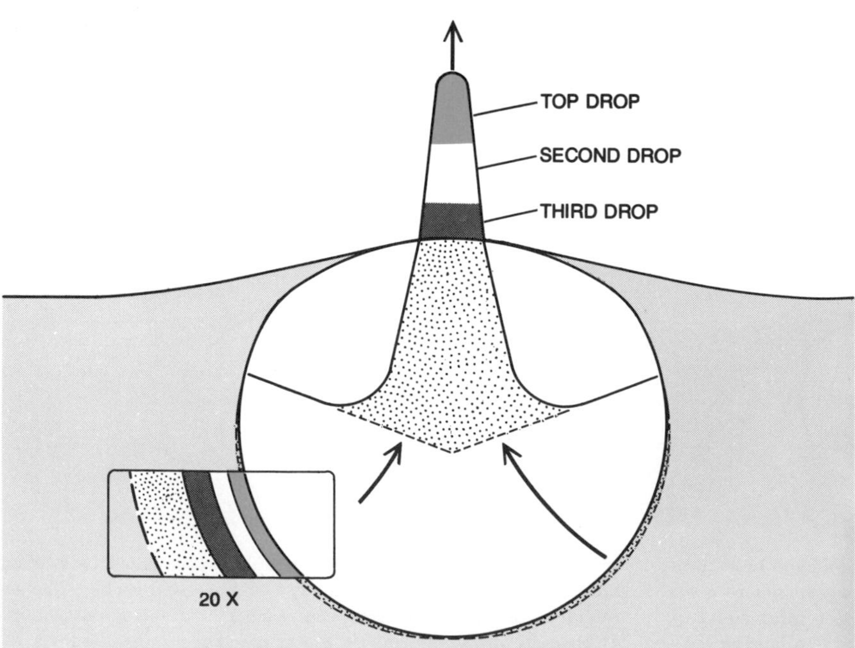

SOURCE OF DROP MATERIAL in a breaking bubble is the thin wall of the bubble just before collapse. The twentyfold enlargement of the bubble wall is color-keyed to show the origin of the top, second and third drops that will ultimately break away from the rising jet.

Workers at the Woods Hole Oceanographic Institution have found that when an inert gas is bubbled through seawater, "dissolved inorganic" phosphate is converted into "particulate organic" phosphate. Brine shrimp will grow slowly when given the phosphate-containing particles as their sole food. Very few marine organisms can extract dissolved material from seawater, and even they do so with great difficulty, but since every particle will be eaten by something, the particle-forming process is important in making phosphate (which is often a limiting nutrient) available to the plankton community.

The mechanism by which bubbles induce the formation of particles was the subject of much controversy. This seems to have been resolved by the finding that very small particles are necessary as nuclei. Apparently if the nuclei are adsorbed onto a bubble, the surface compression that occurs when the bubble breaks results in a permanent transfer of sea surfactant onto the nucleus, whereas in the absence of the nucleus the surfactant simply clumps and redissolves almost immediately. Similar particle growth (and dissolution) can be observed by half-filling a horizontal glass tube with soap solution. When the tube is tipped upright, thereby reducing the surface to a small fraction of its original area, surface compression creates a lump of soap out of what had previously been a monolayer.

A related phenomenon is observed when a gas free of carbon dioxide is bubbled through seawater. Since the gas carries off the dissolved carbon dioxide, equilibrium then demands that calcium carbonate precipitate out of solution. The seawater becomes opalescent with fine particles, a phenomenon known as "whiting" when it occurs over the Bahama Banks (through a different mechanism). The particles precipitated in this fashion are unique among carbonates in that they do not dissolve in acids. The reason seems to be that they form at the bubble surface, where the carbon dioxide concentration is lowest and the carbonate supersaturation is the highest, so that when the bubble breaks, the particles are coated with a tenacious film of organic material that protects them from acid attack. Instant Saran Wrap!

Pollution and Sea Farming

Not only is the oceanic microlayer exposed to all the pollutants in the atmosphere but also it absorbs the brunt of oil spills, including oil released from blowouts on the ocean floor. Some 85 percent of all the organic material now afloat on the Baltic, the Mediterranean and North seas consists of petroleum. The microlayer avidly concentrates such heavy metals as lead, mercury, copper, chromium and zinc. It also retains such long-lasting chlorinated hydrocarbons as DDT and PCB (polychlorinated biphenyls), which reach the sea in precipitation and surface runoff. These substances enter the food chain, where they are progressively concentrated.

Perhaps the most bizarre example of oceanic dispersal of a pollutant is the one deduced to explain the appearance of DDT in Antarctica long before it could be carried there by wind or water currents. The sequence starts with the spraying of DDT on Northern Hemisphere crops. A portion enters the atmosphere by codistilling with the water that is transpired through the leaves of plants. The DDT reaches the North Atlantic in precipitation, where it is trapped by the microlayer. It promptly enters the food chain and is concentrated as it passes from smaller organisms to larger ones. The agent for transmitting the DDT across the Equator is Wilson's petrel (perhaps the most abundant bird in the world), which feeds in the North Atlantic but breeds in the Antarctic. So transported, the DDT is picked up by the food chain in the southern ocean. When it finally reaches the tissues of the crabeater seal, it appears for the first time in concentrations high enough to be detected.

I shall close by mentioning two ways in which increased understanding of microlayer oceanography may be of great practical significance. The first relates to the potential control of hurricanes. It is suspected that both water vapor and salt nuclei from the sea play a crucial role in the spawning and growth of hurricanes. Since both substances must cross the microlayer, it is conceivable that making small and inexpensive changes in the sea surface ahead of an incipient hurricane might reduce the storm's severity. With this in mind one investigator is studying the evaporation rate from the microlayer, having noticed that there are anomalies in the surface properties of seawater at about 30 degrees Celsius, the same ocean temperature known to favor the genesis of hurricanes.

A second way in which the microlayer might be manipulated for man's benefit would be as a source of protein-rich food. One reason agriculture has so greatly outstripped mariculture is the extensive use of fertilizer on land. Fertilizing the sea has never seemed profitable because both fertilizer and crop can escape so easily. (The sustained yield of the semienclosed North Sea fishery depends entirely on the nutrients poured in by the Seine, the Thames, the Rhine and other open sewers—a fortuitous and nonoptimized bit of mariculture.)

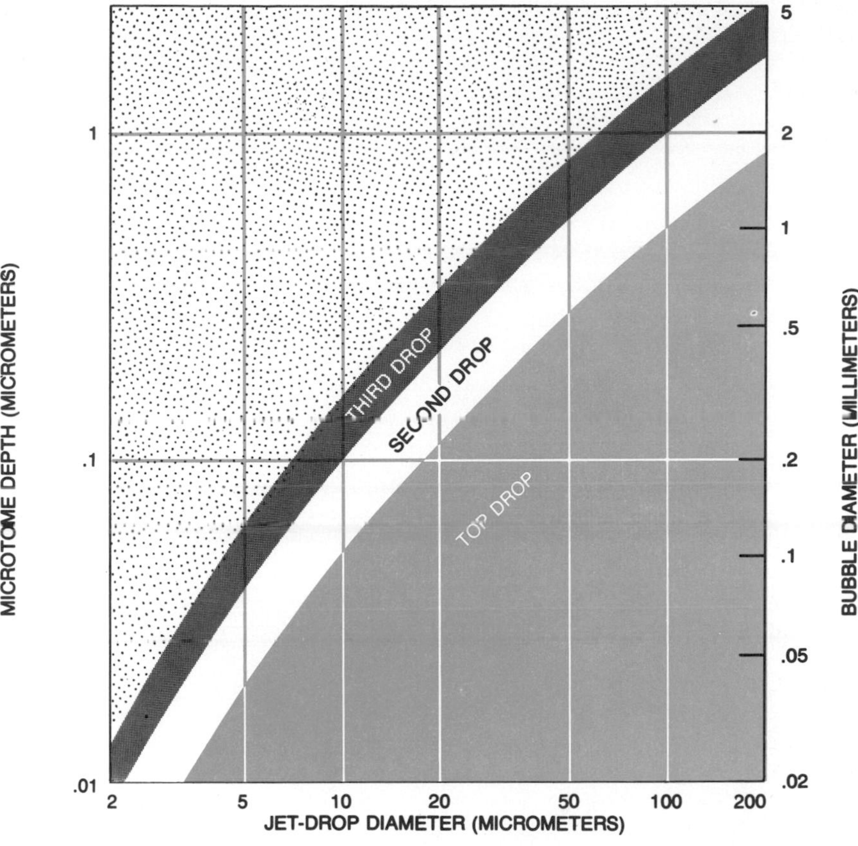

BUBBLE "MICROTOME" can produce a thinner slice of the ocean surface than any mechanical device known. The average thickness of the sample ejected into the atmosphere by a breaking bubble (*scale at left*) is about .05 percent of the bubble diameter. Thus the collapse of a bubble one millimeter in diameter will produce a sample in the top jet representing roughly the top .5 micrometer (5,000 angstroms). Subsequent jet drops, all about 100 micrometers in diameter, originate in onionlike shells successively deeper in the solution.

One can imagine an optimized microlayer farming operation designed along the following lines. Microlayer "weeds," or unwanted organisms, are first removed with a "cultivator," which is simply a Harvey drum skimmer. Surfactant fertilizer, growth hormone and "seed" (desirable varieties of neuston) are next applied. The fertilizer could be designed as a recyclable nutrient carrier that could not escape the microlayer, rather than as a totally consumable molecule. Since each step of the food chain loses 90 percent of the energy contained in the previous step, it is advantageous to harvest the phytoneuston (the plant species) before they are eaten by the zooneuston (the animal ones). Optimum operation of the farm would have most of the nutrients converted into biomass by the time the prevailing wind had moved the microlayer down to the harvesting skimmer. Even so, grazing predators might have to be fenced out. The farm should be about as productive as a Kansas cornfield or a Caribbean sugar plantation.

Obviously many problems need to be solved before a microlayer farm is feasible. Present efforts at collecting oil spills might be directly applicable to harvesting. The operation of a farm would have to be largely automatic, utilizing power available on site from the sun or the wind. The vagaries of weather are no greater at sea than on land. And it might prove feasible to protect the farming apparatus by submerging it 10 meters or so during storms.

Perhaps the biggest advantage of mariculture over agriculture is the ease with which microscopic organisms can be manipulated genetically. Strains of *Chlorella* have already been bred that are 80 percent oil, whereas cottonseed, our richest present source, is only 20 percent oil. Similar results might be achieved with protein content. The food-processing industry should be clever enough to convert neuston into a product both palatable and nutritious.

The Microstructure of the Ocean

14

by Michael C. Gregg
February 1973

The temperature and saltiness of seawater can now be mapped centimeter by centimeter. Such mapping is needed to learn how the sea is so effectively stirred by the winds and the tides

It is just 100 years and two months since H.M.S. *Challenger*, a wooden corvette of 2,306 tons, sailed from Portsmouth, England, on the world's first oceanographic expedition. During the voyage, which lasted four years and covered 69,000 nautical miles, the *Challenger* investigators made observations at 362 locations. As a part of their mission they took tens of thousands of samples of the Atlantic, Pacific and Antarctic oceans by lowering bottles that could be opened at a prescribed depth, filled and sealed before being hoisted back aboard ship. At each depth a temperature reading was taken by breaking the capillary column in a bulb thermometer by remote control. This cumbersome method of collecting samples and taking temperatures prevailed with modest improvements for the next 90 years and has provided much of what we know about the large-scale characteristics of the medium that covers 70 percent of the earth's surface. Although the sample-bottle-and-thermometer method is a good one for establishing the gross seasonal changes in salinity, temperature and density of seawater at various depths, its sampling "mesh" is far too coarse to detect the small-scale transactions in salinity, temperature and density by which energy received from the sun is exchanged between the ocean and the atmosphere and ultimately redistributed over the earth.

Within the past 10 years advances in solid-state electronics have made possible rugged instruments that can supply a continuous record of temperature, salinity, density and velocity from the ocean surface down to depths of a kilometer or more. Because these new instruments are able to sense changes over distances of less than a centimeter they have revealed the widespread existence of coherent structures in the sea that previously escaped detection. In some places steplike profiles of temperature and salinity suggest the presence of homogeneous layers of water separated by thin interfaces. Typical layers 10 to 20 meters thick can often be traced horizontally for as far as 15 kilometers; other layers only 20 to 30 centimeters thick extend laterally for several hundred meters. Vertical measurements in the range from a few millimeters to a few centimeters reveal intense and irregular fluctuations of temperature and salinity, indicating that the top 30 meters of the sea is often completely turbulent. It is here that energy is being exchanged most actively. It is now customary to apply the term "microstructure" to oceanic physical processes on the scale of a few centimeters or less. As we shall see, the rate of diffusion of temperature and salinity along microstructure gradients sets a limit to the smallest significant scales of fluctuation in the ocean.

It has been evident since the voyage of the *Challenger* that the gross features of the principal ocean-current systems and the distribution patterns of temperature and salinity change little from year to year; they represent nearly a steady-state condition. By learning how processes on the microscale lead to this steady-state condition of temperature and salinity we shall be better able to understand how substances such as nutrients and man-made pollutants are distributed in the ocean.

Two closely related processes are responsible for dissipating energy and smoothing variations in the ocean: mixing and stirring. Although the two terms are often used interchangeably in ordinary speech, the oceanographer must make an important distinction between them. This distinction was pointed out by Carl Eckart of the Scripps Institution of Oceanography in a farseeing paper published in 1948. Mixing is the reduction of variations by the action of molecular diffusion along gradients from regions of higher temperature or concentration to regions of lower temperature or concentration. The rate of mixing depends directly on the steepness of the gradients and on the molecular coefficients of diffusivity. Stirring, on the other hand, is the creation of velocity differences by any process that imparts kinetic energy to a liquid. If a liquid is not homogeneous with respect to, say, temperature and salinity, stirring will tend to increase the gradients between non-homogeneous parcels, leading both to greater variability and to faster mixing. For example, if a source of variations is introduced by carefully injecting a blob of salty water into an otherwise uniform and still sample of fresh water, the blob will sink to the bottom and persist for a considerable time because salt diffuses rather slowly. If one stirs the water with a paddle, the resulting velocity differences, or shear, will distort and stretch the blob, thus increasing the area of contact between it and the surrounding water and sharpening the salinity gradients across the interface [*see top illustration on following page*]. Both of these effects will hasten mixing.

An important task of the oceanographer is to identify the forces that stir the ocean. Stirring motions in the deep ocean are usually gentle. Although these weak shears produce marked contrasts in temperature and salinity over distances greater than several meters, they generate little microstructure and therefore do little to enhance the rate of mixing. A common situation arises when two adjacent volumes of water that differ significantly in temperature are sheared sideways by a vertical velocity gradient; the result is a strong temperature contrast in the vertical [*see illustra-*

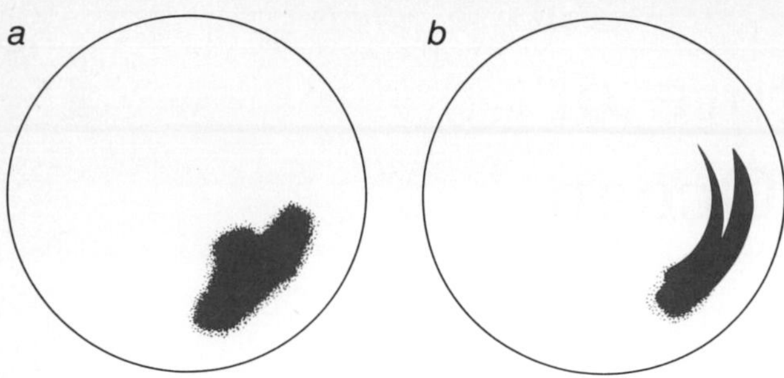

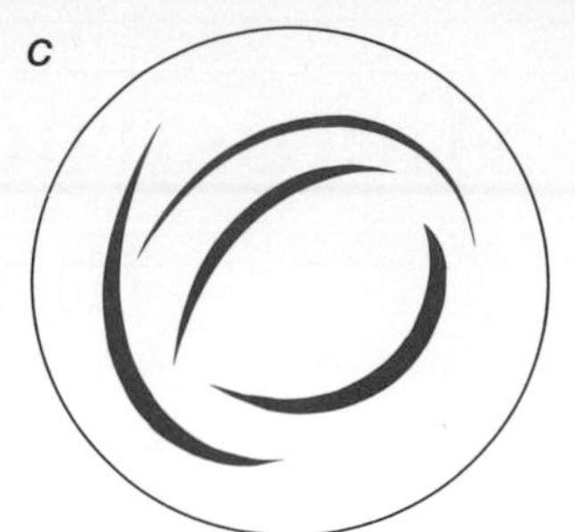

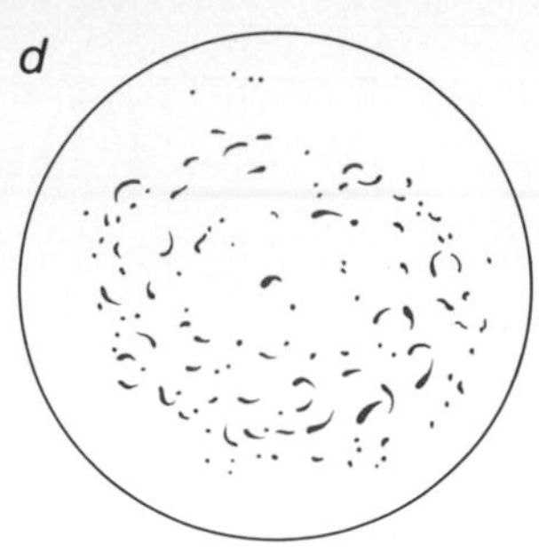

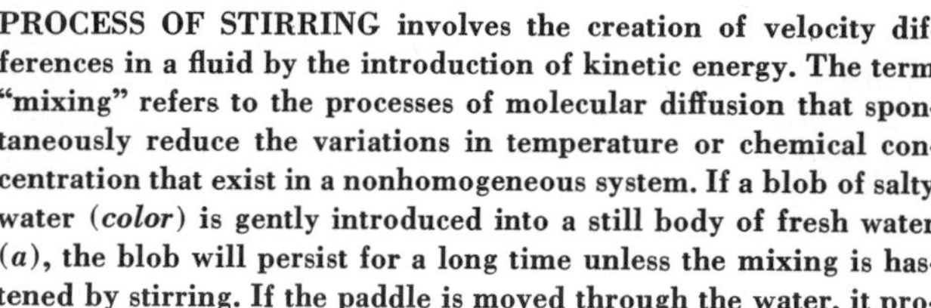

PROCESS OF STIRRING involves the creation of velocity differences in a fluid by the introduction of kinetic energy. The term "mixing" refers to the processes of molecular diffusion that spontaneously reduce the variations in temperature or chemical concentration that exist in a nonhomogeneous system. If a blob of salty water (*color*) is gently introduced into a still body of fresh water (*a*), the blob will persist for a long time unless the mixing is hastened by stirring. If the paddle is moved through the water, it produces differences in velocity that stretch and distort the blob, increasing its area and sharpening the salinity gradients (*b*). If the stirring is now continued (*c*), the blob is broken into long, thin filaments. Both the increase in area and the sharpening of gradients increase the rate of molecular diffusion from regions of higher salinity to regions of lower salinity. With or without stirring, the mixing will eventually approach completion (*d*). The winds and the tides are the principal agencies that act to stir the ocean.

tion below]. This mechanism probably accounts for many features in large-scale vertical temperature profiles and in some of the smaller "stairstep" profiles observed near the surface.

We have substantial evidence for two stirring processes whose strong shearing motions are capable of generating intense microscale activity. The first is turbulent stirring driven either by the winds near the surface or by shear instabilities associated with currents and internal waves at greater depths. The second stirring process arises from vertical convective motions produced by the differing rates of diffusion of heat and salt. This process is a local one whose vertical scale seldom exceeds a few meters; it tends to operate where weak stirring has already increased the vertical gradients of temperature and salinity. Before going further into the processes that generate microstructure I shall discuss how variations on a scale exceeding several meters arise. I shall deal primarily with vertical variations both because they are more pronounced than horizontal ones and because they are better understood.

Since the sea surface is nearly in thermal equilibrium with the atmosphere, variations in surface temperature reflect the local heat balance, which ranges from −1.9 degrees Celsius in polar regions to nearly 30 degrees C. at the Equator. Away from coastal areas, where the runoff of precipitation affects the salt content, the salinity of the ocean varies from 32 to 37 grams of dissolved solid per kilogram of seawater, expressed as 32 to 37 ‰. (The symbol ‰ is read "per mil.") The variations in salinity reflect differences in the relative rates of precipitation and of evaporation from the surface of the ocean. When differences in salinity and temperature produce horizontal differences in density, there is a tendency for the denser waters to sink below those that are less dense.

Although a difference of 1 ‰ in salinity has an effect on the density of seawater about five times greater than a difference of one degree C. in temperature, the greater range of temperature dominates the large-scale density structure of the ocean. Salinity differences, however, have very appreciable effects on smaller scale. The densest waters are believed to be formed under the ice of the Weddell Sea in the Antarctic, in the North Atlantic off Greenland and in the Norwegian Sea. These dense waters sink to the bottom and slowly spread into the major ocean basins. Less dense waters from other high-latitude sites sink to intermediate depths in a similar manner. As they spread into the central oceans the various tongues of water interleave, forming vertical density profiles with a continuous stratification. Since any ocean is very shallow in relation to its breadth, the interleaving of water of various densities gives rise to temperature gradients that are much sharper vertically than horizontally.

The structure of the vertical temperature gradients is seen most simply in the open ocean, far from complicating boundary effects. Here measurements of temperature and salinity show little

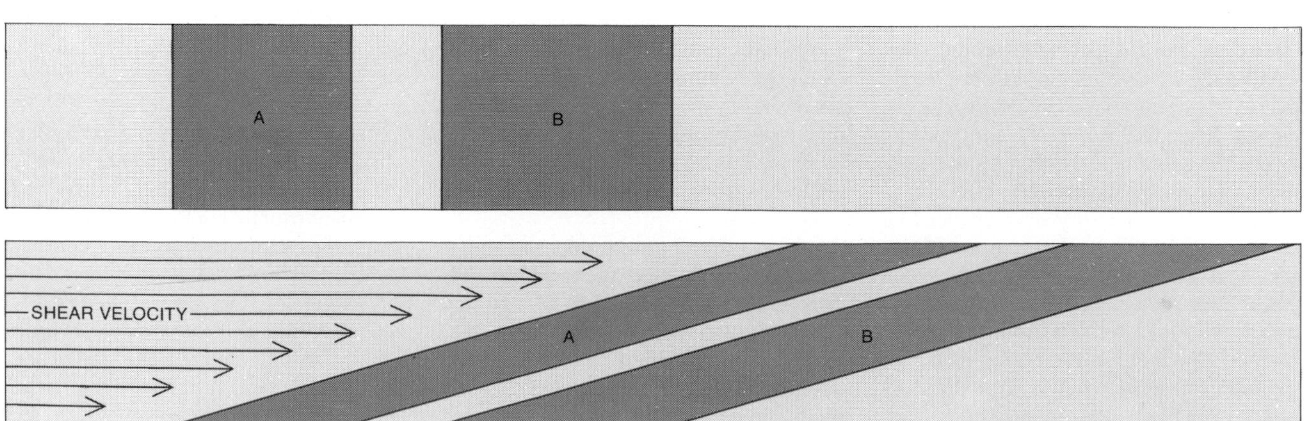

STRONG VERTICAL TEMPERATURE GRADIENTS can be formed when two adjacent volumes of water at different temperatures (*A*, *B*) are tilted sideways by shearing currents. Illustrations on this page follow diagrams first published by Carl Eckart.

change from one location to another or from year to year at the same location. Below the well-mixed layer at the surface the temperature drops very rapidly. Below the region where the temperature gradient is a maximum, known as the thermocline, the temperature profile is very nearly an exponential curve [*see illustration at right*]. The temperature varies smoothly over vertical distances greater than several meters, implying that the tongues of water forming the temperature profile have undergone considerable vertical mixing as they spread toward the Equator from high latitudes. Many theoretical treatments, known as thermocline models, have attempted to establish how this mixing takes place. Such studies require that the amount of water sinking at high latitudes be balanced by the amount rising in low or middle latitudes.

If one assumes in a thermocline model that there is a broad general upward movement of the bottom water over the interior of the ocean basins, one must provide for a moderately intense downward transport of heat. Since molecular heat conduction (that is, simple mixing) along the mean temperature gradients is too slow to match the upward movement of the cold water, some models assume the existence of additional mixing driven by turbulent stirring to increase the heat flow [*see top illustration on page 162*]. Typical values obtained by combining the downward turbulence-driven heat flow and the upward motion of the cold deep water to fit the observed temperature data are an upward velocity of one centimeter per day and a turbulent-mixing coefficient of one square centimeter per second. It is assumed that the rate of mixing is uniform except within a few hundred meters of the surface and the bottom.

An alternative model suggests that enhanced mixing rates may not be necessary if the effects of prevailing winds are considered. The Coriolis force, which is created by the earth's rotation, tends to deflect moving objects to the right in the Northern Hemisphere and to the left in the Southern Hemisphere. Early in this century the Norwegian oceanographer Vagn Walfrid Ekman concluded that the combined effect of a steady wind and the Coriolis force is to produce a net transport of the surface water 90 degrees to the right of the wind in the Northern Hemisphere and an equal amount to the left south of the Equator.

Thus the prevailing westerlies in the middle latitudes and the easterly trade winds in the Tropics combine to drive the surface waters to the center of subtropical gyres (large, slowly rotating masses of water). The net inflow along the surface leads to a slight elevation in the center of the gyres and a tendency for the warm waters under the gyres to sink [*see bottom illustration on page 162*]. The downward heat flux provides a mechanism for the formation of the thermocline without the necessity of invoking turbulence-driven mixing. In this model the rising deep waters cannot penetrate the thermocline over a broad region but are concentrated in restricted areas, such as at the Equator and near the coasts. The two types of model make very different predictions of the rates of mixing to be expected in the deep ocean.

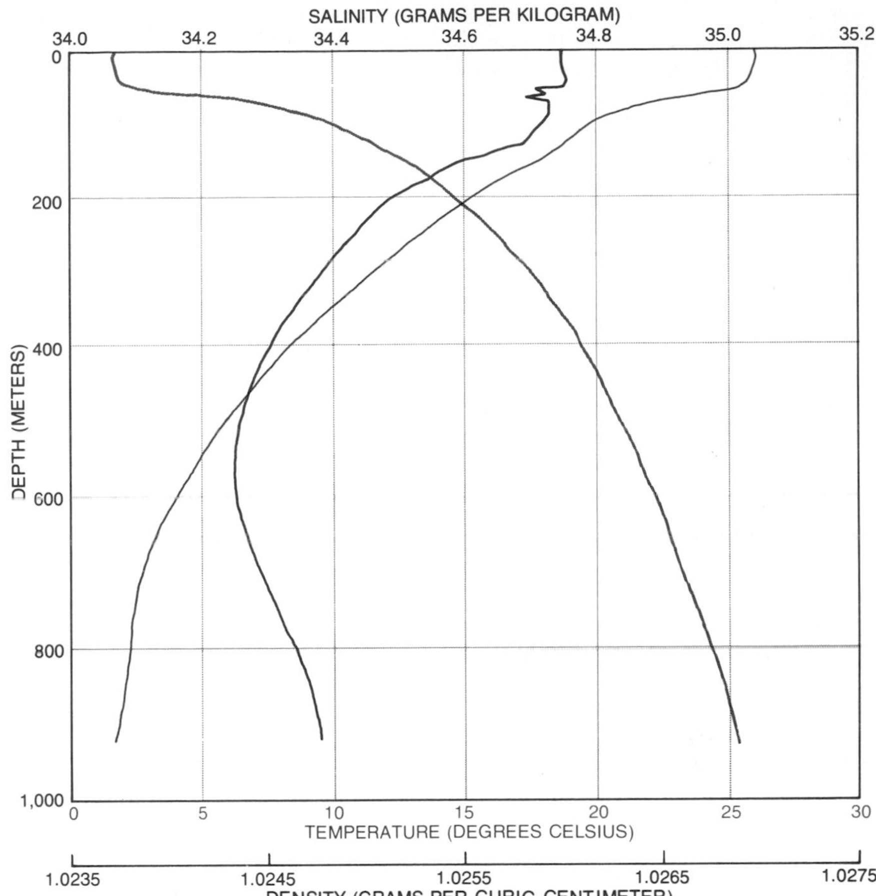

TEMPERATURE AND SALINITY IN MID-PACIFIC were recorded with an instrument having a resolution of a few meters. A well-mixed layer in which the temperature and salinity are nearly constant extends to a depth of about 60 meters. Their temperature (*colored curve*) falls steeply and then more slowly. The region of rapid temperature change extending several hundred meters below the mixed layer is called the thermocline. The salinity (*black curve*) reaches a minimum value at about 550 meters, then rises again. Since the temperature dominates the density structure (*gray curve*), the region of maximum density change coincides with the thermocline. The temperature of the Pacific below a depth of four kilometers is about one degree Celsius; salinity is 34.7 grams per kilogram of seawater.

Temperature variations on an intermediate scale, with vertical dimensions from a few meters to several hundred meters, are most pronounced in coastal regions and near the open-ocean boundaries of major current systems. The variations are due in part to variations in meteorological conditions, which can lead to the formation of local tongues of dense water, and to the stirring effect of the rather high shears associated with the surface and subsurface boundary currents. For example, vertical measurements made a few miles off San Diego show pronounced and irregular layers 10 to 30 meters thick, including numerous temperature inversions, that is, regions where the temperature increases with depth instead of decreasing [*see illustration on next two pages*].

These layers, a characteristic feature below the California Current, are believed to be formed by the interleaving of discordant water types of different densities. The temperature inversions are accompanied by local increases in salinity that compensate for the reduced density associated with the warmer layers of water, so that the net density increases continuously with depth. The warm saline water is formed in the eastern tropical Pacific and flows northward at depths of 200 to 300 meters along the California coast, where it intermingles with cold, less salty water that is flowing south along the surface from the Gulf of Alaska.

High-resolution observations made with continuous recorders have shown that below the near-surface regions high vertical mixing rates tend to be found in small patches where there is intense microstructure activity. The rate of molecular diffusion along these thin but sharp microstructure gradients is at least several factors of 10 higher than the diffusion rate found in the much weaker large-scale gradients. The evidence to date suggests that the patches of microstructure activity and high mixing rates have vertical dimensions of a few meters or less. Most of the gradient features of temperature and salinity profiles greater than 10 meters or so are therefore not the result of mixing but of weak shearing motions. These larger structures are continuously being modified and smoothed by the existence of smaller regions in which there is intense mixing.

Some degree of turbulence is associated with all these local regions of high mixing. In the dynamical instabilities, where motions tend to become disorganized, the role of turbulent stirring is dominant. In the differential diffusion processes, on the other hand, where convective instabilities lead to more orderly velocity patterns, the role of turbulence is more restricted. On scales of a few meters, however, the distinction between features formed by the processes that generate microstructure and features due to weaker motions that do not result in much mixing is complicated by the fact that away from the surface the two kinds of process rarely carry the mixing to completion. In a stratified fluid such partial mixing usually results in the formation of additional layers.

The task of providing a theoretical description of turbulence has been a challenge to generations of hydrodynamicists. Since the random velocity fluctuations that characterize turbulence continue to elude a rigorous theory, the most successful treatments have either been statistical or have relied on physical arguments and dimensional analysis.

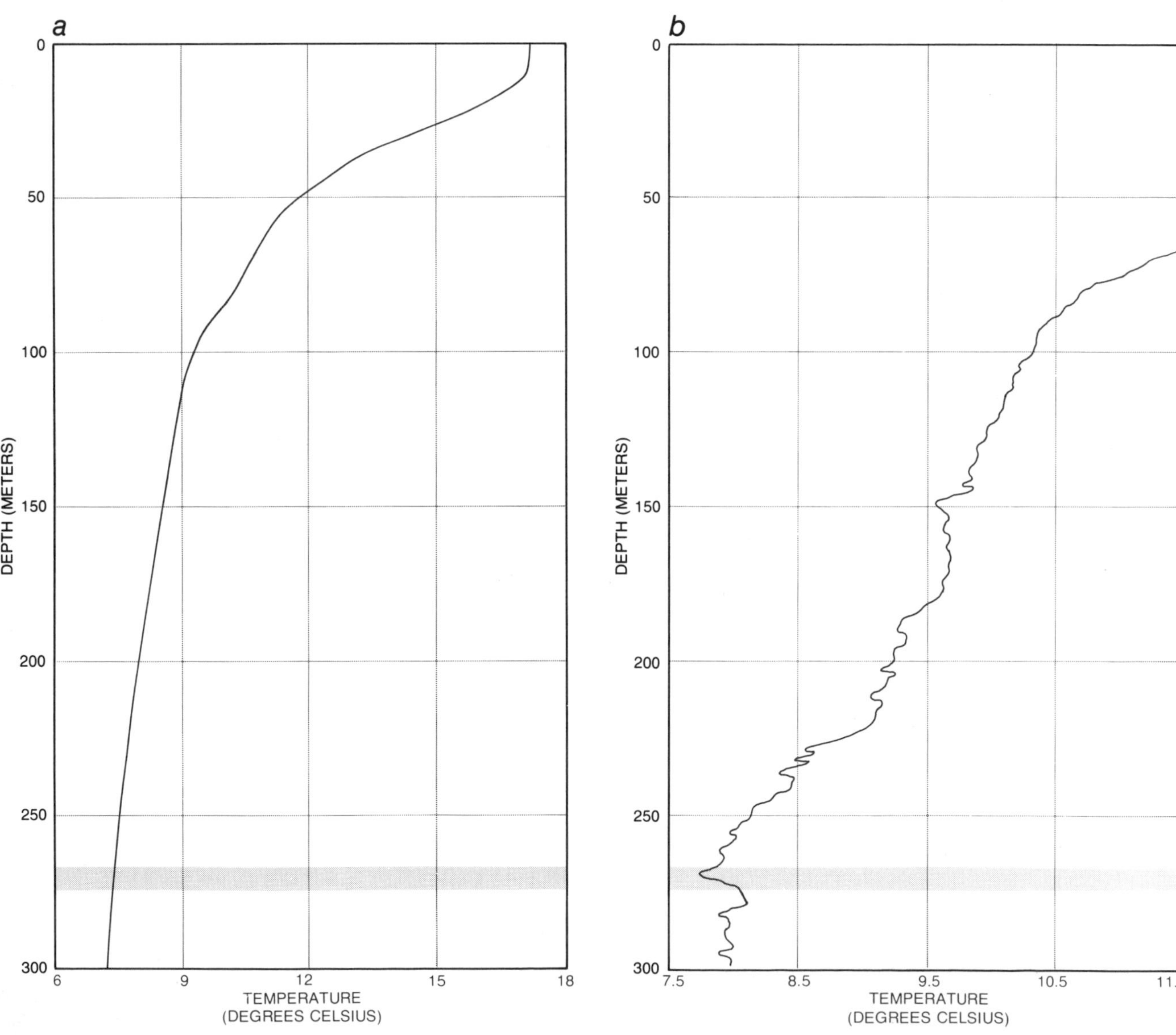

SERIES OF TEMPERATURE READINGS shows how small-scale variations that eluded detection by earlier methods are revealed by electronic recorders. The first curve (*a*) is a hand-plotted record of the ocean temperature off San Diego made by lowering a bulb thermometer, a simple method that dates back more than a century. The second curve (*b*) was made nearby with a commercial electronic instrument, which provides a continuous record with a maximum resolution of a few meters. The third trace (*c*) was made simultaneously with a still more sensitive instrument capable of revealing temperature fluctuations centimeter by centimeter. An

Briefly stated, dimensional analysis involves determining what parameters are present in an equation and then deducing how they must be related to obtain the same dimensions on both sides of the equation. For example, a velocity scale (in centimeters per second) must be the ratio of a length scale (in centimeters) and a time scale (in seconds). It has proved to be useful to think of turbulent flow as a superposition of eddies of different sizes, the largest of which are limited by the dimensions of the region in which the flow takes place or by the scale of the process generating the turbulence.

The energy extracted from the mean flow by the largest eddies cascades through increasingly smaller eddies until it is ultimately dissipated as heat by the action of viscosity. If the turbulence is well developed, so that the smallest eddies are many factors of 10 smaller than the large eddies that hold the bulk of the kinetic energy, it turns out that the motions of smallest scale are statistically independent of the largest eddies. The successive cascading of energy through increasingly smaller scales of motion decouples the smallest eddies from the details of the large-scale flow, so that the energy balance for the small-scale events depends only on the dynamics that are important at that scale.

The Russian mathematician A. N. Kolmogorov postulated some years ago that the dominant energy balance of the smallest eddies depends on only two factors: the net rate at which energy cascades down from large-scale motions and the viscosity of the fluid. Kolmogorov showed by dimensional analysis how the energy-dissipation rate and the viscosity can be combined in an equation that predicts the smallest-scale length for fluctuations of velocity. For the range of dissipation rates prevailing in the ocean the velocity cutoff scale—the dimension below which velocity fluctuations become insignificant—varies from about one centimeter to about six centimeters. Very little turbulent kinetic energy can exist over shorter dimensions.

If variations of salinity or temperature are present, they will be stirred by the same eddies that produce the velocity structure. The important result obtained from dimensional analysis is that the scale cutoffs for temperature and salinity are respectively only about a half and

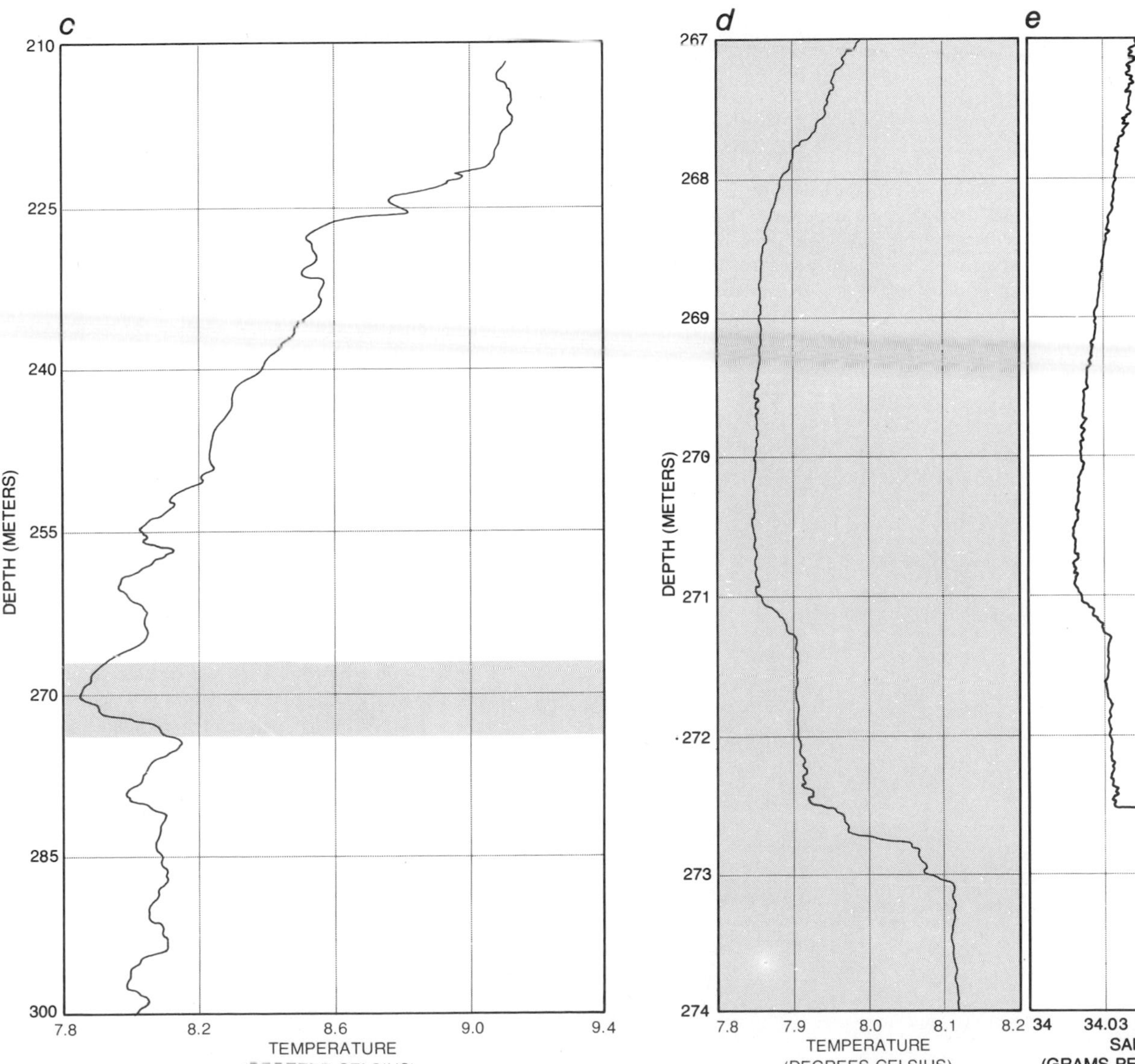

enlarged section of the third record, covering tiny temperature variations over a depth interval of only seven meters, constitutes the fourth trace (*d*). The seven-meter interval from 267 to 274 meters is indicated by a colored band in the first three traces. The last curve (*e*) is a salinity record made simultaneously with the high-resolution temperature record at its left. One can see a steplike layering of temperature and salinity between 271 and 274 meters. Such records provide clues to the transport of heat and chemical elements in the ocean. These records were made by the author and Charles S. Cox of the Scripps Institution of Oceanography.

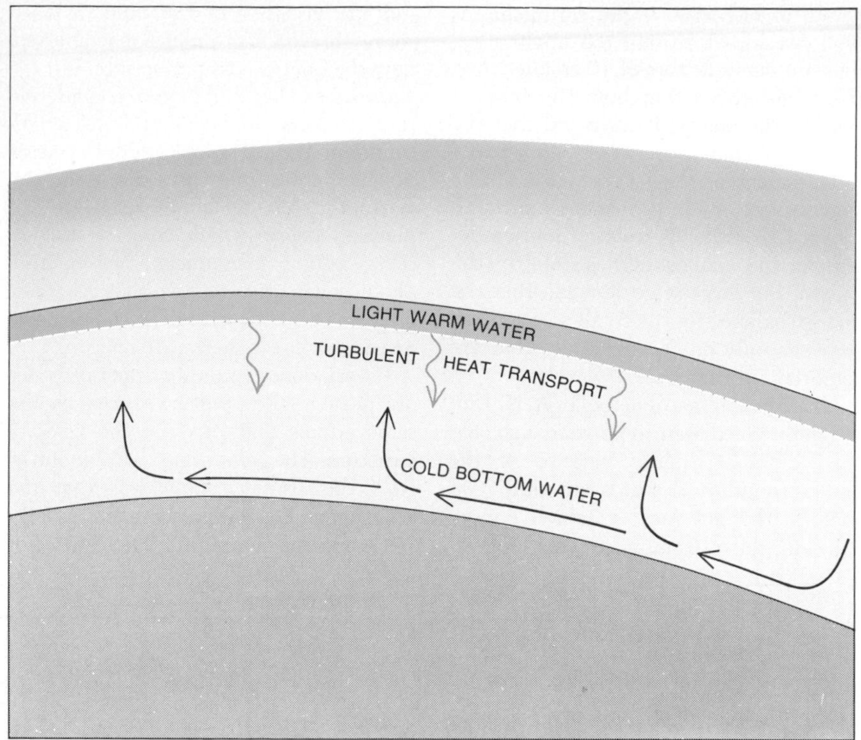

THERMOCLINE MODELS attempt to explain how heat is exchanged between the warm surface waters of the subtropics and the cold deep water that originates in polar regions. The exchange produces the characteristic mid-ocean temperature profile that is shown in the illustration on page 67. This simple model assumes that the downward mixing of heat, driven by turbulent stirring, is balanced by a slow upward drift of cold bottom water.

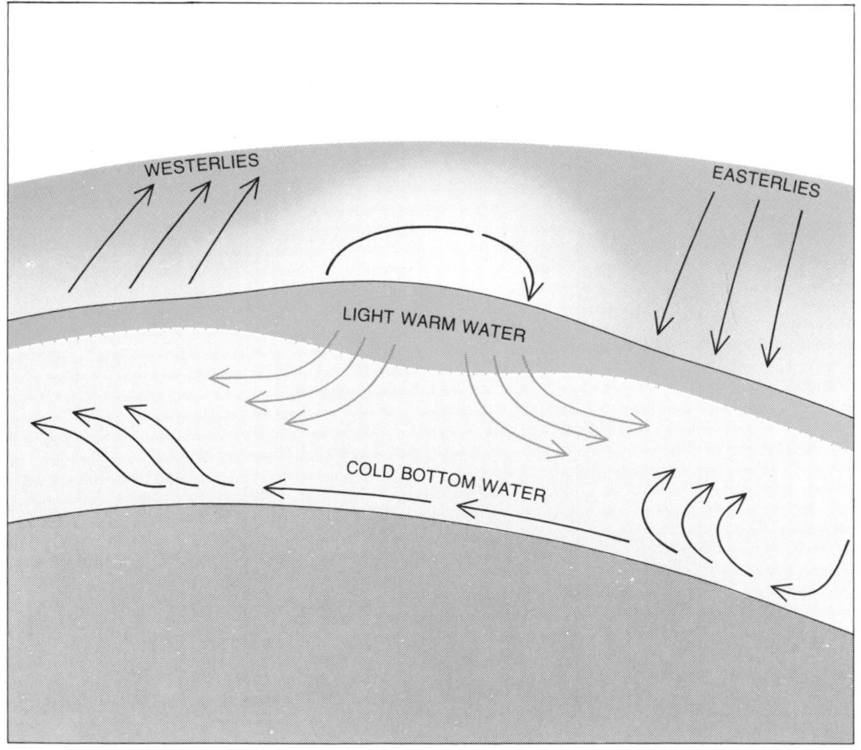

ALTERNATIVE THERMOCLINE MODEL minimizes the effect of turbulent stirring and attributes oceanic heat transport primarily to the development of subtropical gyres: slowly rotating mounds of water hundreds of kilometers across. Gyres are created by the prevailing westerly and easterly winds in conjunction with Coriolis force, which deflects moving bodies to the right in the Northern Hemisphere and to the left in the Southern. There would be a tendency for the warm surface water under a subtropical gyre to sink and flow outward.

WAKE LEFT BY A DYE PELLET is photographed by a diver (*at left in the photograph on the opposite page*) in a study of oceanic microstructure conducted in the Mediterranean by John D. Woods of the University of Southampton. Kinks in the wake of the pellet reveal differences in the motion of adjacent layers in the water. The marker to the right of the dye-pellet wake is held in position by other divers to provide a scale and a reference to the vertical and horizontal. Divers working with Woods were members of a Royal Navy team.

a fifth as large as the scale cutoff for velocity is. This means that in a sample of water in which velocity fluctuations had been smoothed nearly to zero, temperature fluctuations could still exist, and that if mixing continued until temperature fluctuations had been smoothed nearly to zero, salinity fluctuations could still exist. Although there is little kinetic energy associated with motions smaller than the velocity cutoff, the distortion of fluid elements by the remaining velocity shear will stir the temperature and salinity variations over smaller and smaller scales until ultimately they are both smoothed by diffusion. The importance for oceanography is that it is not necessary to resolve scales of variation smaller than the cutoff values in order to compute the rate of mixing in a turbulent flow field.

How does one estimate the energy-dissipation rates for the ocean on which the scale cutoffs are based? There are two principal forces acting on the ocean: the tides and the winds. Each has an average input of from five to 20 ergs per square centimeter of sea surface per second. It is believed that most of the tidal energy is dissipated in the shallow seas near the coasts, whereas most of the motions induced by the winds are dissipated in the upper layers of the open ocean. A typical value for the wind-driven energy input can be obtained by assuming that an average input of 10 ergs per square centimeter per second is dissipated uniformly in a mixed layer 50 meters deep lying above the thermocline. That works out to a dissipation rate of 2×10^{-3} erg per gram-second. A reasonable limit for the dissipation rate for the ocean below the thermocline can be estimated by combining the total tidal and wind energy and dividing it by the total volume of the sea. This gives a dissipation rate of 1×10^{-5} erg per gram-second. One can then compute the scale cut-

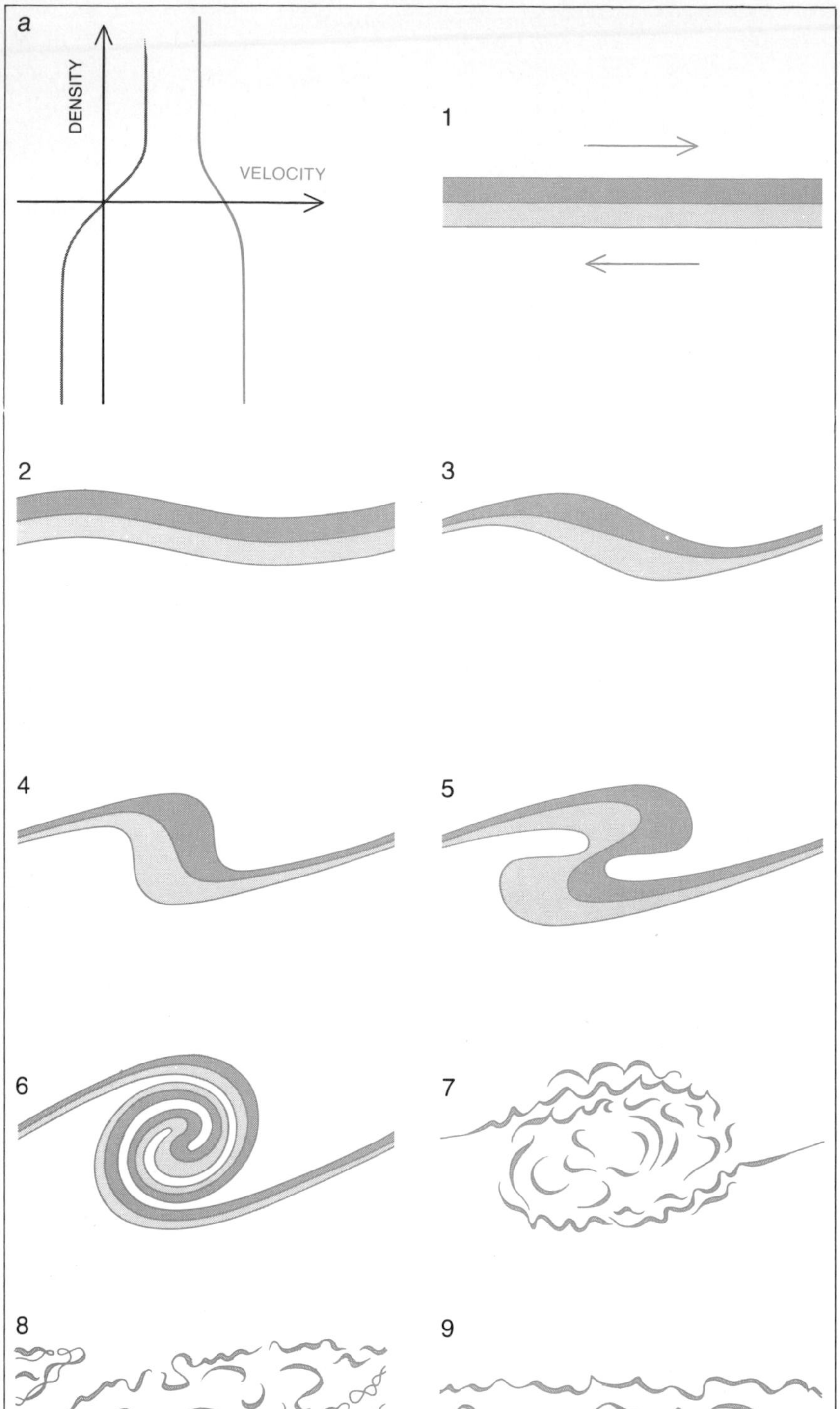

KELVIN-HELMHOLTZ SHEAR INSTABILITY is one form of process believed to stir the ocean in regions where high-frequency internal waves propagate between two layers of significantly different density. The formal situation is shown in *a*, where the horizontal arrow represents the interface between the two layers. The upper layer has a higher velocity and lower density than the lower layer. The following drawings are based on observations of the instability that can be generated in a two-layer laboratory tank by rapidly raising and then lowering one end of the tank. The two layers soon lose their coherence and break up into turbulent patches. The patches are rapidly flattened by stratification, which gives rise to a finely layered microstructure as the sloshing inside the tank subsides. The diagrams are based on a study by S. A. Thorpe of National Institute of Oceanography in England.

off for velocity, temperature and salinity based on these energy-dissipation limits. These dimensions establish the resolution the oceanographer must try to achieve in his measurements. Although a velocity probe must be capable of sensing fluctuations over several centimeters and a temperature probe must be capable of sensing them over about one centimeter, a salinity probe should have a resolution measured in millimeters.

Microstructure measurements capable of resolving the length-scale cutoff for both velocity and temperature have been made in the upper 300 meters of the ocean by a group of Canadian oceanographers, Harold L. Grant, Robert W. Stewart and Anthony Moilliet, who mounted sensitive probes on a vehicle towed horizontally behind a ship. This work has been continued by Patrick W. Nasmyth in the open coastal waters west of Vancouver Island. He has often found the upper 30 meters of the Pacific to be completely turbulent, with energy-dissipation rates as high as 2.5×10^{-2} erg per gram-second. The probes used are sensitive enough to distinguish between the cutoff length for velocity and the shorter cutoff length for temperature in water that is sufficiently turbulent. A probe capable of resolving the still finer cutoff length for salinity has yet to be operated in the ocean.

The source of the nearly uniform turbulent activity documented by the Canadian workers is the wind blowing on the surface of the sea. During strong storms the wind is capable of churning the surface layers to depths of tens of meters. If the intense stirring persists for a few days, it can drive the mixing nearly to completion, producing a highly uniform region, termed the mixed layer, extending 10 to 100 meters below the surface. The weak turbulent motions generated by only moderate winds are then sufficient to maintain the mixed layer for a considerable period.

Below the mixed layer the effects of surface motions are strongly attenuated. Intense turbulence arises only sporadically when there happens to be a local concentration of energy great enough to overcome the normal stratification, creating patches of microstructure activity separated by quieter water [*see illustration on opposite page*]. Dissipation values of 5×10^{-4} erg per gram-second have been recorded in patches at a depth of 90 meters. In addition to the regions of active turbulence one sometimes finds patches of "fossil" turbulence in which the velocity fluctuations have been dissipated by viscosity but the temperature

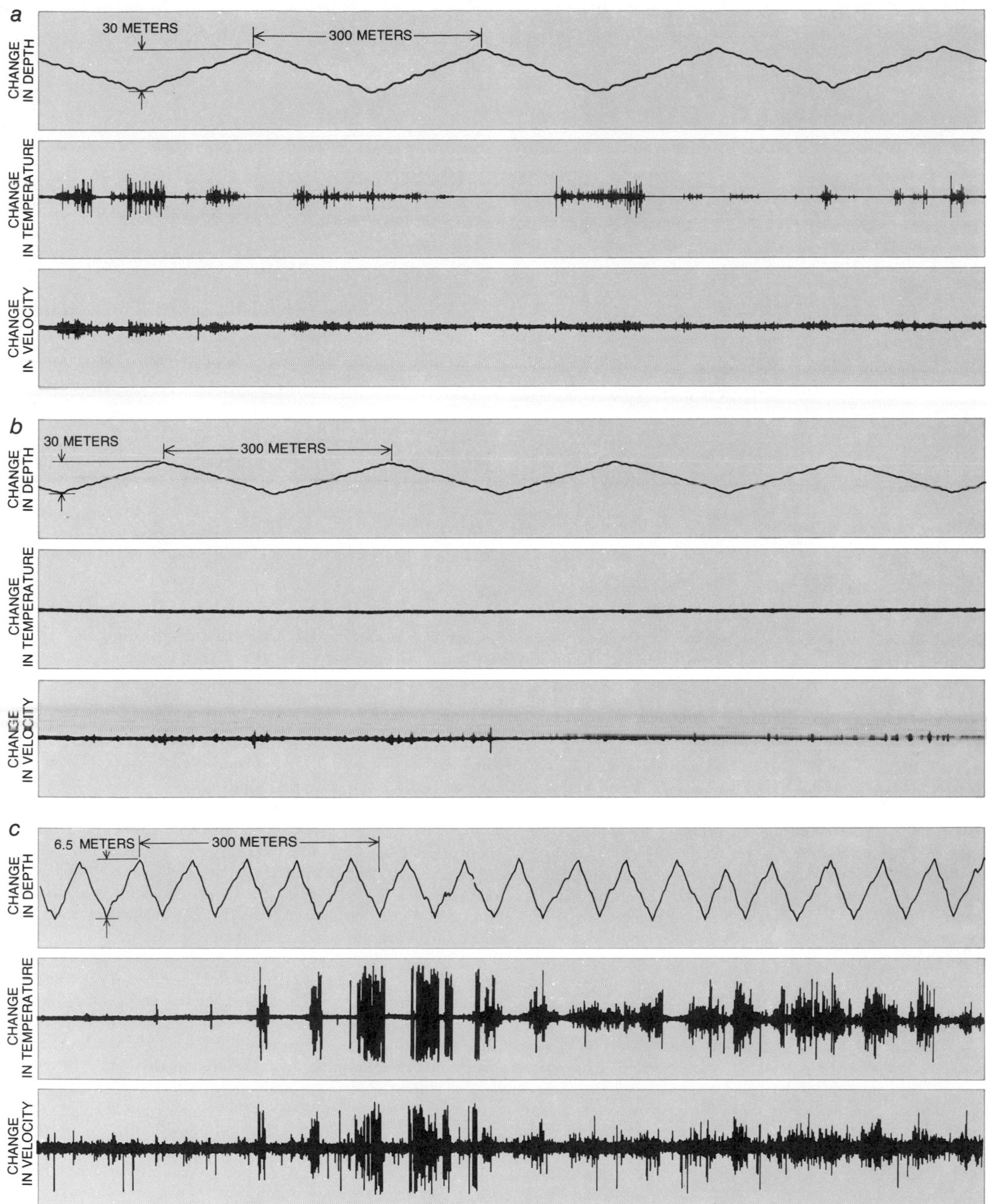

RECORDS OF HORIZONTAL MICROSTRUCTURE were made at three depths in the open Pacific off the west coast of Vancouver Island by Patrick W. Nasmyth as part of his doctoral thesis at the University of British Columbia. Changes in temperature and water velocity were sensed by probes on a submerged vehicle towed behind a ship. As the depth curves show, the vehicle was cycled up and down in porpoising fashion. The first series of curves (*a*) was made in a part of the ocean containing patches of turbulence, as indicated by rapid oscillations of the pens recording changes in temperature and velocity. The two traces generally coincide. The second series of curves (*b*) displays little microstructure activity, indicating that the ocean was quite stable at the recording depth. The last series (*c*) reveals the existence of a sharply defined front, or boundary, between two water layers; the boundary slopes upward about 1½ degrees with respect to the direction of the recording. At the outset the porpoising vehicle remains almost wholly within the upper layer, which is isothermal. Soon, however, it begins dipping into a strongly turbulent region. Owing to the slope of the front, the vehicle spends more and more time in the turbulent layer, until it is almost wholly immersed in it at end of trace.

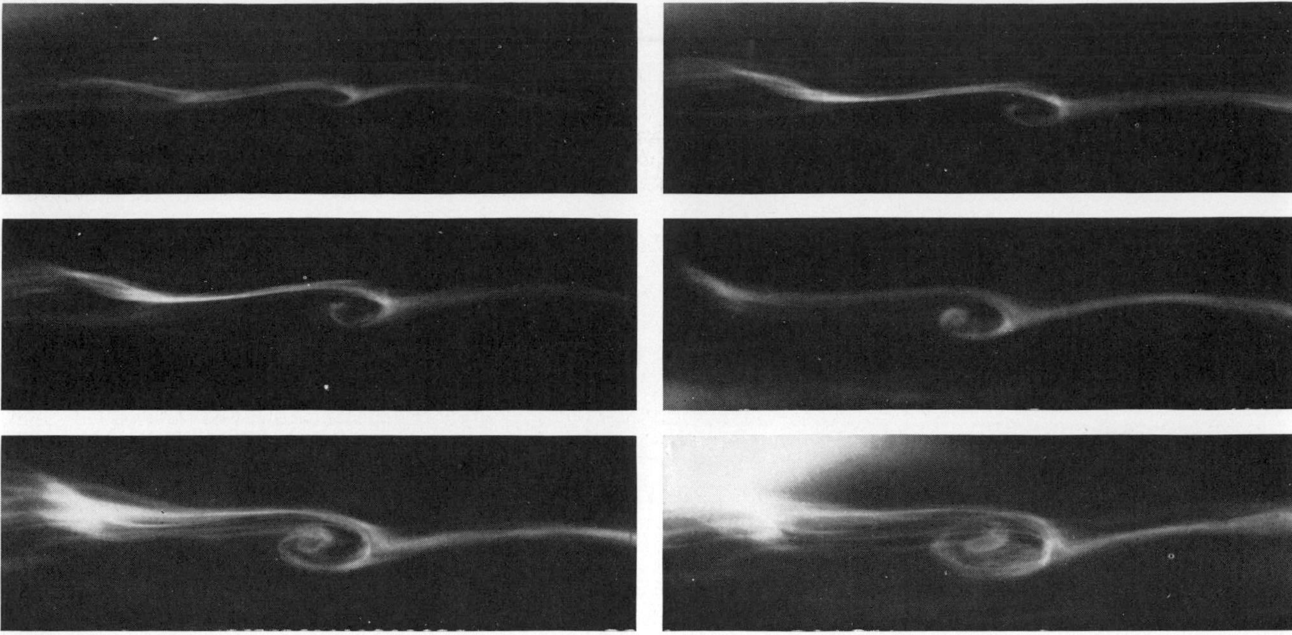

INTERNAL WAVES and Kelvin-Helmholtz shear instabilities were made visible by use of a fluorescein dye in the shallow waters off the island of Malta in the Mediterranean. These photographs, like the one on page 163, were made by Woods and a team of Royal Navy divers. The waves shown in the photographs have a length of 250 centimeters and a crest-to-trough height of 60 centimeters.

gradients have not yet been smoothed. High-resolution measurements of salinity would presumably show even older turbulence fossils in which the temperature as well as the velocity fluctuations have largely vanished.

Vertical observations of temperature and salinity that fully resolve features down to a centimeter or so and respond to fluctuations of millimeter size were first made with a freely sinking probe built by Charles S. Cox in our laboratory at the Scripps Institution of Oceanography. Temperature records obtained in mid-ocean at a depth of 1.5 kilometers appear almost as irregular as those from lesser depths, although both the amplitude and the sharpness of the fluctuations are greatly reduced. One of the common patterns shows intense turbulent fluctuations similar to the patches disclosed by the horizontal measurements. Even though we have not measured velocity fluctuations we can infer from the sharpness of some of the individual temperature gradients, which are less than a centimeter thick, either that they were being maintained by continuous stirring or that they had been formed only a few minutes before the observation. Molecular diffusion will smooth an infinitely sharp temperature step to a thickness of one centimeter in three minutes.

As part of my own graduate research I built a probe that senses electrical conductivity. This device has been used in concert with the more recent observations, making it possible to calculate the salinity and hence the density on a centimeter scale. We have observed local density inversions that cannot exist as long-term features and thus must represent transient overturns of the water. The concentrations of energy capable of producing such turbulent patches often indicate the presence of internal, or subsurface, waves.

Internal waves propagate energy through the body of a stratified fluid in much the same way that energy is propagated by waves at the surface. The ability of the sea surface to sustain undulating motions is a consequence of the buoyant restoring force that results from the density differences between air and water. Density contrasts below the surface are considerably weaker and occur as continuous changes across interfaces that are much more diffuse than the air-water boundary. As a result internal waves usually have greater amplitudes and longer periods than those on the surface. The sharpest density transition in the ocean is at the bottom of the mixed layer; a more diffuse transition coincides with the region of rapid temperature decrease in the thermocline. Internal waves with an amplitude of about 10 meters commonly travel through the thermocline with periods of about 20 minutes.

At least some degree of internal wave motion appears to exist throughout the ocean. Although the energy for these waves is ultimately derived from the tides and the winds, the manner in which the energy is converted into internal oscillations is poorly understood. Two principal mechanisms are thought to be involved. The first is the flow of steady currents and tidal currents over irregular contours on the sea floor, which would induce local oscillations in the mass of water flowing near the bottom. The second likely mechanism is the resonant interaction of surface waves, which transfers some energy to internal waves. Given the weak stratification present in the deep ocean, internal waves can propagate on sloping paths, providing a way to radiate energy from the region of generation into the interior of the ocean.

Internal waves do not result in significant energy dissipation unless they lead to an instability. One form of instability, known as Kelvin-Helmholtz instability, tends to arise when high-frequency waves propagate along rather sharp density discontinuities. Growing slowly at first as small wavelets on the crests of an internal wave, the instabilities rapidly form rolls that show much microstructure activity. Laboratory studies conducted by S. A. Thorpe at the National Institute of Oceanography in England have shown that the structures finally break down into turbulent patches that are rapidly flattened by the surrounding stratified fluid, producing a finely layered region as the stirring motions subside [*see illustration on page 164*].

Using dye tracers in the shallow water

off the island of Malta, a team of Royal Navy divers under the direction of John D. Woods of the University of Southampton were able to observe and photograph internal waves and the occasional shear instabilities they generate [*see illustration on preceding page*]. The instabilities strongly resemble Thorpe's laboratory observations. Although the growth of the instability took only one or two minutes, the turbulence persisted for five or 10 minutes after breaking, and a dye scar remained for many hours. Temperature records that we have made at greater depths in the Pacific show distinctive S-shaped features over vertical distances of half a meter to three meters that suggest similar instabilities in their initial stages.

The mixing effectiveness of an individual instability is hard to determine but is probably rather low because of the rapid damping that follows. Data taken in coastal waters at depths down to several hundred meters suggest that instabilities due to internal waves occur at a given point between once and six times a day, and that their averaged effect is equivalent to a steady turbulent mixing coefficient of between half a square centimeter and one square centimeter per second, or roughly the amount required for mixing in the diffusive thermocline models. By collecting an extensive series of observations from mid-ocean locations it should be possible to estimate ocean-wide mixing rates as one way of testing the various thermocline models.

The other vertical transport process that we know of originates in molecular motions rather than in the meter-scale velocity shears produced by internal waves. The molecular motions arise because heat diffuses much more rapidly from point to point than salt does, with the result that adjacent layers of water can rapidly destabilize under certain conditions. In a two-layer system involving both heat and salt one expects the system to be stable whenever the upper layer is less dense than the lower layer. Depending, however, on the actual distribution of heat and salt, three distinct situations can arise: one inherently stable, one inherently unstable and one where the layers remain separate but where there is convective stirring in each layer.

Laboratory experiments conducted by J. S. Turner of the University of Cambridge and Henry Stommel of the Woods Hole Oceanographic Institution showed that the inherently stable system is produced if the upper layer is both warmer

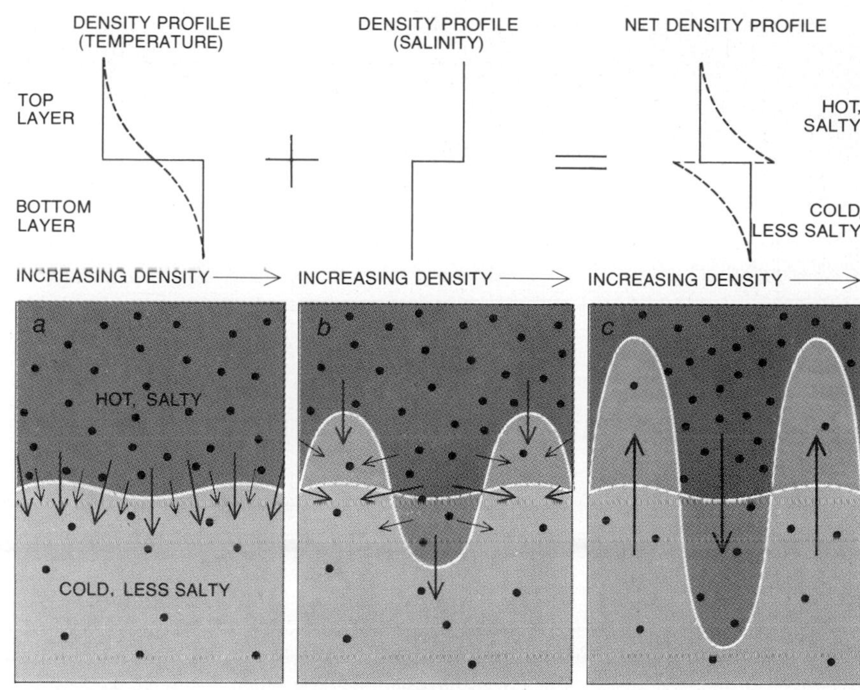

SALT-FINGERING is a special type of vertical-transport process that can arise because heat diffuses about 100 times more rapidly than salt. Salt-fingering occurs in a two-layer system where the upper layer is warmer and saltier than the lower layer, as is shown in the density profiles at the top. Solid lines show the initial density profile, whereas the broken lines show the situation a few minutes later when the more rapid diffusion of heat has produced a potentially unstable situation at the interface. The three lower panels show how small irregularities on the interface lead to salt-fingering. Color represents temperature and heat flow; black dots and arrows indicate salt and salt diffusivity; gray tones indicate density. A downward bulge will cool the water above the interface more rapidly than when the interface is level or curves upward. The rapid loss of heat makes the water above the bulge heavier than its surroundings, so that it tends to sink. Similarly, the water below an upward bulge is warmed more than its surroundings and tends to rise. The result is fingering (*c*).

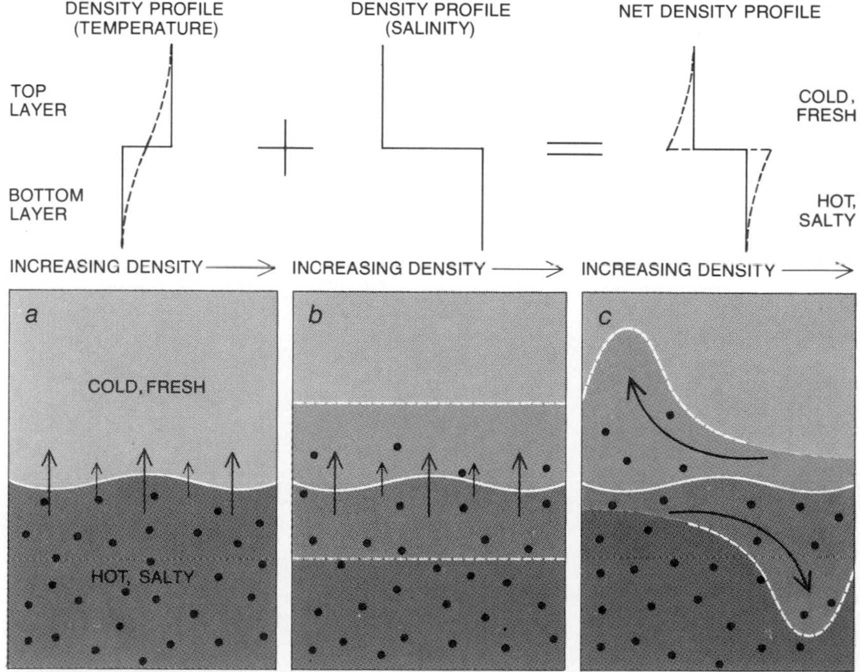

ENHANCED STABILITY of the density step in a two-layer system results when the upper layer of water is cool and fresh and the lower layer is warm and salty. Now as heat diffuses across the interface between the layers more rapidly than salt (*a, b*) the water that is immediately above the interface becomes even lighter than the upper layer as a whole and tends to rise, whereas water just below interface increases in density and tends to sink (*c*).

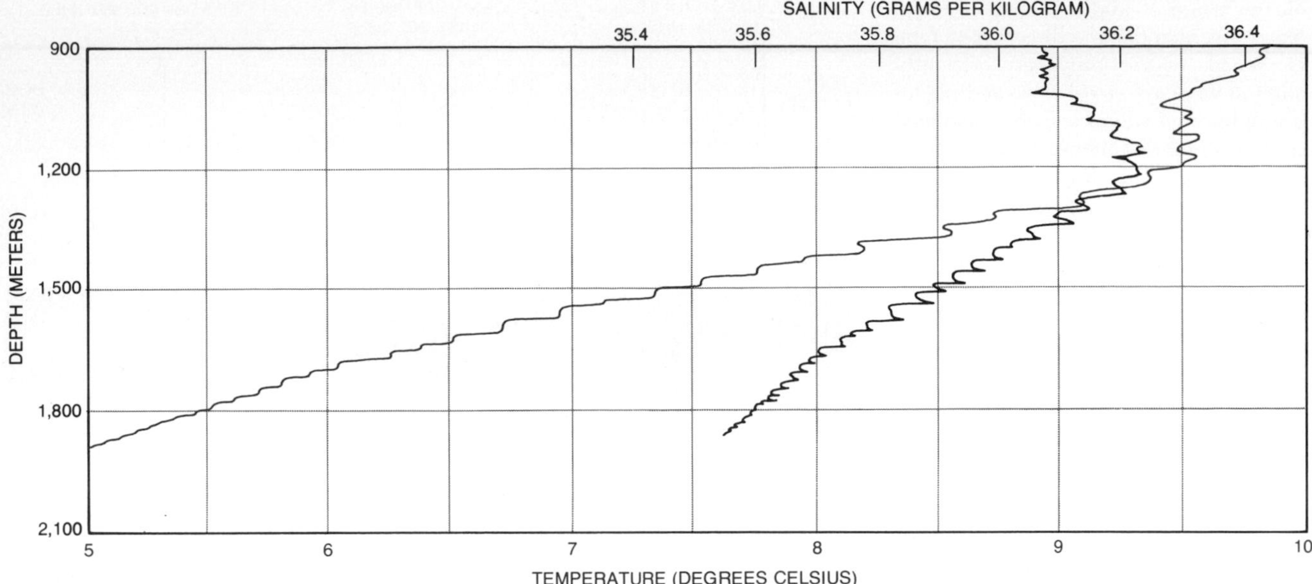

PROBABLE EXAMPLE OF SALT-FINGERING is represented by the series of temperature steps (*color*) and salinity steps (*black*) recorded in the Atlantic Ocean below an intrusion of warm saline water that has flowed westward through the Strait of Gibraltar. Horizontal layers may extend for 30 miles. Recordings were made by R. I. Tait and M. R. Howe of the University of Liverpool.

and less salty than the lower layer; both components act together to produce a stable density step across the interface. The inherently unstable system is produced if the upper layer is warmer and saltier than the lower layer, provided that the lower layer is heavier primarily because of its cooler temperature [*see top illustration on preceding page*]. The third system, which leads to convective stirring in each layer, is produced if the upper layer is cooler and less salty than the lower layer, provided that in this case the lower layer is heavier primarily because of its salinity [*see bottom illustration on preceding page*].

If the last two systems are set up experimentally with sharp interfaces, the temperature step will be smoothed by diffusion in just a few minutes, whereas the salinity step will be virtually unal-

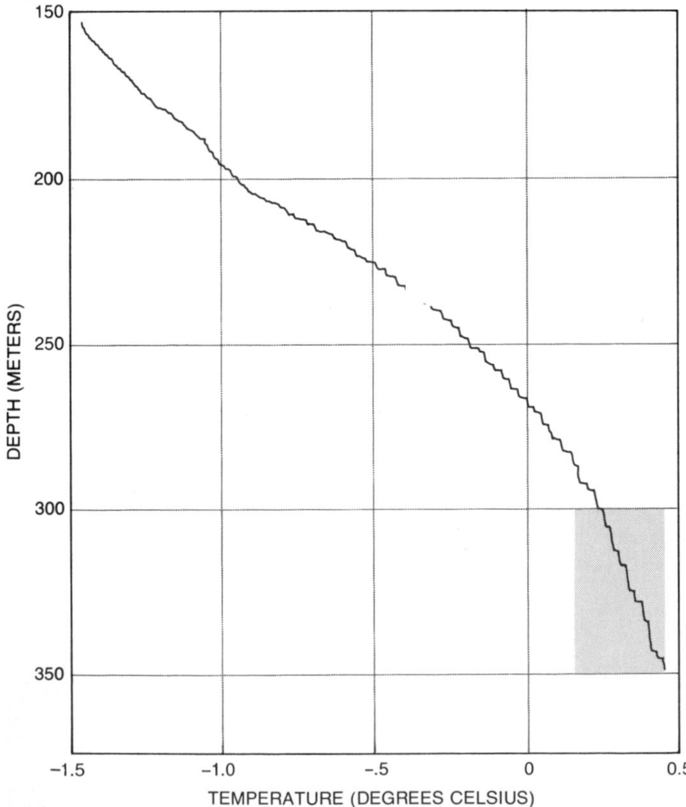

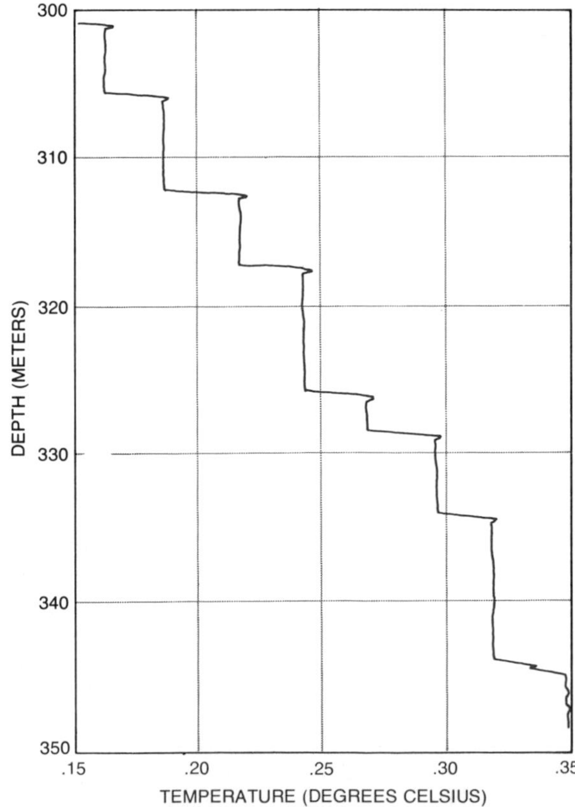

SERIES OF TEMPERATURE STEPS was observed in the Arctic Ocean under the floating ice island designated T-3 by Steve Neshyba and Victor T. Neal of Oregon State University and Warren Denner of the U.S. Naval Postgraduate School at Monterey, Calif. Although the corresponding salinity steps were not recorded, it seems likely that the temperature steps are another example of stratification produced by the unequal diffusivity rates of heat and salt. The layers were formed above an intrusion of warm saline water from the Atlantic. The region in rectangle at left is enlarged in the curve at the right. Seven steps are compressed within .2 degree C.

tered. The reason is that heat diffuses about 100 times more rapidly than salt. In the first case, where the warm, salty water is on top, the rapid loss of heat across the interface makes the water immediately above the interface heavier than the less salty water immediately below it, which is simultaneously absorbing heat and becoming lighter. The result is a convection pattern resembling a tiny checkerboard, in which alternate cells of salty water sink while adjacent cells of less salty water rise; the pattern is known as salt-fingering. In the second case, where warm, salty water is on the bottom, the rapid diffusion of heat upward increases the stability across the interface but produces strong convective stirring within each of the layers.

The salt-fingering pattern of convection persists only for a limited depth below the interface. At a certain point the negative-buoyancy flux due to the descending saline water forms a thick region that undergoes a weak, overturning turbulent motion, thus creating a well-mixed layer whose temperature and salinity are intermediate between those of the two original layers. In a deep column of water a series of thin salt-fingering regions and thick well-mixed layers can form progressively below the original interface, giving rise to stairstep profiles of decreasing temperature and salinity. In a similar manner the convective motions above a layer of warm saline water are self-limiting. In this case a family of steps is progressively formed above the original interface. A vertical profile through such a system shows steps of increasing temperature and salinity with increasing depth. It appears that the same sequence of steps can develop in the absence of a sharp interface if the temperature and salinity gradients are in the proper sense, evidently as a consequence of small irregularities in velocity that distort the original gradients to produce regions of increased heat diffusion.

In the ocean there are many areas over which the temperature and salinity profiles are in the right relation for differential diffusion to occur. As many as 20 steps of decreasing temperature and salinity, apparently the result of salt-fingering, have been observed in the Atlantic under an intrusion of Mediterranean water that moved through the Strait of Gibraltar [*see top illustration on page 168*]. An even greater number of steps of increasing temperature and salinity, each about three meters thick, have been discovered in the Arctic Ocean above a layer of warm saline water that has entered from the Atlantic [*see bottom illustration on page 168*]. This is an example of the second case, in which steps are progressively formed above an intrusion of warm saline water. On a smaller scale the upper sections of temperature inversions 10 meters thick found off San Diego have been resolved into families of layers 20 to 30 centimeters deep separated by descending steps of increasing temperature and salinity that are often less than two centimeters thick.

By producing convective motions, which drive turbulent stirring within layers, and by creating steep gradients of temperature and salinity at layer boundaries, differential diffusion produces vertical transport rates that are much greater than those that would result from simple molecular diffusion along the average gradients. Differential diffusion is clearly a major mixing process in places where warm saline water mingles with cooler fresher water to produce large-scale profiles of high variation. The upward fluxes of temperature and salinity computed for the thin diffusion layers off San Diego are great enough to fill in the minimums of temperature and salinity between successive intrusions in about a day. This suggests that the boundaries of the intrusions are rapidly altered until they are no longer recognizable as separate layers. It remains to be demonstrated whether or not salt-fingering is a general process in the deep ocean. An effort to use optical means for observing centimeter-diameter salt fingers in the open ocean is being made by Albert Williams III of Woods Hole. If salt-fingering is present, it will complement the instabilities created by internal waves and other shearing processes in supplying the downward transport of heat required to maintain the thermocline profile.

Although I have discussed differential diffusion and shear instabilities as separate processes, in the ocean they must interact and modify each other in ways that now can only be guessed at. For example, laboratory work has shown that salt-fingering is altered and partially inhibited by the presence of turbulence or even by weak shearing currents. Conversely, the sharp interfaces that are formed by both types of differential diffusion process must inevitably modify the character of internal waves. By advancing our understanding of the small-scale dynamics of the ocean, observations of oceanic microstructure can provide tests of the validity of the various thermocline models and ultimately improve our ability to predict the dispersal of substances introduced into the sea by both nature and man.

Beaches

by Willard Bascom
August 1960

Where the land meets the sea, the waves usually lay down a strip of sand or other uniform material. The constant shifting of this material has created a conservation problem for beach-loving man

Beaches are natural playgrounds partly under the sun and partly under the sea where people can swim and surfboard, sun themselves and study other people. This human activity tends to obscure the fact that the beach itself is constantly in motion, quietly changing its configuration and restlessly shifting its position, grain by grain, until huge masses of sand have been moved. On a small scale and in a matter of hours the sand castles disappear and the footprints are erased; on a large scale, after days and months, the height of the sand around the rocks changes, the waves break in new places and the beach becomes broader or narrower. Indeed, over a period of years large quantities of sand may arrive or depart, posing complex problems of conservation for the people who want to enjoy the beach. This dynamic quality was incorporated by the late Columbia University physiographer Douglas W. Johnson into a definition: "A beach is a deposit of material which is in transit either alongshore or off-and-on shore." Thus three elements make a beach: a quantity of rocky material, a shore-line area in which it moves and a supply of energy to move it.

Most of the beach material along the coasts of the U. S. consists of light-colored sand—the product of the weathering of granite rock into its two main constituents, quartz and feldspar [see "Sand," by Ph. H. Kuenen; SCIENTIFIC AMERICAN Offprint 803]. Americans therefore tend to think of beaches as stretches of white sand. But white sand is no more required to make a beach than is the sun or sun bathers. Many Tahitians and Hawaiians think that a proper beach is made of black sand—the result of the disintegration of dark volcanic rocks. Along much of the coast of England and much of the French Riviera the beach is composed of small flat stones called shingle. The word beach may originally have meant shore lines made of this shingle. In parts of Labrador, Alaska and Argentina the beaches consist of large cobbles from four to 12 inches in diameter. The beach of the Pacific Coast of Lower California is made of two materials: a flat sandy portion that is exposed only at low tide, while above and behind the sand great cobble ramparts rise in steep steps to a height of 30 feet or more. Nor are stones and sand the only beach materials. At Fort Bragg, Calif., a small pocket beach consists entirely of old tin cans washed in from the city's nearby oceanic dump and arranged by the sea in the usual beach forms. It seems that a beach can be made of almost any material of reasonable size and density that is present in quantity. Because the principles involved in the motion of beaches are much the same regardless of the material, the word sand will be used in this article for all beach materials.

The Work of the Waves

A casual observer thinks of a beach as the sandy surface above water. The student of beaches takes a broader view and includes all of the area in which the sand moves. For him the beach extends from a depth of 30 feet below the water level at the lowest tide to the edge of the permanent coast. The latter may consist of a cliff, sand dunes or man-made structures, none of them really permanent, but all more enduring than the beach as seen at any one time. The offshore boundary of 30 feet below the low-tide water level is the depth beyond which ordinary water motion does not have sufficient energy to move the sand. A beach also has limits in the alongshore direction. A point of land or a stream may make such a boundary.

The waves and currents of the water provide the third element that must be present to make a beach: the energy to keep the sand in transit. At some beaches the wind also moves quite a lot of sand, but its direct effects will be ignored here. Of course it is the wind blowing on the surface of the sea that creates the waves, and so the wind ultimately causes all the movement of beaches. The waves rolling in on the beach may have originated far out at sea. The faster the wind, the longer it blows and the greater the distance over which it blows, the larger the waves it raises. If the storm is near the shore, the waves will be steep and may very quickly change the configuration of the beach. Normally, however, the waves that shape the beach have moved out from under the winds that generated them and are longer, lower and more regular than the wind waves. Such waves are called swell, and they travel away from the storm in all directions with very little loss of energy. Since a storm is almost always taking place somewhere at sea, the swell is constantly molding all the beaches around an ocean.

Swell is described by its height (the vertical distance between the trough and the crest of a wave), by its length (the horizontal distance between crests) and by its period (the time in seconds between crests observed at a given point). As swell moves into shallow water a remarkable change occurs. When the depth equals half a wavelength, the waves are said to "feel bottom." The velocity and length decrease, and the height increases; only the period remains the same. Rolling farther inshore, the waves rise higher and finally topple over and break. The result is surf—a turbulent mass of water [see the article "Ocean Waves," by Willard Bascom, beginning on page 117].

It is the action of waves in shallow

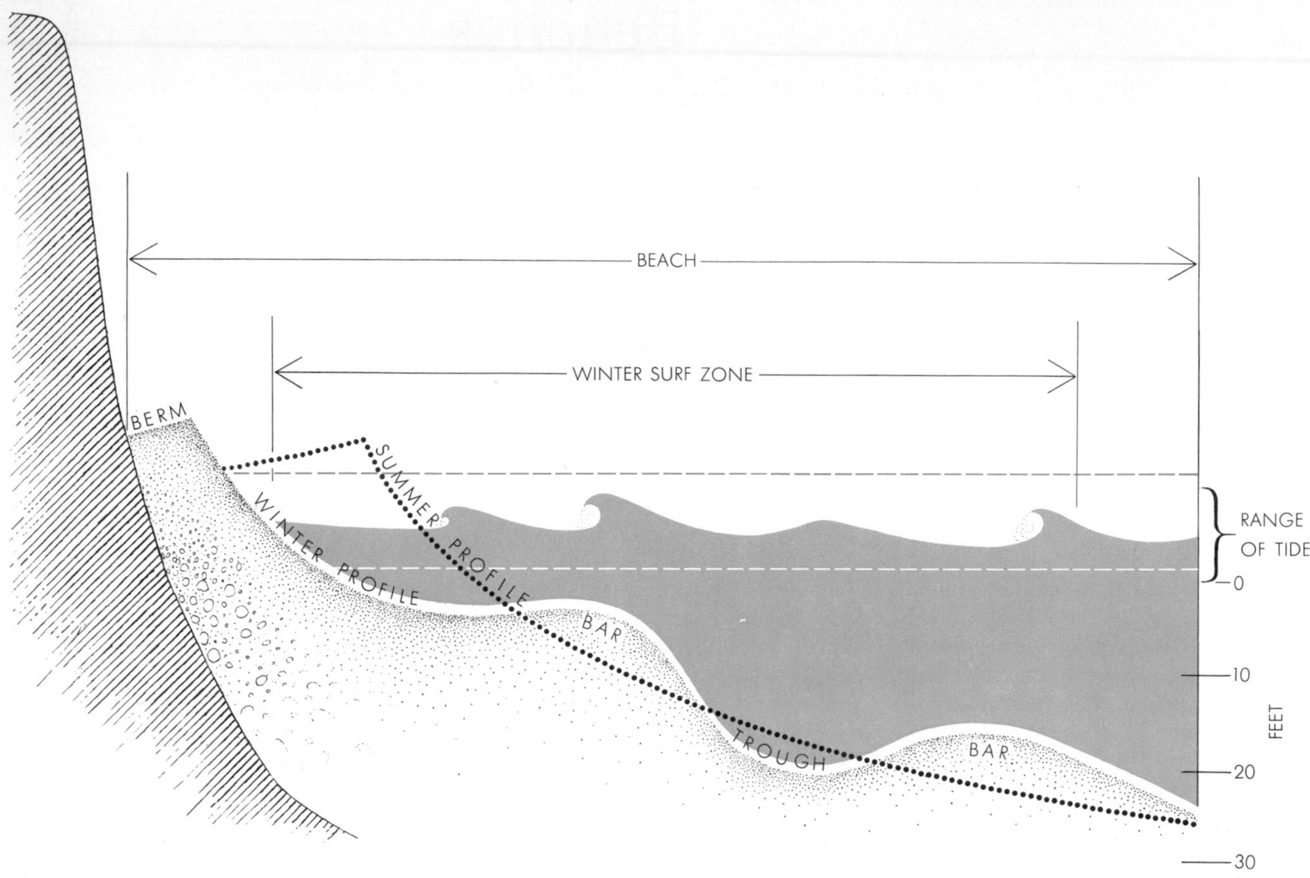

PROFILE OF A BEACH is characterized by a berm (the deposit of material at the top of the beach) and bars. In winter heavy surf removes sand from the berm and deposits it on the bars; in summer, light surf builds the berm. Vertical scale is exaggerated 25 times.

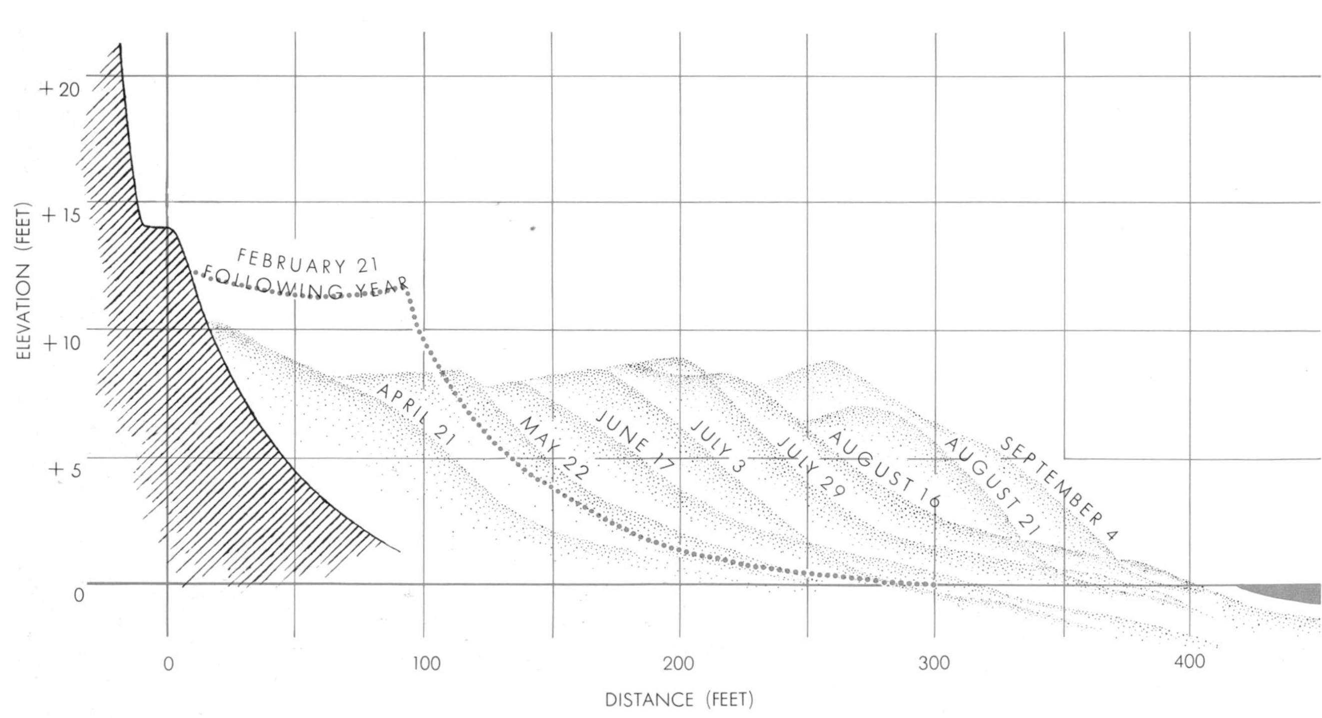

GROWTH OF THE BERM at Carmel, Calif., during the spring and summer is indicated by this series of dated slopes, based on actual measurements. Vertical dimension is exaggerated 10 times. The dotted line shows how berm was cut back during following winter.

water that changes the beach. The basic mechanism is simply the lifting of the sand grain by the turbulence that accompanies the passage of a wave and the free fall of the grain to the bottom as the wave loses its lifting force. Since a sand grain is lighter under water than in air by an amount equal to the weight of water it displaces, the water does not need great energy to lift it. Moreover, the grain settles back rather slowly because of the viscosity and turbulence of the water. While in suspension, a sand grain tends to move with the water, and currents of very low velocity will displace it. Each time a grain is lifted it lands in a slightly different location. Because uncounted millions of sand grains are continually being picked up and relocated, the beach shifts its position.

The Measurement of Beaches

During the years 1945 to 1950 John D. Isaacs, now of Scripps Institution of Oceanography, and I were employed on the "Waves" project of the University of California to study the beaches of the U. S. Pacific Coast. Under the direction of Morrough P. O'Brien, dean of engineering and a member of the Beach Erosion Board of the Army Corps of Engineers, we spent virtually all our time, winter and summer, observing the interaction of waves and beaches. To measure the height and period of the waves, we installed a dozen or so ocean-wave meters offshore, connected by armored submarine cable to recorders on the beach. With radio-controlled cameras we made photographs of the surf simultaneously from the beach, from nearby cliffs and from an aircraft directly above. We threw dye into the surf to determine the nature of its currents.

We spent most of our time, however, making repeated "profiles" of various beaches. This involved going out in an amphibious vehicle, the DUKW, or "duck," of World War II, to beyond the 30-foot low-tide depth and making numerous soundings as we came in to the beach face. Moving at three knots, we would keep the duck lined up with two marker poles set at right angles to the shore line. At short intervals I would heave the sounding lead, read the depth and call the results into a radio transmitter. Isaacs would be listening on shore, about 1,000 feet down the beach from the poles and watching the duck's progress through a surveyor's transit. The poles, the duck and the transit made a right triangle, with Isaacs sighting along the hypotenuse. When he saw the

AMPHIBIOUS TRUCK, the Army "duck" of World War II, rides a 12-foot breaker during a University of California survey of beaches. By piloting the duck straight toward the shore the workers aboard it were able to sound the profile of the beach. This beach is at Carmel.

RIP CHANNELS are marked by the dark lanes in the surf in this aerial photograph of a beach near Monterey, Calif. The channels are formed when a series of waves raises the level of the water inside a bar, and the outrushing water cuts a series of notches in the bar. The water in such channels can flow as fast as four knots.

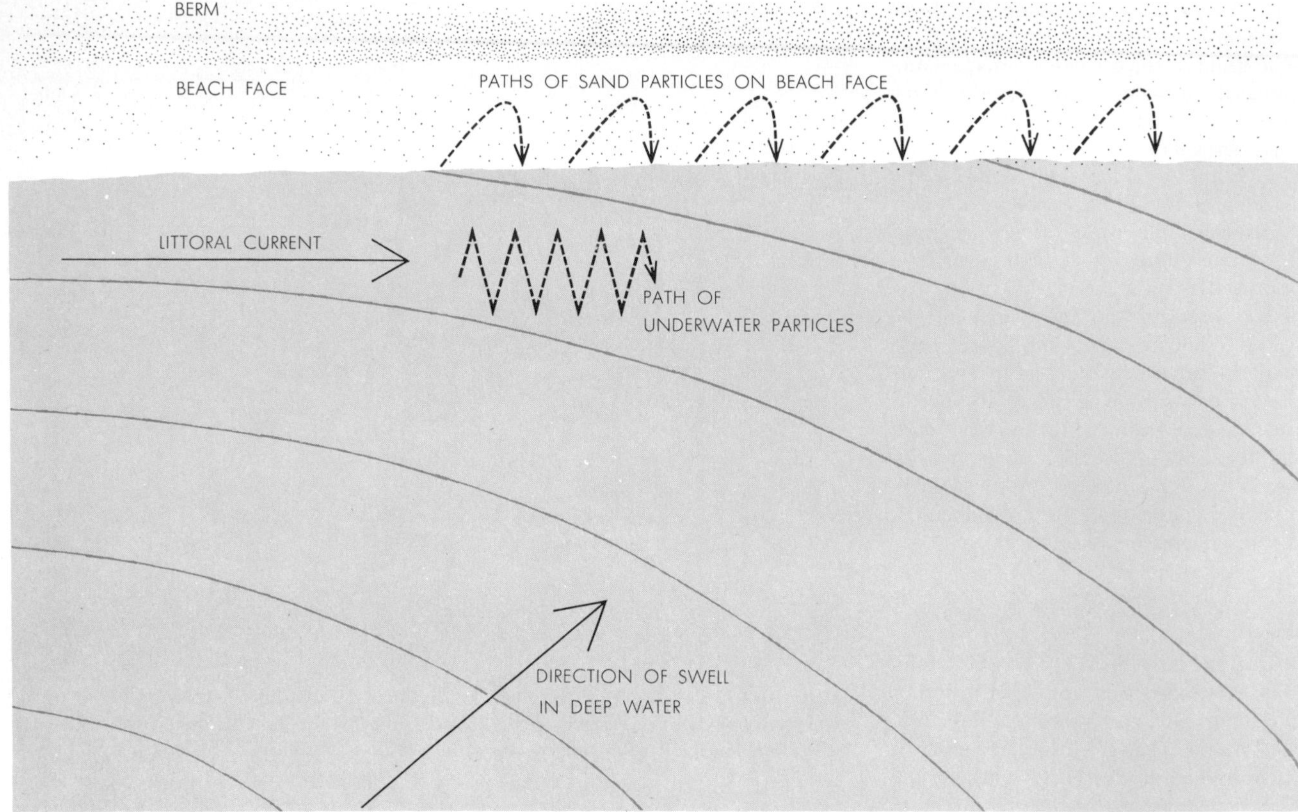

LITTORAL CURRENT, a current running parallel to the beach, is set up when waves move toward the beach at an oblique angle. Under such conditions sand grains lifted by the surf, normally moved at right angles to the beach, are transported with the current.

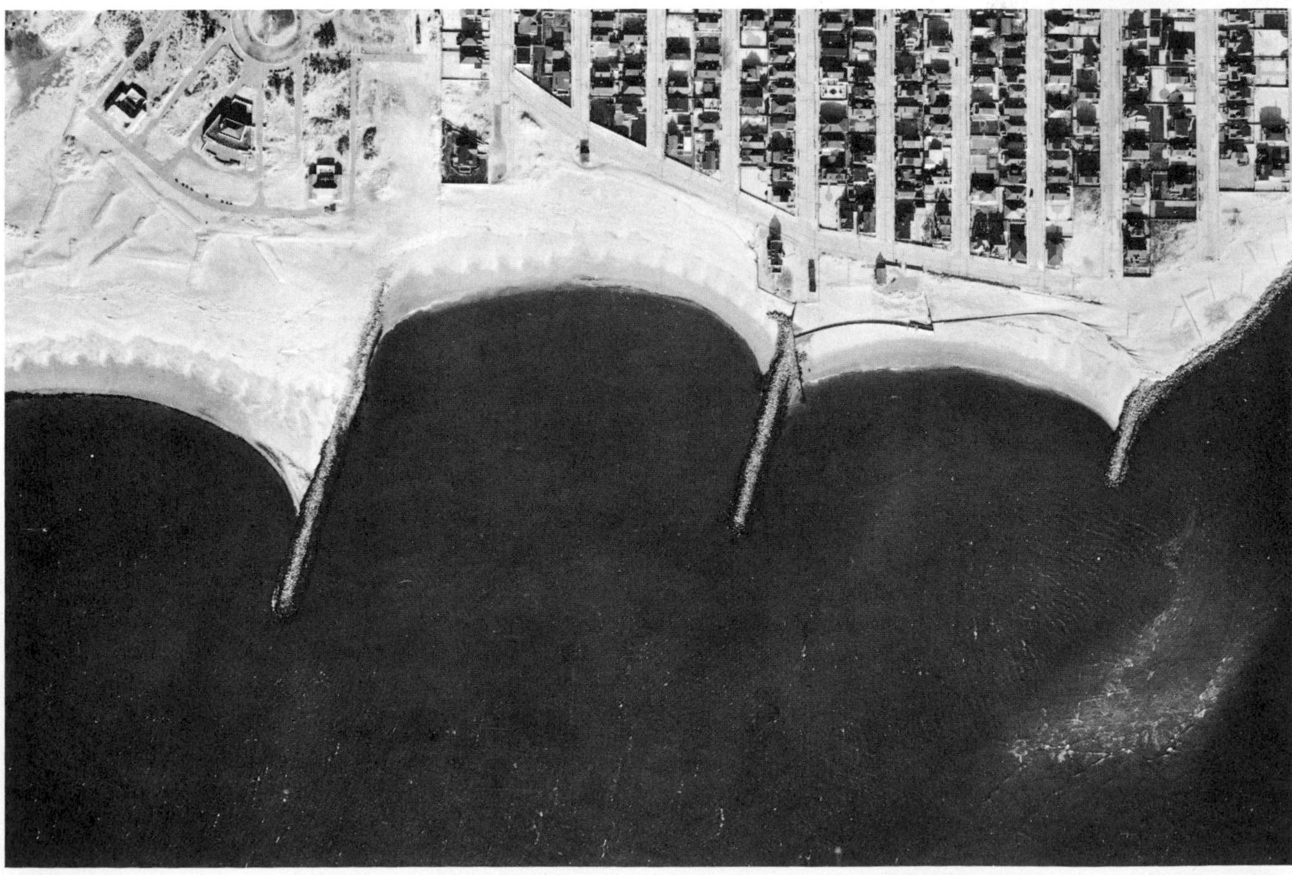

GROINS, dams of stone or wood jutting out from the beach, are widely used to retard the erosion of beaches by littoral currents. The groins in this photograph are on the Atlantic Coast at Point Lookout on Long Island. The current moves from left to right.

lead weight splash, he would read the angle on the transit and call it to an assistant who recorded depths and angles together. By plotting the depth at each distance from the shore line, we could draw a profile of the beach.

Somewhat surprisingly we obtained good profiles even in rough surf, partly because the duck is a fine beach-and-surf craft. It is 32 feet long, has six wheels and was originally designed for moving cargo from ship to shore and inland during amphibious operations. In breakers the front wheels tend to "hook" the crest of the wave, hanging down in front of the shoreward-tumbling water so that the vehicle is carried in like a surfboard. In the zone where it is only partly afloat the wheels and the propeller may drive it at the same time. The air pressure of its big tires is controlled from the cab, and when the vehicle reaches the steep beach-face, the pressure can be adjusted to achieve enough traction for the vehicle to grind its way upward.

On the northern California beaches in winter we often surveyed beaches where the breakers were 12 feet high, and occasionally, having misjudged the waves before starting out, we found ourselves amid breakers half again as high—a remarkable experience in a relatively small craft. Using the reliable duck we recorded changes in beach profiles from winter to summer on more than 30 West Coast beaches. We also kept a record of the ease with which the duck could move about on the part of beach above water. On the hard beaches north of the Columbia River it could travel as fast as it could on a highway—if the beach was being eroded. A day later, however, when the waves had altered the delicate balance of sand transport and deposited a new layer of soft sand, the duck could not exceed 10 miles per hour. We found that the beach face in the zone between high and low tide seemed to be the only place that retained its hardness and its degree of slope over relatively long periods.

Later we correlated our slope and wave measurements with the results of elaborate samplings of sand-grain size. This showed, as even a casual observer will note, that steeper beaches are usually composed of larger sand grains. Our studies also showed that factors not quite so apparent enter into the picture. For example, beaches that are partly protected from the swell will be steeper than beaches composed of sand of the same size that are exposed to it.

The underwater slopes are quite different from those of the beach face, and are usually described in a different way

BREAKWATER AT SANTA BARBARA, CALIF., causes sand transported by a littoral current running from west to east (*i.e.* from left to right) to deposit in the sheltered water. A dredge in center of harbor moves sand beyond the large pier at right.

BREAKWATER AT SANTA MONICA, CALIF., runs parallel to the shore (*bottom right*). Originally it was thought that the littoral current running inside the breakwater would carry sand past it, but instead the sand was deposited in the quiet water.

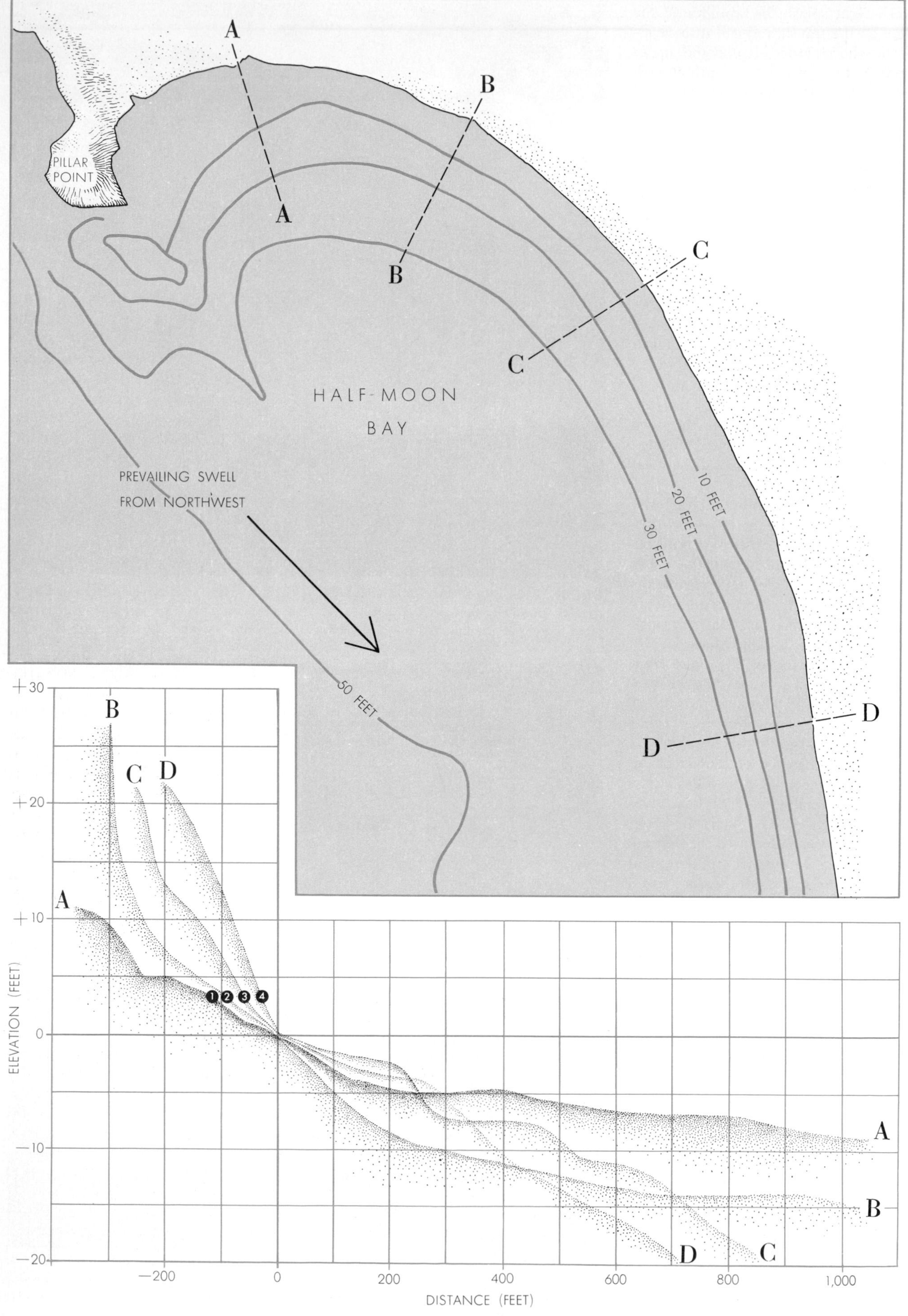

A
A
B
B
C
C
D
D
PILLAR POINT
HALF-MOON BAY
PREVAILING SWELL FROM NORTHWEST
10 FEET
20 FEET
30 FEET
50 FEET
+30
+20
+10
0
−10
−20
ELEVATION (FEET)
−200
0
200
400
600
800
1,000
DISTANCE (FEET)
1
2
3
4

to take account of the substantial irregularities between the waterline and the seaward boundary. In this scheme of classification a flat beach has an average underwater slope with a vertical rise of less than one foot in 75 feet of horizontal distance; a steep beach has a gradient steeper than one foot in 50 feet.

A large part of the movement of beach material consists in an exchange between offshore underwater ridges, or bars, and the berm, the nearly horizontal deposit of material at the top of the beach onshore [*see top illustration on page 172*]. Bars may be considered as products of erosion, since they appear when violent wave action cuts back the berm and deposits the beach material in neat ridges offshore. Because they are associated with storm conditions, and since more storms occur in winter than in summer, bars are regarded as a normal feature of the beach profile in winter. All beaches exposed to the ocean swell (as well as beaches on such large bodies of water as the Great Lakes) have them, and beaches with a slope of less than one foot in 50 frequently have two or more. Essentially continuous bars 10 to 20 miles long are commonplace on the Pacific Coast north of Cape Mendocino.

Beach investigators do not know exactly how bars are formed. They have noted that the creation of bars is somehow related to wave steepness. Using an experimental wave-channel, J. W. Johnson of the University of California found that bars always formed on model beaches when the ratio of wave height to wavelength was steeper than .03, and that bars never appeared when the ratio was less than .03. In nature the numbers seem to be different, but the principle is undoubtedly the same. Since the tiny forces that cause differential motion of individual sand grains are hard to detect amid the general turbulence, the exact manner in which variations in steepness cause the sand to move landward or seaward remains elusive.

The bars in their turn have a decided effect upon waves. The outer slope of a bar is relatively steep, and this abrupt rise of the bottom causes the larger waves to break. The waves often re-form in the trough between bars and proceed toward shore as smaller waves, breaking on the shallower inner bars or on the beach face. The smaller waves in a train of waves of irregular heights will not break on the outer bar. Thus a bar tends to act as a wave filter, breaking and reducing the higher waves and passing waves that are below a certain height. On Pacific Coast beaches that are exposed to the full force of the waves, the top of the innermost bar is usually about a foot below the low-tide water level, the top of the second bar is at a depth of seven feet and the third bar is 13 feet deep. Large swell from a nearby storm will produce violent breakers as high as 30 feet on the outermost bar. On a beach with a gentle slope and a series of bars the waves will re-form and break again and again as they move in, creating a surf zone as much as a mile wide.

After the storm season the steepness of the waves decreases and they begin to move the sand toward the shore. The material from the outer bars fills in the troughs, and soon the beach profile shows no bars. The material from the inner bar migrates to the berm, building it seaward. Except on very flat beaches the berm usually has a well-defined edge, or crest, and its method of growth can be readily observed. As each wave reaches the beach face, it uses up its remaining energy in a thin swash of water that runs up the beach face carrying sand with it. Depending on how permeable and how saturated the sand is, a certain amount of the water sinks into the beach and does not return to the sea as backrush. Thus the transport capacity of the returning water is less than that of the uprushing water, and sand is added to the berm. When conditions are precisely right, a berm may grow as much as six inches an hour, or 10 feet a day. The berm of the beach at Carmel, Calif., which we studied for several years, is about 300 feet wider in September than it is in April, the months which respectively mark the end of the calm season and the storm season.

Large waves build a higher berm than small ones do. At Monterey Bay, Calif., the crest of the berm on the exposed beach at Fort Ord has reached 16 feet above the low-tide water level, whereas the berm of the beach a few miles away, which is protected by a headland from large waves, is six feet lower. Paradoxically the storm seas that remove the summer berm often leave a higher berm of their own at the back of the beach. This berm may remain clearly visible throughout the summer.

The material that stormy seas remove from the berm ordinarily returns during calm seasons. However, an occasional very large storm or a tsunami (a "tidal" wave) may strip a beach face of sand and carry the material to depths so great that the normal waves cannot reach it and return it. At Long Branch, N.J., and Santa Barbara, Calif., the Beach Erosion Board of the Army Corps of Engineers dumped mounds of sand at depths of 38 and 18 feet in the hope that waves would move the sand onto the berm. Unfortunately the sand, like that removed by a great storm, was too deep for normal waves to pick up.

When higher-than-average waves break in quick succession and raise the water level inside a bar, the water rushes back so energetically to sea that it sometimes breaches the bar at a narrow place, producing a so-called rip channel. From then on much of the excess water hurled over the bar by the breakers moves along the beach until it reaches the channel, where it flows out as a rip current [*see bottom illustration on page 173*]. The current can be dangerous, since it flows directly out to sea with a velocity as high as four knots, considerably faster than a man can swim. Fortunately rip currents are confined to relatively narrow channels, and the bather can get out of them simply by swimming a short distance parallel to the shore. On some beaches the lifeguards mark rip currents (they flow on the surface and can be seen from shore by a trained observer), often moving the warning signs several times a day to keep up with the migration of the currents along the beach.

The rip current appears in popular mythology as the fearsome "undertow," which is otherwise a pure invention of the imagination. The undertow is said to flow outward beneath the surf and so pull swimmers out to sea. Experiments with dye markers at beaches marked with undertow warning signs have repeatedly shown that no such current exists. The water does, of course, move in and out along the bottom with every wave, but anyone being pulled seaward would be carried shoreward after six sec-

EFFECT OF A HEADLAND on a beach eroded by a littoral current is depicted in this chart made by the University of California "Waves" project. The headland is Pillar Point and the sheltered area is Half Moon Bay. The profiles of the beach at the locations marked on the map at the top of the opposite page are traced in the graph below the map. The average size of the sand grains at each location and the slope of the bottom are listed in the table below.

REFERENCE POINTS	SAND SIZE (MILLIMETERS)	SLOPE
❶	.17	1:41
❷	.20	1:30
❸	.39	1:13.5
❹	.65	1:8

TWO VERTICAL AERIAL PHOTOGRAPHS show the beach at Table Bluff, Calif., during a light surf (*top*) and during a heavy surf (*bottom*). The first light band below the beach in the top photograph is waves breaking on the beach; the second light band is waves breaking on a sand bar. The light band at the bottom of bottom photograph is larger waves breaking on a deeper bar farther out.

onds or so by the other half of the cycle. Unlike rip currents, undertow can be dismissed as a danger to swimmers.

Cusps and Ripple Marks

A beach feature that offers intriguing problems for the investigator is the cusp. This is a crescent-shaped depression that occasionally forms in regularly spaced series along the beach face. The triangular apex (the horn) at which two crescents join points seaward. A "bay" of sand, which may be deep and narrow or broad and shallow, lies between the horns. Cusps vary in length from a few feet to hundreds of feet and their relief may exceed six feet or be so shallow that they are barely discernible. Although investigators have studied these curious beach forms for years and have constructed many hypotheses about them, no one has produced a generally accepted explanation of how they begin, why they are so regular and why they have the dimensions they have. These factors are almost certainly related to the character of the waves that form them. Perhaps cusps develop when the waves have a "balanced" steepness, so that neither erosion nor deposition of the sand occurs.

Ripple marks, or sand ripples, are roughly parallel arrays of wave-shaped ridges and troughs that are formed in the sand on the sea bottom by the action of the water. Ordinarily their "wavelength" is three or four inches, and they are about an inch deep, but giant ripples something like desert sand dunes and with wavelengths of more than 10 feet have been observed under the surf. Since ripple crests are usually parallel to the wave crests, paleographers have used fossil ripples in sandstone to establish the orientation of ancient beaches.

Ripple marks may begin to form around a pebble or any other small prominence on the bottom. The oscillating wave-motion creates horizontal vortices first on one side of the pebble, then on the other, and a small ridge of sand begins to accumulate. At a certain distance on either side of the original ridge the vortex currents diminish and deposit sand so that another ridge forms and the ripple pattern grows. Water motion due entirely to waves produces symmetrical ripples, but if a current is superimposed on the wave motion, the sides of the ripples are steeper on the lee side. Ripple marks come and go with the changing wave conditions, and they migrate with the currents. Although they, too, have been extensively studied, the relationship between current velocity, sand-grain size and the dimensions of the ripple marks remains unexplained.

The Preservation of Beaches

The major problems of beach conservation are created not by the seasonal movement of sand onshore and offshore, but by the motion of sand parallel to the shore. On some coasts, alongshore or littoral currents, which arise when waves strike the shore at an angle, annually transport millions of tons of sand, eroding one beach and building up another. The largest waves create the strongest littoral currents because they contain the most energy. They "feel" bottom well out from shore, and if they approach

CUSPS are a series of crescent-shaped depressions along a beach. The cusps in this aerial photograph are at El Segundo, Calif.; they are 60 feet across. There is no satisfactory explanation of how cusps are formed, though they are undoubtedly related to wave action.

shore at an angle, they tend to be refracted or bent by the underwater contours so that the wave front becomes parallel to the shore line. Waves are often incompletely refracted, especially if the angle between the deep-water wave fronts and the shore line is great, or if the water becomes shallow abruptly. Consequently waves often strike the shore at an angle.

Along coasts where the prevailing waves arrive in this way, a littoral current flows constantly. These currents usually flow too slowly to move the sand grains by themselves. Turbulence in the surf zone, however, keeps the sand in suspension, and even a low-velocity alongshore current is able to transport large quantities of material. On the beach face the sand particles carried by the uprush describe an arc in the direction of the alongshore current so that each wave moves them along the beach a little way. In shallow water, where the waves can lift sand and move it back and forth, the littoral current gives the sand grains a saw-toothed motion [*see top illustration on page 174*]. With every oscillation the sand moves sideways, and there is no force to return it to its original position. As a result the sand travels downstream as on a conveyor belt, the belt having the width of the surf zone and the velocity of the littoral current.

Many seaside areas afflicted with littoral currents have serious problems of beach erosion. The California coast north of Point Conception faces west, directly into the Pacific Ocean winds and swell, and there is no appreciable alongshore current. But south of this point the shore line turns abruptly to the east, so that these same winds and waves strike the shore at an angle. As a result an almost continuous current moves sand to the east. Any structure that interrupts the flow acts like a dam, and the beaches immediately to the west grow while those to the east are stripped of sand by the waves and currents. Of the several structures that produce this effect, the most interesting is the breakwater at Santa Barbara, for there the amount of sand carried along the coast can be measured. Sand moving from the west past the end of the breakwater abruptly encounters the deeper, quiet water that the breakwater was built to create. There the cessation of turbulence causes all the suspended particles to be deposited just inside the end of the breakwater [*see top illustration on page 175*].

Frequent surveys of the changes in the volume of sand in the spit have revealed the rate of deposition, which equals the rate of transport of sand along the shore. On an average day about 800 cubic yards of sand are dumped in the harbor and under storm conditions four times that much will arrive. To keep the sand from filling the harbor and to prevent damaging erosion on the shore beyond, a dredge pumps the sand from the spit across the quiet water to the downstream beach. Once again it is exposed to wave action, and the littoral current carries it along the coast until eventually it reaches Santa Monica.

That city also needed quiet water for a yacht harbor. Because of the difficulties associated with the damming of the sand by a conventional breakwater, the city built a wave barrier consisting of a straight line of rocks parallel to the shore

RIPPLE MARKS appear in the sand at the bottom of two experimental wave-tanks of the Beach Erosion Board in Washington, D.C. The tank at left contains fine sand; the tank at right, coarse sand. After the sand had been subjected to wave action for 60 hours, the tanks were drained and the configuration of the sand was photographed. Three sand bars have formed in the tank at right.

and several hundred yards out [*see bottom illustration on page 175*]. The sand was expected to flow past in the wide space between the breakwater and the shore. It did not. The sand simply stopped moving in the quiet water, and the beach started to build outward toward the rock wall. Downstream from the structure the beaches retreated. Now Santa Monica must also employ a dredge to put the sand back into circulation.

A similar littoral sand-transport system operates along the southern shore of Long Island on the East Coast. Prevailing winds and the unrefracted waves from the North Atlantic sweep the sand along the shore from Montauk Point at the island's eastern tip to the Rockaway spit at the entrance of New York Harbor. Montauk is rapidly eroding, and Rockaway spit is (or was for a considerable period before the present shore-line structures were built) building at the rate of 200 feet a year.

If no action is taken on erosion problems, everyone shares the erosion. But as soon as one part of the shore is protected, the remainder of the shore must supply the sand. Nevertheless for many years the customary way to stop the retreat of a beach was to build groins, that is, dams made of rocks or wooden piling a few feet high and a few hundred feet long, jutting out from the beach face to stop the passing sand.

Along some coasts groins have been constructed at regular intervals for many miles, each supporting a curving beach that spills over its end, giving the beach a cuspate appearance. The sand still flows, but it is retained temporarily on each little segment of beach. Groins can hold sand if they are properly engineered, but their effect is local and temporary. They are no longer the preferred means of maintaining a beach, for it has been found that they are usually less effective and more expensive than a "beach-nourishment" program. In such programs new sand is supplied to the system from inland dunes or from the bottoms of nearby lagoons. For example, the famous Waikiki Beach in the Hawaiian Islands was recently rebuilt with sand that was trucked in from dunes 14 miles away.

This change of opinion about the best way to maintain beaches is illustrated by the problem now facing the state of New Jersey. The configuration of the coast is such that refracted Atlantic Ocean swell strikes heavily on the New Jersey coast's most prominent point, near Barnegat Inlet. Littoral currents move the sand away from the point in both directions, and the point is eroding rapidly. In the past 50 years nearly $50 million has been spent on shore works in an attempt to stabilize the shore line. The present annual rate of expenditure is more than $2 million, and the results are not entirely satisfactory. Some parts of the shore have long since been stripped of sand; others are still retreating.

The Beach Erosion Board has studied the New Jersey problem and has proposed a project to develop adequate recreational beaches and to prevent further erosion. This project would nourish all the beaches along the coast by supplying new sand to the beaches in the vicinity of Barnegat Inlet. The sand would come partly by truck from inland locations and partly by pumping from Barnegat Bay; wave action and littoral currents would be relied upon to distribute it along the coast. The initial investment would be $28 million, but the program would require less than $1 million per year to maintain the beaches thereafter.

Sixty-six other shore-line construction projects, costing a total of over $100 million, have been planned for the shores of the U. S., and about half are completed or are well under way. The preservation of valuable coastal land, the maintenance of usable harbors and the development of recreational activities require an understanding of the ways of moving sand. In these enterprises the knowledge gained by the scientific study of beaches will play a central role.

IV

MARINE LIFE AND LIVING RESOURCES

IV MARINE LIFE AND LIVING RESOURCES

INTRODUCTION

The greatest revolution in biology was generated by the theory of evolution, almost exactly a century before the plate-tectonics revolution in geology. Scientists responded in very similar ways to the two revolutions, despite the great advances in science itself during that century. Thus it appears that the way they act manifests a combination of human nature and their training and will be repeated in the future. Marine biology was not central to Charles Darwin's evidence for evolution, which is hardly surprising because little was known about the subject in the early nineteenth century. However, marine biology was stimulated by the Darwinian revolution and became embroiled in the arguments that followed. Thus, the early history of the subject provides illuminating examples of how the work and thought of scientists are influenced by new equipment and new ideas, and of the pernicious effects of old prejudices, which cannot be discarded like outmoded nets or dredges.

Prior to the rise of science, the sea was thought to be full of familiar fish in shallow water and strange monsters in the deep. Both the fish and the monsters are now generally thought to be much rarer, but the dawn of science in the eighteenth century generated a belief that life was totally absent except near the surface. This belief was a consequence of faulty measurements and mistaken science. Some of the earliest measurements of temperature below the sea surface indicated that the deeper water was isothermal, but this was merely the result of using thermometers that were not protected from pressure. Unfortunately, the isothermal temperature was very close to 3.98°C, at which fresh water reaches its maximum density. Thus, although sea water has different properties, the mistaken belief arose that the deep sea was stagnant. It was popular at the same time as a curious popular notion, that increasing pressure caused seawater to become denser with depth, and that, as a consequence, sunken ships and even cannon balls floated at levels above the bottom.

Although scientists gradually became aware that the temperature of the sea increases with depth and that water is almost incompressible, they still thought of the deep sea as dark, cold, stagnant, and subject to pressures intolerable for living organisms. It is small wonder that, lacking information, speculation about life in the deep sea was strongly influenced by the general prejudices of the time. A few biologists were beginning to collect at sea at the time, but only beginning. For example, it was not until 1828 that J. Vaughn Thomson, a British naturalist, conceived of the idea of sampling sea water by towing a conical gauze net with a glass bottle at the end. He thereupon discovered the existence of microscopic marine animals and plants, the previously unsuspected but most abundant life on earth.

In 1818 Sir John Ross, not to be confused with Sir James his nephew, set out to explore Arctic waters with an array of newly-designed equipment on board. The bottom sampler would not work, so he designed and built a "deep-sea clamm" aboard the ship and succeeded in grabbing mud in 1800 meters of water. This bottom mud contained many worms, and a living starfish was attached to the line at a depth of 1400 meters. These facts were steadfastly ignored in the speculation about life in the deep sea.

In the 1840s the young British naturalist Edward Forbes decided to study the distribution of marine animals and to establish the depth of the azoic zone, in which the sea became lifeless. It took him some time to organize financial and institutional support, but in 1842 he put to sea on H.M.S. *Beacon*, equipped with suitable dredges. He sampled the Aegean Sea to a depth of 410 meters, and found that the organisms were grouped according to depth and were sparser downward. He extrapolated his findings to conclude that the azoic zone began at 540 meters (300 fathoms). It is now known that the Aegean is rather barren compared to other seas—it is typical of nature to delight in fooling us as we try to unravel her secrets.

Dredging continued, perhaps stimulated by Forbes' success, perhaps by the enthusiasm of his publications. Meanwhile J. M. Brooke, a midshipman in the American navy, invented a sounding machine that collected a sample every time it touched bottom. Even the samples from 2000 meters contained the remains of microscopic organisms. A controversy arose about whether they had lived on the bottom or rained down from the surface waters. In 1860, information came from yet another source. A submarine telegraph cable resting at 1800 meters in the Mediterranean was lifted for repairs and found to be covered with a host of living animals.

Scientists continued to dredge, looking for the upper limit of the azoic zone in which conditions were impossible for life. In 1866 the Norwegian Michael Sars dredged to depths of 800 meters, and retrieved stemmed crinoids, which had previously been known only as fossils and which have been rare for 100 million years. The year before, Charles Darwin had published *On the Origin of Species*, which provoked an enormous controversy. One of the questions raised concerned the fact that the links in the evolutionary chain were missing. Perhaps they were living in the sea like the crinoids.

It might be expected that the concept of a shallow azoic zone would have vanished by then. Indeed, a recent history of oceanography contains a reproduction of a drawing of the famous telegraph cable teeming with life, and says that this evidence "dispelled all doubt" about the prevalence of life in the deep sea. However, as I have already remarked in connection with the revolution in the earth sciences, scientists often cannot be convinced except by evidence of a type that is familiar to them. Thus, Wyville Thomson wrote in 1877 that "in 1868 . . . the question was thus undecided," and went on to describe the organization of the oceanographic cruise of H.M.S. *Lightning* in 1868, under the auspices of the Royal Society. That expedition found life abundant off Britain at all depths to 1100 meters, and for some reason that seemed to settle the matter—at least for Wyville Thomson.

At that point, biologists did a flip-flop, presumably on the philosophical grounds that, whatever the deep sea was like, it must be odd. If it was not sterile, perhaps it was teeming with unknown creatures. This phase began with a discovery by T. H. Huxley, the great defender of Darwin's theory of evolution and one of the premier biologists of his day. In 1868 he was examining marine samples collected eleven years before, and he found to his surprise that there was a thin layer of viscid jelly containing granules resembling the nuclei of cells. He concluded that it was an organism, so far as the term was applicable to a creature without organs, and he named it *Bathybius haeckelii*, in honor of the famous German biologist.

Ernst Haeckel himself naturally took an interest in this remarkable creature, and he decided that it was the clue to understanding life in the sea, the nature of evolution, and much else. *Bathybius* was nothing less than a primordial slime that covered the entire floor of the sea. It provided the mysterious food required to support the observed life in the deep. It remained in its primordial state because it was in an absolutely stable environment. At shallower depths the environment was increasingly changeable; consequently, living fossils were stratified at different levels in the sea.

When the *Challenger* expedition put to sea four years after Huxley's discovery, one of its objectives was to discover more about this extremely important creature and chart its distribution. Not a trace was ever found. After the ship returned, careful experiments finally demonstrated that "*Bathybius*" was produced by a slow chemical reaction in sea water when alcohol was added to preserve specimens. After that, marine biologists, possibly somewhat subdued, busily explored the sea for real organisms and discovered the real complexity of interactions, which is described in "The Nature of Oceanic Life," by John Isaacs. Even now this exploration is far from complete. It is difficult to sample the rocks of the sea floor because they are hard and far away. It is difficult to study ocean currents, because the water moves around. However, the problems of other ocean scientists are minor compared to those of marine biologists. In "Active Animals of the Deep-Sea Floor," Isaacs and Richard Schwartzlose describe how marine biologists completely misunderstood some aspects of abyssal life. For example, it was thought that there were no large active creatures in the abyss, simply because no one realized that you cannot take a shark's picture unless you first give him a present.

The discovery of abundant life in the deep bears on the question whether the sea is capable of feeding the burgeoning population of the land. This is the subject of the last three articles in this section. From the beginning, marine biological research has been closely associated with commercial fisheries. Indeed, many aspects of the research depended on analyses of the contents of fishermen's nets. Likewise, the fishermen, who in this country are small independent businessmen, depend on the government to support research related to their industry. Now our coastline is peppered with laboratories of the National Marine Fisheries Service, and their ships sample the world ocean. Many other nations are equally active in fisheries research, with the result that a vast amount of information has accumulated and is being analyzed. It is summarized in "The Food Resources of the Ocean," by S. J. Holt.

Much has been learned, but it is increasingly clear that although scientific information is necessary it is not sufficient to produce rational fisheries. At present the response of most coastal nations to living resource problems is simply to grab more water. Lacking any agreements from the interminable United Nations conference on the law of the sea, the United States and other nations are extending their jurisdiction from 12 miles to 200 miles from shore. The scramble for fisheries has hardly begun. Inland nations are already asking for secondary rights to the fisheries of adjacent coastal nations. Meanwhile, nations with global fishing fleets, such as Japan and Russia, defend their rights to fish anywhere.

Many treaties exist to control marine fisheries, but some are difficult to enforce. One reason is the natural variation in the size of fish populations. An outstanding example of this comes from the enormous anchovy fishery of Peru. The article by Holt leaves this fishery in 1967, still expanding rapidly, with the remark that it "sorely needed management." The next article, "The Anchovy Crisis," by C. P. Idyll, carries the story forward a few years to when the world's greatest fishery reached its peak catch of 12.3 million metric tons in 1970, which then plunged to 2.9 million tons in 1972. Overfishing and natural changes combined with disastrous results.

If fishing wild stocks has limitations, why not farm the sea as we have the land? "Marine Farming," by Gifford Pinchot, addresses this question. Shellfish farming is ancient and widespread, and it varies according to local ecology and need. It is quite possible to fatten oysters in polluted waters near urban sewage outlets and then transport them to pure water, where they purge themselves in a few months. Unfortunately, such oysters are not cheap; thus, they are no longer common in the diet of the poor, as they once were on the eastern seaboard. Much research needs to be done before we learn to extract the maximum sustainable yield from the sea.

The final article in this section is about the integration of the recent geological revolution with the biological revolution that is a century old. Natural selection seems to require a varied environment, which marine biologists first sought by visualizing a stratified ocean carpeted with *Bathybius*. In "Plate Tectonics and the History of Life in the Oceans," James Valentine and Eldridge Moores show how the present distribution and diversity of life may be related to the drift of continents and the formation of new oceans. The stratification of time is not vertical in the water of the sea, it is horizontal in the magnetic stripes of the sea floor.

The Nature of Oceanic Life

by John D. Isaacs
September 1969

The conditions of the marine environment have given rise to a food web in which the dominant primary production of organic matter is carried out by microscopic plants

I plan to take the reader on a brief tour of marine life from the surface layers of the open sea, down through the intermediate layers to the deep-sea floor, and from there to the living communities on continental shelves and coral reefs. Like Dante, I shall be able to record only a scattered sampling of the races and inhabitants of each region and to point out only the general dominant factors that typify each domain; in particular I shall review some of the conditions, principles and interactions that appear to have molded the forms of life in the sea and to have established their range and compass.

The organisms of the sea are born, live, breathe, feed, excrete, move, grow, mate, reproduce and die within a single interconnected medium. Thus interactions among the marine organisms and interactions of the organisms with the chemical and physical processes of the sea range across the entire spectrum from simple, adamant constraints to complex effects of many subtle interactions.

Far more, of course, is known about the life of the sea than I shall be able even to suggest, and there are yet to be achieved great steps in our knowledge of the living entities of the sea. I shall mention some of these possibilities in my concluding remarks.

A general discussion of a living system should consider the ways in which plants elaborate basic organic material from inorganic substances and the successive and often highly intricate steps by which organisms then return this material to the inorganic reservoir. The discussion should also show the forms of life by which such processes are conducted. I shall briefly trace these processes through the regions I have indicated, returning later to a more detailed discussion of the living forms and their constraints.

Some organic material is carried to the sea by rivers, and some is manufactured in shallow water by attached plants. More than 90 percent of the basic organic material that fuels and builds the life in the sea, however, is synthesized within the lighted surface layers of open water by the many varieties of phytoplankton. These sunny pastures of plant cells are grazed by the herbivorous zooplankton (small planktonic animals) and by some small fishes. These in turn are prey to various carnivorous creatures, large and small, who have their predators also.

The debris from the activities in the surface layers settles into the dimly lighted and unlighted midlayers of the sea, the twilight mesopelagic zone and the midnight bathypelagic zone, to serve as one source of food for their strange inhabitants. This process depletes the surface layers of some food and particularly of the vital plant nutrients, or fertilizers, that become trapped below the surface layers, where they are unavailable to the plants. Food and nutrients are also actively carried downward from the surface by vertically migrating animals.

The depleted remnants of this constant "rain" of detritus continue to the sea floor and support those animals that live just above the bottom (epibenthic animals), on the bottom (benthic animals) and burrowed into the bottom. Here filter-feeding and burrowing (deposit-feeding) animals and bacteria rework the remaining refractory particles. The more active animals also find repast in mid-water creatures and in the occasional falls of carcasses and other larger debris. Except in unusual small areas there is an abundance of oxygen in the deep water, and the solid bottom presents advantages that allow the support of a denser population of larger creatures than can exist in deep mid-water.

In shallower water such as banks, atolls, continental shelves and shallow seas conditions associated with a solid bottom and other regional modifications of the general regime enable rich populations to develop. Such areas constitute about 7 percent of the total area of the ocean. In some of these regions added food results from the growth of larger fixed plants and from land drainage.

With the above bare recitation for general orientation, I shall now discuss these matters in more detail.

The cycle of life in the sea, like that on land, is fueled by the sun's visible light acting on green plants. Of every million photons of sunlight reaching the earth's surface, some 90 enter into the net production of basic food. Perhaps 50

NEW EVIDENCE that an abundance of large active fishes inhabit the deep-sea floor was obtained recently by the author and his colleagues in the form of photographs such as the one on the opposite page. The photograph was made by a camera hovering over a five-gallon bait can at a depth of 1,400 meters off Lower California. The diagonal of the bait can measures a foot. The larger fish are mostly rat-tailed grenadiers and sablefish. The fact that large numbers of such fish are attracted almost immediately to the bait suggests that two rather independent branches of the marine food web coexist in support of the deep-bottom creatures by dead material: the rain of fine detritus, which supports a variety of attached filter-feeding and burrowing organisms, and rare, widely separated falls of large food fragments, which support active creatures adapted to the discovery and utilization of such food.

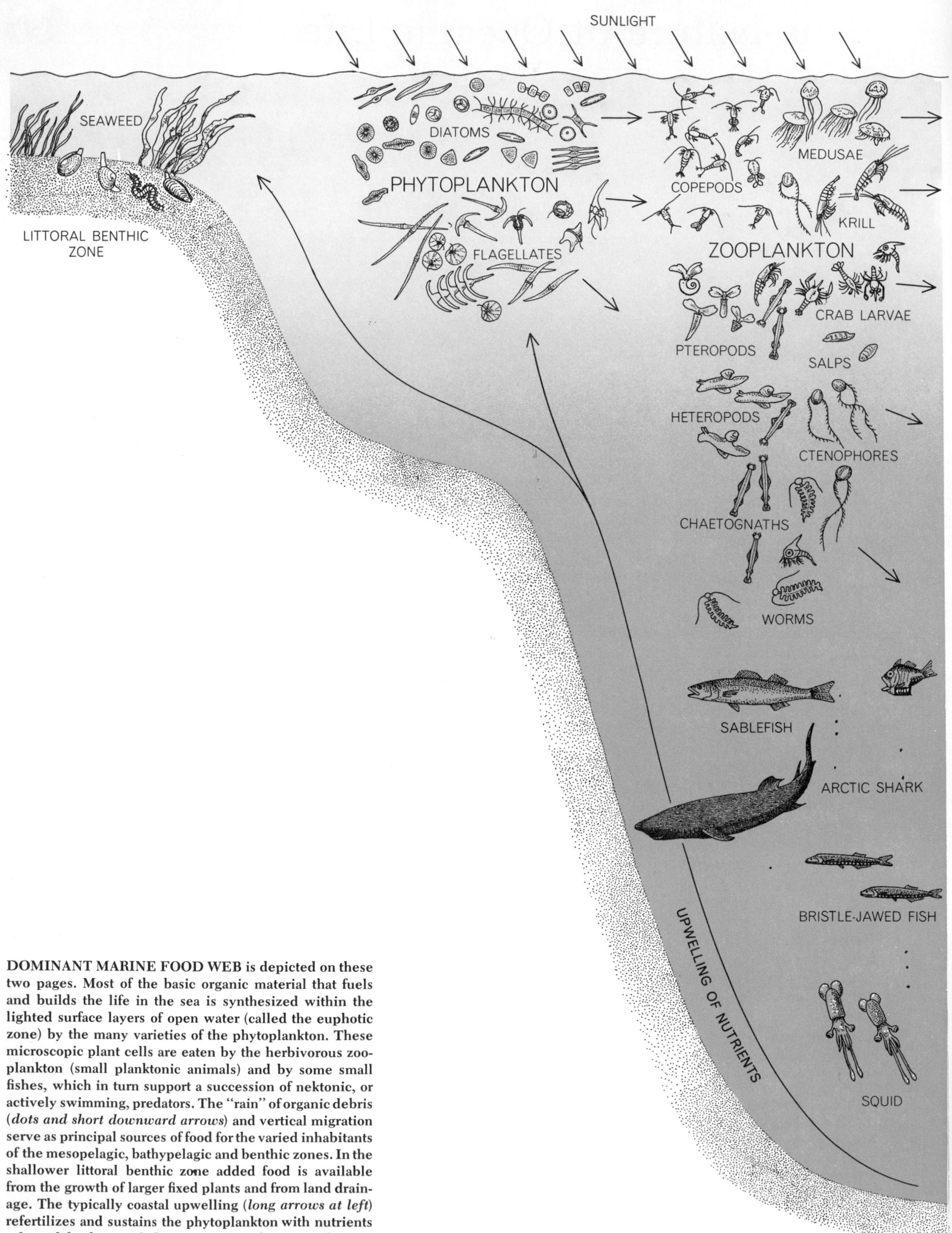

DOMINANT MARINE FOOD WEB is depicted on these two pages. Most of the basic organic material that fuels and builds the life in the sea is synthesized within the lighted surface layers of open water (called the euphotic zone) by the many varieties of the phytoplankton. These microscopic plant cells are eaten by the herbivorous zooplankton (small planktonic animals) and by some small fishes, which in turn support a succession of nektonic, or actively swimming, predators. The "rain" of organic debris (*dots and short downward arrows*) and vertical migration serve as principal sources of food for the varied inhabitants of the mesopelagic, bathypelagic and benthic zones. In the shallower littoral benthic zone added food is available from the growth of larger fixed plants and from land drainage. The typically coastal upwelling (*long arrows at left*) refertilizes and sustains the phytoplankton with nutrients released by bacterial decomposition of organic detritus on the bottom. The organisms are not drawn to same scale.

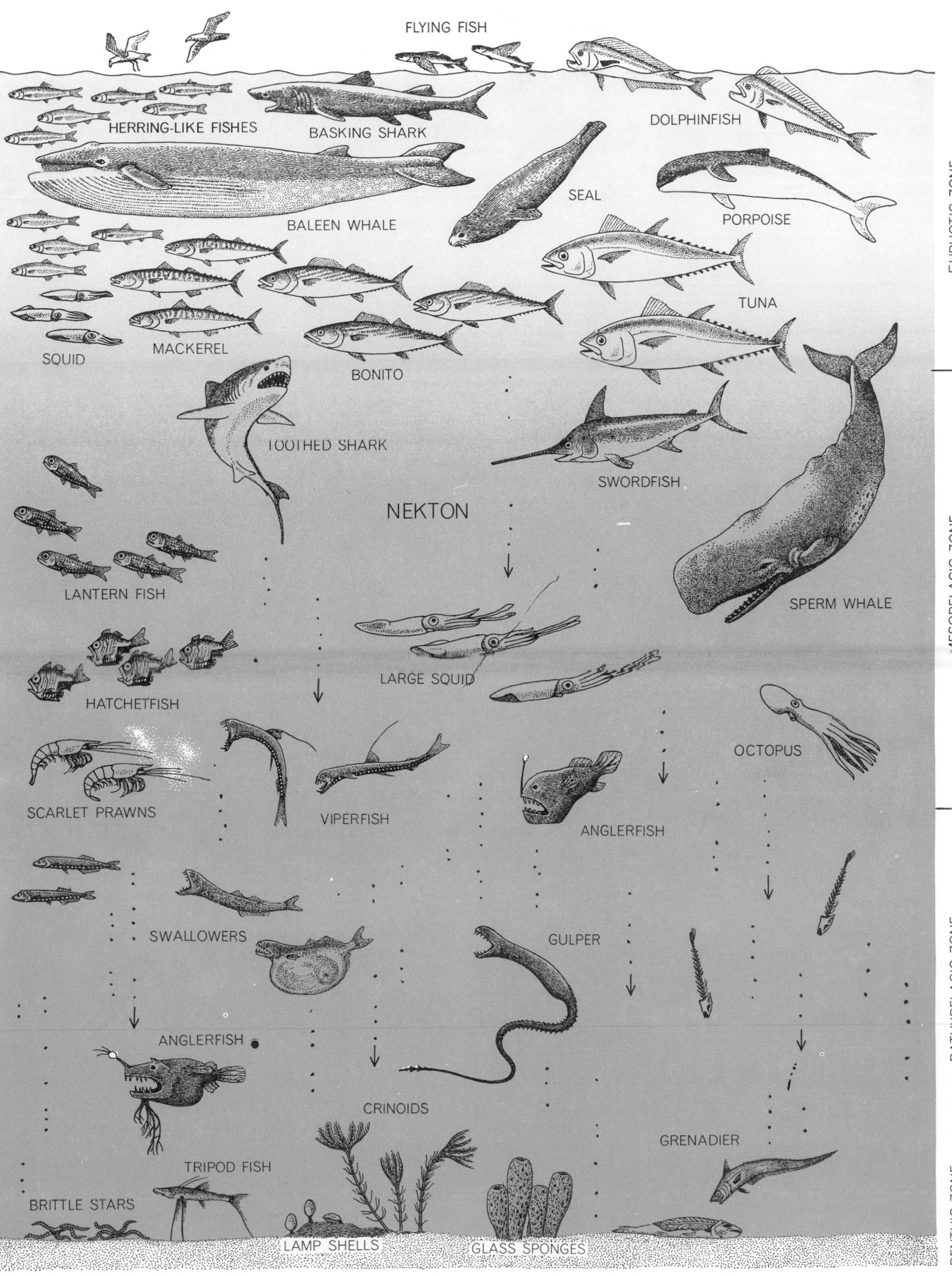
FLYING FISH
HERRING-LIKE FISHES
BASKING SHARK
DOLPHINFISH
SEAL
BALEEN WHALE
PORPOISE
TUNA
SQUID
MACKEREL
BONITO
TOOTHED SHARK
SWORDFISH
NEKTON
LANTERN FISH
SPERM WHALE
LARGE SQUID
HATCHETFISH
OCTOPUS
SCARLET PRAWNS
VIPERFISH
ANGLERFISH
SWALLOWERS
GULPER
ANGLERFISH
CRINOIDS
GRENADIER
TRIPOD FISH
BRITTLE STARS
LAMP SHELLS
GLASS SPONGES
EUPHOTIC ZONE
MESOPELAGIC ZONE
BATHYPELAGIC ZONE
BENTHIC ZONE

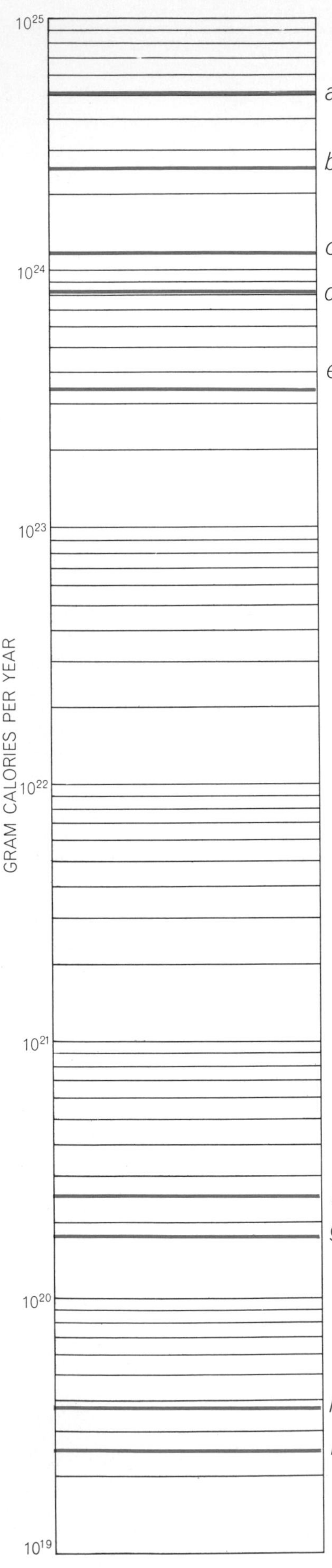

of the 90 contribute to the growth of land plants and about 40 to the growth of the single-celled green plants of the sea, the phytoplankton [*see illustration at left*]. It is this minute fraction of the sun's radiant energy that supplies the living organisms of this planet not only with their food but also with a breathable atmosphere.

The terrestrial and marine plants and animals arose from the same sources, through similar evolutionary sequences and by the action of the same natural laws. Yet these two living systems differ greatly at the stage in which we now view them. Were we to imagine a terrestrial food web that had developed in a form limited to that of the open sea, we would envision the land populated predominantly by short-lived simple plant cells grazed by small insects, worms and snails, which in turn would support a sparse predaceous population of larger insects, birds, frogs and lizards. The population of still larger carnivores would be a small fraction of the populations of large creatures that the existing land food web can nurture, because organisms in each of these steps pass on not more than 15 percent of the organic substance.

In some important respects this imaginary condition is not unlike that of the dominant food web of the sea, where almost all marine life is sustained by microscopic plants and near-microscopic herbivores and carnivores, which pass on only a greatly diminished supply of food to sustain the larger, more active and more complex creatures. In other respects the analogy is substantially inaccurate, because the primary marine food production is carried out by cells dispersed widely in a dense fluid medium.

This fact of an initial dispersal imposes a set of profound general conditions on all forms of life in the sea. For comparison, the concentration of plant food in a moderately rich grassland is of the order of a thousandth of the volume of the gross space it occupies and of the order of half of the mass of the air in which it is immersed. In moderately rich areas of the sea, on the other hand, food is hundreds of times more dilute in volume and hundreds of thousands of times more dilute in relative mass. To crop this meager broth a blind herbivore or a simple pore in a filtering structure would need to process a weight of water hundreds of thousands of times the weight of the cell it eventually captures. In even the densest concentrations the factor exceeds several thousands, and with each further step in the food web dilution increases. Thus from the beginnings of the marine food web we see many adaptations accommodating to this dilution: eyes in microscopic herbivorous animals, filters of exquisite design, mechanisms and behavior for discovering local concentrations, complex search gear and, on the bottom, attachments to elicit the aid of moving water in carrying out the task of filtration. All these adaptations stem from the conditions that limit plant life in the open sea to microscopic dimensions.

It is in the sunlit near-surface of the open sea that the unique nature of the dominant system of marine life is irrevocably molded. The near-surface, or mixed, layer of the sea varies in thickness from tens of feet to hundreds depending on the nature of the general circulation, mixing by winds and heating [see the article "The Atmosphere and the Ocean," by R. W. Stewart, beginning on page 129]. Here the basic food production of the sea is accomplished by single-celled plants. One common group of small phytoplankton are the coccolithophores, with calcareous plates, a swimming ability and often an oil droplet for food storage and buoyancy. The larger microscopic phytoplankton are composed of many species belonging to several groups: naked algal cells, diatoms with complex shells of silica and actively swimming and rotating flagellates. Very small forms of many groups are also abundant and collectively are called nannoplankton.

The species composition of the phytoplankton is everywhere complex and varies from place to place, season to season and year to year. The various regions of the ocean are typified, however, by

PRODUCTIVITY of the land and the sea are compared in terms of the net amount of energy that is converted from sunlight to organic matter by the green cells of land and sea plants. Colored lines denote total energy reaching the earth's upper atmosphere (*a*), total energy reaching earth's surface (*b*), total energy usable for photosynthesis (*c*), total energy usable for photosynthesis at sea (*d*), total energy usable for photosynthesis on land (*e*), net energy used for photosynthesis on land (*f*), net energy used for photosynthesis at sea (*g*), net energy used by land herbivores (*h*) and net energy used by sea herbivores (*i*). Although more sunlight falls on the sea than on the land (by virtue of the sea's larger surface area), the total land area is estimated to outproduce the total sea area by 25 to 50 percent. This is primarily due to low nutrient concentrations in the euphotic zone and high metabolism in marine plants. The data are from Walter R. Schmitt of Scripps Institution of Oceanography.

dominant major groups and particular species. Seasonal effects are often strong, with dense blooms of phytoplankton occurring when high levels of plant nutrients suddenly become usable or available, such as in high latitudes in spring or along coasts at the onset of upwelling. The concentration of phytoplankton varies on all dimensional scales, even down to small patches.

It is not immediately obvious why the dominant primary production of organic matter in the sea is carried out by microscopic single-celled plants instead of free-floating higher plants or other intermediate plant forms. The question arises: Why are there no pelagic "trees" in the ocean? One can easily compute the advantages such a tree would enjoy, with its canopy near the surface in the lighted levels and its trunk and roots extending down to the nutrient-rich waters under the mixed layer. The answer to this fundamental question probably has several parts. The evolution of plants in the pelagic realm favored smallness rather than expansion because the mixed layer in which these plants live is quite homogeneous; hence small incremental extensions from a plant cell cannot aid it in bridging to richer sources in order to satisfy its several needs.

On land, light is immediately above the soil and nutrients are immediately below; thus any extension is of immediate benefit, and the development of single cells into higher erect plants is able to follow in a stepwise evolutionary sequence. At sea the same richer sources exist but are so far apart that only a very large ready-made plant could act as a bridge between them. Although such plants could develop in some other environment and then adapt to the pelagic conditions, this has not come about. It is difficult to see how such a plant would propagate anyway; certainly it could not propagate in the open sea, because the young plants there would be at a severe disadvantage. In the sea small-scale differential motions of water are rapidly damped out, and any free-floating plant must often depend on molecular diffusion in the water for the uptake of nutrients and excretion of wastes. Smallness and self-motion are then advantageous, and a gross structure of cells cannot exchange nutrients or wastes as well as the same cells can separately or in open aggregations.

In addition the large-scale circulation of the ocean continuously sweeps the pelagic plants out of the region to which they are best adapted. It is essential that some individuals be returned to renew

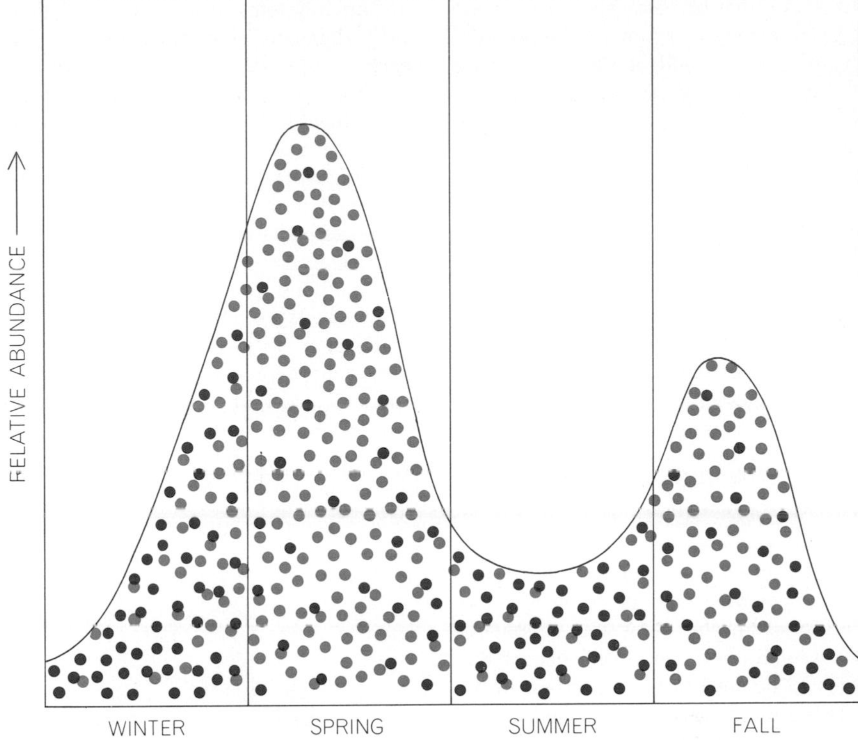

SPECIES COMPOSITION AND ABUNDANCE of the phytoplankton varies from season to season, particularly at high latitudes. During the winter the turbulence caused by storms replenishes the supply of nutrients in the surface layers. During this period flagellates (*black dots*) tend to dominate. In early spring the increase in the amount of sunlight reaching the surface stimulates plant growth, and diatoms (*colored dots*) are stimulated to grow. Later in spring grazing by zooplankton and a decrease in the supply of nutrients caused by calmer weather result in a general reduction in the phytoplankton population, which reaches a secondary minimum in midsummer, during which time flagellates again dominate. The increased mixing caused by early autumn storms causes a rise in the supply of nutrients and a corresponding minor surge in the population of diatoms. The decreasing sunlight of late fall and grazing by zooplankton again reduce the general level of the plant population.

the populations. More mechanisms for this essential return exist for single-celled plants than exist for large plants, or even for any conventional spores, seeds or juveniles. Any of these can be carried by oceanic gyres or diffused by large-scale motions of surface eddies and periodic counterflow, but single-celled plants can also ride submerged countercurrents while temporarily feeding on food particles or perhaps on dissolved organic material. Other mechanisms of distribution undoubtedly are also occasionally important. For example, living marine plant cells are carried by stormborne spray, in bird feathers and by well-fed fish and birds in their undigested food.

No large plant has solved the many problems of development, dispersal and reproduction. There *are* no pelagic trees, and these several factors in concert therefore restrict the open sea in a profound way. They confine it to an initial food web composed of microscopic forms, whereas larger plants live attached only to shallow bottoms (which comprise some 2 percent of the ocean area). Attached plants, unlike free-floating plants, are not subject to the aforementioned limitations. For attached plants all degrees of water motion enhance the exchange of nutrients and wastes. Moreover, their normal population does not drift, much of their reproduction is by budding, and their spores are adapted for rapid development and settlement. Larger plants too are sometimes found in nonreproducing terminal accumulations of drifting shore plants in a few special convergent deep-sea areas such as the Sargasso Sea.

Although species of phytoplankton will populate only regions with conditions to which they are adapted, factors other than temperature, nutrients and light levels undoubtedly are important in determining the species composition of phytoplankton populations. Little is understood of the mechanisms that give rise to an abundance of particular species under certain conditions. Grazing herbivores may consume only a part of the size range of cells, allowing certain sizes and types to dominate temporarily. Little is understood of the mechanisms that give rise to an abundance of particular species under certain conditions. Chemical by-products of certain species probably exclude certain other species. Often details of individual cell behavior are probably also important in the introduction and success of a species in a particular area. In some cases we can glimpse what these mechanisms are.

For example, both the larger diatoms and the larger flagellates can move at appreciable velocities through the water. The diatoms commonly sink downward, whereas the flagellates actively swim upward toward light. These are probably patterns of behavior primarily for increasing exchange, but the interaction of such unidirectional motions with random turbulence or systematic convective motion is not simple, as it is with an inactive particle. Rather, we would expect diatoms to be statistically abundant in upward-moving water and to sink out of the near-surface layers when turbulence or upward convection is low.

Conversely, flagellates should be statistically more abundant in downwelling water and should concentrate near the surface in low turbulence and slow downward water motions. These effects seem to exist. Off some continental coasts in summer flagellates may eventually collect in high concentrations. As they begin to shade one another from the light, each individual struggles closer to the lighted surface, producing such a high density that large areas of the water are turned red or brown by their pigments. The concentration of flagellates in these "red tides" sometimes becomes too great for their own survival. Several species of flagellates also become highly toxic as they grow older. Thus they sometimes both produce and participate in a mass death of fish and invertebrates that has been known to give rise to such a high yield of hydrogen sulfide as to blacken the white houses of coastal cities.

Large diatom cells, on the other hand, spend a disproportionately greater time in upward-moving regions of the water and an unlimited time in any region where the upward motion about equals their own downward motion. (The support of unidirectionally moving objects by contrary environmental motion is observed in other phenomena, such as the production of rain and hail.) Diatom cells are thus statistically abundant in upwelling water, and the distribution of diatoms probably is often a reflection of the turbulent-convective regime of the water. Sinking and the dependence of the larger diatoms on upward convection and turbulence for support aids them in reaching upwelling regions, where nutrients are high; it helps to explain their dominance in such regions and such other features of their distribution as their high proportion in rich ocean regions and their frequent inverse occurrence with flagellates. Differences in adaptations to the physical and chemical conditions, and the release of chemical products, probably reinforce such relations.

In some areas, such as parts of the equatorial current system and shallow seas, where lateral and vertical circulation is rapid, the species composition of phytoplankton is perhaps more simply a result of the inherent ability of the species to grow, survive and reproduce under the local conditions of temperature, light, nutrients, competitors and herbivores. Elsewhere second-order effects of the detailed cell behavior often dominate. Those details of behavior that give rise to concentrations on any dimensional scale are particularly important to all subsequent steps in the food chain.

All phytoplankton cells eventually settle from the surface layers. The depletion of nutrients and food from the surface layers takes place continuously through the loss of organic material, plant cells, molts, bodies of animals, fecal pellets and so forth, which release their content of chemical nutrients at

FAVORABLE CONDITIONS for the growth of phytoplankton occur wherever upwelling or mixing tends to bring subsurface nutrients up to the euphotic layer of the ocean. This map,

various depths through the action of bacteria and other organisms. The periodic downward migration of zooplankton further contributes to this loss.

These nutrients are "trapped" below the level of light adequate to sustain photosynthesis, and therefore the water in which plants must grow generally contains very low concentrations of such vital substances. It is this condition that is principally responsible for the comparatively low total net productivity of the sea compared with that of the land. The regions where trapping is broken down or does not exist—where there is upwelling of nutrient-rich water along coasts, in parts of the equatorial regions, in the wakes of islands and banks and in high latitudes, and where there is rapid recirculation of nutrients over shallow shelves and seas—locally bear the sea's richest fund of life.

The initial factors discussed so far have placed an inescapable stamp on the form of all life in the open sea, as irrevocably no doubt as the properties and distribution of hydrogen have dictated the form of the universe. These factors have limited the dominant form of life in the sea to an initial microscopic sequence that is relatively unproductive, is stimulated by upwelling and mixing and is otherwise altered in species composition and distribution by physical, chemical and biological processes on all dimensional scales. The same factors also limit the populations of higher animals and have led to unexpectedly simple adaptations, such as the sinking of the larger diatoms as a tactic to solve the manifold problems of enhancing nutrient and waste exchange, finding nutrients, remaining in the surface waters and repopulating.

The grazing of the phytoplankton is principally conducted by the herbivorous members of the zooplankton, a heterogeneous group of small animals that carry out several steps in the food web as herbivores, carnivores and detrital (debris-eating) feeders. Among the important members of the zooplankton are the arthropods, animals with external

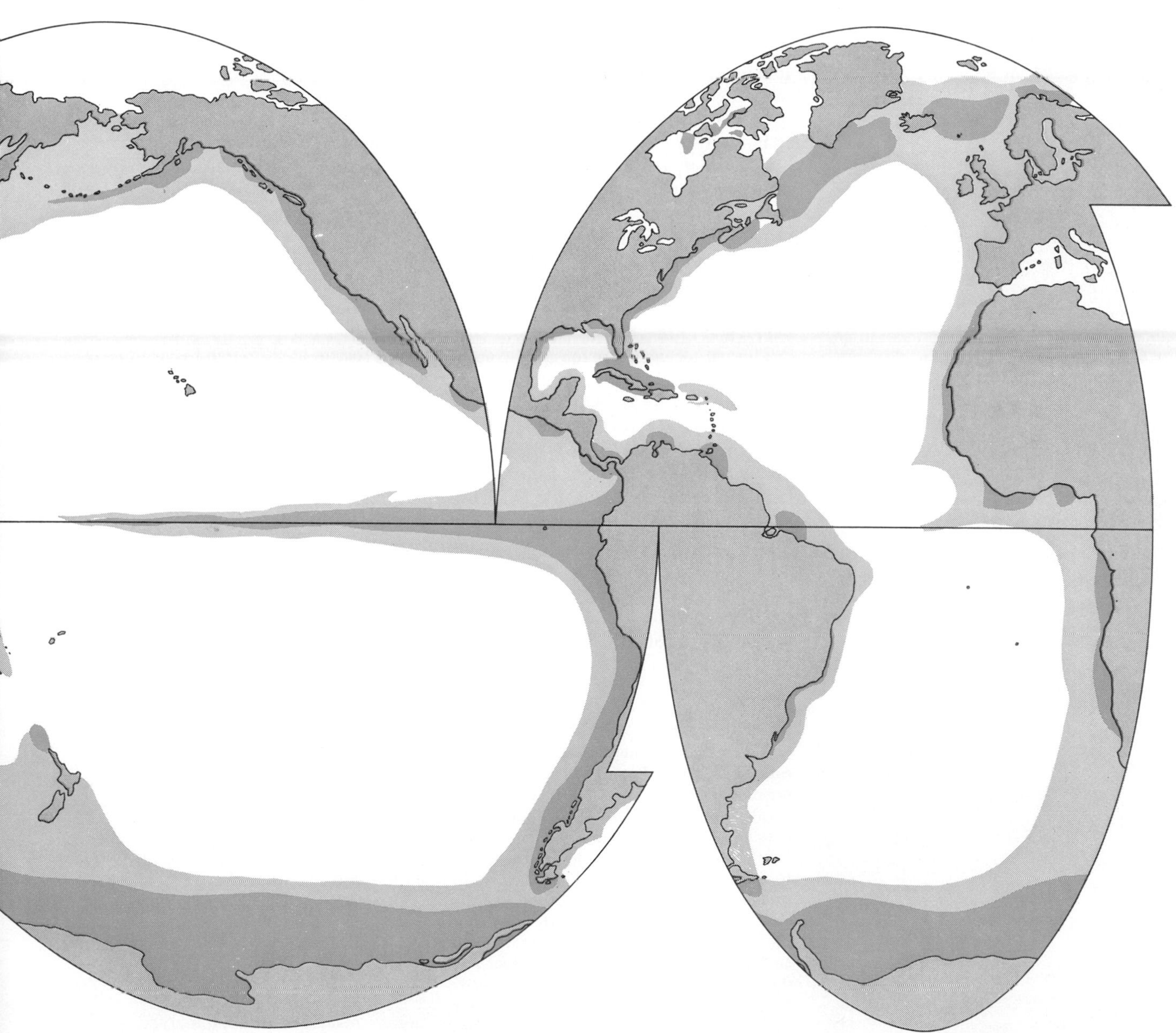

which is adapted from one compiled by the Norwegian oceanographer Harald U. Sverdrup, shows the global distribution of such waters, in which the productivity of marine life would be expected to be very high (*dark color*) and moderately high (*light color*).

skeletons that belong to the same broad group as insects, crabs and shrimps. The planktonic arthropods include the abundant copepods, which are in a sense the marine equivalent of insects. Copepods are represented in the sea by some 10,000 or more species that act not only as herbivores, carnivores or detrital feeders but also as external or even internal parasites! Two or three thousand of these species live in the open sea. Other important arthropods are the shrimplike euphausiids, the strongest vertical migrators of the zooplankton. They compose the vast shoals of krill that occur in high latitudes and that constitute one of the principal foods of the baleen whales. The zooplankton also include the strange bristle-jawed chaetognaths, or arrowworms, carnivores of mysterious origin and affinities known only in the marine environment. Widely distributed and abundant, the chaetognaths are represented by a surprisingly small number of species, perhaps fewer than 50. Larvae of many types, worms, medusae (jellyfish), ctenophores (comb jellies), gastropods (snails), pteropods and heteropods (other pelagic mollusks), salps, unpigmented flagellates and many others are also important components of this milieu, each with its own remarkably complex and often bizarre life history, behavior and form.

The larger zooplankton are mainly carnivores, and those of herbivorous habit are restricted to feeding on the larger plant cells. Much of the food supply, however, exists in the form of very small particles such as the nannoplankton, and these appear to be available almost solely to microscopic creatures. The immense distances between plant cells, many thousands of times their diameter, place a great premium on the development of feeding mechanisms that avoid the simple filtering of water through fine pores. The power necessary to maintain a certain rate of flow through pores or nets increases inversely at an exponential rate with respect to the pore or mesh diameter, and the small planktonic herbivores, detrital feeders and carnivores show many adaptations to avoid this energy loss. Eyesight has developed in many minute animals to make possible selective capture. A variety of webs, bristles, rakes, combs, cilia and other structures are found, and they are often sticky. Stickiness allows the capture of food that is finer than the interspaces in the filtering structures, and it greatly reduces the expenditure of energy.

A few groups have developed extremely fine and apparently quite effective nets. One group that has accomplished this is the Larvacea. A larvacian produces and inhabits a complex external "house," much larger than its owner, that contains a system of very finely constructed nets through which the creature maintains a gentle flow [*see illustration on page 198*]. The Larvacea have apparently solved the problem of energy loss in filtering by having proportionately large nets, fine strong threads and a low rate of flow.

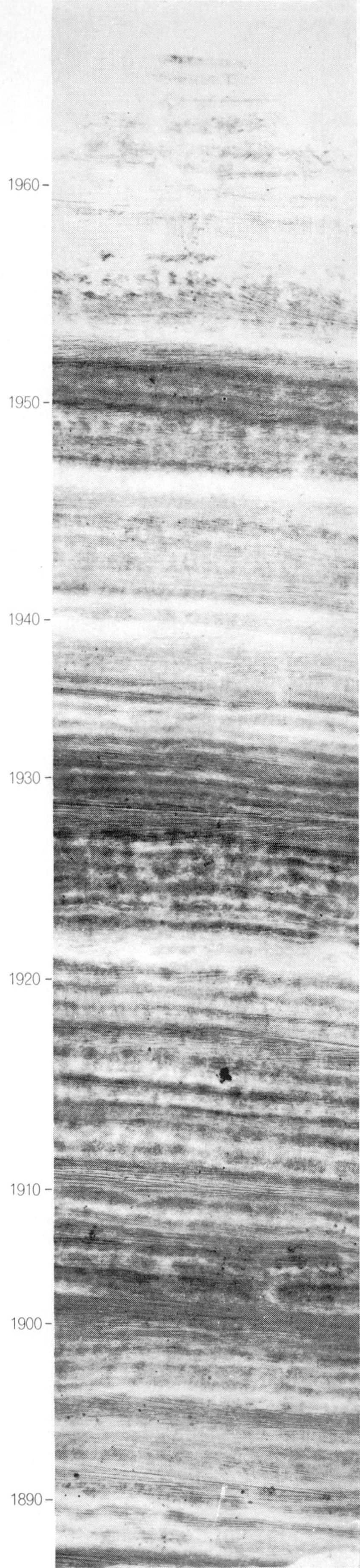

RARE SEDIMENTARY RECORD of the recent annual oceanographic, meteorological and biological history of part of a major oceanic system is revealed in this radiograph of a section of an ocean-bottom core obtained by Andrew Soutar of the Scripps Institution of Oceanography in the Santa Barbara Basin off the California coast. In some near-shore basins such as this one the absence of oxygen causes refractory parts of the organic debris to be left undecomposed and the sediment to remain undisturbed in the annual layers called varves. The dark layers are the densest and represent winter sedimentation. The lighter and less dense layers are composed mostly of diatoms and represent spring and summer sedimentation.

The composition of the zooplankton differs from place to place, day to night, season to season and year to year, yet most species are limited in distribution, and the members of the planktonic communities commonly show a rather stable representation of the modes of life.

The zooplankton are, of course, faced with the necessity of maintaining breeding assemblages and, like the phytoplankton, with the necessity of establishing a reinoculation of parent waters. In addition, their behavior must lead to a correspondence with their food and to the pattern of large-scale and small-scale spottiness already imposed on the marine realm by the phytoplankton. The swimming powers of the larger zooplankton are quite adequate for finding local small-scale patches of food. That this task is accomplished on a large scale is indirectly demonstrated by the observed correspondence between the quantities of zooplankton and the plant nutrients in the surface waters. How this large-scale task is accomplished is understood for some groups. For example, some zooplankton species have been shown to descend near the end of suitable conditions at the surface and to take temporary residence in a submerged countercurrent that returns them upstream.

There are many large and small puz-

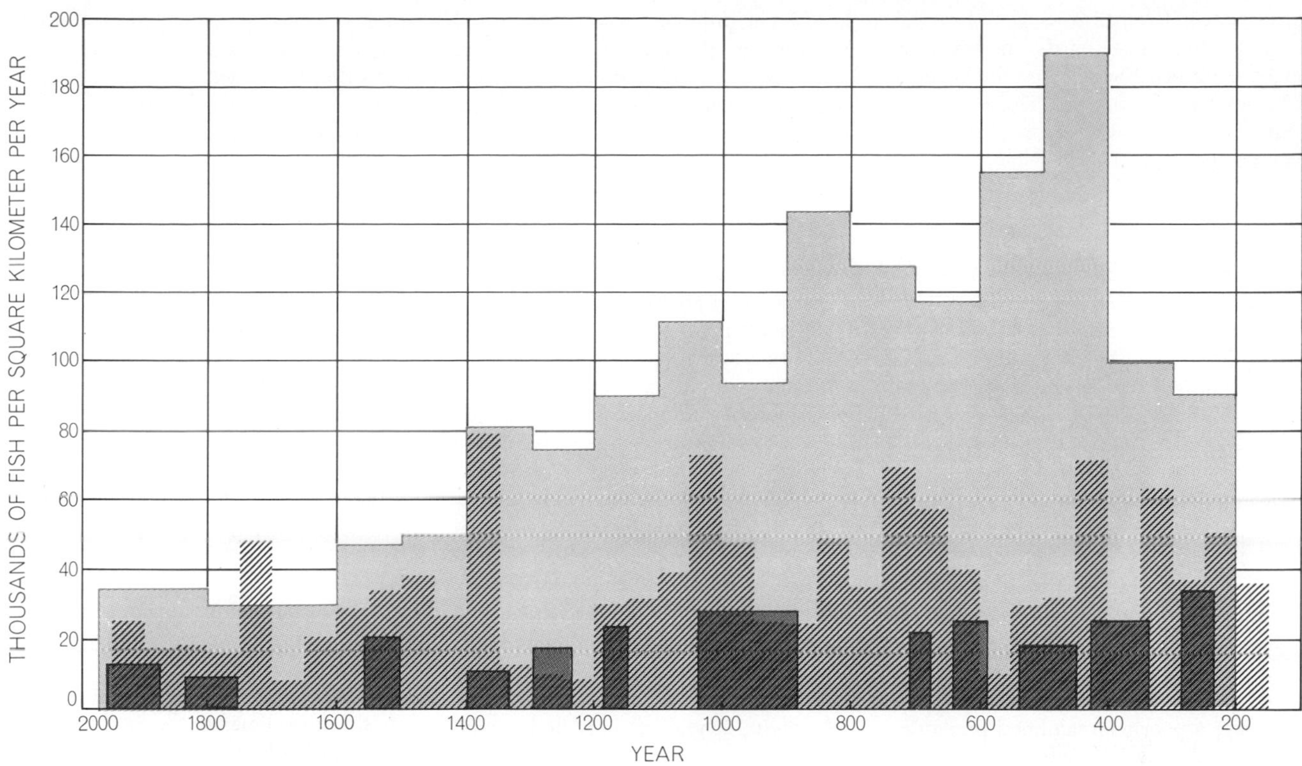

ESTIMATED FISH POPULATIONS in the Santa Barbara Basin over the past 1,800 years were obtained for three species by counting the average number of scales of each species in the varves of the core shown on the opposite page. Minimum population estimates for fish one year old and older are given for Pacific sardines (*gray*), northern sardines (*colored areas*) and Pacific hake (*hatched*).

zles in the distribution of zooplankton. As an example, dense concentrations of phytoplankton are often associated with low populations of zooplankton. These are probably rapidly growing blooms that zooplankton have not yet invaded and grazed on, but it is not completely clear that this is so. Chemical repulsion may be involved.

The concentration of larger zooplankton and small fish in the surface layers is much greater at night than during the day, because of a group of strongly swimming members that share their time between the surface and the mesopelagic region. This behavior is probably primarily a tactic to enjoy the best of two worlds: to crop the richer food developing in the surface layers and to minimize mortality from predation by remaining always in the dark, like timid rabbits emerging from the thicket to graze the nighttime fields, although still in the presence of foxes and ferrets. Many small zooplankton organisms also make a daily migration of some vertical extent.

In addition to its primary purpose daily vertical migration undoubtedly serves the migrating organisms in a number of other ways. It enables the creatures to adjust their mean temperature, so that by spending the days in cooler water the amount of food used during rest is reduced. Perhaps such processes as the rate of egg development are also controlled by these tactics. Many land animals employ hiding behavior for similar kinds of adjustment. Convincing arguments have also been presented to show that vertical migration serves to maintain a wide range of tolerance in the migrating species, so that they will be more successful under many more conditions than if they lived solely in the surface layers. This migration must also play an important part in the distribution of many species. Interaction of the daily migrants with the water motion produced by daily land-sea breeze alternation can hold the migrants offshore by a kind of "rectification" of the oscillating water motion. More generally, descent into the lower layers increases the influence of submerged countercurrents, thereby enhancing the opportunity to return upstream, to enter upwelling regions and hence to find high nutrient levels and associated high phytoplankton productivity.

Even minor details of behavior may strongly contribute to success. Migrants spend the day at a depth corresponding to relatively constant low light levels, where the movement of the water commonly is different from that at the surface. Most of the members rise somewhat even at the passage of a cloud shadow. Should they be carried under the shadow of an area rich in phytoplankton, they migrate to shallower depths, thereby often decreasing or even halting their drift with respect to this rich region to which they will ascend at night. Conversely, when the surface waters are clear and lean, they will migrate deeper and most often drift relatively faster.

We might simplistically view the distribution of zooplankton, and phytoplankton for that matter, as the consequence of a broad inoculation of the oceans with a spectrum of species, each with a certain adaptive range of tolerances and a certain variable range of feeding, reproducing and migrating behavior. At some places and at some times the behavior of a species, interacting even in detailed secondary ways with the variable conditions of the ocean and its other inhabitants, results in temporary, seasonal or persistent success.

There are a few exceptions to the microscopic dimensions of the herbivores in the pelagic food web. Among these the herrings and herring-like fishes are able to consume phytoplankton as a substantial component of their diet. Such an adaptation gives these fishes access to many times the food supply of the more carnivorous groups. It is therefore no surprise that the partly herbivorous fishes comprise the bulk of the world's fisheries [see the article "The Food Resources of the Ocean," by S. J. Holt, beginning on page 210].

The principal food supplies of the pelagic populations are passed on in incremental steps and rapidly depleted quantity to the larger carnivorous zooplankton, then to small fishes and squids, and ultimately to the wide range of larger carnivores of the pelagic realm. In this region without refuge, either powerful static defenses, such as the stinging cells of the medusae and men-o'-war, or increasing size, acuity, alertness, speed and strength are the requirements for survival at each step. Streamlining of form here reaches a high point of development, and in tropical waters it is conspicuous even in small fishes, since the lower viscosity of the warmer waters will enable a highly streamlined small prey to escape a poorly streamlined predator, an effect that exists only for fishes of twice the length in cold, viscous, arctic or deep waters.

The pelagic region contains some of the largest and most superbly designed creatures ever to inhabit this earth: the exquisitely constructed pelagic tunas; the multicolored dolphinfishes, capturers of flying fishes; the conversational porpoises; the shallow- and deep-feeding swordfishes and toothed whales, and the greatest carnivores of all, the baleen whales and some plankton-eating sharks, whose prey are entire schools of krill or small fishes. Seals and sea lions feed far into the pelagic realm. In concert with these great predators, large carnivorous sharks await injured prey. Marine birds, some adapted to almost continuous pelagic life, consume surprising quantities of ocean food, diving, plunging, skimming and gulping in pursuit. Creatures of this region have developed such faculties as advanced sonar, unexplained senses of orientation and homing, and extreme olfactory sensitivity.

These larger creatures of the sea commonly move in schools, shoals and herds. In addition to meeting the needs of mating such grouping is advantageous in both defensive and predatory strategy, much like the cargo-ship convoy and submarine "wolf pack" of World War II. Both defensive and predatory assemblages are often complex. Small fishes of several species commonly school together. Diverse predators also form loosely cooperative groups, and many species of marine birds depend almost wholly on prey driven to the surface by submerged predators.

At night, schools of prey and predators are almost always spectacularly illuminated by bioluminescence produced by the microscopic and larger plankton. The reason for the ubiquitous production of light by the microorganisms of the sea remains obscure, and suggested explanations are controversial. It has been suggested that light is a kind of inadvertent by-product of life in transparent organisms. It has also been hypothesized that the emission of light on disturbance is advantageous to the plankton in making the predators of the plankton conspicuous to *their* predators! Unquestionably it does act this way. Indeed, some fisheries base the detection of their prey on the bioluminescence that the fish excite. It is difficult, however, to defend the thesis that this effect was the direct factor in the original development of bioluminescence, since the effect was of no advantage to the individual microorganism that first developed it. Perhaps the luminescence of a microorganism also discourages attack by the light-avoiding zooplankton and is of initial survival benefit to the individual. As it then became general in the population, the effect of revealing plankton predators to their predators would also become important.

The fallout of organic material into the deep, dimly lighted mid-water supports a sparse population of fishes and invertebrates. Within the mesopelagic and bathypelagic zones are found some of the most curious and bizarre creatures of this earth. These range from the highly developed and powerfully predaceous intruders, toothed whales and swordfishes, at the climax of the food chain, to the remarkable squids, octopuses, euphausiids, lantern fishes, gulpers and anglerfishes that inhabit the bathypelagic region.

In the mesopelagic region, where some sunlight penetrates, fishes are often countershaded, that is, they are darker above and lighter below, as are surface fishes. Many of the creatures of this dimly lighted region participate in the daily migration, swimming to the upper layers at evening like bats emerging from their caves. At greater depths, over a half-mile or so, the common inhabitants are often darkly pigmented, weak-bodied and frequently adapted to unusual feeding techniques. Attraction of prey by luminescent lures or by mimicry of small prey, greatly extensible jaws and expansible abdomens are common. It is, however, a region of Lilliputian monsters, usually not more than six inches in length, with most larger fishes greatly reduced in musculature and weakly constructed.

There are some much larger, stronger and more active fishes and squids in this region, although they are not taken in trawls or seen from submersibles. Knowledge of their existence comes mainly from specimens found in the stomach of sperm whales and swordfish. They must be rare, however, since the slow, conservative creatures that are taken in trawls could hardly coexist with large numbers of active predators. Nevertheless, populations must be sufficiently large to attract the sperm whales and swordfish. There is evidence that the sperm whales possess highly developed long-range hunting sonar. They may lo-

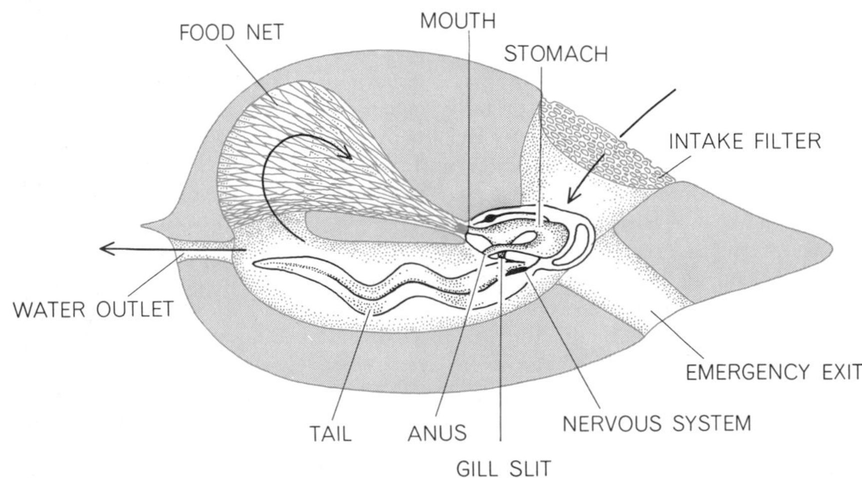

LARVACIAN is representative of a group of small planktonic herbivores that has solved the problem of energy loss in filtering, apparently without utilizing "stickiness," by having proportionately large nets, strong fine threads and a low rate of water flow. The larvacian (*black*) produces and inhabits a complex external "house" (*color*), much larger than its owner, which contains a system of nets through which the organism maintains a gentle flow. In almost all other groups simple filters are employed only to exclude large particles.

cate their prey over relatively great distances, perhaps miles, from just such an extremely sparse population of active bathypelagic animals.

Although many near-surface organisms are luminescent, it is in the bathypelagic region that bioluminescence has reached a surprising level of development, with at least two-thirds of the species producing light. Were we truly marine-oriented, we would perhaps be more surprised by the almost complete absence of biological light in the land environment, with its few rare cases of fireflies, glowworms and luminous bacteria. Clearly bioluminescence can be valuable to higher organisms, and the creatures of the bathypelagic realm have developed light-producing organs and structures to a high degree. In many cases the organs have obvious functions. Some fishes, squids and euphausiids possess searchlights with reflector, lens and iris almost as complex as the eye. Others have complex patterns of small lights that may serve the functions of recognition, schooling control and even mimicry of a small group of luminous plankton. Strong flashes may confuse predators by "target alteration" effects, or by producing residual images in the predators' vision. Some squids and shrimps are more direct and discharge luminous clouds to cover their escape. The luminous organs are arranged on some fishes so that they can be used to countershade their silhouettes against faint light coming from the surface. Luminous baits are well developed. Lights may also be used for locating a mate, a problem of this vast, sparsely populated domain that has been solved by some anglerfishes by the development of tiny males that live parasitically attached to their relatively huge mates.

It has been shown that the vertebrate eye has been adapted to detect objects in the lowest light level on the earth's surface—a moonless, overcast night under a dense forest canopy—but not lower. Light levels in the bathypelagic region can be much lower. This is most probably the primary difference that accounts for the absence of bioluminescence in higher land animals and the richness of its development in the ocean forms.

The densest populations of bathypelagic creatures lie below the most productive surface regions, except at high latitudes, where the dearth of winter food probably would exhaust the meager reserves of these creatures. All the bathypelagic populations are sparse, and in this region living creatures are less than one hundred-millionth of the water volume. Nevertheless, the zone is of immense dimensions and the total populations may be large. Some genera, such as the feeble, tiny bristle-jawed fishes, are probably the most numerous fishes in the world and constitute a gigantic total biomass. There are some 2,000 species of fishes and as many species of the larger invertebrates known to inhabit the bathypelagic zone, but only a few of these species appear to be widespread. The barriers to distribution in this widely interconnected mid-water region are not obvious.

The floor of the deep sea constitutes an environment quite unlike the mid-water and surface environments. Here are sites for the attachment of the larger

CHAMPION FILTER FEEDER of the world ocean in terms of volume is the blue whale, a mature specimen of which lies freshly butchered on the deck of a whaling vessel in this photograph. The whale's stomach has been cut open with a flensing knife to reveal its last meal: an immense quantity of euphausiids, or krill, each measuring about three inches in length. The baleen whales are not plankton-filterers in the ordinary sense but rather are great carnivores that seek out and engulf entire schools of small fish or invertebrates. The photograph was made by Robert Clarke of the National Institute of Oceanography in Wormley, England.

invertebrates that filter detritus from the water. Among these animals are representatives of some of the earliest multicelled creatures to exist on the earth, glass sponges, sea lilies (crinoids)—once thought to have been long extinct—and lamp shells (brachiopods).

At one time it was also thought that the abyssal floor was sparsely inhabited and that the populations of the deep-ocean floor were supplied with food only by the slow, meager rain of terminal detrital food material that has passed through the surface and bathypelagic populations. Such refractory material requires further passage into filter feeders or through slow bacterial action in the sediment, followed by consumption by larger burrowing organisms, before it becomes available to active free-living animals. This remnant portion of the food web could support only a very small active population.

Recent exploration of the abyssal realm with a baited camera throws doubt on the view that this is the exclusive mechanism of food transfer to the deep bottom. Large numbers of active fishes and other creatures are attracted to the bait almost immediately [*see illustration on page 180*]. It is probably true that several rather independent branches of the food web coexist in support of the deep-bottom creatures: one the familiar rain of fine detritus, and the other the rare, widely separated falls of large food particles that are in excess of the local feeding capacity of the broadly diffuse bathypelagic population. Such falls would include dead whales, large sharks or other large fishes and fragments of these, the multitude of remnants that are left when predators attack a school of surface fish and now, undoubtedly, garbage from ships and kills from underwater explosions. These sources result in an influx of high-grade food to the sea floor, and we would expect to find a population of active creatures adapted to its prompt discovery and utilization. The baited cameras have demonstrated that this is so.

Other sources of food materials are braided into these two extremes of the abyssal food web. There is the rather subtle downward diffusion of living and dead food that results initially from the daily vertical migration of small fish and zooplankton near the surface. This migration appears to impress a sympathetic daily migration on the mid-water populations down to great depths, far below the levels that light penetrates. Not only may such vertical migration bring feeble bathypelagic creatures near the bottom but also it accelerates in itself the flux of dead food material to the bottom of the deep sea.

There must also be some unassignable flux of food to the abyssal population resulting from the return of juveniles to their habitat. The larvae and young of many abyssal creatures develop at much shallower levels. To the extent that the biomass of juveniles returning to the deep regions exceeds the biomass of spawn released from it, this process, which might be called "Faginism," constitutes an input of food.

Benthic animals are much more abundant in the shallower waters off continents, particularly offshore from large rivers. Here there is often not only a richer near-surface production and a less hazardous journey of food to the sea floor but also a considerable input of food conveyed by rivers to the bottom. The deep slopes of river sediment wedges are typified by a comparatively rich population of burrowing and filtering animals that utilize this fine organic material. All the great rivers of the world save one, the Congo, have built sedimentary wedges along broad reaches of their coast, and in many instances these wedges extend into deep water. The shallow regions of such wedges are highly productive of active and often valuable marine organisms. At all depths the wedges bear larger populations than are common at similar depths elsewhere. Thus one wonders what inhabits the fan of the Congo. That great river, because of a strange invasion of a submarine canyon into its month, has built no wedge but rather is depositing a vast alluvial fan in the two-mile depths of the Angola Basin. This great deep region of the sea floor may harbor an unexplored population that is wholly unique.

In itself the pressure of the water at great depths appears to constitute no insurmountable barrier to water-breathing animal life. The depth limitations of many creatures are the associated conditions of low temperature, darkness, sparse food and so on. It should perhaps come as no surprise, therefore, that some of the fishes of high latitudes, which are of course adapted to cold dark waters, extend far into the deep cold waters in much more southern latitudes. Off the coast of Lower California, in water 1,200 to 6,000 feet deep, baited cameras have found an abundance of several species of fishes that are known at the near surface only far to the north. These include giant arctic sharks, sablefish and others. It appears that some of the fishes that have been called arctic species are actually fishes of the dark cold waters of the seas, which only "outcrop" in the Arctic, where cold water is at the surface.

I have discussed several of the benthic and epibenthic environments without pointing out some of the unique features the presence of a solid interface entails. The bottom is much more variable than the mid-water zone is. There are as a result more environmental niches for an organism to occupy, and hence we see organisms that are of a wider range of form and habit. Adaptations develop for hiding and ambuscade, for mimicry and controlled patterns. Nests and burrows can be built, lairs occupied and defended and booby traps set.

Aside from the wide range of form and function the benthic environment elicits from its inhabitants, there are more fundamental conditions that influence the nature and form of life there. For example, the dispersed food material settling from the upper layers becomes much concentrated against the sea floor. Indeed, it may become further concentrated by lateral currents moving it into depressions or the troughs of ripples.

In the mid-water environment most creatures must move by their own energies to seek food, using their own food stores for this motion. On the bottom, however, substantial water currents are present at all depths, and creatures can await the passage of their food. Although this saving only amounts to an added effectiveness for predators, it is of critical importance to those organisms that filter water for the fine food material it contains, and it is against the bottom interface that a major bypass to the microscopic steps of the dominant food web is achieved. Here large organisms can grow by consuming microscopic or even submicroscopic food particles. Clams, scallops, mussels, tube worms, barnacles and a host of other creatures that inhabit this zone have developed a wide range of extremely effective filtering mechanisms. In one step, aided by their attachment, the constant currents and the concentration of detritus against the interface, they perform the feat, most unusual in the sea, of growing large organisms directly from microscopic food.

Although the benthic environment enables the creatures of the sea to develop a major branch of the food web that is emancipated from successive microscopic steps, this makes little difference to the food economy of the sea. The sea is quite content with a large population of tiny organisms. From man's standpoint, however, the shallow benthic environment is an unusually effective producer of larger creatures for his food, and he widely utilizes these resources.

Man may not have created an ideal environment for himself, but of all the environments of the sea it is difficult to conceive of one better for its inhabitants than the one marine creatures have created almost exclusively for themselves: the coral islands and coral reefs. In these exquisite, immense and well-nigh unbelievable structures form and adaptation reach a zenith.

An adequate description of the coral reef and coral atoll structure, environments and living communities is beyond the scope of this article. The general history and structure of atolls is well known, not only because of an inherent fascination with the magic and beauty of coral islands but also because of the wide admiration and publicity given to the prescient deductions on the origin of atolls by Charles Darwin, who foresaw much of what modern exploration has affirmed.

From their slowly sinking foundations of ancient volcanic mountains, the creatures of the coral shoals have erected the greatest organic structures that exist. Even the smallest atoll far surpasses any of man's greatest building feats, and a large atoll structure in actual mass approaches the total of all man's building that now exists.

These are living monuments to the success of an extremely intricate but balanced society of fish, invertebrates and plants, capitalizing on the basic advantages of benthic populations already discussed. Here, however, each of the reef structures acts almost like a single great isolated and complex benthic organism that has extended itself from the deep poor waters to the sunlit richer surface. The trapping of the advected food from the surface currents enriches the entire community. Attached plants further add to the economy, and there is considerable direct consumption of plant life by large invertebrates and fish. Some of the creatures and relationships that have developed in this environment are among the most highly adapted found on the earth. For example, a number of the important reef-building animals, the corals, the great tridacna clams and others not only feed but also harbor within their tissues dense populations of single-celled green plants. These plants photosynthesize food that is then directly available within the bodies of the animals; the plants in turn depend on the animal waste products within the body fluids, with which they are bathed, to derive their basic nutrients. Thus within the small environment of these plant-animal composites both the entire laborious nutrient cycle and the microscopic food web of the sea appear to be substantially bypassed.

There is much unknown and much to be discovered in the structure and ecology of coral atolls. Besides the task of unraveling the complex relationships of its inhabitants there are many questions such as: Why have many potential atolls never initiated effective growth and remained submerged almost a mile below the surface? Why have others lost the race with submergence in recent times and now become shallowly submerged, dying banks? Can the nature of the circulation of the ancient ocean be deduced from the distribution of successful and unsuccessful atolls? Is there circulation within the coral limestone structure that adds to the nutrient supply, and is this related to the curious development of coral knolls, or coral heads, within the lagoons? Finally, what is the potential of cultivation within these vast, shallow-water bodies of the deep open sea?

There is, of course, much to learn about all marine life: the basic processes of the food web, productivity, populations, distributions and the mechanisms of reinoculation, and the effects of intervention into these processes, such as pollution, artificial upwelling, transplantation, cultivation and fisheries. To learn of these processes and effects we must understand the nature not only of strong simple actions but also of weak complex interactions, since the forms of life or the success of a species may be determined by extremely small second- and third-order effects. In natural affairs, unlike human codes, *de minimis curat lex*—the law *is* concerned with trivia!

Little is understood of the manner in which speciation (that is, the evolution of new species) occurs in the broadly intercommunicating pelagic environment with so few obvious barriers. Important yet unexpected environmental niches may exist in which temporary isolation may enable a new pelagic species to evolve. For example, the top few millimeters of the open sea have recently been shown to constitute a demanding environment with unique inhabitants. Further knowledge of such microcosms may well yield insight into speciation.

As it has in the past, further exploration of the abyssal realm will undoubtedly reveal undescribed creatures including members of groups thought long extinct, as well as commercially valuable populations. As we learn more of the conditions that control the distribution of species of pelagic organisms, we shall become increasingly competent to read the pages of the earth's marine-biological, oceanographic and meteorological history that are recorded in the sediments by organic remains. We shall know more of primordial history, the early production of a breathable atmosphere and petroleum production. Some of these deposits of sediment cover even the period of man's recorded history with a fine time resolution. From such great records we should eventually be able to increase greatly our understanding of the range and interrelations of weather, ocean conditions and biology for sophisticated and enlightened guidance of a broad spectrum of man's activities extending from meteorology and hydrology to oceanography and fisheries.

Learning and guidance of a more specific nature can also be of great practical importance. The diving physiology of marine mammals throws much light on the same physiological processes in land animals in oxygen stress (during birth, for example). The higher flowering plants that inhabit the marine salt marshes are able to tolerate salt at high concentration, desalinating seawater with the sun's energy. Perhaps the tiny molecule of DNA that commands this process is the most precious of marine-life resources for man's uses. Bred into existing crop plants, it may bring salt-water agriculture to reality and nullify the creeping scourge of salinization of agricultural soils.

Routine upstream reinoculation of preferred species of phytoplankton and zooplankton might stabilize some pelagic marine populations at high effectiveness. Transplanted marine plants and animals may also animate the dead saline lakes of continental interiors, as they have the Salton Sea of California.

The possible benefits of broad marine-biological understanding are endless. Man's aesthetic, adventurous, recreational and practical proclivities can be richly served. Most important, undoubtedly, is the intellectual promise: to learn how to approach and understand a complex system of strongly interacting biological, physical and chemical entities that is vastly more than the sum of its parts, and thus how better to understand complex man and his interactions with his complex planet, and to explore with intelligence and open eyes a huge portion of this earth, which continuously teaches that when understanding and insight are sought for their own sake, the rewards are more substantial and enduring than when they are sought for more limited goals.

Active Animals of the Deep-Sea Floor

by John D. Isaacs and Richard A. Schwartzlose
October 1975

Baited automatic cameras dropped to the bottom of the ocean reveal a surprising population of large fishes and other scavengers that find and consume dead animals that fall from the waters far above

Photographs made by automatic cameras that have been dropped to the deep-ocean floor have confirmed a finding that was first suggested by evidence gathered from baited traps some years ago. A deep-sea population of large, active animals thrives in what was generally assumed to be a province inhabited mainly by small, feeble creatures such as worms, snails and sponges. The thousands of pictures make it clear that much of the deep-sea floor teems with numerous species of scavengers: vigorous invertebrates and fishes, including some gigantic sharks, that are supported by a marine food web whose extent and complexity is only beginning to be perceived.

The celebrated expedition of H.M.S. *Challenger* in the 1870's laid to rest the old idea that the deep waters of the open ocean were a lifeless desert. The *Challenger's* trawls and dredges brought to the surface and to the attention of biologists a vast collection of deep-midwater and bottom-dwelling creatures from even the deepest ocean trenches [see the article "The Voyage of the 'Challenger,' " by Herbert S. Bailey, Jr., beginning on page 8]. The inventory included some of the most grotesque forms into which higher animals have ever been molded by adaptation to extreme conditions, and several animals that had been thought to be long extinct.

FIFTEEN-FOOT SHARK attacks crushed bait can in the color photograph on the opposite page, obtained by the authors from a depth of some 750 meters in the eastern Mediterranean Sea. Such large fish typically arrive at the bait from three to eight hours after the camera system reaches bottom, frightening off most other feeding creatures. The line (*left*) leads to the camera.

In this century much has been added to that inventory and to our understanding of deep-living animals. The investigations were conducted, however, by means of trawls and dredges, by direct observation from deep-diving submersible vehicles and by inspection of photographs made by cameras lowered on cables from ships, and it is characteristic of these methods that they ordinarily sample only the stationary or slow-moving animals of the deep-sea communities; the deeper the investigation is, the more difficult and the less effective it is. That gave rise to the prevalent assumption of a few years ago that the ocean bottom was sparsely inhabited by weak creatures specially adapted to live on the only food material then thought to be available at great depths: a terminal food web supported by the thin but constant rain of detritus that sifts down from the surface layers and that is metabolized by primitive filter feeders or bacteria and deposit feeders in the bottom ooze. A proper reconnaissance of the active deep-sea creatures whose presence was revealed by the baited traps required new tools.

Over the past seven years Meredith Sessions, Richard Shutts and the authors, working at the Scripps Institution of Oceanography, have designed and constructed robot motion-picture and still cameras and other instruments with which to explore the bottom and to study the nature, distribution and behavior of its inhabitants. We find, first of all, that the population of sea-floor invertebrates is by no means sparse and that many of its species are far from weak. And they are not alone. Their domain is shared by a population of scavenging fishes and crustaceans adapted for the prompt discovery and consumption of larger falls of food: the bodies of dead animals descending from above, mid-water creatures that happen to approach the bottom and juveniles of their own species that return to the deeps from the shallower water where they undergo their early development.

The cameras we have devised for the ocean-floor study are free-fall devices connected to a recovery buoy, a floodlight and a bait holder. Released at the surface, the camera falls to the bottom and remains there, anchored by the bait holder for between 12 and 48 hours, making still photographs or short bursts of motion pictures at five- to 15-minute intervals. The bait is in the foreground of each frame and the camera's view is either vertical or oblique. At the end of the mission the bait ballast is released and the camera and its buoy rise to the surface, where a radio transmitter broadcasts a signal that aids in recovery. The free-fall technique allows a research ship to conduct various other missions while a number of cameras it has distributed keep functioning on the sea floor. And the work can be done from rather simple, inexpensive vessels, provided only that they have reasonable navigation and sonic-sounding equipment.

In a typical drop the camera is released in water that is between 400 and 7,000 meters deep. Greater depths call for a special camera housing and flotation gear because of the immense pressures. The camera operates during the descent, which may take as long as several hours, but ordinarily not much is seen in photographs made on the way down except for the "snow" of small particles, which are ubiquitous in the midwaters of the sea, and a few fleeting small crustaceans.

Sometimes the very first photographs

on the bottom, made a few minutes after the camera reaches it, show great activity, with fishes and invertebrates already tearing at the bait. More commonly, however, the first scenes show only brittle stars, large shrimps, amphipods (small crustaceans) and perhaps a fish or two. On more than half of the missions at least one fish has been photographed within 30 minutes of the camera's arrival on the bottom, even at the greater depths. (Curiously the first photographs sometimes record a fish of a species that is not seen again in the sequence, as though the camera had invaded the territory of a creature that was not a part of the population of active scavengers and wanted nothing to do with the subsequent activity.)

Usually the number of fish gathered around the bait increases slowly, reaching a maximum after a few hours. Often the scene develops into one of furious activity, with several species of fish competing for the bait, thrashing and tearing at it and sometimes attacking one another. Shrimps, brittle stars, amphipods and other invertebrates encroach on the melee. In almost half of the sequences from drops down to 2,000 meters the party ends abruptly after three to eight hours, when some creature, usually a large shark, moves in, frightens off the other fish and consumes the bulk of the bait. In any case the time comes when most of the bait has been eaten. The fish depart, and slowly crabs, sea urchins, snails and other such creatures arrive to complete the task of sanitizing the sea floor.

Sometimes the direction of the near-bottom currents has been determined and it can be seen that these latecomers to the banquet plod upstream toward the bait—an indication that they are probably following a scent. The fishes also probably depend on scent for close-in detection of the bait; on some occasions we could relate the number of fish gathered around the bait to the strength of the current, which suggests that an increased current had carried the scent more widely. Like wide-ranging scavengers on land, however, fish that are farther away are probably led to the area of the bait by other cues. They may sense the successive collapse of loosely established territories as the scavengers that held them move closer to the bait, each invading the area once held by an absent neighbor. The western prairie wolf, vultures and other terrestrial scavengers respond to just such a territorial collapse to converge on large kills.

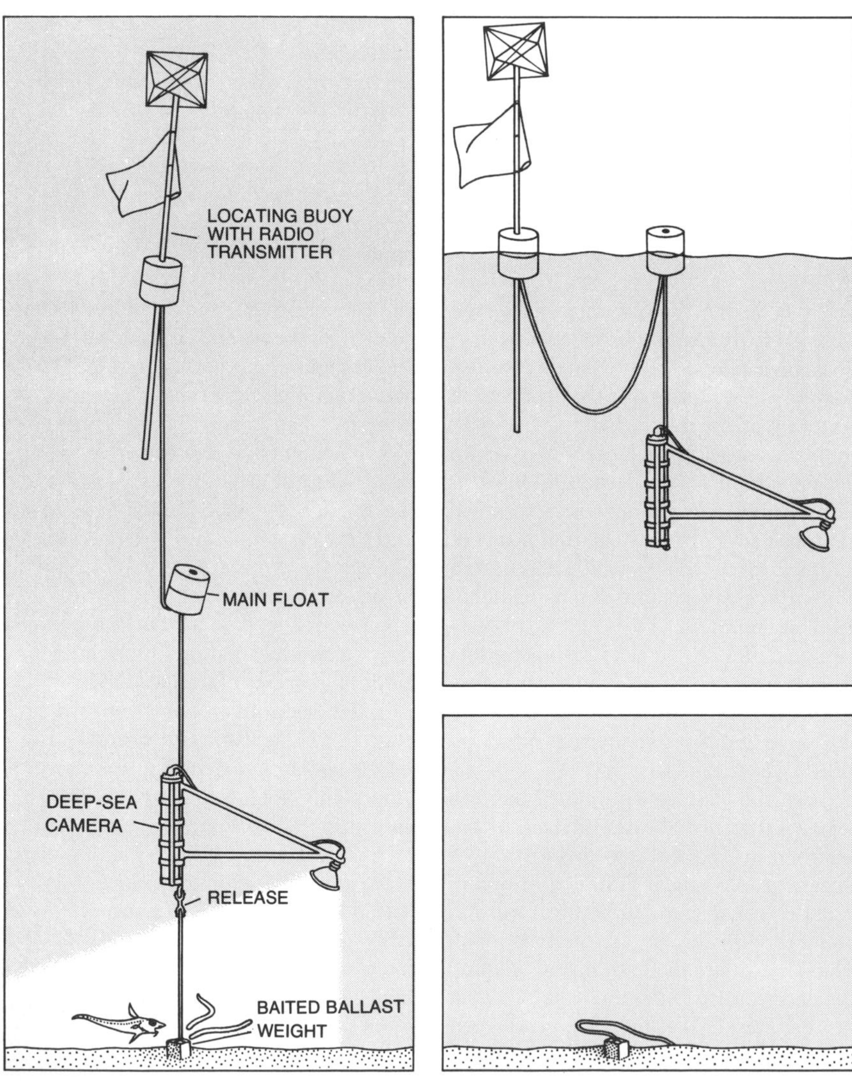

FREE-FALL CAMERA is dropped to the ocean bottom, where it is held in position by a baited ballast weight and a float made of foamed plastic (*left*). The camera can be suspended above the bait as shown here or can be positioned off to the side. At a preset time a clock mechanism cuts the camera system loose from the ballast and the system rises to the surface (*right*). An attached buoy with a flag, a radar target and a radio transmitter marks the site.

Surprisingly, we find particularly abundant bottom populations in regions of the Pacific where the surface populations are the least abundant, and vice versa. For example, under the least productive surface waters of the North Pacific Gyre, on an austere bottom paved with manganese nodules nearly 6,000 meters down, more than 40 large fish and shrimps were attracted by the bait within a few hours; at least four species were represented. Moderately productive areas of the Indian Ocean, the California Current system and biologically poor areas around the Hawaiian Islands have also shown remarkably large numbers of deep-sea creatures. On the other hand, only 4,000 meters down under the most productive oceanic waters of the world, in the Antarctic over a bottom of soft organic ooze, the bait was visited by only a few eelpouts, brotulids and rattail fish (grenadiers), along with some small crustaceans, the louselike isopods. The brotulids (fishes distantly related to cods) remained almost motionless for hours, nibbling gently at the bait. Camera drops along the underwater ridges of the Line Islands, a highly productive equatorial area of the Pacific, revealed only a few eels. Photographs from the bottom below the rich Peru Current showed very few fish but large numbers of invertebrates: furious masses of amphipods that stripped the bait in a few hours.

The explanation for such a puzzling distribution must lie in the nature of this part of the marine food web. These abyssal roving scavengers must depend for their sustenance in substantial part on

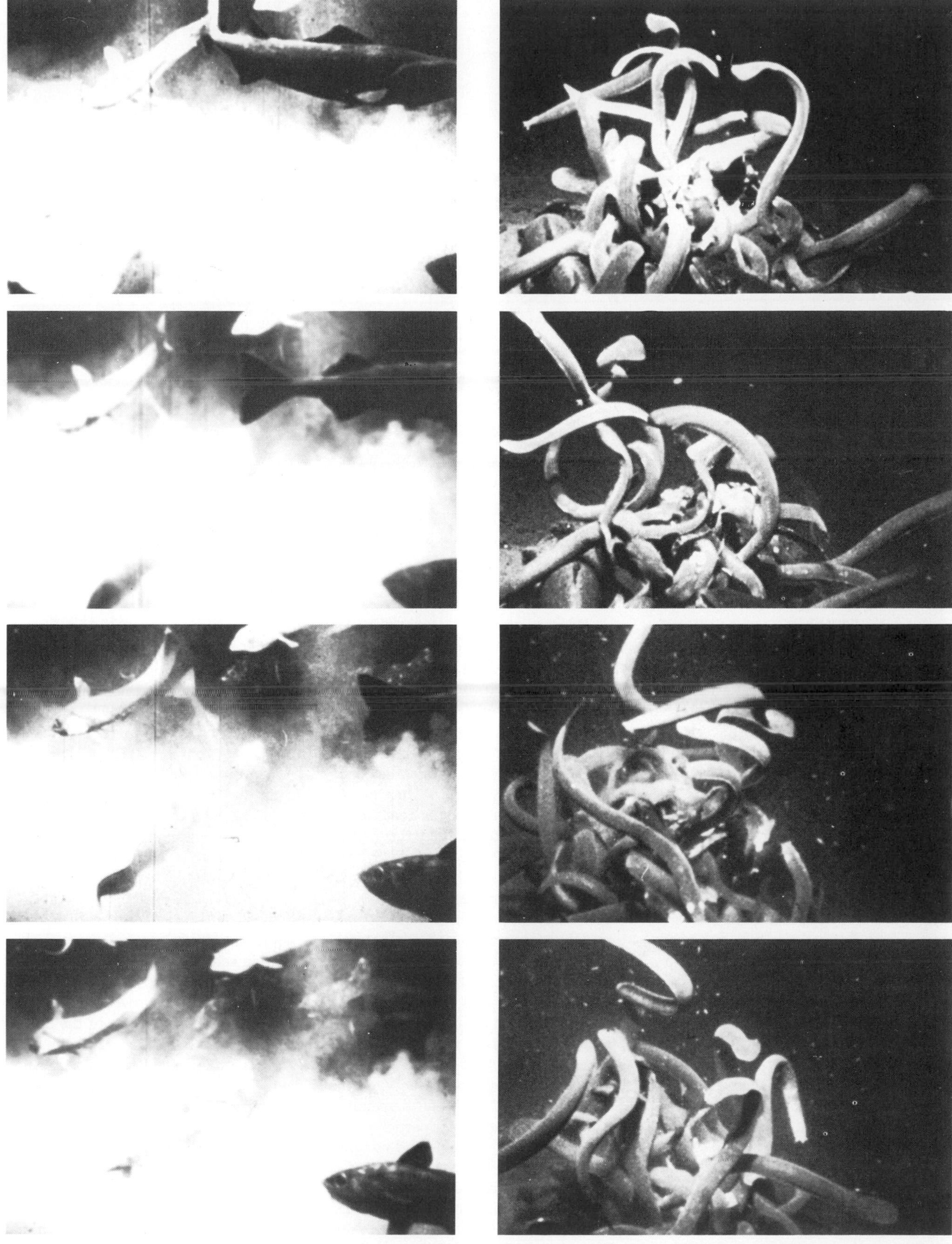

TWO MOTION-PICTURE SEQUENCES show the frenzied activity that develops around the bait. These pictures were made with a 16-millimeter camera dropped to the Pacific floor at a depth of 1,300 meters some 50 miles off the coast of southern California. In sequence at left a variety of fishes attack bait, stirring up the sediment on the bottom. In the sequence at the right the bait is entirely hidden by a mass of hagfish: primitive eyeless and jawless chordates that drill into the bait and consume it from the inside out.

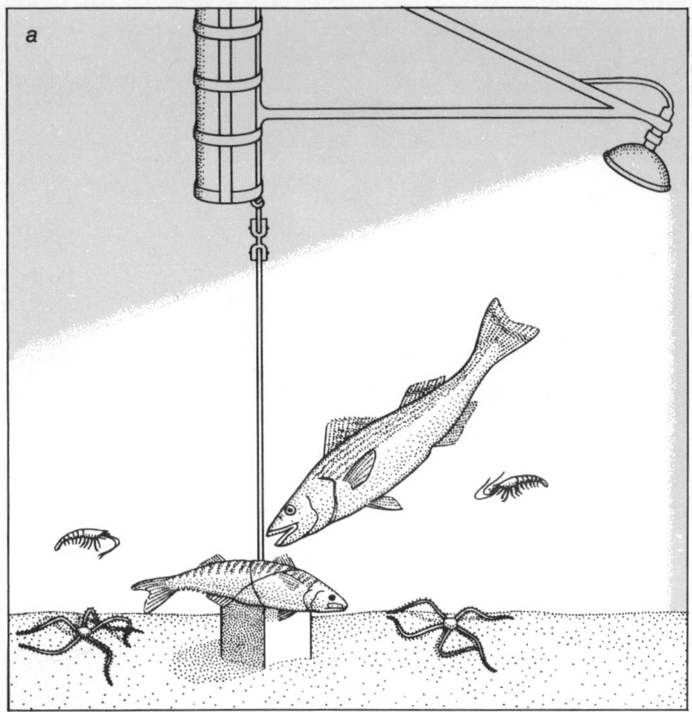

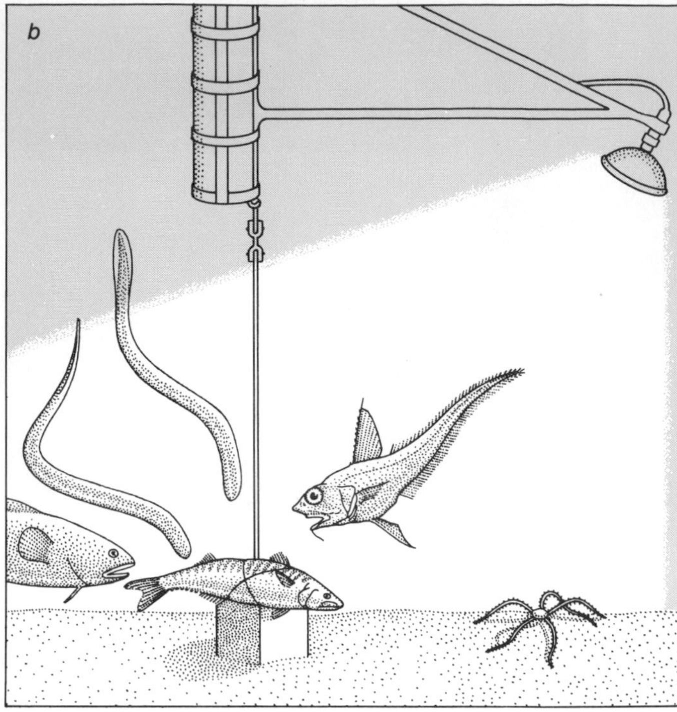

ONE LIKELY SEQUENCE OF EVENTS following the arrival of a baited camera system on the ocean floor is shown in the idealized drawings on these two pages. The first few photographs made after touchdown usually show only brittle stars, shrimps and other small crustaceans (*a*). Within 30 minutes or so the larger fishes begin to arrive on the scene (*b*). The number of fish gathered around the

falls of large food particles. Such particles include fish and marine mammals that have died, fragments of forage from surface-feeding schools of fish, vertically migrating creatures, garbage from ships and (nowadays) animals killed by underwater explosions, ships' propellers and whaling. (The North Pacific Gyre photographs were made below an area patrolled by a weather ship, and her garbage may be partly responsible for the large numbers of animals observed there.) Falling food material can support a population of deep scavengers only if the food descends to the bottom, and it may be that the mid-water population below highly productive regions is sufficiently dense and continuous to consume such food on the way down. Regions of low productivity, on the other hand, may support such sparse and discontinuous mid-water populations that a substantial proportion of dead surface creatures do fall all the way to the sea floor. And there on the floor even a meager fall can support a sizable population, whose scavenging can be much more intensive than that of fish living on the relatively brief passage of falls through the immense volume of the mid-waters.

Somewhat different explanations are possible, perhaps in combination with this mid-water-population effect. The western north-central Pacific, where so many fish were photographed over a manganese-nodule bottom, is an area through which tuna and other large surface fishes and whales and other cetaceans periodically migrate. Just as terrestrial deserts are traps for aged, infirm or injured creatures, so may this marine desert be a trap. The number of large migrants that succumb in traversing its unproductive surface waters may be sufficient to support the bottom population. We know nothing of the "natural" end of the largest fishes and cetaceans; it could well be that these vast areas of low productivity inflict the final stress on aging and infirm members of the populations of great marine creatures, exhausting their ultimate reserves. Deep below areas of high productivity, on the other hand, the principal descending food may be rather small particles, sufficient to support a sizable and continuous population of mid-water creatures as well as throngs of small bottom scavengers that do not need to rove far for sustenance. Moreover, in such waters the large near-surface animals have plentiful food and are not in such a precarious situation as the schools of large migrants crossing the great marine deserts. Areas of high productivity might therefore fail to support the larger scavengers of the deep-sea floor; few active, roving fish would be searching for large food falls such as our bait represents.

Some of the fishes now being observed for the first time at great depths at low latitudes turn out to be species that are well known in near-surface waters at high latitudes. The flatnose codling, the sablefish and the arctic sleeper shark are common inhabitants of the bottom off the coast of southern California and Lower California; the sablefish, which is found in commercially valuable numbers off the coast of Washington, British Columbia and Alaska, is particularly abundant on the deep-sea floor of the Southern California Bight thousands of miles to the south. On the basis of the incidence of this species in a random series of pictures made with unbaited cameras, we have estimated that there are 800,000 tons of sablefish at depths of between 800 and 1,500 meters in the southern California waters; they seem to be of the same race as the commercially harvested variety far to the north. At least along the Pacific coast of North America, then, some species that are generally considered arctic fishes apparently represent mere near-surface outcroppings of populations that inhabit

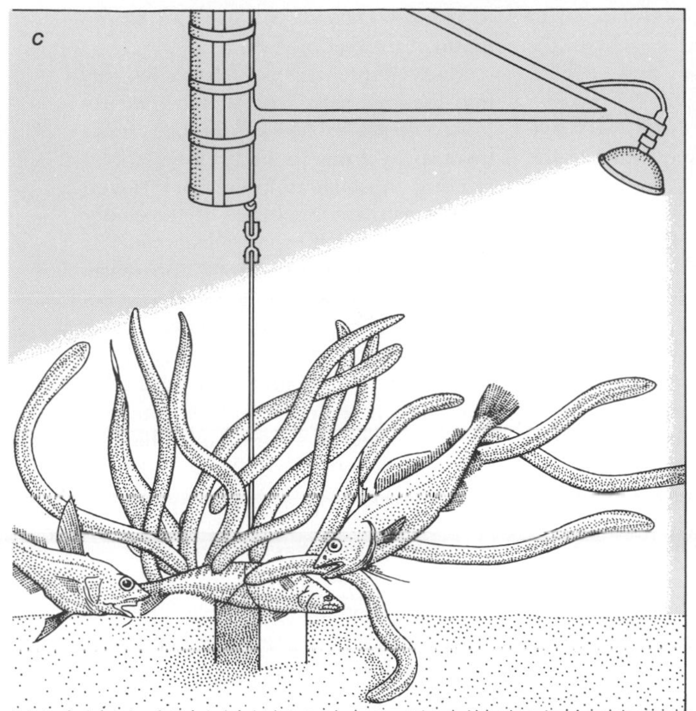

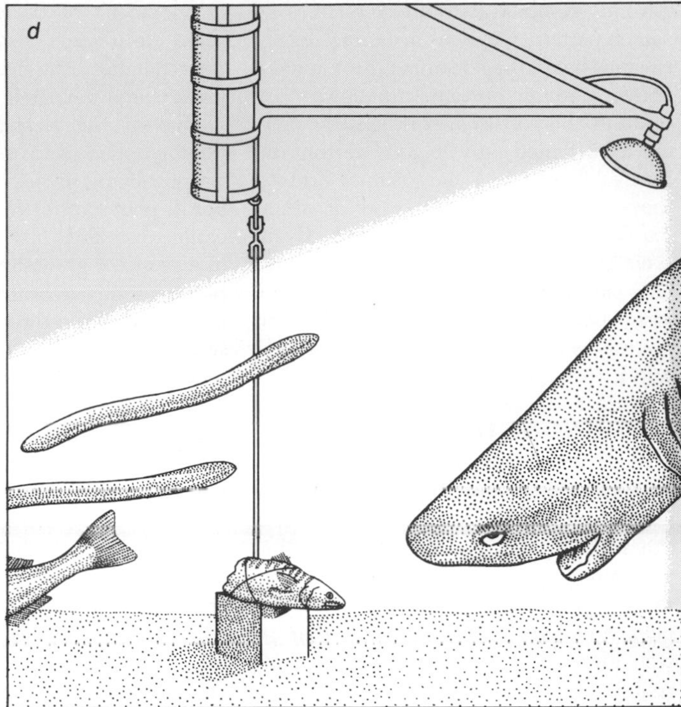

bait increases slowly, reaching a maximum after a few hours (*c*). After three to eight hours some much larger creature, usually a shark, moves in and consumes the bulk of the bait (*d*). Afterward sea urchins and snails accumulate to finish off the scraps. The latecomers to the feast appear to sense some kind of territorial collapse as successive waves of scavengers converge on large kills.

the cold, deep waters of continental borders to the south. If that is true of other coasts as well, the size and extent of a number of fish populations may be dramatically larger than has been thought. The total numbers of large shrimps in particular must be immense, since they are almost always attracted to the bait at all depths and in all regions we have sampled.

In one of our sequences a small brown shark appears in the bottom waters of the Santa Barbara Basin, which contain virtually no dissolved oxygen. We have also photographed dead mid-water fish on the bottom there. The combination suggests another source of food in some deep waters. Perhaps anoxic basins serve as traps for unsuspecting vertically migrating animals that have never encountered pockets of suffocating water. A scavenger adapted for short forays into such a basin could find a rich harvest. Perhaps the brown shark is so adapted, able to survive a brief period of oxygen deficit.

Sometimes no fish appear in our pictures; the bait is completely consumed by vast numbers of invertebrates such as shrimps and swarming amphipods. Motion pictures from the Santa Cruz Basin off southern California show the amphipods accumulating until the bait is totally covered and the surrounding water is nearly saturated by a roiling mass of these crustaceans. In that sequence one sablefish came near the bait but left immediately, gaping as if to rid itself of amphipods. Is it possible that where there are swarms of aggressive invertebrates such as amphipods, fish are not able to compete for the small amount of food that reaches the bottom? Amphipods can sustain themselves on small food particles, but they are also capable of quickly and completely devouring the much larger pieces that fall to the ocean floor. Deep-living octopuses are other invertebrates that tend to monopolize the bait. In one motion-picture sequence made at about 4,000 meters a small octopus squats on the bait, keeping the grenadiers at tentacle's length, and in an unusual sequence made with a still camera off Cedros Island in Lower California two octopuses fend off a single grenadier from the bait.

The bait is often taken over aggressively and completely by innumerable hagfish, primitive eyeless and jawless chordates that drill into dead creatures and consume them from the inside out [see "The Hagfish," by David Jensen; Scientific American Offprint 1035]. Hagfish thrive at a depth of from 200 meters to nearly 2,000 meters. At first we were puzzled by the reluctance of other fishes to penetrate the Gorgon's-head tangle of hagfish and feed on the bait. Closeup motion pictures gave the answer: the hagfish enclose the bait in a thick cocoon of slime that other fishes apparently find distressing. On a number of occasions fish emerge from the feeding mass making frantic efforts to clear their gills of slime. The spectacular ability of the hagfish to exude slime has long been known; the exudation of a single hagfish can convert a large container of water into a slimy gel. Their employment of this defense mechanism to sequester food, however, had not been suggested.

Some of the scavenging bottom fishes display a feeding capability that could scarcely be predicted from an examination of their jaw and tooth structure. The grenadiers are quite capable of tearing out the abdomen of a bait fish; sablefish shake large baits as terriers do and spin furiously to twist off mouthfuls of food.

The large sharks that frequent the deep-ocean floor have been photographed to depths of about 2,000 meters. Their behavior in approaching the bait appears to be mediated in part by a sense of smell: the fish execute slow, deliberate geometrical maneuvers that apparently combine to establish a complex search pattern. On some occasions, when the bait holder has been hung somewhat above a rough bottom, the sharks are quite unable to discover it. Clearly they are accustomed only to food resting on the sea floor; when they cannot find it, they nudge and bite at rocks or other sea-floor objects under the bait. Even their search for bait that is on the ocean floor usually requires more than one sortie before the bait is found. Several picture sequences end just as the shark is in a position to seize the bait, with the bait dead ahead within the width of the shark's jaws. Yet the next sequence, made five minutes later, may show the bait untouched and the shark still engaged in a slow, deliberate search. Like sharks, hagfish are unable to discover the bait when it is a meter or so above the bottom, whereas eels and grenadiers have no difficulty locating bait that is well above the bottom. Knowledge of such feeding limitations will help in the development of better techniques for some deep-water fisheries, where the depredations of sharks and hagfish greatly limit the catch.

One of the shyest fishes we have photographed, and perhaps one of the deepest-living, is the brotulid. In a number of photographs from the deepest locations brotulids lurk at the outer edge of a group of grenadiers. In one motion-picture series a brotulid hovers like a motionless blimp through sequence after sequence, facing the feeding grenadiers. Unlike most of the deep-water fishes, the brotulids have very reduced eyes, and it may be that their behavior is related to diminished visual acuity. Creatures living more than several hundred meters below the surface are maneuvering in a profound darkness that is lighted only by the faint glow or brighter brief flashes of bioluminescent organisms. Only fishes with a highly developed visual system can be expected to be guided by visual cues. Brotulids may be responding not to the presence of the bait but to the sound of the feeding fish.

Many sequences yield fresh insights into feeding behavior. A crab gingerly lifts some sea urchins off the bait, holds them away from its body like a spider ejecting a distasteful insect from its web and drops them. Sheltered by the empty bait holder, a small spiny lobster flails its antennae in a strong current, apparently grasps a small swimming crustacean between the antennae and conveys it to its mouth by some movement too quick for us to make out. A grenadier goes after small food particles in the sediment with a sudden explosive thrust into the bottom, throwing a cloud of sediment through its gills.

SWIRLING MASS of shrimps and amphipods (deep-sea relatives of sand fleas) completely covers the bait and fills the surrounding water in this oblique photograph, made at a depth of 7,000 meters in the Peru-Chile Trench. Exhibiting surprisingly aggressive behavior, the small crustaceans stripped the bait in a few hours. Curiously no fishes appeared in this sequence of pictures, filmed under one of the most highly productive fisheries in the oceans.

The uniformly large size of most of the fish photographed at great depths presents something of a puzzle. Very few small fish are ever seen. The grenadiers photographed at from 750 to 6,000 meters in the Pacific, Antarctic and Indian oceans are all large, mature fish, some measuring more than a meter in length. We have so far recovered no free grenadier eggs or juvenile fish from collections made off California anywhere between the surface and the bottom, although females with ripe ovaries have been collected near the bottom. All the arctic sleeper sharks photographed off southern California and Lower California have been very large, but most of the photographs have shown only a small portion of their total length, which we can only estimate as being between five and eight meters. They also must be quite common, since they have been photographed in nearly half of our missions off California down to 2,000 meters; on a number of occasions when more than one large shark was photographed during a mission, we could tell from distinguishing scars and other markings that the sharks were different individuals.

It is probable that the juveniles of these deep-living fishes inhabit much shallower depths than the adults. The young of the sablefish, for example, are numerous in many places at depths of 100 meters or so. It may be that the rarity of juveniles is merely the result of great adult longevity and low fecundity. On the other hand, juveniles of many species must return to the deep bottom environment; indeed, their return may constitute a meaningful importation of food from the more productive upper layers for the nourishment of the total adult population. We usually think of the relation between juveniles and adult stocks only in terms of replacement. Juveniles may also be important as prey; indeed, in some freshwater environments the young of a species are a principal prey of adults of the same species. To the degree that this process, which has been called Faginism, is important to the food economy of the deep-sea populations, we would of course expect to observe a paucity of juveniles in the populations: most of them are quickly consumed by their elders!

The baited camera suffers from some limitations common to many simple exploratory tools. Its sample is highly selective. Quantification and even identification can be doubtful. There are several ways to deal with these problems. One can, for example, set out fishlines and traps in order to retrieve specimens for sure identification. A particularly promising means of quantification is an unbaited drifting camera that lightens itself until it rises just a few meters above the bottom. It drifts with the current, making overlapping photographs and meanwhile recording the direction and distance of drift. Such a camera has been successfully operated in waters about 1,000 meters deep. We have plans to develop a drifting camera that will remain submerged for several months, making photographs along a drift track as long as 200 kilometers. The drifting camera should yield a meaningful census of the creatures that have been observed by stationary baited cameras and thus help us to assess the potential food harvest from the deep-ocean floor and to understand the ramifications of this remarkable branch of the marine food web.

Meanwhile vast areas await investigation with the baited camera. A series of photographic "sections" across the continental shelves along all the major land masses and down the slopes to the abyss would fill many gaps in our knowledge of bottom-dwelling animals and should reveal entirely new fishery resources. Among the environments that are of special interest to oceanographers and marine biologists are the floor of the Arctic Ocean, the deep delta of the Congo River, the Antarctic slope, the deeps of the Mediterranean, the top of seamounts and mid-ocean ridges and the slope and bottom of deep oceanic trenches. Since the free-fall cameras can be operated from inexpensive, unspecialized craft, they are particularly suitable for studies by investigators in underdeveloped countries who want to know more about the creatures and the potential deep fisheries off their coast.

18

The Food Resources of the Ocean

by S. J. Holt
September 1969

The present harvest of the oceans is roughly 55 million tons a year, half of which is consumed directly and half converted into fish meal. A well-managed world fishery could yield more than 200 million tons

I suppose we shall never know what was man's first use of the ocean. It may have been as a medium of transport or as a source of food. It is certain, however, that from early times up to the present the most important human uses of the ocean have been these same two: shipping and fishing. Today, when so much is being said and written about our new interests in the ocean, it is particularly important to retain our perspective. The annual income to the world's fishermen from marine catches is now roughly $8 billion. The world ocean-freight bill is nearly twice that. In contrast, the wellhead value of oil and gas from the seabed is barely half the value of the fish catch, and all the other ocean mineral production adds little more than another $250 million.

Of course, the present pattern is likely to change, although how rapidly or dramatically we do not know. What is certain is that we shall use the ocean more intensively and in a greater variety of ways. Our greatest need is to use it wisely. This necessarily means that we use it in a regulated way, so that each ocean resource, according to its nature, is efficiently exploited but also conserved. Such regulation must be in large measure of an international kind, particularly insofar as living resources are concerned. This will be so whatever may be the eventual legal regime of the high seas and the underlying bed. The obvious fact about most of the ocean's living resources is their mobility. For the most part they are lively animals, caring nothing about the lines we draw on charts.

The general goal of ecological research, to which marine biology makes an important contribution, is to achieve an understanding of and to turn to our advantage all the biological processes that give our planet its special character. Marine biology is focused on the problems of biological production, which are closely related to the problems of production in the economic sense. Our most compelling interest is narrower. It lies in ocean life as a renewable resource: primarily of protein-rich foods and food supplements for ourselves and our domestic animals, and secondarily of materials and drugs. I hope to show how in this field science, industry and government need each other now and will do so even more in the future. First, however, let me establish some facts about present fishing industries, the state of the art governing them and the state of the relevant science.

The present ocean harvest is about 55 million metric tons per year. More than 90 percent of this harvest is finfish; the rest consists of whales, crustaceans and mollusks and some other invertebrates. Although significant catches are reported by virtually all coastal countries, three-quarters of the total harvest is taken by only 14 countries, each of which produces more than a million tons annually and some much more. In the century from 1850 to 1950 the world catch increased tenfold—an average rate of about 25 percent per decade. In the next decade it nearly doubled, and this rapid growth is continuing [*see illustration on page 215*]. It is now a commonplace that fish is one of the few major foodstuffs showing an increase in global production that continues to exceed the growth rate of the human population.

This increase has been accompanied by a changing pattern of use. Although some products of high unit value as luxury foods, such as shellfish, have maintained or even enhanced their relative economic importance, the trend has been for less of the catch to be used directly as human food and for more to be reduced to meal for animal feed. Just before World War II less than 10 percent of the world catch was turned into meal; by 1967 half of it was so used. Over the same period the proportion of the catch preserved by drying or smoking declined from 28 to 13 percent and the proportion sold fresh from 53 to 31 percent. The relative consumption of canned fish has hardly changed but that of frozen fish has grown from practically nothing to 12 percent.

While we are comparing the prewar or immediate postwar situation with the present, we might take a look at the composition of the catch by groups of species. In 1948 the clupeoid fishes (herrings, pilchards, anchovies and so on), which live mainly in the upper levels of the ocean, already dominated the scene (33 percent of the total by weight) and provided most of the material for fish meal. Today they bulk even larger (45 percent) in spite of the decline of several great stocks of them (in the North Sea and off California, for example). The next most important group, the gadoid fishes (cod, haddock, hake and so on), which live mainly on or near the bottom, comprised a quarter of the total in 1948. Although the catch of these fishes has continued to increase absolutely, the

SCHOOL OF FISH is spotted from the air at night by detecting the bioluminescent glow caused by the school's movement through the water. As the survey aircraft flew over the Gulf of Mexico at an altitude of 3,500 feet, the faint illumination in the water was amplified some 55,000 times by an image intensifier before appearing on the television screen seen in the photograph on the opposite page. The fish are Atlantic thread herring. Detection of fish from the air is one of several means of increasing fishery efficiency being tested at the Pascagoula, Miss., research base of the U.S. Bureau of Commercial Fisheries.

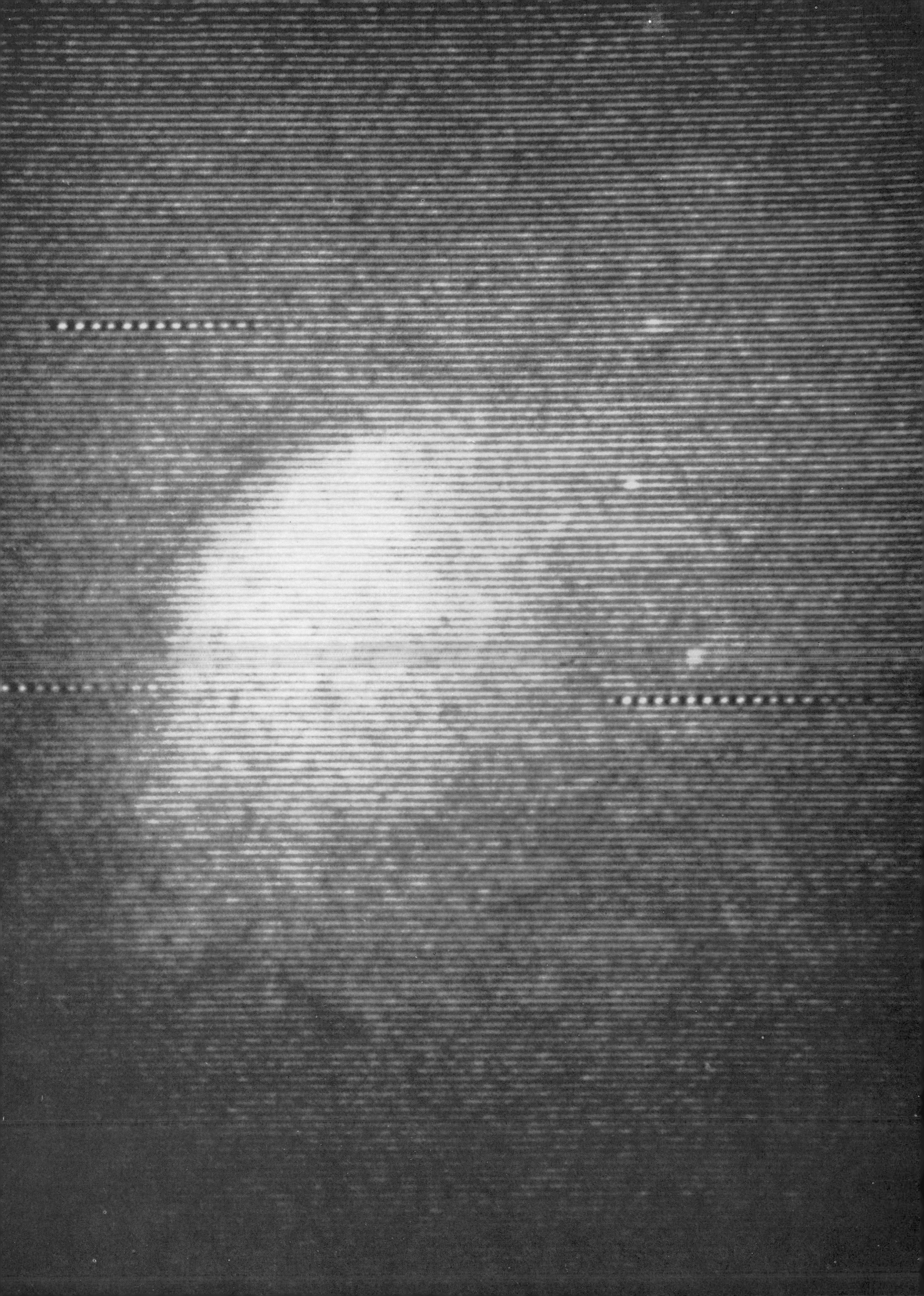

proportion is now reduced to 15 percent. The flounders and other flatfishes, the rosefish and other sea perches and the mullets and jacks have collectively stayed at about 15 percent; the tunas and mackerels, at 7 percent. Nearly a fifth of the total catch continues to be recorded in statistics as "Unsorted and other"—a vast number of species and groups, each contributing a small amount to a considerable whole.

The rise of shrimp and fish meal production together account for another major trend in the pattern of fisheries development. A fifth of the 1957 catch was sold in foreign markets; by 1967, two-fifths were entering international trade and export values totaled $2.5 billion. Furthermore, during this same period the participation of the less developed countries in the export trade grew from a sixth to well over 25 percent. Most of these shipments were destined for markets in the richer countries, particularly shrimp for North America and fish meal for North America, Europe and Japan. More recently several of the less developed countries have also become importers of fish meal, for example Mexico and Venezuela, South Korea and the Republic of China.

The U.S. catch has stayed for many years in the region of two million tons, a low figure considering the size of the country, the length of the coastline and the ready accessibility of large resources on the Atlantic, Gulf and Pacific seaboards. The high level of consumption in the U.S. (about 70 pounds per capita) has been achieved through a steady growth in imports of fish and fish meal: from 25 percent of the total in 1950 to more than 70 percent in 1967. In North America 6 percent of the world's human population uses 12 percent of the world's catch, yet fishermen other than Americans take nearly twice the amount of fish that Americans take from the waters most readily accessible to the U.S.

There has not been a marked change in the broad geography of fishing [*see illustration on these two pages*]. The Pacific Ocean provides the biggest

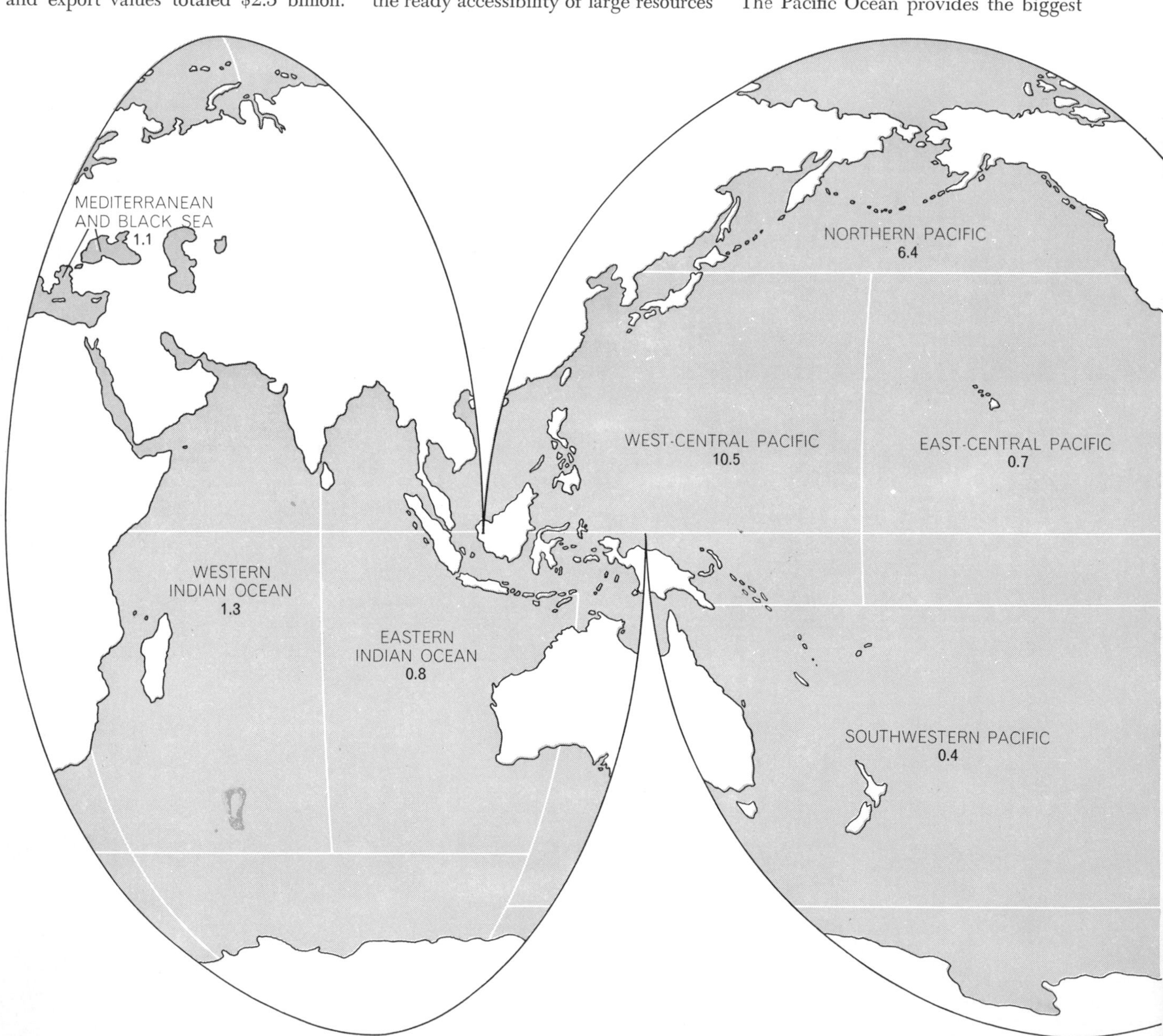

MAJOR MARINE FISHERY AREAS are 14 in number: two in the Indian Ocean (*left*), five in the Pacific Ocean (*center*) and six in the Atlantic (*right*). Due to the phenomenal expansion of the Peru fishery, the total Pacific yield is now a third larger than the Atlantic total. The bulk of Atlantic and Pacific catches, however, is still taken well north of the Equator. The Indian Ocean, with a

share (53 percent) but the Atlantic (40 percent, to which we may add 2 percent for the Mediterranean) is yielding considerably more per unit area. The Indian Ocean is still the source of less than 5 percent of the catch, and since it is not a biologically poor ocean it is an obvious target for future development. Within the major ocean areas, however, there have been significant changes. In the Pacific particular areas such as the waters off Peru and Chile and the Gulf of Thailand have rapidly acquired importance. The central and southern parts of the Atlantic, both east and west, are of growing interest to more nations. Although, with certain exceptions, the traditional fisheries in the colder waters of the Northern Hemisphere still dominate the statistics, the emergence of some of the less developed countries as modern fishing nations and the introduction of long-range fleets mean that tropical and subtropical waters are beginning to contribute significantly to world production.

Finally, in this brief review of the trends of the past decade or so we must mention the changing importance of countries as fishing powers. Peru has become the leading country in terms of sheer magnitude of catch (although not of value or diversity) through the development of the world's greatest one-species fishery: 10 million tons of anchovies per year, almost all of which is reduced to meal [*see illustration on page 217*]. The U.S.S.R. has also emerged as a fishing power of global dimension, fishing for a large variety of products throughout the oceans of the world, particularly with large factory ships and freezer-trawlers.

At this point it is time to inquire about the future expectations of the ocean as a major source of protein. In spite of the growth I have described, fisheries still contribute only a tenth of the animal protein in our diet, although this proportion varies considerably from one part of the world to another. Before such an inquiry can be pursued, however, it is necessary to say something about the problem of overfishing.

A stock of fish is, generally speaking, at its most abundant when it is not being exploited; in that virgin state it will include a relatively high proportion of the larger and older individuals of the species. Every year a number of young recruits enter the stock, and all the fish—but particularly the younger ones—put on weight. This overall growth is balanced by the natural death of fish of all ages from disease, predation and perhaps senility. When fishing begins, the large stock yields large catches to each fishing vessel, but because the pioneering vessels are few, the total catch is small.

Increased fishing tends to reduce the level of abundance of the stock progressively. At these reduced levels the losses accountable to "natural" death will be less than the gains accountable to recruitment and individual growth. If, then, the catch is less than the difference between natural gains and losses, the stock will tend to increase again; if the catch is more, the stock will decrease. When the stock neither decreases nor increases, we achieve a sustained yield. This sustained yield is small when the stock is large and also when the stock is small; it is at its greatest when the stock is at an intermediate level—somewhere between two-thirds and one-third of the virgin abundance. In this intermediate stage the average size of the individuals will be smaller and the age will be younger than in the unfished condition, and individual growth will be highest in relation to the natural mortality.

The largest catch that on the average can be taken year after year without causing a shift in abundance, either up or down, is called the maximum sustainable yield. It can best be obtained by leaving the younger fish alone and fishing the older ones heavily, but we can also get near to it by fishing moderately, taking fish of all sizes and ages. This

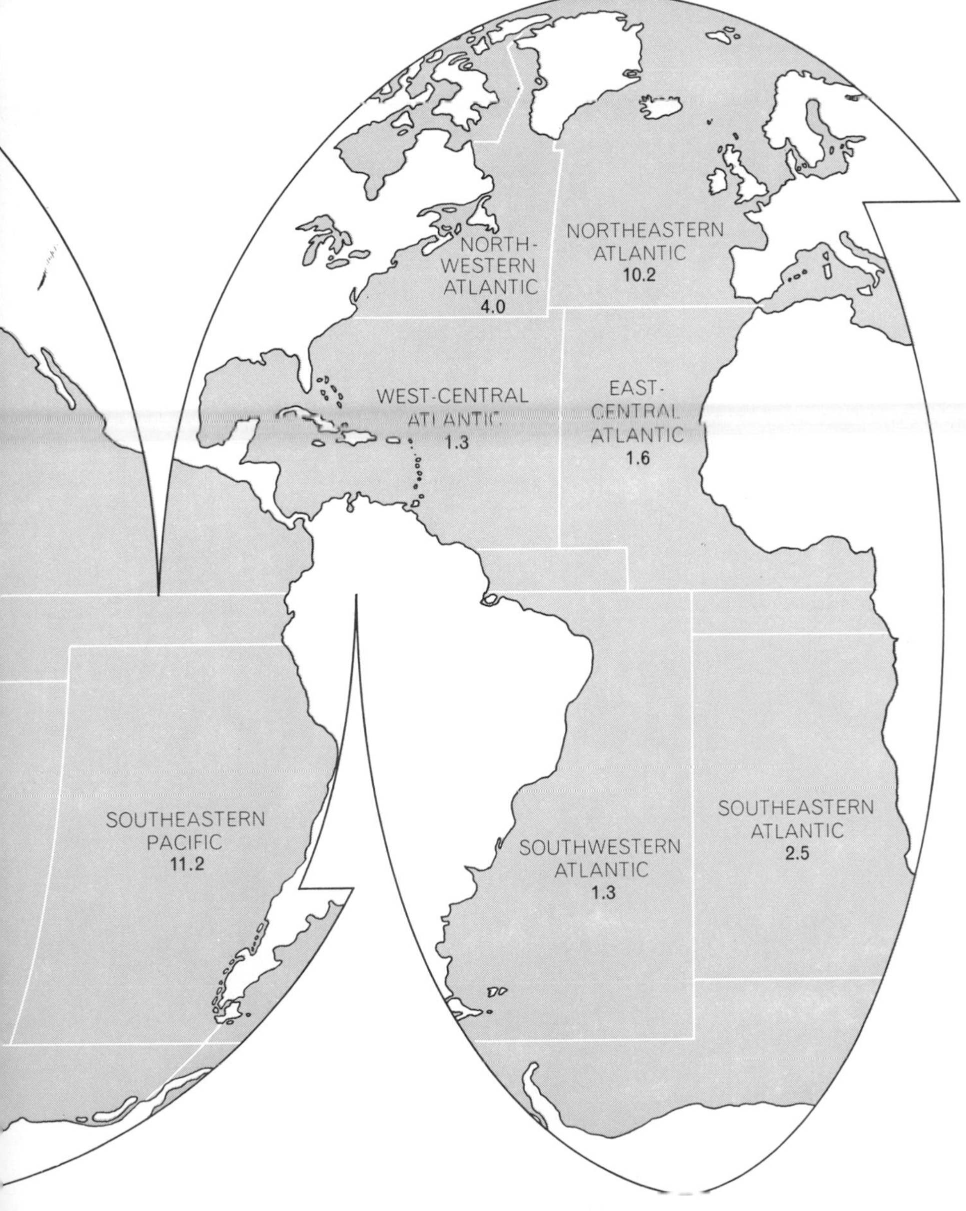

total catch of little more than two million metric tons, live weight, is the world's major underexploited region. The number below each area name shows the millions of metric tons landed during 1967, as reported by the UN Food and Agriculture Organization.

phenomenon—catches that first increase and then decrease as the intensity of fishing increases—does not depend on any correlation between the number of parent fish and the number of recruits they produce for the following generation. In fact, many kinds of fish lay so many eggs, and the factors governing survival of the eggs to the recruit stage are so many and so complex, that it is not easy to observe any dependence of the number of recruits on the number of their parents over a wide range of stock levels.

Only when fishing is intense, and the stock is accordingly reduced to a small fraction of its virgin size, do we see a decline in the number of recruits coming in each year. Even then there is often a wide annual fluctuation in this number. Indeed, such fluctuation, which causes the stock as a whole to vary greatly in abundance from year to year, is one of the most significant characteristics of living marine resources. Fluctuation in number, together with the considerable variation in "availability" (the change in the geographic location of the fish with respect to the normal fishing area), largely account for the notorious riskiness of fishing as an industry.

For some species the characteristics of growth, natural mortality and recruit-

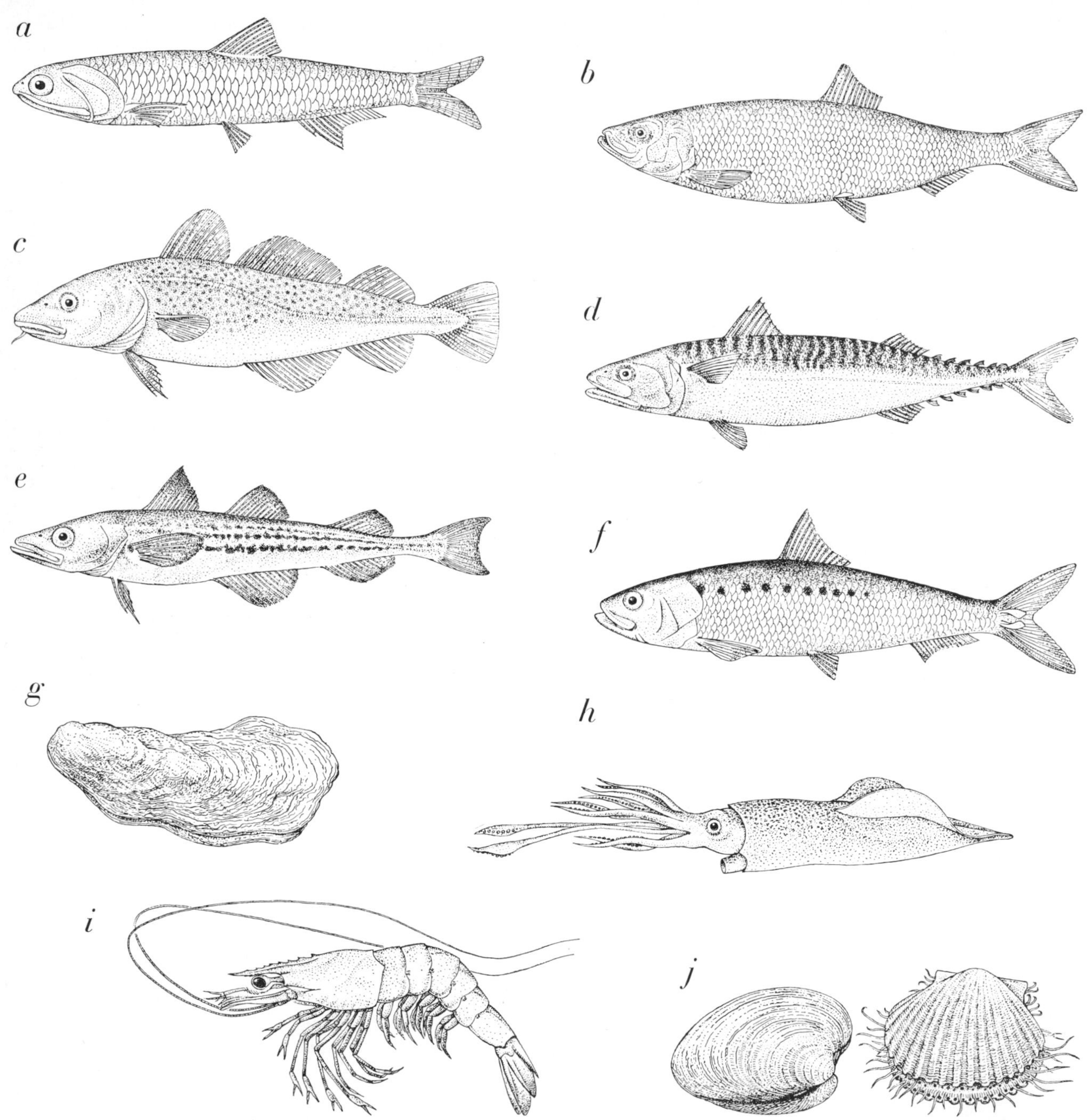

LARGEST CATCHES of individual fish species include the five fishes shown here (*left*). They are, according to the most recent detailed FAO fishery statistics (1967), the Peruvian anchoveta (*a*), with a catch of more than 10.5 million metric tons; the Atlantic herring (*b*), with a catch of more than 3.8 million tons; the Atlantic cod (*c*), with a catch of 3.1 million tons; the Alaska walleye pollack (*d*), with a catch of 1.7 million metric tons, and the South African pilchard (*e*), with a catch of 1.1 million tons. No single invertebrate species (*right*) is harvested in similar quantities. Taken as a group, however, various oyster species (*f*) totaled .83 million tons in 1967; squids (*g*), .75 million tons; shrimps and prawns (*h*), .69 million tons; clams and cockles (*i*), .48 million tons.

ment are such that the maximum sustainable yield is sharply defined. The catch will decline quite steeply with a change in the amount of fishing (measured in terms of the number of vessels, the tonnage of the fleet, the days spent at sea or other appropriate index) to either below or above an optimum. In other species the maximum is not so sharply defined; as fishing intensifies above an optimum level the sustained catch will not significantly decline, but it will not rise much either.

Such differences in the dynamics of different types of fish stock contribute to the differences in the historical development of various fisheries. If it is unregulated, however, each fishery tends to expand beyond its optimum point unless something such as inadequate demand hinders its expansion. The reason is painfully simple. It will usually still be profitable for an individual fisherman or ship to continue fishing after the *total* catch from the stock is no longer increasing or is declining, and even though his own rate of catch may also be declining. By the same token, it may continue to be profitable for the individual fisherman to use a small-meshed net and thereby catch young as well as older fish, but in doing so he will reduce both his own possible catch and that of others in future years. Naturally if the total catch is declining, or not increasing much, as the amount of fishing continues to increase, the net economic yield from the fishery—that is, the difference between the total costs of fishing and the value of the entire catch—will be well past its maximum. The well-known case of the decline of the Antarctic baleen whales provides a dramatic example of overfishing and, one would hope, a strong incentive for the more rational conduct of ocean fisheries in the future.

There is, then, a limit to the amount that can be taken year after year from each natural stock of fish. The extent to which we can expect to increase our fish catches in the future will depend on three considerations. First, how many as yet unfished stocks await exploitation, and how big are they in terms of potential sustainable yield? Second, how many of the stocks on which the existing fisheries are based are already reaching or have passed their limit of yield? Third, how successful will we be in managing our fisheries to ensure maximum sustainable yields from the stocks?

The first major conference to examine the state of marine fish stocks on a global basis was the United Nations Scientific Conference on the Conservation and Utilization of Resources, held in 1949 at Lake Success, N.Y. The small group of fishery scientists gathered there concluded that the only overfished stocks at that time were those of a few high-priced species in the North Atlantic and North Pacific, particularly plaice, halibut and salmon. They produced a map showing 30 other known major stocks they believed to be underfished. The situation was reexamined in 1968. Fishing on half of those 30 stocks is now close to or beyond that required for maximum yield. The fully fished or overfished stocks include some tunas in most ocean areas, the herring, the cod and ocean perch in the North Atlantic and the anchovy in the southeastern Pacific. The point is that the history of development of a fishery from small beginnings to the stage of full utilization or overutilization can, in the modern world, be compressed into a very few years. This happened with the anchovy off Peru, as a result of a massive local fishery growth, and it has happened to some demersal, or bottom-dwelling, fishes elsewhere through the large-scale redeployment of long-distance trawlers from one ocean area to another.

It is clear that the classical process of fleets moving from an overfished area to another area, usually more distant and less fished, cannot continue indefinitely. It is true that since the Lake Success

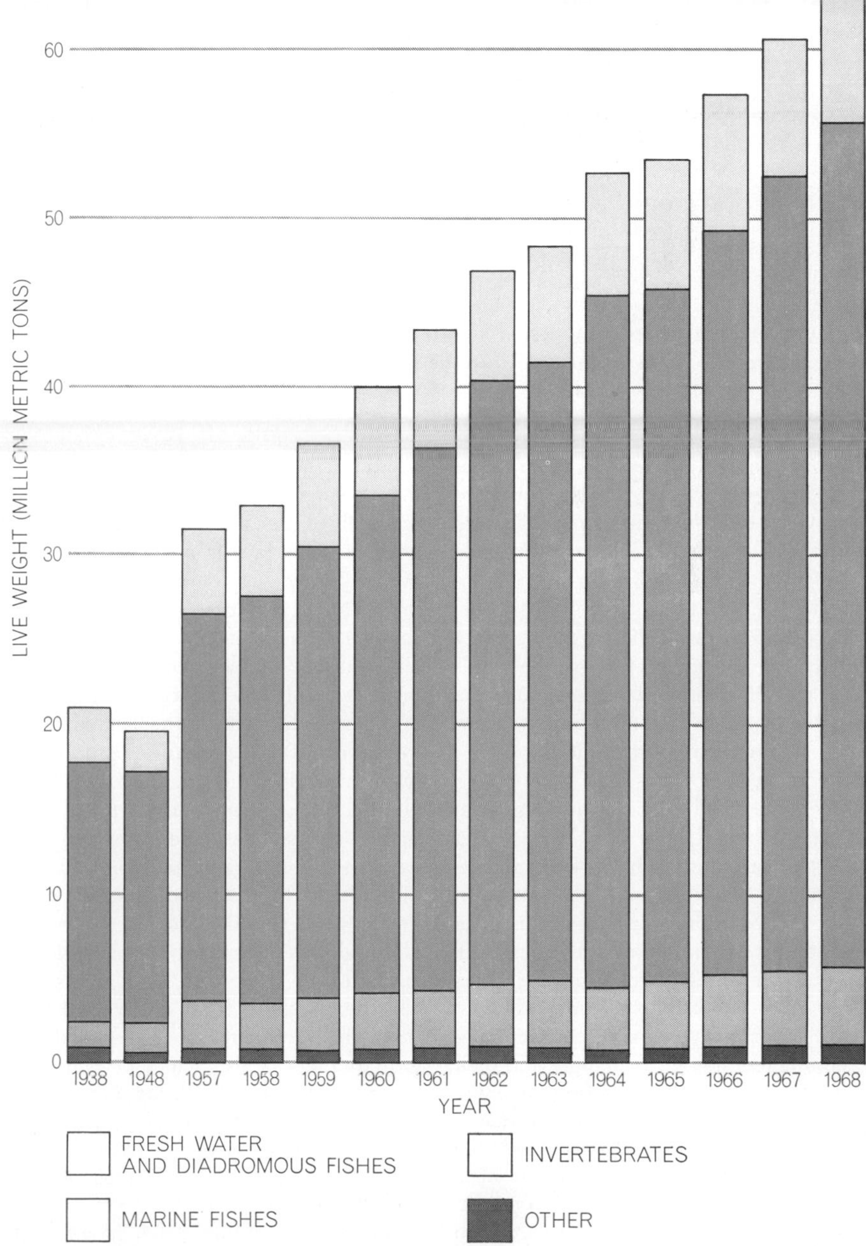

WORLD FISH CATCH has more than tripled in the three decades since 1938; the FAO estimate of the 1968 total is 64 million metric tons. The largest part consists of marine fishes. Humans directly consume only half of the catch; the rest becomes livestock feed.

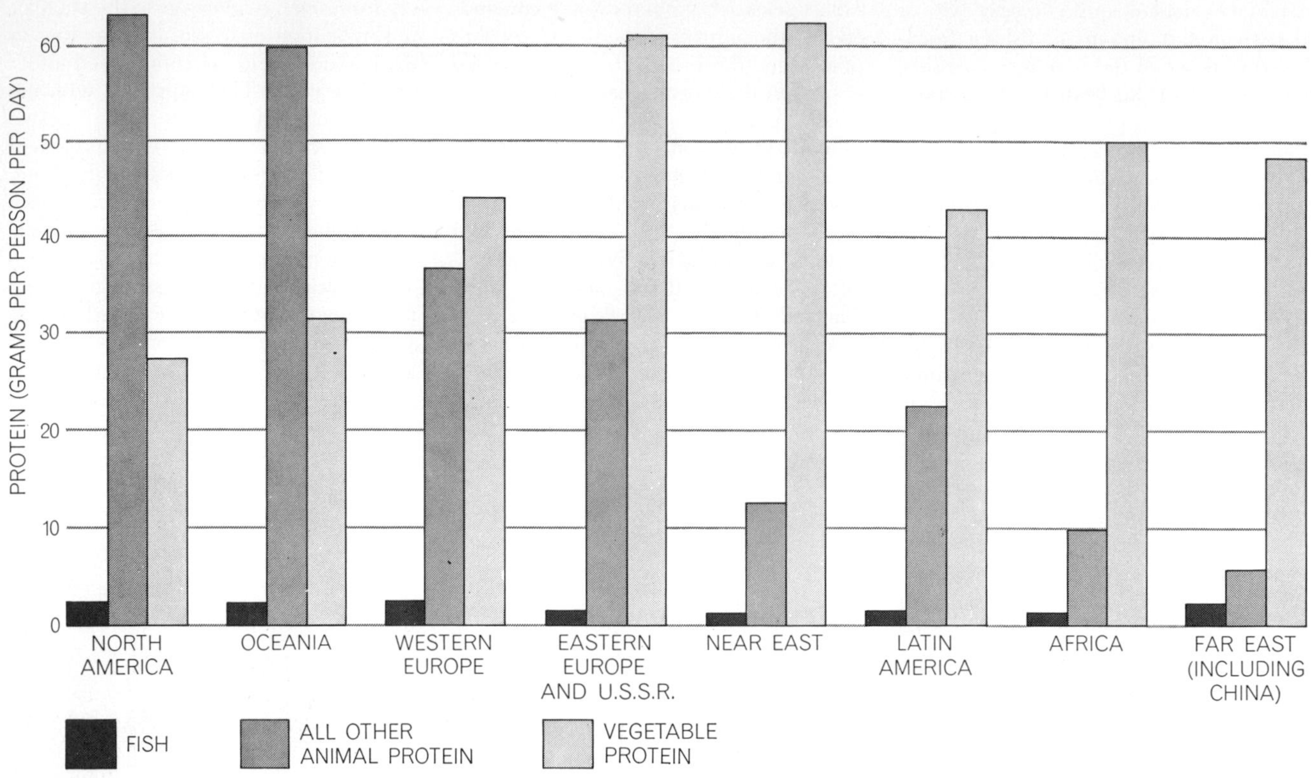

RELATIVELY MINOR ROLE played by fish in the world's total consumption of protein is apparent when the grams of fish eaten per person per day in various parts of the world (*left column in each group*) is compared with the consumption of other animal protein (*middle column*) and vegetable protein (*right column*). The supply is nonetheless growing more rapidly than world population.

meeting several other large resources have been discovered, mostly in the Indian Ocean and the eastern Pacific, and additional stocks have been utilized in fishing areas with a long history of intensive fishing, such as the North Sea. In another 20 years, however, very few substantial stocks of fish of the kinds and sizes of commercial interest and accessible to the fishing methods we know now will remain underexploited.

The Food and Agriculture Organization of the UN is now in the later stages of preparing what is known as its Indicative World Plan (IWP) for agricultural development. Under this plan an attempt is being made to forecast the production of foodstuffs in the years 1975 and 1985. For fisheries this involves appraising resource potential, envisioning technological changes and their consequences, and predicting demand. The latter predictions are not yet available, but the resource appraisals are well advanced. With the cooperation of a large number of scientists and organizations estimates are being prepared in great detail on an area basis. They deal with the potential of known stocks, both those fished actively at present and those exploited little or not at all. Some of these estimates are reliable; others are naturally little more than reasonable guesses. One fact is abundantly clear: We still have very scrappy knowledge, in quantitative terms, of the living resources of the ocean. We can, however, check orders of magnitude by comparing the results of different methods of appraisal. Thus where there is good information on the growth and mortality rates of fishes and measures of their numbers in absolute terms, quite good projections can be made. Most types of fish can now in fact virtually be counted individually by the use of specially calibrated echo sounders for area surveys, although this technique is not yet widely applied. The size of fish populations can also be deduced from catch statistics, from measurements of age based on growth rings in fish scales or bands in fish ear stones, and from tagging experiments. Counts and maps of the distribution of fish eggs in the plankton can in some cases give us a fair idea of fish abundance in relative terms. We can try to predict the future catch in an area little fished at present by comparing the present catch with the catch in another area that has similar oceanographic characteristics and basic biological productivity and that is already yielding near its maximum. Finally, we have estimates of the food supply available to the fish in a particular area, or of the primary production there, and from what we know about metabolic and ecological efficiency we can try to deduce fish production.

So far as the data permit these methods are being applied to major groups of fishes area by area. Although individual area and group predictions will not all be reliable, the global totals and subtotals may be. The best figure seems to be that the potential catch is about three times the present one; it might be as little as twice or as much as four times. A similar range has been given in estimates of the potential yield from waters adjacent to the U.S.: 20 million tons compared with the present catch of rather less than six million tons. This is more than enough to meet the U.S. demand, which is expected to reach 10 million tons by 1975 and 12 million by 1985.

Judging from the rate of fishery development in the recent past, it would be entirely reasonable to suppose that the maximum sustainable world catch of between 100 and 200 million tons could be reached by the second IWP target date, 1985, or at least by the end of the century. The real question is whether or not this will be economically worth the effort. Here any forecast is, in my view, on soft ground. First, to double the catch we have to more than double the amount of fishing, because the stocks decline in abundance as they are exploited. Moreover, as we approach the global maximum more of the stocks that are lightly fished at present will be brought down to intermediate levels. Second, fishing will become even more competitive and costly if the nations fail to agree, and agree soon, on regulations to cure overfishing situations. Third, it is quite uncertain what will happen in the long run to the costs of production and the price of protein of marine origin in relation to other protein sources, particularly from mineral or vegetable bases.

In putting forward these arguments I am not trying to damp enthusiasm for the sea as a major source of food for coming generations; quite the contrary. I do insist, however, that it would be dangerous for those of us who are interested in such development to assume that past growth will be maintained along familiar lines. We need to rationalize present types of fishing while preparing ourselves actively for a "great leap forward." Fishing as we now know it will need to be made even more efficient; we shall need to consider the direct use of the smaller organisms in the ocean that mostly constitute the diet of the fish we now catch; we shall need to try harder to improve on nature by breeding, rearing and husbanding useful marine animals and cultivating their pasture. To achieve this will require a much larger scale and range of scientific research, wedded to engineering progress; expansion by perhaps an order of magnitude in investment and in the employment of highly skilled labor, and a modified legal regime for the ocean and its bed not only to protect the investments but also to ensure orderly development and provide for the safety of men and their installations.

To many people the improvement of present fishing activities will mean increasing the efficiency of fishing gear and ships. There is surely much that could be done to this end. We are only just beginning to understand how trawls, traps, lines and seines really work. For example, every few years someone tries a new design or rigging for a deep-sea trawl, often based on sound engineering and hydrodynamic studies. Rarely do these "improved" rigs catch more than the old ones; sometimes they catch much less. The error has been in thinking that the trawl is simply a bag, collecting more or less passive fish, or at least predictably active ones. This is not so at all. We really have to deal with a complex, dynamic relation between the lively animals and their environment, which includes in addition to the physical and biological environment the fishing gear itself. We can expect success in understanding and exploiting this relation now that we can telemeter the fishing gear, study its hydrodynamics at full scale as well as with models in towing tanks, monitor it (and the fish) by means of underwater television, acoustic equipment and divers, and observe and experiment with fish behavior both in the sea and in large tanks. We also probably have something to learn from studying, before they become extinct, some kinds of traditional "primitive" fishing gear still used in Asia, South America and elsewhere—mainly traps that take advantage of subtleties of fish behavior observed over many centuries.

Successful fishing depends not so much on the size of fish stocks as on their concentration in space and time. All fishermen use knowledge of such concentrations; they catch fish where they have gathered to feed or to reproduce, or where they are on the move in streams or schools. Future fishing methods will surely involve a more active role for the fishermen in causing the fish to congregate. In many parts of the world lights or sound are already used to attract fish. We can expect more sophistication in the employment of these and other stimuli, alone and in combination.

Fishing operations as a whole also depend on locating areas of concentration and on the efficient prediction, or at least the prompt observation, of changes in these areas. The large stocks of pelagic, or open-sea, fishes are produced mainly in areas of "divergencies," where water is rising from deeper levels toward the surface and hence where surface waters are flowing outward. Many such areas are the "upwellings" off the western coasts of continental masses, for example off western and southwestern Africa, western India and western South America. Here seasonal winds, currents and continental configurations combine to cause a periodic enrichment of the surface waters.

Divergencies are also associated with

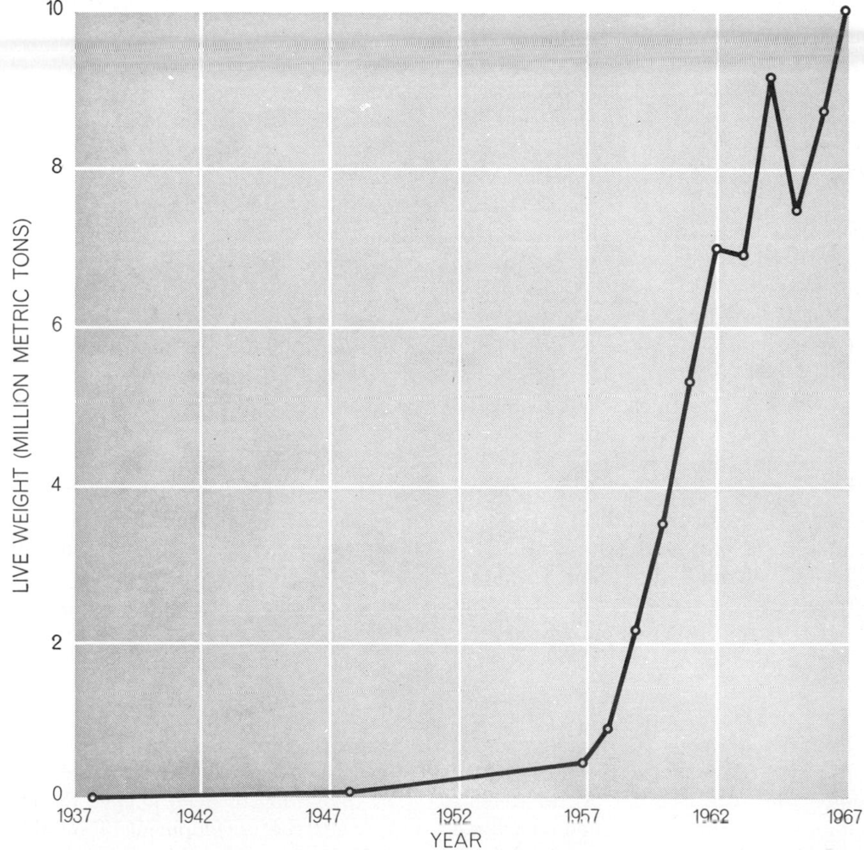

EXPLOSIVE GROWTH of the Peruvian anchoveta fishery is seen in rising number of fish taken between 1938 and 1967. Until 1958 the catch remained below half a million tons. By 1967, with more than 10.5 million tons taken, the fishery sorely needed management.

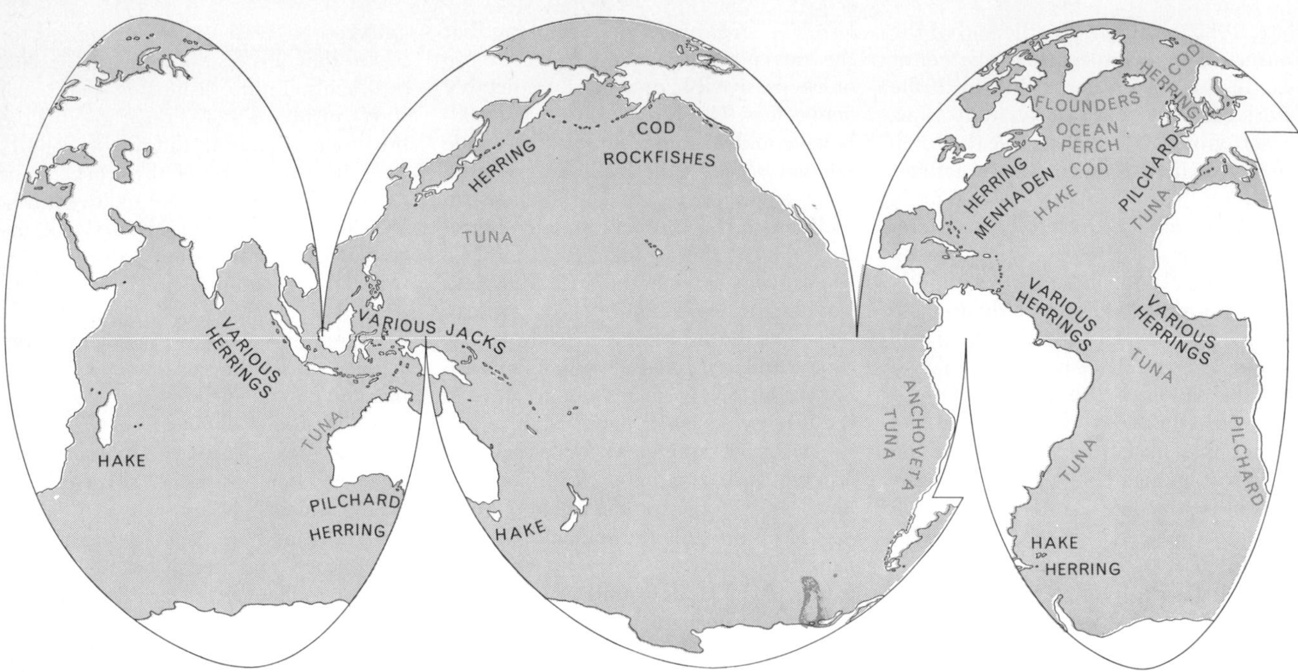

EXPLOITATION OF FISHERIES during the past 20 years is evident from this map, which locates 30 major fish stocks that were thought to be underfished in 1949. Today 14 of the stocks (*color*) are probably fully exploited or in danger of being overfished.

certain current systems in the open sea. The classical notion is that biological production is high in such areas because nutrient salts, needed for plant growth and in limited supply, are thereby renewed in the surface layers of the water. On the other hand, there is a view that the blooming of the phytoplankton is associated more with the fact that the water coming to the surface is cooler than it is associated with its richness in nutrients. A cool-water regime is characterized by seasonal peaks of primary production; the phytoplankton blooms are followed, after a time lag, by an abundance of herbivorous zooplankton that provides concentrations of food for large schools of fish. Fish, like fishermen, thrive best not so much where their prey are abundant as where they are most aggregated. In any event, the times and places of aggregation vary from year to year. The size of the herbivore crop also varies according to the success of synchronization with the primary production cycle.

There would be great practical advantage to our being able to predict these variations. Since the weather regime plays such a large part in creating the physical conditions for high biological production, the World Weather Watch, under the auspices of the World Meteorological Organization, should contribute much to fishery operations through both long-range forecasting and better short-term forecasting. Of course our interest is not merely in atmospheric forecasts, nor in the state of the sea surface, but in the deeper interaction of atmosphere and ocean. Thus, from the point of view of fisheries, an equal and complementary partner in the World Weather Watch will be the Integrated Global Ocean Station System (IGOSS) now being developed by the Intergovernmental Oceanographic Commission. The IGOSS will give us the physical data, from networks of satellite-interrogated automatic buoys and other advanced ocean data acquisition systems (collectively called ODAS), by which the ocean circulation can be observed in "real time" and the parameters relevant to fisheries forecast. A last and much more difficult link will be the observation and prediction of the basic biological processes.

So far we have been considering mainly the stocks of pelagic fishes in the upper layers of the open ocean and the shallower waters over the continental shelves. There are also large aggregations of pelagic animals that live farther down and are associated particularly with the "deep scattering layer," the sound-reflecting stratum observed in all oceans. The more widespread use of submersible research vessels will tell us more about the layer's biological nature, but the exploitation of deep pelagic resources awaits the development of suitable fishing apparatus for this purpose.

Important advances have been made in recent years in the design of pelagic trawls and in means of guiding them in three dimensions and "locking" them onto fish concentrations. We shall perhaps have such gear not only for fishing much more deeply than at present but also for automatically homing on deep-dwelling concentrations of fishes, squids and so on, using acoustic links for the purpose. The Indian Ocean might become the part of the world where such methods are first deployed on a large scale; certainly there is evidence of a great but scarcely utilized pelagic resource in that ocean, and around its edge are human populations sorely in need of protein. The Gulf of Guinea is another place where oceanographic knowledge and new fishing methods should make accessible more of the large sardine stock that is now effectively exploited only during the short season of upwelling off Ghana and nearby countries, when the schools come near the surface and can be taken by purse seines.

The bottom-living fishes and the shellfishes (both mollusks and crustaceans) are already more fully utilized than the smaller pelagic fishes. On the whole they are the species to which man attaches a particularly high value, but they cannot have as high a global abundance as the pelagic fishes. The reason is that they are living at the end of a longer food chain. All the rest of ocean life depends on an annual primary production of 150 billion tons of phytoplankton in the 2 to 3 percent of the water mass into which light penetrates and photosynthesis can occur. Below this "photic" zone dead and dying organisms sink as a continual rain of organic matter and are eaten or decompose. Out in the deep ocean little, if any, of this organic matter reaches the bottom, but nearer land a substantial quantity does; it nourishes an entire community of marine life, the benthos, which itself provides food for animals such as cod, ocean perch, flounder and shrimp that dwell or visit there.

Thus virtually everywhere on the bed of the continental shelf there is a thriving demersal resource, but it does not end there. Where the shelf is narrow but primary production above is high, as in the upwelling areas, or where the zone of high primary production stretches well away from the coast, we may find considerable demersal resources on the continental slopes beyond the shelf, far deeper than the 200 meters that is the average limiting depth of the shelf itself. Present bottom-trawling methods will work down to 1,000 meters or more, and it seems that, at least on some slopes, useful resources of shrimps and bottom-dwelling fishes will be found even down to 1,500 meters. We still know very little about the nature and abundance of these resources, and current techniques of acoustic surveying are not of much use in evaluating them. The total area of the continental slope from, say, 200 to 1,500 meters is roughly the same as that of the entire continental shelf, so that when we have extended our preliminary surveys there we might need to revise our IWP ceiling upward somewhat.

Another problem is posed for us by the way that, as fishing is intensified throughout the world, it becomes at the same time less selective. This may not apply to a particular type of fishing operation, which may be highly selective with regard to the species captured. Partly as a result of the developments in processing and trade, and partly because of the decline of some species, however, we are using more and more of the species that abound. This holds particularly for species in warmer waters, and also for some species previously neglected in cool waters, such as the sand eel in the North Sea. This means that it is no longer so reasonable to calculate the potential of each important species stock separately, as we used to do. Instead we need new theoretical models for that part of the marine ecosystem which consists of animals in the wide range of sizes we now utilize: from an inch or so up to several feet. As we move toward fuller utilization of all these animals we shall need to take proper account of the interactions among them. This will mean devising quantitative methods for evaluating the competition among them for a common food supply and also examining the dynamic relations between the predators and the prey among them.

These changes in the degree and quality of exploitation will add one more dimension to the problems we already face in creating an effective international system of management of fishing activities, particularly on the high seas. This system consists at present of a large number—more than 20—of regional or specialized intergovernmental organizations established under bilateral or multilateral treaties, or under the constitution of the FAO. The purpose of each is to conduct and coordinate research leading to resource assessments, or to promulgate regulations for the better conduct of the fisheries, or both. The organizations are supplemented by the 1958 Geneva Convention on Fishing and Conservation of the Living Resources of the High Seas. The oldest of them, the International Council for the Exploration of the Sea, based in Copenhagen and concerned particularly with fishery research in the northeastern Atlantic and the Arctic, has had more than half a century of activity. The youngest is the International Commission for the Conservation of Atlantic Tunas; the convention that establishes it comes into force this year.

For the past two decades many have hoped that such treaty bodies would ensure a smooth and reasonably rapid approach to an international regime for ocean fisheries. Indeed, a few of the organizations have fair successes to their credit. The fact is, however, that the fisheries have been changing faster than the international machinery to deal with them. National fishery research budgets and organizational arrangements for guiding research, collecting proper statistics and so on have been largely inadequate to the task of assessing resources. Nations have given, and continue to give, ludicrously low-level support to the bodies of which they are members, and the bodies themselves do not have the powers they need properly to manage the fisheries and conserve the resources. Add to this the trend to high mobility and range of today's fishing fleets, the problems of species interaction and the growing number of nations at various stages of economic development participating in international fisheries, and the regional bodies are indeed in trouble! There is some awareness of this, yet the FAO, having for years been unable to give adequate financial support to the fishery bodies it set up years ago in the Indo-Pacific area, the Mediterranean and the southwestern Atlantic, has been pushed, mainly through the enthusiasm of its new intergovernmental Committee on Fisheries, to establish still other bodies (in the Indian Ocean and in the east-central and southeastern Atlantic) that will be no better supported than the ex-

RUSSIAN FACTORY SHIP *Polar Star* lies hove to in the Barents Sea in June, 1968, as two vessels from its fleet of trawlers unload their catch for processing. The worldwide activities of the Russian fishing fleet have made the U.S.S.R. the third-largest fishing nation.

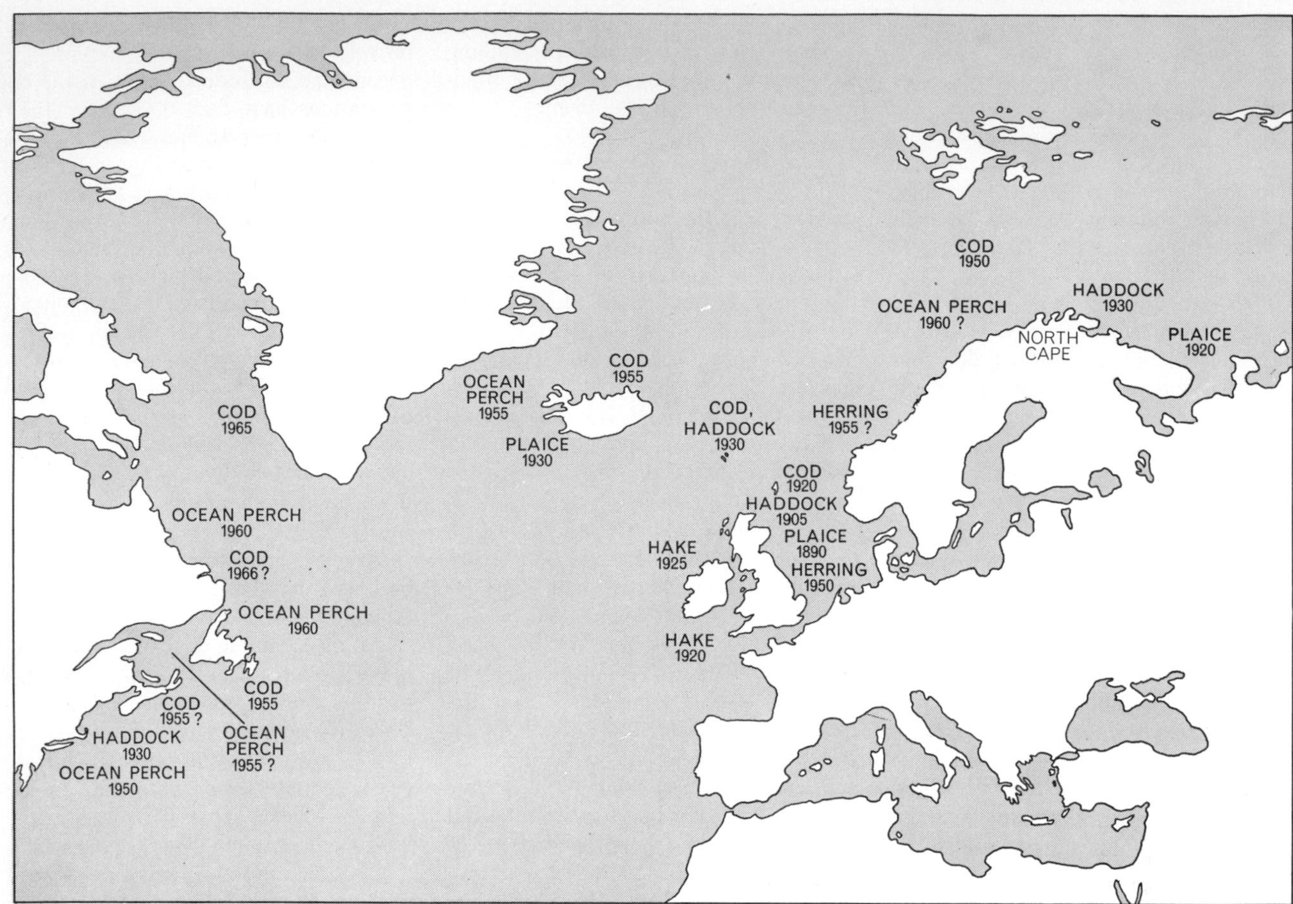

OVERFISHING in the North Atlantic and adjacent waters began some 80 years ago in the North Sea, when further increases in fishing the plaice stock no longer produced an increase in the catch of that fish. By 1950 the same was true of North Sea cod, haddock and herring, of cod, haddock and plaice off the North Cape and in the Barents Sea, of plaice, haddock and cod south and east of Iceland and of the ocean perch and haddock in the Gulf of Maine. In the period between 1956 and 1966 the same became true of ocean perch off Newfoundland and off Labrador and of cod west of Greenland. It may also be true of North Cape ocean perch and Labrador cod.

isting ones. A grand plan to double the finance and staff of the FAO's Department of Fisheries (including the secretariats and working budgets of the associated regional fishery bodies) over the six-year period 1966–1971, which member nations endorsed in principle in 1965, will be barely half-fulfilled in that time, and the various nations concerned are meanwhile being equally parsimonious in financing the other international fishery bodies.

Several of these bodies are now facing a crucial, and essentially political, problem: How are sustainable yields to be shared equitably among participating nations? It is now quite evident that there is really no escape from the paramount need, if high yields are to be sustained; this is to limit the fishing effort deployed in the intensive fisheries. This could be achieved by setting total quotas for each species in each type of fishery, but this only leads to an unseemly scramble by each nation for as large a share as possible of the quota. This can only be avoided by agreement on national allocations of the quotas. On what basis can such agreement be reached? On the historical trends of national participation? If so, over what period: the past two years, the past five, the past 20? On the need for protein, on the size or wealth of the population or on the proximity of coasts to fishing grounds? Might we try to devise a system for maximizing economic efficiency in securing an optimum net economic yield? How can this be measured in an international fishery? Would some form of license auction be equitable, or inevitably loaded in favor of wealthy nations? The total number or tonnage of fishing vessels might be fixed, as the United Kingdom suggested in 1946 should be done in the North Sea, but what flags should the ships fly and in what proportion? Might we even consider "internationalizing" the resources, granting fishing concessions and using at least a part of the economic yield from the concessions to finance marine research, develop fish-farming, police the seas and aid the participation of less developed nations?

Some of my scientific colleagues are optimistic about the outcome of current negotiations on these questions, and indeed when the countries participating are a handful of nations at a similar stage of economic and technical development, as was the case for Antarctic whaling, agreement can sometimes be reached by hard bargaining. What happens, however, when the participating countries are numerous, widely varying in their interests and ranging from the most powerful nations on earth to states most newly emerged to independence? I must confess that many of us were optimistic when 20 years ago we began proposing quite reasonable net-mesh regulations to conserve the young of certain fish stocks. Then we saw these simple—I suppose oversimple—ideas bog down in consideration of precisely how to measure a mesh of a particular kind of twine, and how to take account of the innumerable special situations that countries pleaded for, so that fishery research sometimes seemed to be becoming perverted from its earlier clarity and broad perspective.

Apprehension and doubt about the ultimate value of the concept of regulation through regional commissions of the present type have, I think, contributed to the interest in recent years in alternative regimes: either the "appropriation" of high-seas resources to some form of international "ownership" instead of today's condition of no ownership or, at the other extreme, the appropriation of increasingly wide ocean areas to national ownership by coastal states. As is well known, a similar dialectic is in progress in connection with the seabed and its mineral resources. Either solution would have both advantages and disadvantages, depending on one's viewpoint, on the time scale considered and on political philosophy. I do not propose to discuss these matters here, although personally I am increasingly firm in the conclusion that mankind has much more to gain in the long run from the "international" solution, with both seabed and fishery resources being considered as our common heritage. We now at least have a fair idea of what is economically at stake.

Here are some examples. The wasted effort in capture of cod alone in the northeastern Atlantic and salmon alone in the northern Pacific could, if rationally deployed elsewhere, increase the total world catch by 5 percent. The present catch of cod, valued at $350 million per year, could be taken with only half the effort currently expended, and the annual saving in fishing effort would amount to $150 million or more. The cost of harvesting salmon off the West Coast of North America could be reduced by three-quarters if management policy permitted use of the most efficient fishing gear; the introduction of such a policy would increase net economic returns by $750,000 annually.

The annual benefit that would accrue from the introduction and enforcement of mesh regulations in the demersal fishery—mainly the hake fishery—in the east-central Atlantic off West Africa is of the order of $1 million. Failure to regulate the Antarctic whaling industry effectively in earlier years, when stocks of blue whales and fin whales were near their optimum size, is now costing us tens of millions of dollars annually in loss of this valuable but only slowly renewable resource. Even under stringent regulation this loss will continue for the decades these stocks will need to recover. Yellowfin tuna in the eastern tropical Pacific are almost fully exploited. There is an annual catch quota, but it is not allocated to nations or ships, with the classic inevitable results: an increase in the catching capacity of fleets, their use in shorter and shorter "open" seasons and an annual waste of perhaps 30 percent of the net value of this important fishery.

Such regulations as exist are extremely difficult to enforce (or to be seen to be enforced, which is almost as important). The tighter the squeeze on the natural resources, the greater the suspicion of fishermen that "the others" are not abiding by the regulations, and the greater the incentive to flout the regulations oneself. There has been occasional provision in treaties, or in *ad hoc* arrangements, to place neutral inspectors or internationally accredited observers aboard fishing vessels and mother ships (as in Antarctic whaling, where arrangements were completed but never implemented!). Such arrangements are exceptional. In point of fact the effective supervision of a fishing fleet is an enormously difficult undertaking. Even to know where the vessels are going, let alone what they are catching, is quite a problem. Perhaps one day artificial satellites will monitor sealed transmitters compulsorily carried on each vessel. But how to ensure compliance with minimum landing-size regulations when increasing quantities of the catch are being processed at sea? With factory ships roaming the entire ocean, even the statistics reporting catches by species and area can become more rather than less difficult to obtain.

Some of these considerations and pessimism about their early solution have, I think, played their part in stimulating other approaches to harvesting the sea. One of these is the theory of "working back down the food chain." For every ton of fish we catch, the theory goes, we might instead catch say 10 tons of the organisms on which those fish feed. Thus by harvesting the smaller organisms we could move away from the fish ceiling of 100 million or 200 million tons and closer to the 150 billion tons of primary production. The snag is the question of concentration. The billion tons or so of "fish food" is neither in a form of direct interest to man nor is it so concentrated in space as the animals it nourishes. In fact, the 10-to-one ratio of fish food to fish represents a use of energy—perhaps a rather efficient use by which biomass is concentrated; if the fish did not expend this energy in feeding, man might have to expend a similar amount of energy—in fuel, for example—in order to collect the dispersed fish food. I am sure the technological problems of our using fish food will be solved, but only careful analysis will reveal whether or not it is better to turn fish food, by way of fish meal, into chickens or rainbow trout than to harvest the marine fish instead.

There are a few situations, however, where the concentration, abundance and homogeneity of fish food are sufficient to be of interest in the near future. The best-known of these is the euphausiid "krill" in Antarctic waters: small shrimplike crustaceans that form the main food of the baleen whales. Russian investigators and some others are seriously charting krill distribution and production, relating them to the oceanographic features of the Southern Ocean, experiment-

JAPANESE MARICULTURE includes the raising of several kinds of marine algae. This array of posts and netting in the Inland Sea supports a crop of an edible seaweed, *Porphyra*.

AUSTRALIAN MARICULTURE includes the production of some 60 million oysters per year in the brackish estuaries of New South Wales. The long racks in the photograph have been exposed by low tide; they support thousands of sticks covered with maturing oysters.

ing with special gear for catching the krill (something between a mid-water trawl and a magnified plankton net) and developing methods for turning them into meal and acceptable pastes. The krill alone could produce a weight of yield, although surely not a value, at least as great as the present world fish catch, but we might have to forgo the whales. Similarly, the deep scattering layers in other oceans might provide very large quantities of smaller marine animals in harvestable concentration.

An approach opposite to working down the food chain is to look to the improvement of the natural fish resources, and particularly to the cultivation of highly valued species. Schemes for transplanting young fish to good high-seas feeding areas, or for increasing recruitment by rearing young fish to viable size, are hampered by the problem of protecting what would need to be quite large investments. What farmer would bother to breed domestic animals if he were not assured by the law of the land that others would not come and take them as soon as they were nicely fattened? Thus mariculture in the open sea awaits a regime of law there, and effective management as well as more research.

Meanwhile attention is increasingly given to the possibilities of raising more fish and shellfish in coastal waters, where the effort would at least have the protection of national law. Old traditions of shellfish culture are being reexamined, and one can be confident that scientific bases for further growth will be found. All such activities depend ultimately on what I call "productivity traps": the utilization of natural or artificially modified features of the marine environment to trap biological production originating in a wider area, and by such a biological route that more of the production is embodied in organisms of direct interest to man. In this way we open the immense possibilities of using mangrove swamps and productive estuarine areas, building artificial reefs, breeding even more efficient homing species such as the salmon, enhancing natural production with nutrients or warm water from coastal power stations, controlling predators and competitors, shortening food chains and so on. Progress in such endeavors will require a better predictive ecology than we now have, and also many pilot experiments with corresponding risks of failure as well as chances of success.

The greatest threat to mariculture is perhaps the growing pollution of the sea. This is becoming a real problem for fisheries generally, particularly coastal ones, and mariculture would thrive best in just those regions that are most threatened by pollution, namely the ones near large coastal populations and technological centers. We should not expect, I think, that the ocean would not be used at all as a receptacle for waste—it is in some ways so good for such a purpose: its large volume, its deep holes, the hydrolyzing, corrosive and biologically degrading properties of seawater and the microbes in it. We should expect, however, that this use will not be an indiscriminate one, that this use of the ocean would be internationally registered, controlled and monitored, and that there would be strict regulation of any dumping of noxious substances (obsolete weapons of chemical and biological warfare, for example), including the injection of such substances by pipelines extending from the coast. There are signs that nations are becoming ready to accept such responsibilities, and to act in concert to overcome the problems. Let us hope that progress in this respect will be faster than it has been in arranging for the management of some fisheries, or in a few decades there may be few coastal fisheries left worth managing.

I have stressed the need for scientific research to ensure the future use of the sea as a source of food. This need seems to me self-evident, but it is undervalued by many persons and organizations concerned with economic development. It is relatively easy to secure a million dollars of international development funds for the worthy purpose of assisting a country to participate in an international fishery or to set up a training school for its fishermen and explore the country's continental shelf for fish or shrimps. It is more difficult to justify a similar or lesser expenditure on the scientific assessment of the new fishery's resources and the investigation of its ocean environment. It is much more difficult to secure even quite limited support for international measures that might ensure the continued profitability of the new fishery for all participants.

Looking back a decade instead of forward, we recall that Lionel A. Walford of the U.S. Fish and Wildlife Service wrote, in a study he made for the Conservation Foundation: "The sea is a mysterious wilderness, full of secrets. It is inhabited only by wild animals and, with the exception of a few special situations, is uncultivated. Most of what we know about it we have had to learn indirectly with mechanical contrivances to probe, feel, sample, fish." There are presumably fewer wild animals now than there were then—at least fewer useful ones—but there seems to be a good chance that by the turn of the century the sea will be less a wilderness and more cultivated. Much remains for us and our children to do to make sure that by then it is not a contaminated wilderness or a battlefield for ever sharper clashes between nations and between the different users of its resources.

The Anchovy Crisis

by C. P. Idyll
June 1973

Over the past decade the world's largest fishery has been in the Peru Current. A periodic ecological disturbance, combined with the heavy fishing, now threatens to destroy the industry

The world of the Peruvian anchovy is the sweep of a great cold ocean current. In a slow northward drift the current carries the little fishes along in company with countless tiny plants and animals that the anchovies avidly devour and with larger fishes, squids and a host of other marine animals. The anchovies form thronging legions that wheel and dart in the current. Their world is often entered by aliens: birds that plunge from above, snatching up the anchovies by the hundreds of thousands, and men who cast great net enclosures around the fishes, carrying them off to shore by the millions.

In the brief life-span of the anchovy, rarely longer than three years, its cold-current environment usually changes only within narrow limits. During the lifetime of some generations, however, their world may be put out of joint. The slow northward drift of the current, only two-tenths to three-tenths of a knot (compared with the six knots of the Gulf Stream off Florida), becomes still slower and may even reverse itself. The water grows warmer and less salty; the makeup of its populations changes and many of the usually abundant microscopic plants and animals dwindle in number. Finding their world poorer, the little anchovies scatter; many may die prematurely and many in the successor generation may not be born at all. The sea change known as El Niño has arrived.

Coastal Peru is normally a cool and misty land, quite unlike the steamy Tropics that occupy the same latitudes on the eastern coast of South America. The Peru coast is kept that way by the temperature of the ocean current, which in the south can be as cool as 10 degrees Celsius (50 degrees Fahrenheit) and only reaches about 22 degrees C. (71.6 degrees F.) in the north. Although the northernmost part of Peru is a mere three degrees of latitude below the Equator, the average air temperature is a moderate 18 to 22 degrees C. (64.4 to 71.6 degrees F.). Along the 1,475 miles of Peruvian coastline, a distance 100 miles longer than the Pacific coast of the U.S., there are no marshes, mud flats and estuaries—only arid desert, most of it a treeless, monotonous, barren brown. This bleak strip of sand extends a short distance inland, rarely more than 40 miles or so, to the upthrust Andes, one of the most awesome mountain ranges in the world. The high rain shadow of the Andes robs the prevailing southeast winds of their moisture and keeps the coastal region arid. A few streams that rise in the mountains cross the desert; their narrow valleys support what little agriculture exists along the coast.

The Peru Current consists of four components, the interaction of which creates, molds and changes the world of the anchovy. Two of the components travel in a northerly direction: the Coastal Current, flowing next to the shore, and the Oceanic Current, located farther out to sea. Between the two, on and near the surface, runs the Peru Countercurrent. Beneath all three runs the Peru Undercurrent. Both the Countercurrent and the Undercurrent flow southward.

The Coastal Current runs deep and hugs the land from about Valparaiso in Chile in the south to north of Chimbote in Peru, a stretch of some 2,000 miles. The anchovies live mostly in the northern part of this great band of water, which constantly changes shape and size, becoming wider or narrower, deeper or shallower, altering and twisting like an elongated amoeba. The Oceanic Current is longer than the Coastal Current. It is often several hundred miles wide, and it runs as deep as 700 meters. It flows north to a point about opposite the Gulf of Guayaquil before bending west.

The great northward sweep of water is often called the Humboldt Current, after the German naturalist who described the phenomenon following a visit to South America in 1803. Humboldt thought that the cold of the water was the chill of the Antarctic. His conjecture was partly right: the current does include subantarctic water. Much of the cold, however, is the cold of subsurface water. As the water on the ocean surface is swept away by the prevailing winds, deeper low-temperature water wells up slowly to replace it. The trade winds in this part of the world, channeled and bent by the Andes, blow from the south and southeast, mostly parallel to the shore. This prevailing wind urges the surface water northward at the same time that another influence, the Coriolis force, deflects it to the west. As the resulting steady offshore drift skims off the surface layer the cold subsurface water rises with stately slowness to replace it, traveling vertically at a rate ranging from 20 to 100 meters per month, depending on the location and the season.

The biological effect of the upwelling is enormous. That stretch of water, only a tiny fraction of the ocean surface, produces fully 22 percent of all the fish caught throughout the world. Its richness springs from a constantly renewed supply of the chemical nutrients—principally phosphates and nitrates—that stimulate plant growth. Accumulated gradually in the deep layers of the ocean as the debris of dead marine plants and animals sinks to the bottom, the nutrients travel with the upwelling water to the top levels. There the light is sufficient to drive photosynthesis, and the nutrients help the marine plants to flourish. The concentration of nutrients in the Peru upwelling is many times greater than that in the open ocean. In terms of

ANCHOVY TRAWLERS by the score lie at anchor in the fishing port of Pisco in Peru. Structures in the foreground are part of a fish-meal factory. Tonnage of anchovies alone landed in Peru in a normal year outweighs the combined fish catch of any other nation.

CASCADE OF ANCHOVIES spills from a draining grate into the cargo hold of a trawler as crewmen keep watch. The fish gather by the billions near shore and close to the surface, particularly in summer months; trawlers can then net anchovies 100 tons at a time.

the amount of carbon fixed photosynthetically per cubic meter of water per day, the range in the upwelling region is from 45 to 200 milligrams, compared with less than 15 milligrams in the waters immediately adjacent. Perhaps only one other part of the world ocean is richer: the Benguela Current off the southwestern coast of Africa.

The Peru upwelling sustains an enormous flow of living matter. The food chain begins with the microscopic diatoms and other members of the phytoplankton that comprise the pasturage of the sea. The plants absorb the nutrients and grow in rich profusion, providing fodder for billions of grazing animals, principally minute crustaceans such as copepods but including arrowworms and a wide variety of other small marine herbivores. The food chain can then go on to several more links, progressing from the small fishes that eat the herbivores to the larger fishes and squids that prey on the small fishes and perhaps continuing to include one or more further advanced levels of marine predation. The food chain in the Peru Current does go on in this fashion to some degree, but most of its energy flow stops with the anchovies. This single fish species has succeeded in capturing an extraordinarily high proportion of the total energy available in the ecosystem and in converting it into enormous quantities of living matter. At the height of the anchovies' annual cycle the total bulk of the species is probably of the order of 15 to 20 million metric tons.

The Peruvian anchovy belongs to the same genus (*Engraulis*) as the common anchovy of the eastern Atlantic and the Mediterranean (*E. encrasicolus*), but it comprises a separate species (*E. ringens*). Its life begins in the form of an egg, a tiny oval spot of nearly transparent protoplasm adrift in the sea. Eggs can be spawned at almost any time of the year but there are two periods when the anchovies' reproductive activity is highest. The major spawning occurs in August and September, during the southern winter, and it is repeated on a lesser scale in January and February. Anchovies are precocious: most females are capable of spawning when they are a year old. By then each female, a little over four inches long, may produce 10,000 eggs. If she survives to the age of two and reaches a length of six inches, her output increases to some 20,000 eggs.

The delicate larvae that hatch from the eggs lead a perilous existence. Many species of fish produce eggs that contain a considerable store of yolk; the reserve

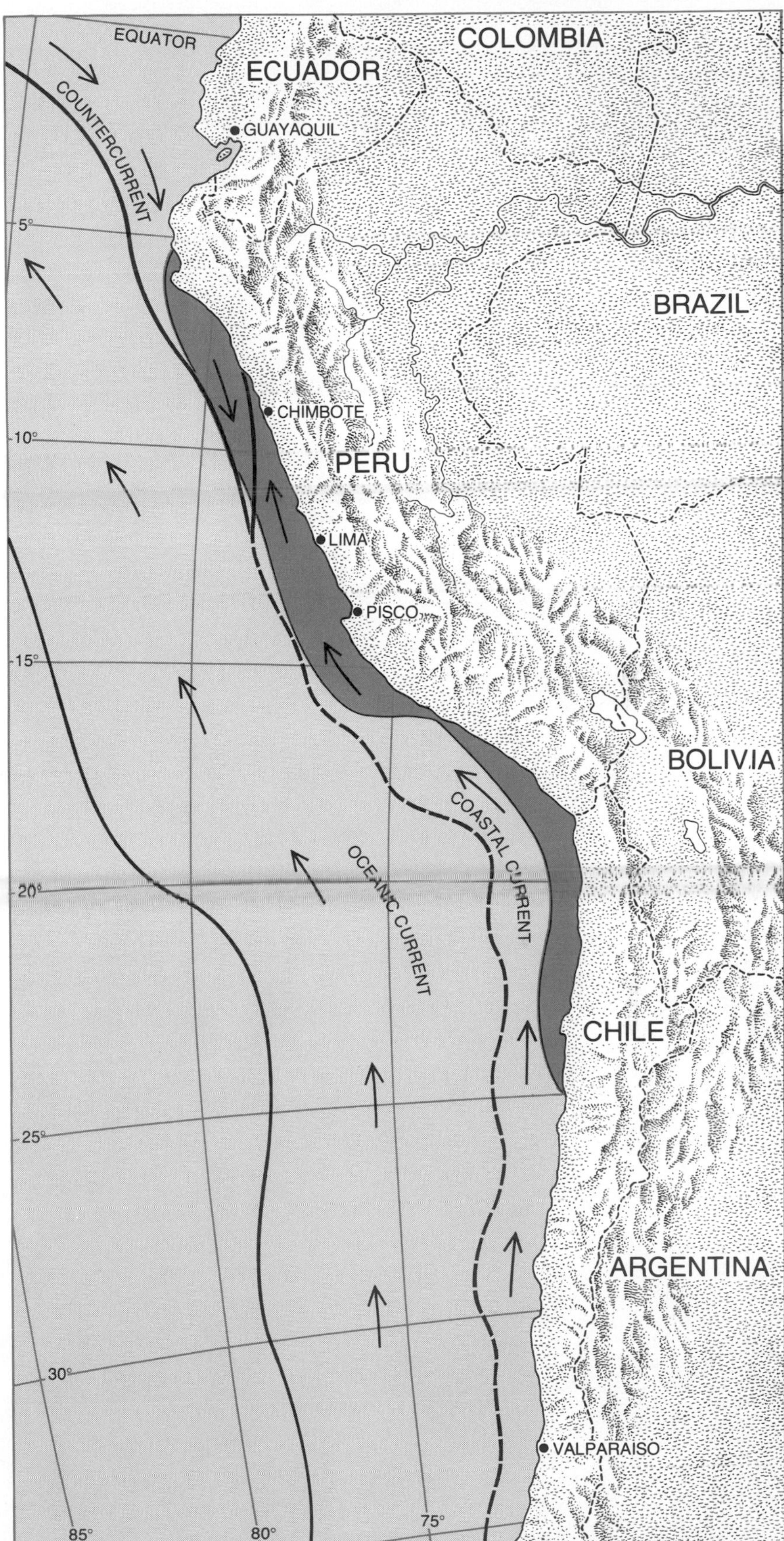

TWO NORTH-FLOWING COMPONENTS of the Peru Current are the deep, narrow Coastal Current that hugs the land from Valparaiso in Chile to north of Chimbote in Peru and the deeper and wider Oceanic Current that reaches the latitude of the Gulf of Guayaquil. The anchovies (*dark color*) are normally found in the Coastal Current between 25 and five degrees south latitude; they may consist of a northern and a southern population.

of nutrient helps to sustain the newly hatched young until they adjust to finding their own food. The anchovy egg has a negligible yolk store, and so the larva must locate food quickly or starve. To make matters worse, the larva has limited swimming powers and a high rate of metabolism. If more than a few wiggles are required to obtain the food it needs, it will not survive. Because every larva consumes plankton in substantial amounts and because the peak hatches produce larvae numbering in multiples of billions, only enormous swarms of microscopic plants and larval crustaceans can sustain the anchovy stock. Nor is starvation the only peril: the larval anchovies feed swarms of predators. They are eaten by the same copepods that will, if the little fish survive, be the anchovies' own main sustenance. Arrowworms also devour them, and so do their own parents.

One month after the time the anchovy larvae are hatched more than 99 percent of them have perished. Even with such a high mortality rate, a process that begins with billions of spawning fish, each casting 10,000 to 20,000 eggs, produces enormous quantities of anchovy larvae. The little fish grow rapidly; in the course of the first year they attain a length of 4.2 to 4.3 inches. The year-old fish are so slender, however, that they weigh a scant third of an ounce.

The anchovy schools do not move at random; apparently because of a strong preference for the cold water of the Coastal Current they remain within a comparatively restricted zone. The Coastal Current is at its narrowest during the southern summer, running close to shore and seldom exceeding 200 meters in depth. Within this shrunken world the anchovies press together in enormous concentrations near the shore and close to the surface. It is now that predators fare best. The larger fishes and the squids feed well; the several species of guano birds have to fly only short distances from their island nesting grounds and need not dive deep to reach their prey. The greatest of the predators, the fisherman, finds summer work the easiest. He can often set his purse seine within sight of port and gather in anchovies 100 tons at a time.

El Niño occurs at irregular intervals. There is said to be a seven-year cycle, but in actuality the phenomenon is far

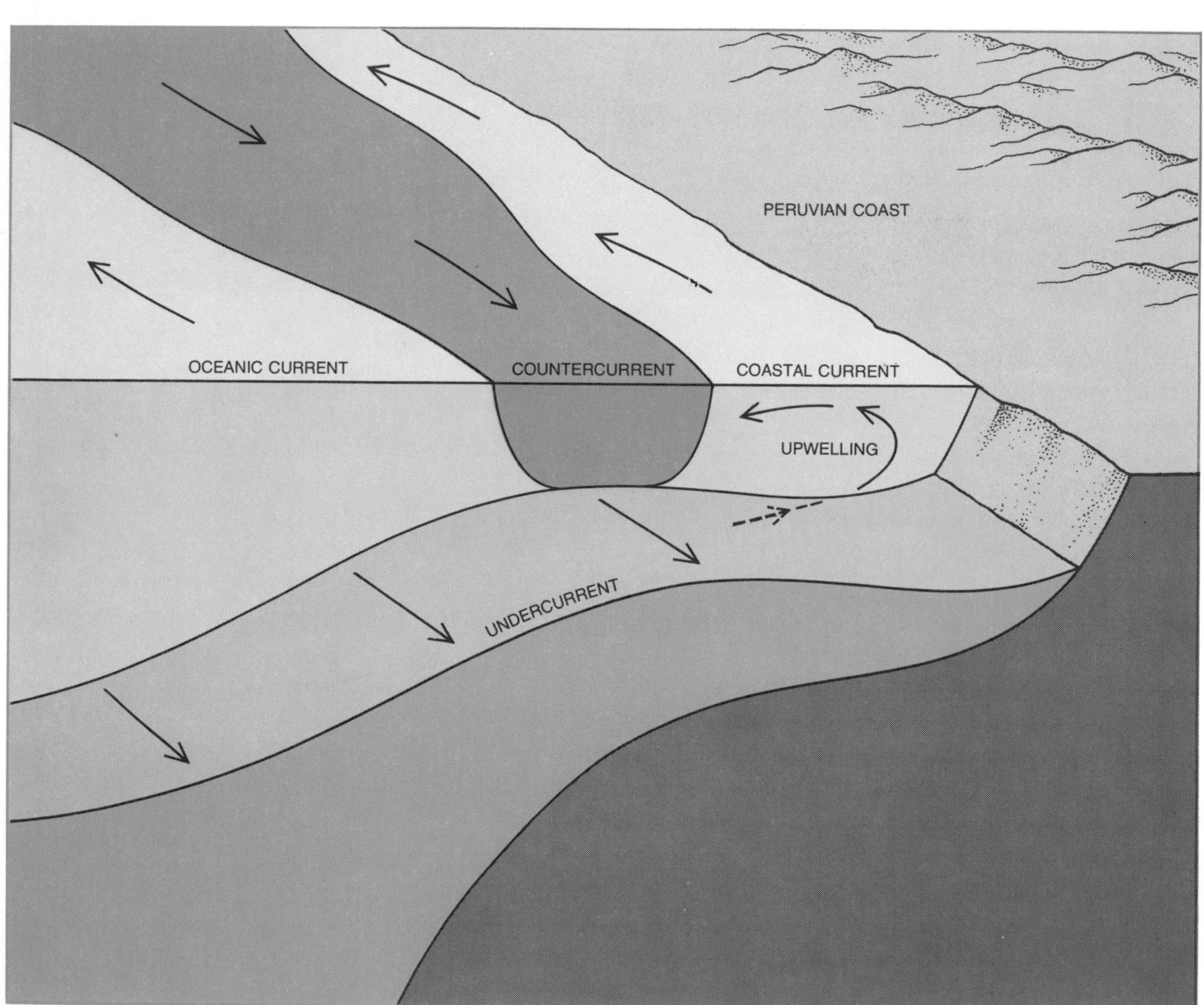

TWO SOUTH-FLOWING COMPONENTS of the Peru Current are seen schematically in relation to the two north-flowing components in this diagram. The Countercurrent (*dark color*), a surface or near-surface stream of water, intrudes between the north-flowing cold Coastal and Oceanic currents. Normally the Countercurrent does not extend much south of the Equator, but when the wind that moves the north-flowing currents along falters or changes direction, its warm water pushes far to the south with disastrous biological consequences. Deep below all three currents is the second, far larger south-flowing component, the Undercurrent (*light color*).

less precise in its appearance. The severe environmental dislocations may be repeated for two or more years in a row or may not recur for a decade or longer. Another kind of regularity, however, has given El Niño its name. The change usually begins around Christmastime and so is given the Spanish name for the Christ Child. The complicated chain of events in a year of El Niño disturbs the anchovies, sometimes profoundly. The wind now comes from the west rather than from the southeast, and it is laden with moisture from the Pacific. With no mountains to rob the air of its burden of water the arid coast is often subjected to torrential rains and severe windstorms. In a desert region where even a heavy mist can cause problems the floods of El Niño are often devastating. Oddly enough, however, in some Niño years no rain falls.

A warning that the sea change may be on its way is given when the temperature of the coastal water begins to rise. If the increase in temperature persists and spreads, the delicately adjusted world of the anchovy tilts. With the warm water come unfamiliar inhabitants of the northern Tropics: the yellowfin tuna, the dolphinfish, the manta ray and the hammerhead shark. Some of them feed on the anchovies. A greater threat to the anchovies' survival, however, is a slowing of the northbound Coastal Current and a decline or even a halt in the usual upwelling of subsurface waters. As the supply of nutrients diminishes, the planktonic plant life that provides the base of the ocean food chain becomes less abundant. As a result herbivorous planktonic animals become scarcer, and so it goes link by link up the chain. Furthermore, the water temperature is now too high to suit the anchovies themselves. Even if the shortage of food has not yet greatly reduced their numbers, the fish scatter, no longer forming the enormous schools that normally afford the guano birds and fishermen such rewarding targets.

The effect on the guano birds and marine animals is among the most serious of the changes wrought by El Niño. The birds starve or fly away, deserting their nestlings. Fishes, squids and even turtles and small sea mammals die. Their decaying bodies release evil-smelling hydrogen sulfide that bubbles up through the water and blackens the paint on the boats in the harbors. This unpleasant phenomenon is called El Pintor (The Painter). Patches of reddish, brownish or yellow water similar to the "red tides" that upset the Florida tourist industry become relatively common. They are caused by prodigious blooms of dinoflagellates: microscopic planktonic plants that are toxic in high concentrations. The greater frequency of the blooms during Niño years may be because the nutrient composition of the seawater suits the organisms better then or because the less vigorous currents fail to disperse accumulating clusters of dinoflagellates as quickly as usual.

The causes of El Niño are wind changes and sea changes on a very large scale. When the steady southeast trade winds weaken or when the wind blows from the west, the ocean currents that run to the northwest are no longer pushed along with the same vigor. Under normal circumstances the south-flowing Peru Countercurrent is weak, but when the prevailing winds fail or are reversed, the Countercurrent thrusts a tongue of warm water into the cleft between the now less vigorous Coastal and Oceanic currents. As it meets less resistance the Countercurrent penetrates farther south, pushing the weak north-flowing currents aside and covering their cold waters with a 30-meter layer of warm tropical water. The water may have come from as far away as the Panama Bight, north of the Equator, and part of it may even have originated as land runoff in Central America. It can be as much as seven degrees C. warmer than the north-flowing Coastal Current and is lower in salinity, deficient in oxygen and poor in nutrients. In some Niño years the Countercurrent pushes the tropical water as far as 600 miles south of the Equator.

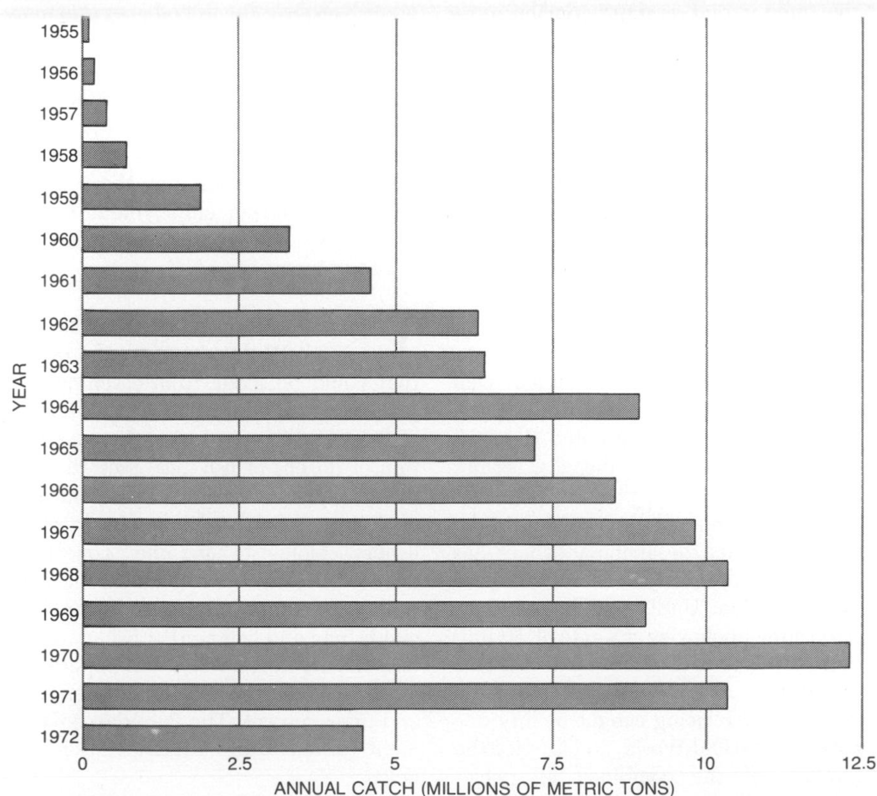

ANCHOVY CATCH remained below two million metric tons until the 1960's. Exploitation of the fishery skyrocketed thereafter; the annual rate approached or exceeded nine million tons for six of eight consecutive years. The peak year, 1970, saw a catch of 12.3 million tons.

The organisms most obviously affected by a Niño year are the guano birds, various species that are colloquially lumped together under the same name that is applied to the droppings that accumulate in large quantities on the rocky islands where they nest. There are three principal species: the guanay, or cormorant (*Phalacrocorax bougainvillii*); the piquero, or booby (*Sula variegata*), and the alcatras, or pelican (*Pelicanus thagus*). Over the millenniums the bird droppings have accumulated in piles as high as 150 feet on some islands. Because guano is perhaps the finest natural fertilizer known, the guano islands have provided the foundation for a valuable industry. Of the guano birds' diet between 80 and 95 percent is made up of anchovies, and the coastal waters of Peru support what is probably the largest population of oceanic birds anywhere in the world. In recent years estimates of the birds' total number have gone as

SACKS OF FISH MEAL awaiting shipment abroad line a wharf at a Peruvian port. The meal is used to enrich feeds for poultry and other livestock. In 1970 the export of fish meal and fish oil earned some $340 million, a third of Peru's total export revenue.

high as 30 million. The five million individuals that inhabited one particular guano island are believed to have consumed 1,000 tons of anchovies a day. The guano birds' annual anchovy catch in recent years is calculated to average 2.5 million metric tons, or between a fourth and a fifth of the commercial-fishery catch.

Following every Niño year of any consequence the bird population declines just as the anchovy population does. When the warmer water scatters the dense surface schools of fish, the birds find it harder to feed themselves, let alone their nestlings. Adult birds fly to other areas. Juvenile birds, less efficient fishers than their parents, perish in large numbers. The deserted nestlings are doomed to starvation. After the severe Niño year of 1957 the guano-bird population, then estimated to be 27 million, plummeted to six million and dropped to a low of 5.5 million the following year. Numbers slowly increased thereafter, so that there were 17 million birds when the Niño year of 1965 arrived. That year the population fell to 4.3 million.

Since then the guano birds have failed to recover at the normal rate. There is concern that the commercial anchovy fishery, which has expanded greatly in the same period, is depriving the birds of so much food that their numbers may fall below the level that is critical to their survival as social species. The late Robert Cushman Murphy of the American Museum of Natural History devoted some years to the study of these populations, and it was his opinion that the birds and the fishermen were essentially incompatible. It seems likely, however, that in spite of Murphy's contrary view the two competitors will be able to coexist at some suitable level of commercial fishing. At the same time it may well be that the size of the commercial catch in recent years has prevented the bird population from regaining its former numbers.

It is not commonly known that in the past few years the anchovy fishery has made Peru the world's leading fish-producing nation. Until recently Peru was harvesting anchovies at a rate of 10 million metric tons or more per year. This is a greater weight than that of all the species of fish being caught by any one nation in the Old World, and is twice the tonnage of the combined all-species catch of all the nations of North and Central America. The fish meal made from the Peruvian catch is sold around the world to enrich feeds for poultry and other livestock; the fish oil goes into margarine, paint, lipstick and a score of other products. In 1970 the export of fishery products brought Peru some $340 million, nearly a third of the nation's foreign-exchange earnings. In addition to this the tax revenues from the industry and the domestic employment it provides have become major elements in Peru's economy.

The anchovy industry began in earnest in 1957. Within 10 years the profits that could be made from catching and processing the fish attracted hundreds of fishing boats and led to the construction of dozens of fish-meal factories. No fish stock, however, can stand unchecked exploitation. Government authorities and fishery biologists became concerned about the future of the resource. Peru was a newcomer to large-scale commercial fishing and had neither fishery scientists nor administrators with experience in the complexities of fishery research and management. The Peruvian government turned to the United Nations for help.

In 1960, with a grant from the UN Development Programme and a matching amount in Peruvian funds, the Insti-

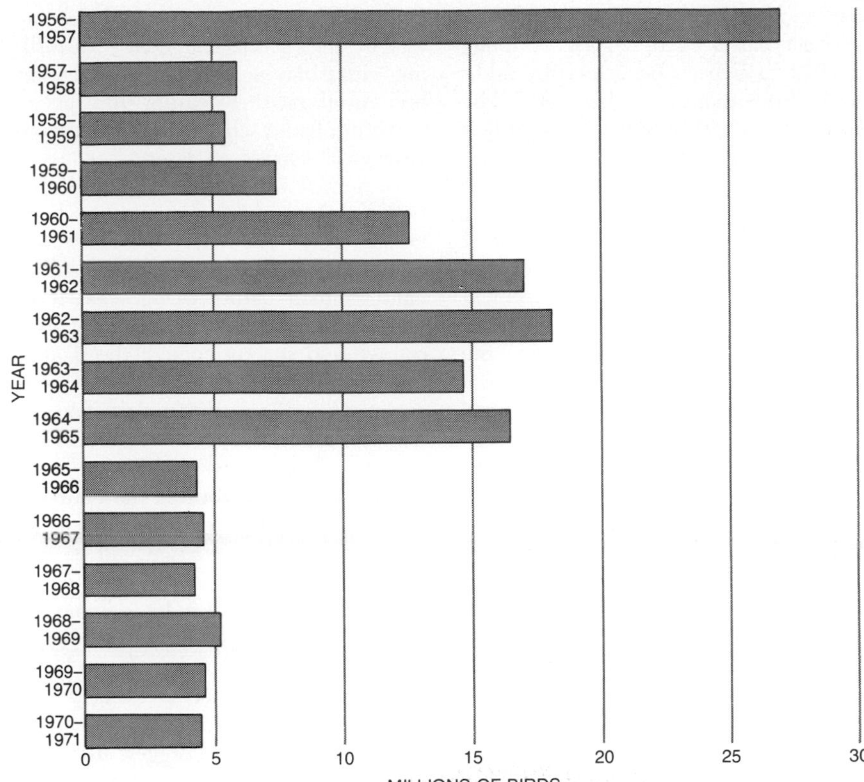

GUANO-BIRD POPULATION, hard hit by El Niño in 1957, had regained more than half its former numbers when El Niño reappeared in 1965. The population did not recover at the normal rate thereafter, possibly because the large commercial anchovy catches cut down the birds' food supply. Only a million or so guano birds may have survived the 1972 Niño.

tuto del Mar del Peru was set up to conduct research on the anchovy stocks and to advise the government on management of the fishery. The Food and Agriculture Organization of the UN (FAO) recruited experienced fishery scientists from around the world to work at the institute, conduct research on the anchovy stocks and train a Peruvian staff. Located in Lima, the institute is now a firmly established fishery-research center where more than 50 young Peruvian scientists are conducting the studies needed to establish conservation regulations for the anchovy fishery. The Peruvian staff is advised by a few resident FAO scientists and by a panel of distinguished experts from around the world, organized by the FAO, that meets twice a year.

It has taken biologists nearly a century to unravel the intricacies of fish populations and the complexities of their response to the dual stresses of environmental change and human exploitation. Until last year fishery biologists could point to the management of the Peruvian fishery as an exemplary application of this hard-won knowledge. A major stock had been put under rational control before exploitation had depleted it, and conservation measures seemed to be ensuring an enormously high yield at the limit of the biological capacity of the Peru Current ecosystem. Then in 1972 such pride was chastened, if not utterly humbled.

After the Niño year of 1965 the fishery had enjoyed several very successful seasons, culminating in 1970 with an anchovy catch of 12.3 million metric tons. Then, toward the end of April, 1972, and only a few weeks after the start of the season, fishing suddenly faltered. By the end of June catches had dwindled to almost nothing, and at the close of the 1972 season only 4.5 million tons of anchovies had been harvested. The catch this year threatens to be even poorer. Indeed, there is some reason to fear that the world's greatest stock of fish may have been irreversibly damaged, in which case the Peru fishery would be destined to collapse altogether.

That is the gloomiest outlook. It is based on two disturbing circumstances. First, the size of the "standing stock" (the total anchovy population) now appears to be far smaller than normal. It may be as low as one or two million metric tons, compared with an average of 15 million tons in recent years and an estimated 20 million tons in 1971. Second, "recruitment" (the numbers of fish grown big enough to enter the commercial fishery in any year) has been by far the smallest ever observed. It is scarcely 13 percent of the recruitment in a normal year.

What is causing the trouble? Very possibly El Niño has been a major factor, but some puzzling circumstances have made the scientists closest to the subject uncertain about the extent of the relation between the sea change and the reduction in the anchovy stock. It is clear, however, that the Niño year of 1972 was one of the most severe ever observed. Instead of remaining at the normal level of 22 degrees C. the surface temperature of the Coastal Current rose to 30.3 degrees in February. Although the temperature fluctuated thereafter, it remained high for the rest of the year and was still above normal in January, 1973. It did not fall to near-normal temperature until March of this year. At the same time that the temperature rose the salinity of the water declined from a normal 35 or 36 parts per 1,000 to 32.7 parts per 1,000.

Tropical and subtropical marine plants and animals began to appear far south of their usual limits: dolphinfish, skipjack tuna and the tropical crab *Euphilax*. The guano birds fled from their nesting islands, abandoning their young; their population may now number no more than a million. The warm Countercurrent forced the cold-seeking anchovies so close to the shore that the fishermen often found the water too shallow for their nets. Moreover, the crowded fish did not spawn as abundantly as usual and the eggs and larvae, already reduced in numbers, did not survive at the usual rate because of greater predation by their own close-packed parents.

Biologists are nonetheless unwilling to blame El Niño for all these occurrences. For example, with respect to the anchovies they note that recruitment of young fish to the adult population was observed to fall below normal levels before it became obvious that the surface temperature of the Coastal Current had risen. In seeming contradiction to this, however, the tropical crabs too made their appearance before the surface temperature rose, apparently indicating that a body of tropical water had by then already invaded the world of the anchovies. In the light of such oddities the experts frankly admit that they do not know how much influence El Niño exerted on the anchovies in 1972.

Quite apart from the sea change, however, the Peruvian commercial fishery must accept a share of the blame. The 1970 catch of 12.3 million tons considerably exceeded the 10-million-ton level that fishery biologists had estimated to be the maximum sustainable yield of the Peruvian stock. Several economic and political stresses were responsible for the excessive catch. Foremost among these harsh realities is that there are many more fishing boats and fish-meal factories in Peru than are needed to harvest and process the catch. The anchovy fleet is so large that it could harvest the equivalent of the annual U.S. catch of yellowfin tuna in a single day or the annual U.S. salmon catch in two and a half days. The fleet could be reduced by more than 25 percent and still comfortably harvest a rational quota of 10 million tons of anchovies a year. Moreover, the record 1970 catch figure does not measure the full toll the fishery took of the anchovy stock that year. Conservative estimates of losses from spoilage at sea and unloading and processing ashore raise the commercial total to some 13 or 14 million tons.

At this writing the future of the Peruvian anchovy is uncertain. The gloomy forecasts based on biological sampling in 1972 have been confirmed by additional observations early this year and by the results of trial fishing allowed at that time. For three weeks in March the anchovy fleet went to sea and about one million tons of fish were caught. During that brief period the catch per unit of fishing effort (a statistic that provides a measure of the size of the stock) declined rapidly. This suggests that the fleet had caught a significant proportion of all the fish that were available. Most of the catch consisted of fish recruited since July, 1972. There will not be any substantial additions to the stock until this coming October, when the progeny of the present population, much reduced by the fishing in March, have grown big enough to enter the fishery. Even so, fishing was authorized again in April with the quota set at 800,000 tons. Only 400,000 tons were taken. At present the 1973 catch is forecast at no more than three million tons.

If things are as bad as the worst prognostications indicate, the anchovy fishery may, like the California sardine industry and the Hokkaido herring industry, collapse forever. Many aspects of the history of the Peru fishery bear a disturbing resemblance to the events that brought about these earlier disasters.

Nature being what it is, the Peruvian coast will sooner or later once again have normal winds and ocean currents. If the anchovy population has not been too severely reduced, the fishery will then begin to recover. On the other hand, human nature being what it is, difficulties may arise in enforcing soon enough and strictly enough the moderate catch quotas required to avoid overexploitation of the diminished population. Unless such a policy of moderation is achieved, not only will the fish stock suffer but also the world of the Peruvian anchovy will be permanently changed. The guano-bird population will be further reduced and perhaps even eliminated. There will also be enormously complex effects among the many other animals that depend to a greater or lesser degree on the presence of the little fish. Finally, if the anchovies' world is allowed to go awry, the biggest loser will be man. He will have lost not only a rich natural resource but also some of the quality of his own world.

GUANO ISLAND off the coast of Peru is a nesting ground for a vast population of piqueros, or boobies. Cormorants, pelicans and boobies are the principal guano birds. Although now reduced in numbers, they once ate some 2.5 million tons of anchovies per year.

Marine Farming

by Gifford B. Pinchot
December 1970

Man gets food from the sea essentially by hunting and gathering. Yet the farming of fish and shellfish has been pursued for some 2,000 years, and its potentialities are far from being exhausted

A major concern of modern man is the possibility that the earth will not be able to produce enough food to nourish its expanding population. A particularly controversial issue is the question of how much food can ultimately be obtained from the sea. It is argued on the one hand that, on the basis of area, the oceans receive more than twice as much solar energy—the prime source of all biological productivity—as the land. This suggests that the oceans' potential productivity should greatly exceed the land's. On the other hand, most of the sea is biologically a desert. Its fertile areas are found where runoff from the land or the upwelling of nutrient-rich deep water fertilizes the surface water and stimulates the growth of marine plants, the photosynthetic organisms on which all other marine life depends. Even at today's high level of exploitation the fisheries of the world provide only a small fraction of human food needs, and there is some danger that they may supply even less in the future because of overfishing.

Does this mean that there is no hope of increasing our yield of food from the sea? I do not think so. It does mean, however, that instead of concentrating exclusively on more efficient means of fishing we must also learn to develop the potential of the oceans by farming them, just as early man learned that farming rather than hunting was the more effective method of feeding a human population. The purpose of this article is to examine briefly the contribution marine farming now makes to our food supply, and to consider some possibilities for its future role.

Marine farming has a long history. The earliest type of farming was the raising of oysters. Laws concerning oyster-raising in Japan go back to well before the time of Christ. Aristotle discusses the cultivation of oysters in Greece, and Pliny gives details of Roman oyster-farming in the early decades of the Christian Era. By the 18th century the natural oyster beds in France were beginning to be overexploited and were saved only by extensive developments in rearing practices.

Carp (*Cyprinus carpio*) were commonly raised in European freshwater ponds in both Roman and medieval times. Records concerning the regulation of salt or brackish ponds for raising milkfish (*Chanos chanos*) in Java date back to the 15th century. Carp and milkfish are both herbivores that thrive on a diet of aquatic plants. Oysters, as filter-feeders, can also be loosely classified as herbivores.

Oysters are particularly appropriate for marine farming because their spawn can be collected and used for "seeding" new areas of cultivation. An oyster produces more than 100 million eggs at a single spawning. The egg soon develops into a free-swimming larval form, known as a veliger, which settles to the bottom after two or three weeks. Veligers attach themselves to any clean surface and develop into miniature adult oysters, called "spat" because oystermen once believed the adult oysters spat them out. At this point the oyster farmer enters the picture. He distributes a supply of "cultch": clean material with a smooth, hard surface, such as old oystershell or ceramic tile. The cultch receives a "set" of spat and is then used to seed new oyster beds.

The bottom is prepared for seeding by removing as many natural enemies of the oyster as possible. In the eastern U.S. this is usually done by dragging a rope mat along the bottom to sweep the area clear of starfish, one of the major predators. In France, where more intensive labor is employed, the spat are usually planted on the exposed bottom of an estuary at low tide. The predators are removed by hand and the oyster bed is fenced to prevent their return. The oysters are moved after a few years to *claires,* special fattening areas where the water is rich in diatoms. This produces oysters of improved taste and color. When the oysters have reached marketable size, they are moved again to shallower water, where they must stay closed for longer periods at low tide. The French oystermen believe this treatment prepares the oysters for their trip to the market.

A significant advance in oyster-farming is the use of suspension cultures. This method, pioneered in Japan, is now spreading to the rest of the world. The spat are collected on shells that are

NUTRIENT-RICH WATER from the depths of the Pacific appears dark blue in the photograph on the following page. Land (*left*) is the coast of Taiwan as seen from *Gemini X*. Natural upwellings such as this are caused by winds or currents. Areas of continuous upwelling are extremely productive fishing grounds because the organisms eaten by fish flourish in them.

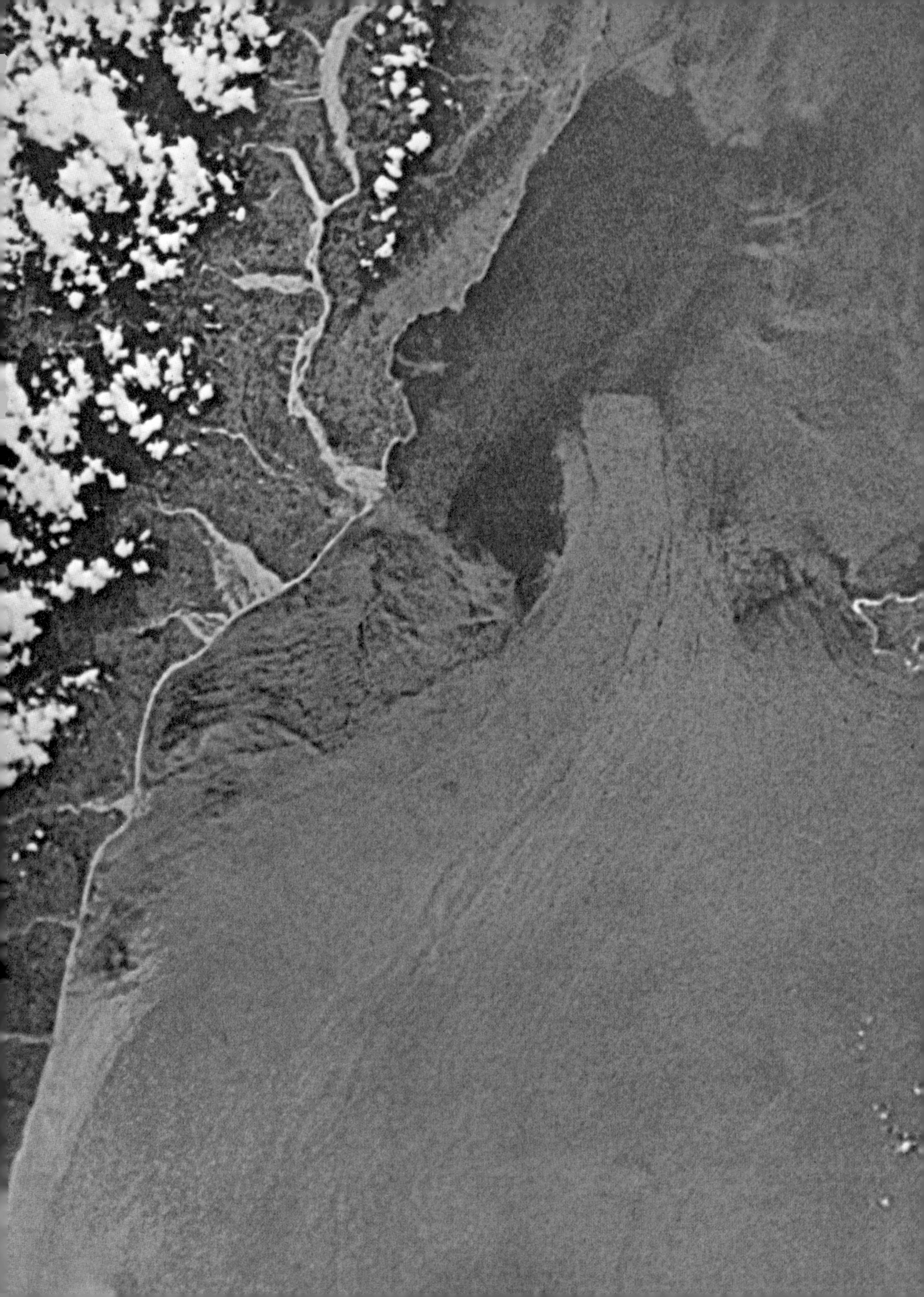

strung in long bundles and immersed in tidal water. The strings, which do not touch the bottom, are sometimes attached to stakes but more generally are attached to rafts. The suspension method has a number of advantages over growth on the bottom. The oysters are protected from predators and from silting, and they feed on the suspended food in the entire column of water rather than being limited to what reaches the bottom. The result is faster growth, rounder shape and superior flavor.

In small areas of Japan's Inland Sea suspension cultures of oysters annually yield 46,000 pounds of shucked meats per acre of cultivated area. This does not mean that one can multiply the total acreage of the Inland Sea by this figure to estimate the potential productivity of the area. Tidal flow allows the anchored oysters to filter much larger volumes of water than surround them at any given time. In addition, inshore waters are generally more productive than those farther from land. The figure does illustrate, however, the production of meat that is possible with our present farming practices in inshore waters.

Luther Blount of Warren, R.I., has tested oyster suspension cultures in Rhode Island waters over the past several years, using spat set on scallop shells. Blount spaces seven scallop shells well apart on each suspension string. At the end of seven months' growth he harvested one group of suspended oysters from 3,200 square feet of float area. The oysters weighed nearly 40,000 pounds and yielded 2,500 pounds of oyster meat. His experience suggests that the coastal waters of the eastern U.S. might yield more than 16,000 tons of meat per square mile of float per year.

Although the farming of oysters in suspension cultures is a comparatively recent development, the same technique has long been used in Europe to raise mussels. The Bay of Vigo is one of the many Spanish ports where acres of mussel floats are a common sight. French and Italian mussel growers are less inclined to use rafts. Their mussel strings are usually suspended from stakes set in the estuary bottom.

John H. Ryther and G. C. Matthiessen of the Woods Hole Oceanographic Institution have studied the yields obtained by the mussel farmers of Vigo. The annual harvest produces an average of 240,000 pounds of mussel meat per acre. This is equivalent to 70,000 tons of meat per square mile of float, or better than four times the yield of oysters in suspension cultures in the U.S. and Japan.

The farming of fish is more difficult than the farming of bivalves for at least two reasons. First, the fish, being motile, must be held in ponds. Second, the salt-water species that are most commonly farmed—milkfish and mullet (*Mugil*)—breed only at sea. This means that the fry have to be caught where and when they occur naturally, and in some years the supply is not adequate. Furthermore, unwanted species and predators have to be sorted out by hand, with the inevitable result that some of both are introduced into the ponds along with the desired species.

In spite of such handicaps pond farm-

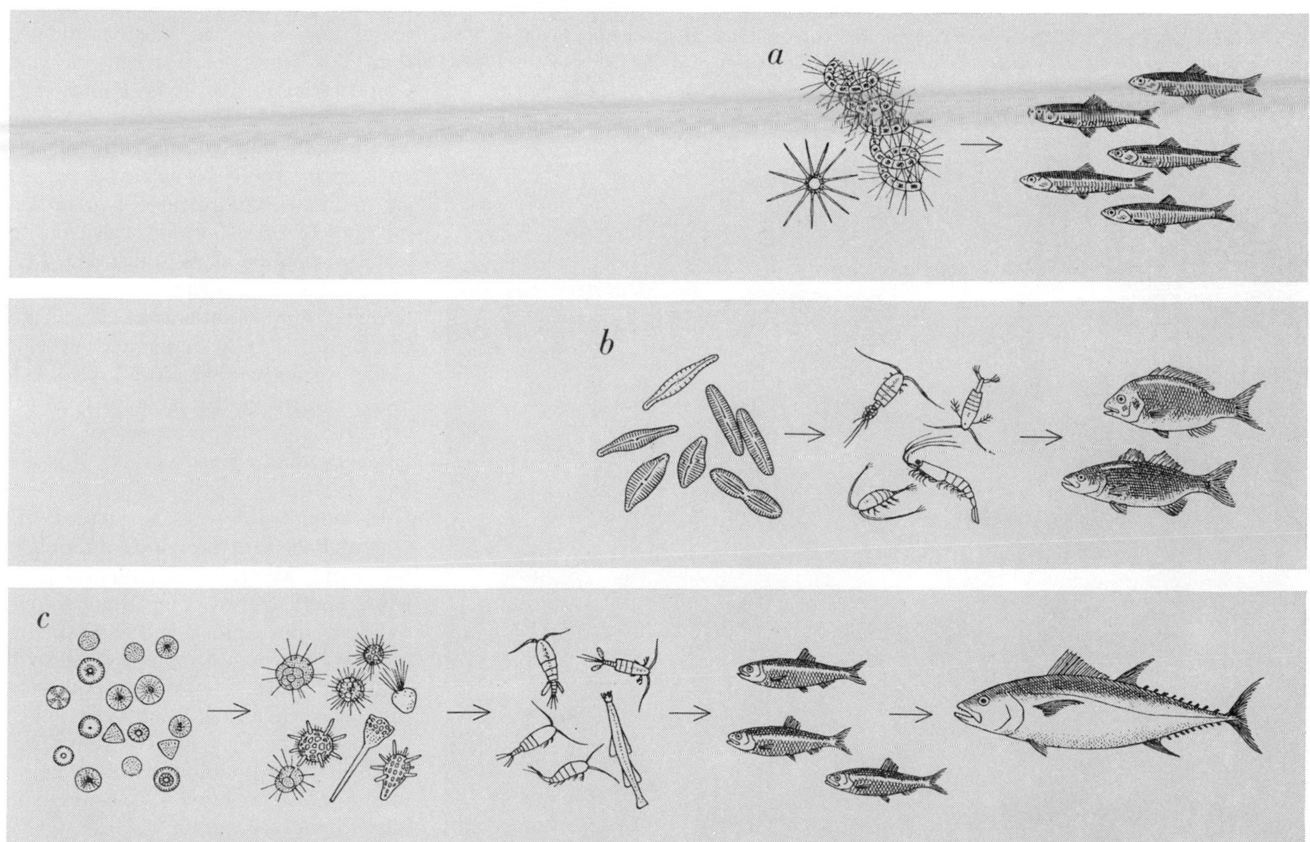

HIGH PRODUCTIVITY of upwelling areas and coastal waters, in contrast to the low productivity of the open sea, is not due only to greater mineral enrichment. In upwelling areas (*a*) the phytoplankton at the bottom of the food chain are usually aggregates of colonial diatoms that are large enough to feed fish of exploitable size. As a result the food chain is very short, with an average of 1.5 steps. The food chain in coastal water (*b*) is longer, averaging 3.5 steps. In the open sea (*c*), where phytoplankton at the bottom of the chain are widely scattered, single-celled diatoms, five steps are needed to produce exploitable fish and the energy transfer at each step is low in efficiency. The length of the chains was calculated by John H. Ryther of the Woods Hole Oceanographic Institution.

OYSTER FARM at Port Stephens in New South Wales in Australia is seen at low tide. The rack-and-stick cultivation system was adopted after depletion of oyster beds in the 1870's.

SUSPENSION CULTURE of young oysters dangles from a float in a Rhode Island oyster pond. Each string supports a series of scallop shells set with oyster spat, a product used commercially for seeding conventional oyster beds. Luther Blount, who is farming the oyster suspension cultures experimentally, has recorded four-year weight gains of 1,000 percent.

ing is remarkably productive. In the Philippine Republic, for example, the annual milkfish harvest is estimated at some 21,000 tons and the productivity of the ponds averages 78 tons per square mile. A comparable estimate for the annual productivity of free-swimming fish in coastal waters, as calculated by Ryther and Matthiessen, falls between six and 17 tons per square mile. In the Philippines, moreover, it is not customary to enrich the pond waters artificially, a process that accelerates the growth of the fishes' plant food. In Taiwan, where milkfish ponds are fertilized, the average annual yield is 520 tons per square mile, and in Indonesia, where sewage is diverted into the ponds in place of commercial fertilizer, the annual yield reaches 1,300 tons per square mile.

Fish farming in Asia is still a long way from reaching its maximum potential. The United Nations Food and Agriculture Organization has calculated that more than 140,000 square miles of land in southern and eastern Asia could be added to the area already devoted to milkfish husbandry. Even if this additional area were no more productive than the ponds of Taiwan, its yield would be more than today's total catch from all the world's oceans. Assuming an adequate supply of milkfish fry, such an increase could be achieved without any technological advance over present methods of pond farming. Even the fry problem may be close to solution. Mullet, a largely herbivorous fish, is now extensively farmed not only in Hawaii and China but also in India and even in Israel. Recently it has proved possible to breed mullet in the laboratory, which brings closer the prospect of mullet hatcheries and a steady supply of mullet fry.

In looking for ways to increase the potential yield of fishponds throughout the world, we are faced with two problems. The first is whether or not we can overcome the sanitary and aesthetic objections to using sewage as a growth stimulant. This is a complex question, but it is worth noting that some practical progress is being made by transferring shellfish from polluted areas to unpolluted ones for a period of "cleaning" before shipment to market.

An equally important question is to what degree commercial fertilizer could increase productivity. Oysters or mussels suspended from rafts in small ponds should provide a simple test organism for such experiments, and they are particularly appropriate because of their high natural yields.

The effect of adding commercial fertilizer to Long Island Sound water has been studied by Victor L. Loosanoff of the U.S. Fish and Wildlife Service. He wanted to produce large amounts of marine plants as food for experiments in rearing oysters and clams. He found marked stimulation of plant growth, but the zooplankton—the marine animals in the water—also grew and ate the plants, thus competing with the shellfish for food. After trying various methods of inhibiting the zooplankton's growth, Loosanoff finally came to the use of pure cultures of the plants, but this would be a very expensive practice on a commercial scale.

The growing of both marine plants and marine animals in a pond could be rewarding, and the zooplankton could be converted from a pest to an asset by adding an organism that feeds on them. Rainbow trout might fill this requirement: they are carnivorous, adapt readily to salt water and are said to grow faster and have a better flavor than when they live in fresh water. In addition they are readily available from hatcheries and have a good market value. To dispose of the inevitable organic debris sinking to the bottom of the pond one might add clams and a few lobsters, since both are in demand and their young are being reared in hatcheries and could be obtained.

It seems to me of the utmost importance that we follow the principles of ecology in our efforts to develop marine farming, by working with nature to establish balanced, stable communities rather than by supporting large single crops artificially, as we do on land, with what are now becoming recognized as disastrous side effects. Perhaps the single most exciting challenge we have in marine farming is this opportunity to make a new start in the production of food, utilizing the ecological knowledge now available.

If the results of the pond experiments are satisfactory, it is technically feasible to consider applying fertilizer to estuaries or even to the open ocean. The mechanical problem here is that the applied fertilizer sinks to the bottom in estuaries and tends to become absorbed by mud, and in the open ocean it simply sinks below the zone where the marine plant life grows. A solution for this problem would be to combine the fertilizer with some floating material that would disintegrate and liberate it slowly. The political and legal problems of controlling the harvest of the crop seem more difficult than the technical one of developing floating fertilizer.

Beyond the continental shelf in the open ocean the surface water is normally poor in nutrients and as barren as any desert on land. As irrigation projects have frequently shown, the addition of water makes the desert bloom. Adding nutrients to the ocean has much the same effect. There is, however, a significant difference between the two measures. The availability of fresh water may ultimately limit our agricultural output on land, but the deep ocean holds an immense supply of nutrients that is constantly being renewed.

The concentration of nitrogen and phosphorus compounds in the ocean reaches its maximum value at depths of from 2,000 to 3,000 feet below the surface. That is well below the region penetrated by sunlight, making the nutrients unavailable for plant growth. What then

ARTIFICIAL PONDS are widely used in Asia to raise fish fry, netted at sea, until they reach edible size. The ponds seen here are in Indonesia, where the use of sewage as fertilizer for pond algae brings a harvest that is equal to 1,300 tons of fish per square mile.

is the practical possibility of bringing these nutrients to the surface? If we were able to do it, would the number of fish increase? The answer to the second question can be found in nature. The upwelling of nutrient-rich deep water occurs naturally in some parts of the ocean, and the world's most productive fisheries are found in these areas. One of the best-known of these is the Peru Current on the west coast of South America. Along the shores of Chile and Peru the southeast trade winds blow the surface water away from the land, with the result that it is replaced by deeper water containing the nutrients needed for plant growth. In 1968, 10.5 million tons of fish—mostly anchovies—were harvested in an area 800 miles long and 30 miles wide along this coast. That is a yield of 440 tons per square mile of ocean surface. If, as seems likely, an equal quantity of fish was taken by predators, it means that this area of natural upwelling approaches the productivity of the heavily fertilized Asiatic fishponds. Incidentally, it far surpasses the production of protein by the raising of cattle on pastureland, which Ryther and Matthiessen give as between 1.5 and 80 tons per square mile.

The Peru Current demonstrates that bringing nutrient-rich water to the surface leads to an enormous increase in fish growth. In fact, the areas of natural upwelling, which comprise only .1 percent of the oceans' surface, supply almost half of the total fish catch, whereas the open oceans, where upwelling does not occur, account for 90 percent of the surface and yield only about 1 percent of the catch. In other words, natural upwelling increases the productivity of the open ocean almost 50,000-fold in terms of fish actually landed.

It would obviously be worthwhile to stimulate upwelling artificially, not only because of the probability of high fish yields but also because the stable ecological communities that inhabit the natural areas of ocean upwelling are models of efficient food production for man, with none of the drawbacks—such as herbicides, pesticides, pollution and excessive human intervention—that such highly productive systems usually entail ashore.

To achieve artificial upwelling we need first some kind of container. Deep water is cold and therefore dense, and without a container it would sink again, taking its nutrients with it. We also need to surround the fish we hope to grow, not only to protect them from predators and to simplify harvesting, but even more important to keep them from being caught by fishermen who have not paid for the upwelling. We also need a land area where the pumping and processing activities can be located and of course a supply of deep water nearby.

There are hundreds of coral atolls in the Pacific and the Indian Ocean that meet these specifications. Rings of coral reef surrounding shallow lagoons, atolls vary in area from less than a square mile to more than 800 square miles. Low islands are often found on the encircling reef, and since atolls are the remains of sunken volcanic peaks topped with coral they are steepsided, with deep water generally less than a mile from the reef itself. The trade winds blow over many of these islands and carry energy enough to bring deep water to the surface; that, after all, is the mechanism in many natural upwelling areas. There is even a built-in stirring system: the trade winds produce a downwind current on the lagoon surface that is matched by a return current near the bottom of the lagoon.

FRESHWATER FISH, the herbivorous carp, has been reared in ponds around the world for more than two millenniums. The carp in the photograph are from a pond in Burma. The 1968 crop of carp and carplike fishes in neighboring China totaled 1.5 million tons.

Given an atoll lagoon filled with nutrient-rich water from the deeps, what kinds of marine plants and animals should be grown in it? Perhaps the simplest procedure would be to leave the passages between lagoon and ocean open and allow nature to take its course in introducing new species. A balanced and stable community such as the one found in the Peru Current might establish itself. If it did not, colonies of plants and fishes from the Peru Current could be introduced in the hope that they could establish themselves in the new location.

Perhaps a crop of suitably large zooplankton such as krill—the shrimplike animals that are the principal food of the baleen whales in the Antarctic—could be raised in a fertilized lagoon. In that case another particularly interesting experiment might be possible. This would be to determine whether or not baleen whales, particularly the now almost vanished blue whales, could adapt to such a restricted environment. A school of blue whales raised in captivity could be regularly culled for a significant yield in meat and edible oil, and at the same time its existence would protect the species from what now seems to be certain extinction.

We know that blue whales migrate into the tropical Pacific to bear their young. The migrants could be followed by attaching radio transmitters to them in Antarctica. Techniques for capturing, transporting and keeping smaller whales have already been worked out. Humpback whales, which are about half the length of blue whales, have been captured at sea by investigators at the Sea Life Park in Hawaii by dropping a net over the whale's head. There is a real possibility that whale farms could be started by capturing pregnant female blue whales and confining them in fertilized atolls.

Artificial upwelling, on a small scale at least, has already been achieved by Oswald A. Roels, Robert D. Gerard and J. Lamar Worzel of the Lamont-Doherty Geological Observatory of Columbia University. They have installed on St. Croix in the Virgin Islands a 3½-inch plastic pipe that extends nearly a mile into the Caribbean, enabling them to pump deep water with a temperature of 40 degrees Fahrenheit into small ponds on shore. They find that selected plant life from the seawater off St. Croix grows 27 times faster in water from the pipe than in water from the surface. They are now exploring the possibilities of feeding a variety of marine herbivores on these artificial blooms.

The Lamont-Doherty group has also

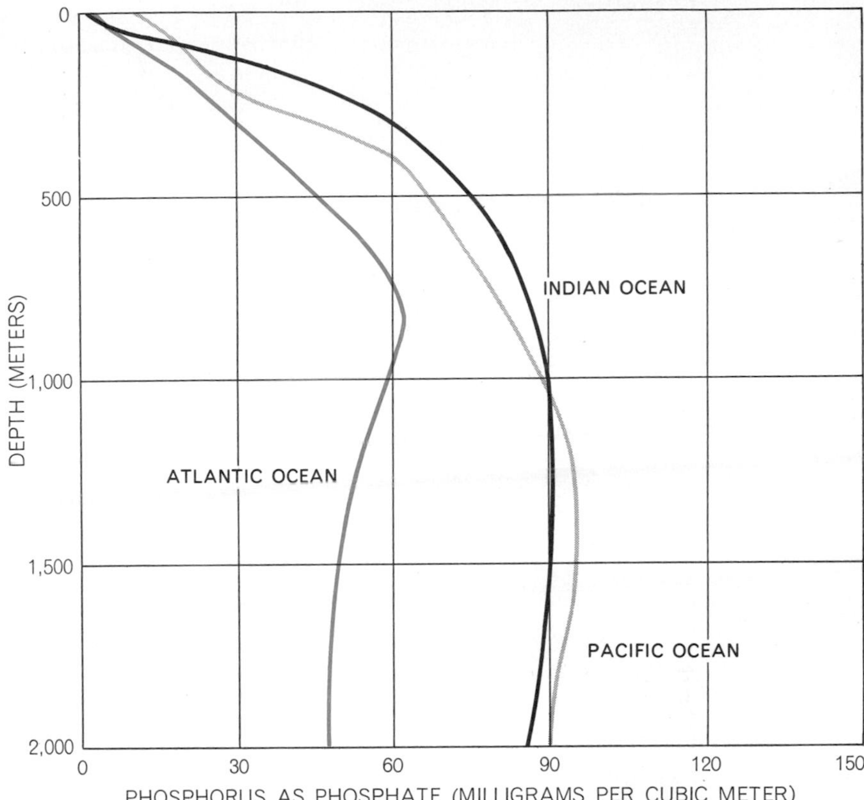

PHOSPHATE IN SEAWATER, present only in small amounts at the surface, increases in the deeper zones and reaches a near-maximum concentration of 90 milligrams per cubic meter in the Pacific and Indian Ocean and about 60 milligrams in the Atlantic at a depth of 1,000 meters. Data are from a study by Lela M. Jeffrey of the University of Nottingham.

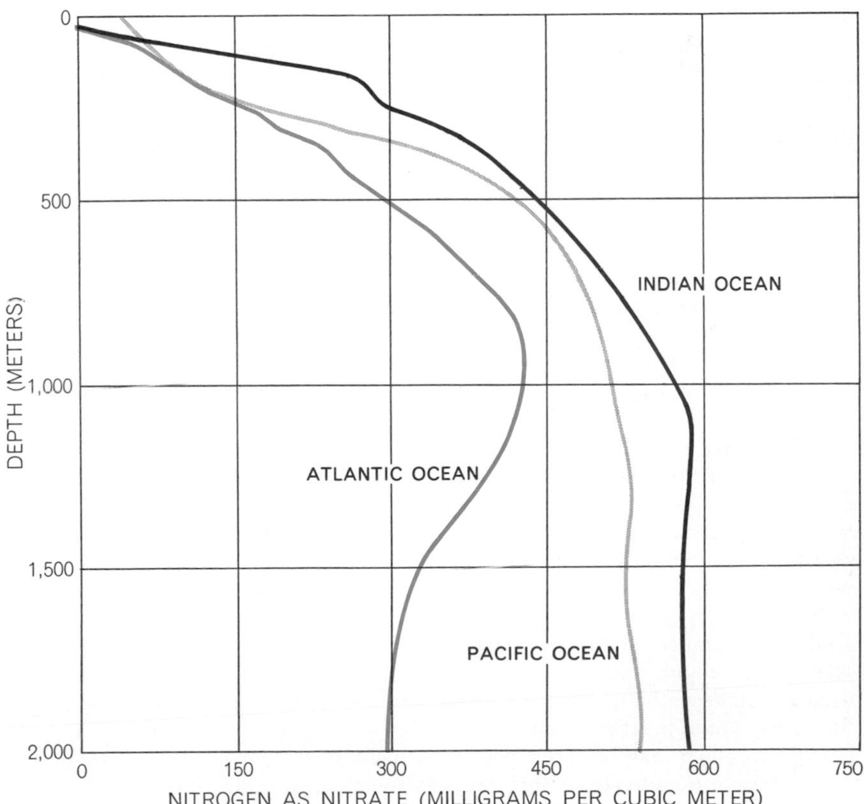

NITRATE IN SEAWATER is also scarce at the surface but approaches its maximum concentration at 1,000 meters. Again the Atlantic has the least; data are from Miss Jeffrey.

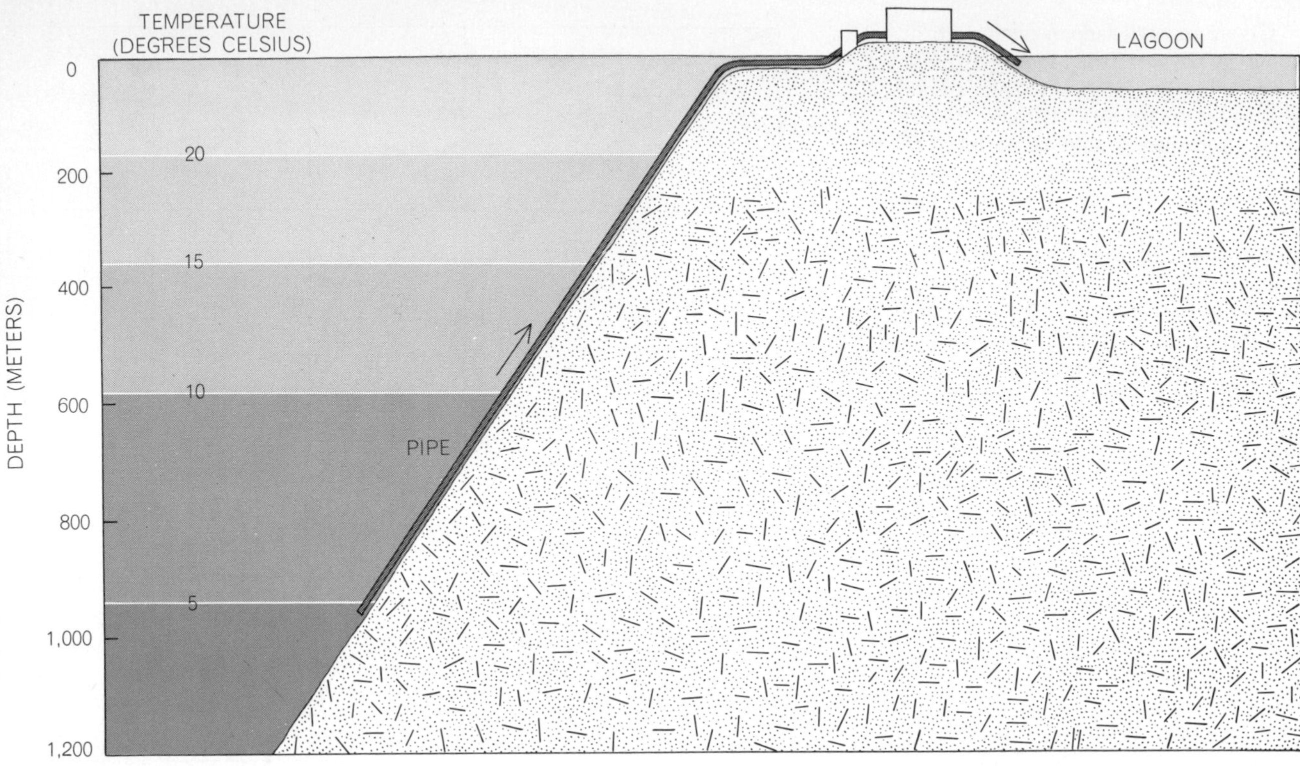

ARTIFICIAL UPWELLING of deep water might be contrived in an atoll setting, as this diagram suggests. The atoll's steep drop to seaward means that the wanted water would be pumped the least possible distance. The central lagoon would provide a catchment basin for the pumped water, retaining its nutrients at or near the surface. The difference in temperature between the surface and the deep water might be used to generate more power than is required for the pumps. A pilot version of this experiment is being conducted by workers in the Virgin Islands, who are pumping deep-sea water ashore and accelerating the growth of phytoplankton in ponds.

pointed out in a recent report that the energy represented by the nearly 40-degree difference in temperature between deep water and surface water can be used for air conditioning, the generation of power and the condensation of fresh water from the atmosphere. (The last idea emerged after observation of the condensation of atmospheric moisture on a Martini glass in a St. Croix bar.) In addition the low temperature of deep water offers the possibility of using the water to cool power plants, including nuclear reactors, without causing thermal pollution.

These fringe benefits, particularly the possibility of producing more than enough power to pump the deep water to the surface, may at first seem to suggest the dream of getting something for nothing. No physical laws would be violated, however; the water-temperature gradient is simply another product of solar-energy input, just as the energy fixed by photosynthetic plants is. From the standpoint of practical economics artificial upwelling may be too expensive to be feasible exclusively for fish farming at the present time. The system seems entirely practical, however, if its cost can be shared with some additional service such as air conditioning or the cooling of power plants.

A less elegant but much cheaper means of enriching the lagoons of atolls would be the addition of commercial fertilizer. To bring an atoll one square mile in area and 30 feet deep to a level of phosphate concentration equal to the level of nutrient-rich deep water would require only about 10 tons of fertilizer and might cost less than $500. In principle, if the lagoon were entirely enclosed, fertilizer would be removed only as the end product of the farming operation. In actual practice, of course, there would be other losses. Even assuming that one recovered only 10 percent of the fertilizer in marketable fish, however, the cost would be only half a cent per pound of fish produced. From the economic point of view this would seem to be a highly practical experiment.

Advances in technology frequently generate further threats to the quality of our already overburdened environment. It is encouraging to realize that the use of deep water from the sea both to stimulate food production and to obtain power or fresh water is a pollution-free process. The deep water returns to the sea at the same temperature and with about the same nutrient concentration as the waters that receive it, without having an adverse effect on either the atmosphere or the ocean. The same is true of the use of commercial fertilizer in atoll lagoons, since the fertilizer is almost wholly consumed in the process. Yet at the same time animal-protein production could be stimulated to a level not yet approached by conventional agriculture. Large areas of our planet could be developed into highly productive marine farms. The time seems ripe for applying the fundamental knowledge we already possess to the practical problems of developing them.

Plate Tectonics and the History of Life in the Oceans

by James W. Valentine and Eldridge M. Moores
April 1974

The breakup of the ancient supercontinent of Pangaea triggered a long-term evolutionary trend that has led to the unprecedented variety of the present biosphere

During the 1960's a conceptual revolution swept the earth sciences. The new world view fundamentally altered long established notions about the permanency of the continents and the ocean basins and provided fresh perceptions of the underlying causes and significance of many major features of the earth's mantle and crust. As a consequence of this revolution it is now generally accepted that the continents have greatly altered their geographic position, their pattern of dispersal and even their size and number. These processes of continental drift, fragmentation and assembly have been going on for at least 700 million years and perhaps for more than two billion years.

Changes of such magnitude in the relative configuration of the continents and the oceans must have had far-reaching effects on the environment, repatterning the world's climate and influencing the composition and distribution of life in the biosphere. These more or less continual changes in the environment must also have had profound effects on the course of evolution and accordingly on the history of life.

Natural selection, the chief mechanism by which evolution proceeds, is a very complex process. Although it is constrained by the machinery of inheritance, natural selection is chiefly an ecological process based on the relation between organisms and their environment. For any species certain heritable variations are favored because they are particularly well suited to survive and to reproduce in their prevailing environment. To answer the question of why any given group of organisms has evolved, then, one needs to understand two main factors. First, it is necessary to know what the ancestral organisms were that formed the "raw material" on which selection worked. And second, one must have some idea of the sequence of environmental conditions that led the ancestral stock to evolve along a particular pathway to a descendant group. Given these factors, one can then infer the organism-environment interactions that gave rise to the evolutionary events. The study of the relations between ancient organisms and their environment is called paleoecology.

The new ideas of continental drift that came into prominence in the 1960's revolve around the theory of plate tectonics. According to this theory, new sea floor and underlying mantle are currently being added to the crust of the earth at spreading centers under deep-sea ridges and in small ocean basins at rates of up to 10 centimeters per year. The sea floor spreads laterally away from these centers and eventually sinks into the earth's interior at subduction zones, which are marked by deep-sea trenches. Volcanoes are created by the consumption process and flank the trenches. The lithosphere, or rocky outer shell of the earth, therefore comprises several major plates that are generated at spreading centers and consumed at subduction zones. Most lithospheric plates bear one continent or more, which passively move with the plate on which they rest. Because the continents are too light to sink into the trenches they remain on the surface. Continents can fragment at new ridges, however, and hence oceans may appear across them. Conversely, continents can be welded together when they collide at the site of a trench. Thus continents may be assembled into supercontinents, fragmented into small continents and generally moved about the earth's surface as passive riders on plates. In tens or hundreds of millions of years entire oceans may be created or destroyed, and the number, size and dispersal pattern of continents may be vastly altered.

The record of such continental fragmentation and reassembly is evident as deformed regions in the earth's mountain belts, particularly those mountain belts that contain the rock formations known as ophiolites. These formations are characterized by a certain sequence of rocks consisting (from bottom to top) of ultramafic rock (a magnesium-rich rock composed mostly of olivine), gabbro (a coarse-grained basaltic rock), volcanic rocks and sedimentary rocks. The major ophiolite belts of the earth are believed to represent preserved fragments of vanished ocean basins [*see illustration on pages 240 and 241*]. The existence of such a belt within a continent (for example the Uralian belt in the U.S.S.R.) is evidence for the former presence there of an ocean basin separating two continental fragments that at some time in the past collided with each other and were welded into the single larger continent. The timing of such events as the opening of ocean basins, the dispersal of continents and the closing of oceans by continental collisions can accordingly be "read" from the geology of a given mountain system.

Of course, the biological environment is constantly being altered as well. For example, the changes in continental configuration will greatly affect the ocean currents, the temperature, the nature of seasonal fluctuations, the distribution of nutrients, the patterns of productivity and many other factors of fundamental importance to living organisms. Therefore evolutionary trends in marine animals must have varied through geologic time in response to the major environmental changes, as natural selection acted to adapt organisms to the new conditions.

It should in principle be possible to detect these changes in the fossil record. Indeed, paleontologists have long recognized that vast changes in the composition, distribution and diversity of marine life are well documented by the fossil record. Now for the first time, however, it is possible to reconstruct the sequence of environmental changes based on the theory of plate tectonics, to determine their environmental consequences and to attempt to correlate them with the sequence of faunal changes that is seen in the fossil record. Such a thorough reconstruction ultimately may explain many of the enigmatic faunal changes known for many years. Even at this early stage paleontologists have succeeded in shedding much new light on a number of major extinctions and diversifications of the past.

As a first step toward understanding the relation between plate tectonics and the history of life it is helpful to investigate the relations that exist today between marine life, the present pattern of continental drift and plate-tectonic theory. The vast majority of marine species (about 90 percent) live on the continental shelves or on shallow-water portions of islands or subsurface "rises" at depths of less than about 200 meters

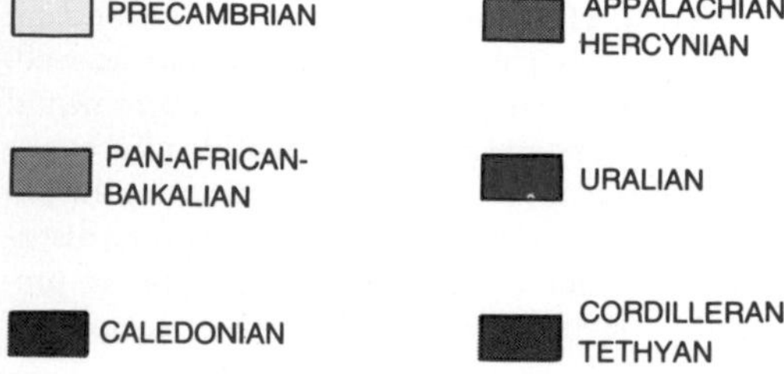

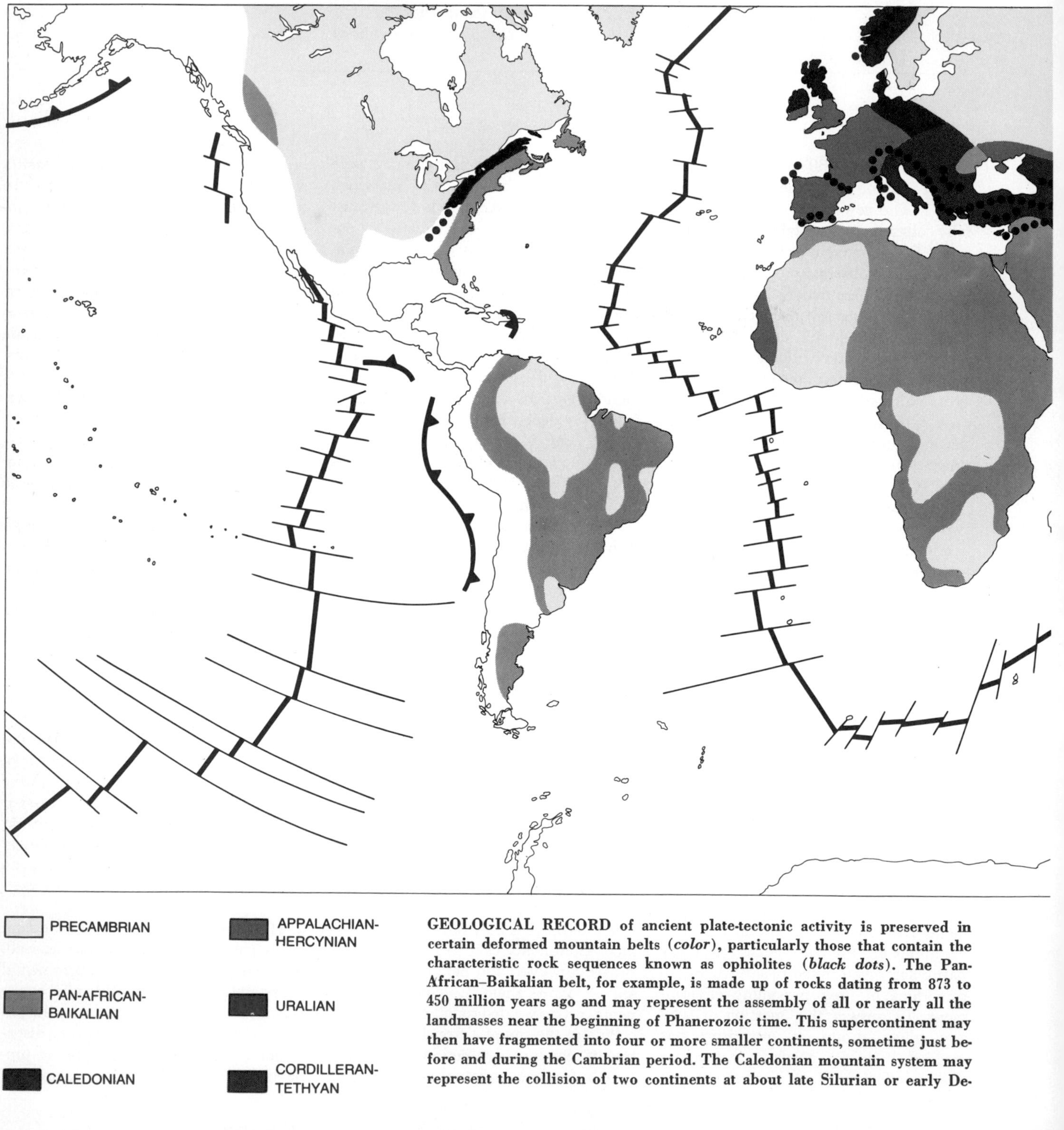

GEOLOGICAL RECORD of ancient plate-tectonic activity is preserved in certain deformed mountain belts (*color*), particularly those that contain the characteristic rock sequences known as ophiolites (*black dots*). The Pan-African–Baikalian belt, for example, is made up of rocks dating from 873 to 450 million years ago and may represent the assembly of all or nearly all the landmasses near the beginning of Phanerozoic time. This supercontinent may then have fragmented into four or more smaller continents, sometime just before and during the Cambrian period. The Caledonian mountain system may represent the collision of two continents at about late Silurian or early De-

(660 feet); most of the fossil record also consists of these faunas. Therefore it is the pattern of shallow-water sea-floor animal life that is of particular interest here.

The richest shallow-water faunas are found today at low latitudes in the Tropics, where communities are packed with vast numbers of highly specialized species. Proceeding to higher latitudes, diversity gradually falls; in the Arctic or Antarctic regions less than a tenth as many animals are living as in the Tropics, when comparable regions are considered [*see illustration on pages 242 and 243*]. The diversity gradient correlates well with a gradient in the stability of food supplies; as the seasons become more pronounced, fluctuations in primary productivity become greater. Although this strong latitudinal gradient dominates the earth's overall diversity pattern, there are important longitudinal diversity trends as well. In regions of similar latitude, for example, diversity is lower where there are sharp seasonal changes (such as variations in the surface-current pattern or in the upwelling of cold water) that affect the nutrient supply by causing large fluctuations in productivity.

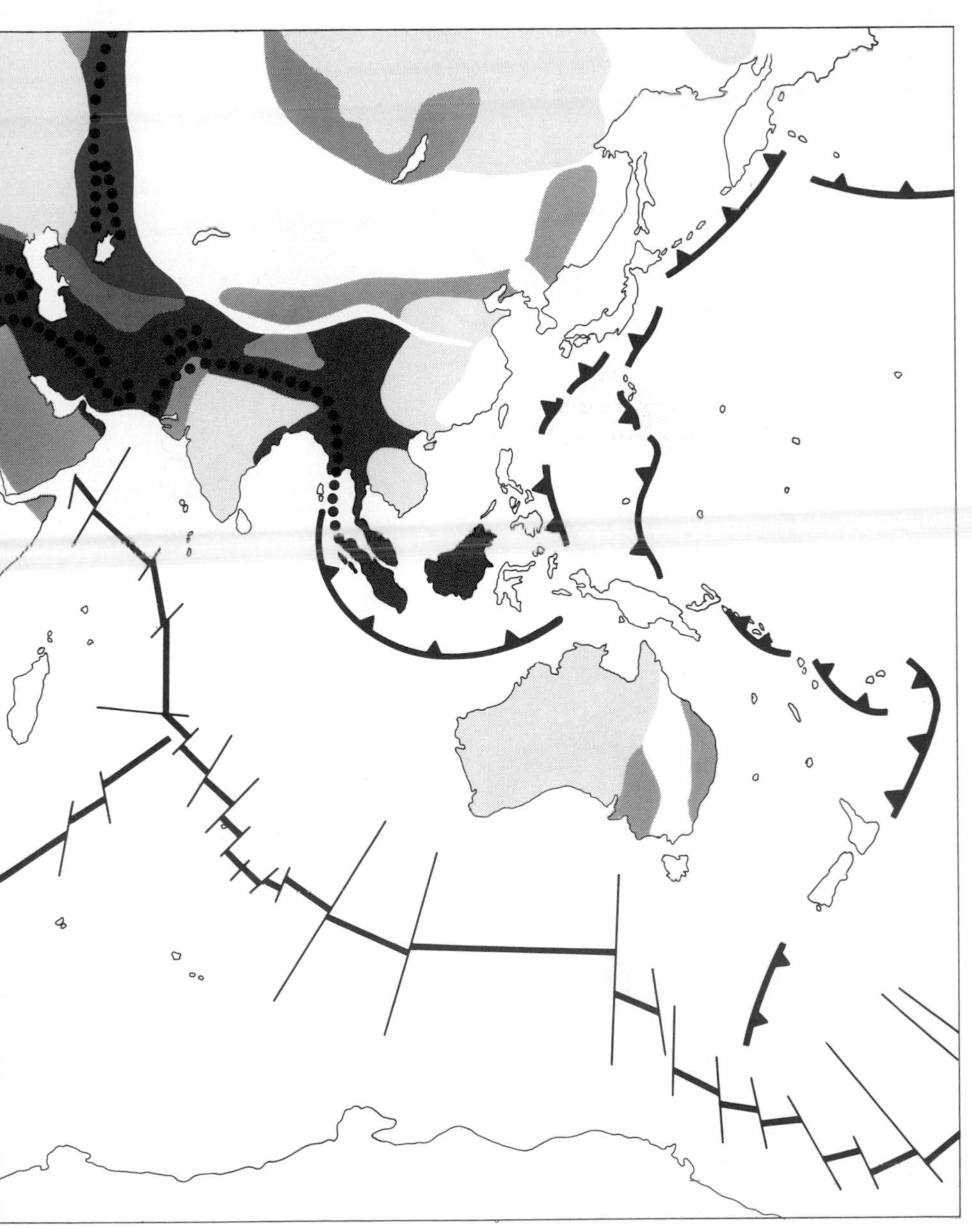

vonian time (approximately 400 million years ago). The Appalachian-Hercynian system may represent a two-continent collision during the late Carboniferous period (300 million years ago). The Uralian mountains may represent a similar collision at about Permo-Triassic time (220 million years ago). The Cordilleran-Tethyan system represents regions of Mesozoic mountain-building and includes the continental collisions that resulted in the Alpine-Himalayan mountain system. The ophiolite belts shown are the preserved remnants of ocean floor exposed in the mountain systems in question. Spreading ridges such as the Mid-Atlantic Ridge are indicated by heavy lines cut by lighter lines, which correspond to transform faults. Subduction zones are marked by heavy black curved lines with triangles.

At any given latitude, therefore, diversity is highest off the shores of small islands or small continents in large oceans, where fluctuations in nutrient supplies are least affected by the seasonal effects of landmasses, whereas diversity is lowest off large continents, particularly when they face small oceans, where shallow-water seasonal variations are greatest. In short, whereas latitudinal diversity increases generally from high latitudes to low, longitudinal diversity increases generally with distance from large continental landmasses. In both of these trends the increase in diversity is correlated with increasing stability of food resources. The resource-stability pattern depends largely on the shape of the continents and should also be sensitive to the extent of inland seas and to the presence of coastal mountains. Seas lying on continental platforms are particularly important: not only do extensive shallow seas provide much habitat area for shallow-water faunas but also such seas tend to damp seasonal climatic changes and to have an ameliorating influence on the local environment.

Today shallow marine faunas are highly provincial, that is, the species living in different oceans or on opposite sides of the same ocean tend to be quite different. Even along continuous coastlines there are major changes in species composition from place to place that generally correspond to climatic changes. The deep-sea floor, generated at oceanic ridges, forms a significant barrier to the dispersal of shallow-water organisms, and latitudinal climatic changes clearly form other barriers. The present dominantly north-south series of ridges forms a pattern of longitudinally alternating oceans and continents, thereby creating a series of barriers to shallow-water marine organisms. The steep latitudinal climatic gradient, on the other hand, creates chains of provinces along north-south coastlines. As a result the marine faunas today are partitioned into more than 30 provinces, among which there is in general only a low percentage of common species [*see illustration on pages 244 and 245*]. It is estimated that the shallow-water marine fauna represents more than 10 times as many species today as would be present in a world with only a single province, even a highly diverse one.

The volcanic arcs that appear over subduction zones form fairly continuous island chains and provide excellent dispersal routes. When long island chains are arranged in an east-west pattern so as to lie within the same climatic zone, they are inhabited by wide-ranging faunas that are highly diverse for their latitude. Indeed, the widest ranging marine province, and also by far the most diverse, is the Indo-Pacific province, which is based on island arcs in its central regions. The faunal life of this province spills from these arcs onto tropical continental shelves in the west (India and East Africa) and also onto tropical intraplate volcanoes (the Polynesian and Micronesian islands) that are reasonably close to them. This vast tropical biota is cut off from the western American mainland by the East Pacific Barrier, a zoogeographic obstruction formed by a spreading ridge.

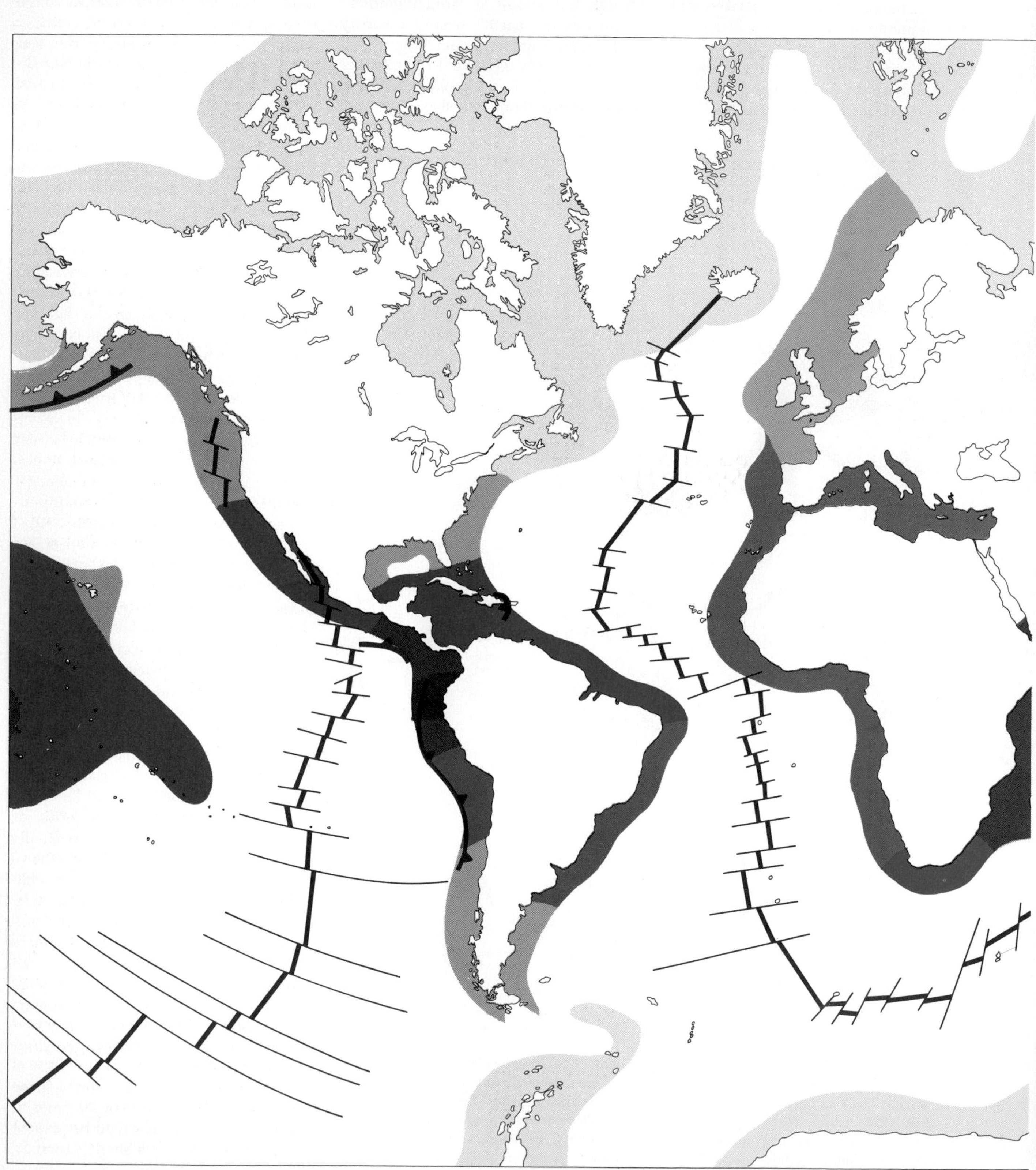

RELATIVE DIVERSITY of shallow-water, bottom-dwelling species in the present oceans is suggested by the colored patterns in this world map. The diversity classes are not based on absolute counts but are inferred from the diversity patterns of the best-

Since current patterns of marine provinciality and diversity fit closely with the present oceanic and continental geography and the resulting environmental patterns, one would expect ancient provinces and ancient diversity patterns also to fit past geographies. One of the best-established of ancient geographies is the one that existed near the beginning of the Triassic period, about 225 million years ago. The continents were then assembled into a single supercontinent named Pangaea, which must have had a continuous shallow-water margin running all the way around it, with no major physical barriers to the dispersal of shallow-water marine animals [*see illustration on page 246*]. Therefore provinciality must have been low compared with today, and it must have been attributable entirely to climatic effects. It is likely that the marine climate was quite mild and that even in high latitudes water temperatures were much warmer than they are today. As a result climatic provinciality must have been greatly reduced also. Furthermore, the seas at that time were largely confined to the ocean basins and did not extend significantly over the continental shelves. Thus the habitat area for shallow-water marine organisms was greatly reduced, first by the diminution of coastline that accompanies the creation of a supercontinent from smaller continents, and second by the general withdrawal of seas from continental platforms. The reduced habitat area would make for low species diversity. Finally, the extreme emergence of such a supercontinent would provide unstable nearshore conditions, with the result that food resources would have been very unstable compared with those of today. All these factors tend to reduce species diversity; hence one would expect to find that Triassic biotas were widespread and were made up of comparatively few species. That is precisely what the fossil record indicates.

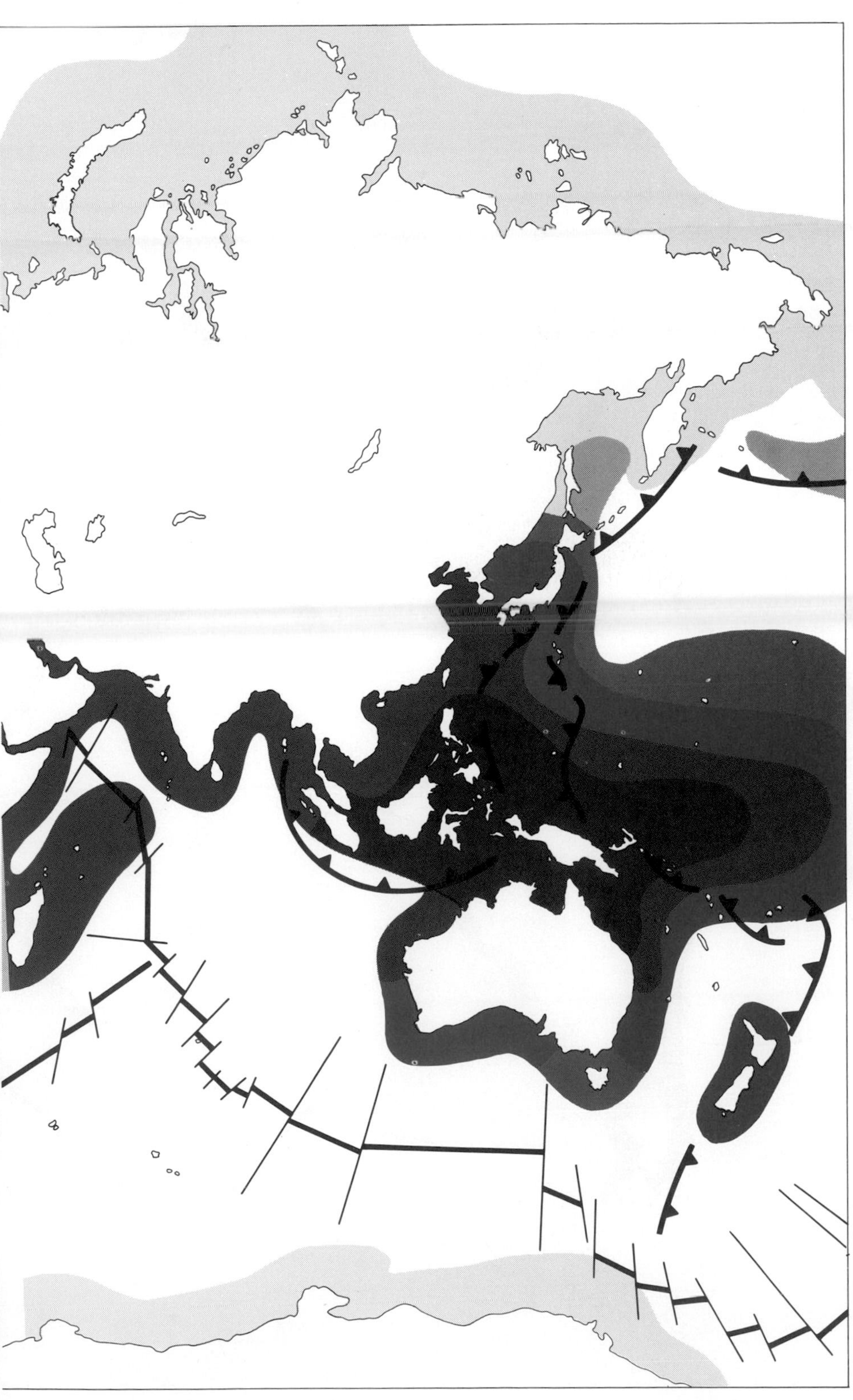

known skeletonized groups, chiefly the bivalves, gastropods, echinoids and corals. The highest class (*darkest color*) is about 20 times as diverse as the lowest (*lightest color*).

Prior to the Triassic period, during the late Paleozoic, diversity appears to have been much higher [*see top illustration on page 248*]. It was sharply reduced again near the close of the Permian period during a vast wave of extinction that on balance is the most severe known to have been suffered by the marine fauna. The late Paleozoic species that were the more elaborately adapted specialists became virtually extinct, whereas the surviving descendants tended to have simple skeletons. A high proportion of these survivors appear to have been detritus feeders or suspension feeders that harvested the water layers just above the sea floor. These successful types seem to be ecologically similar to the populations found today in unstable environments, for instance in high latitudes; the unsuccessful specialists, on the other hand, seem ecologically similar to the populations found in stable environments, for instance in the Tropics. Thus the extinctions appear to have been caused by the reduced potential for diversity of the shallow seas, a trend associated with less provinciality, less habitat area and less stable environmental conditions.

In the period following the great extinction, as Pangaea broke up and the resulting continents themselves gradually fragmented and migrated to their present positions, provinciality increased, communities in stabilized regions became filled with numerous specialized animals and the overall diversity of species in the world ocean rose to unprecedented heights, even though occasional waves of extinctions interrupted this long-range trend.

There is another time in the past besides the early Triassic period when low provinciality and low diversity were coupled with the presence of a high proportion of detritus feeders and near-bottom suspension feeders. That is in the late Precambrian and Cambrian periods, when a widespread, soft-bodied fauna of low diversity gave way to a slightly provincialized, skeletonized fauna of some-

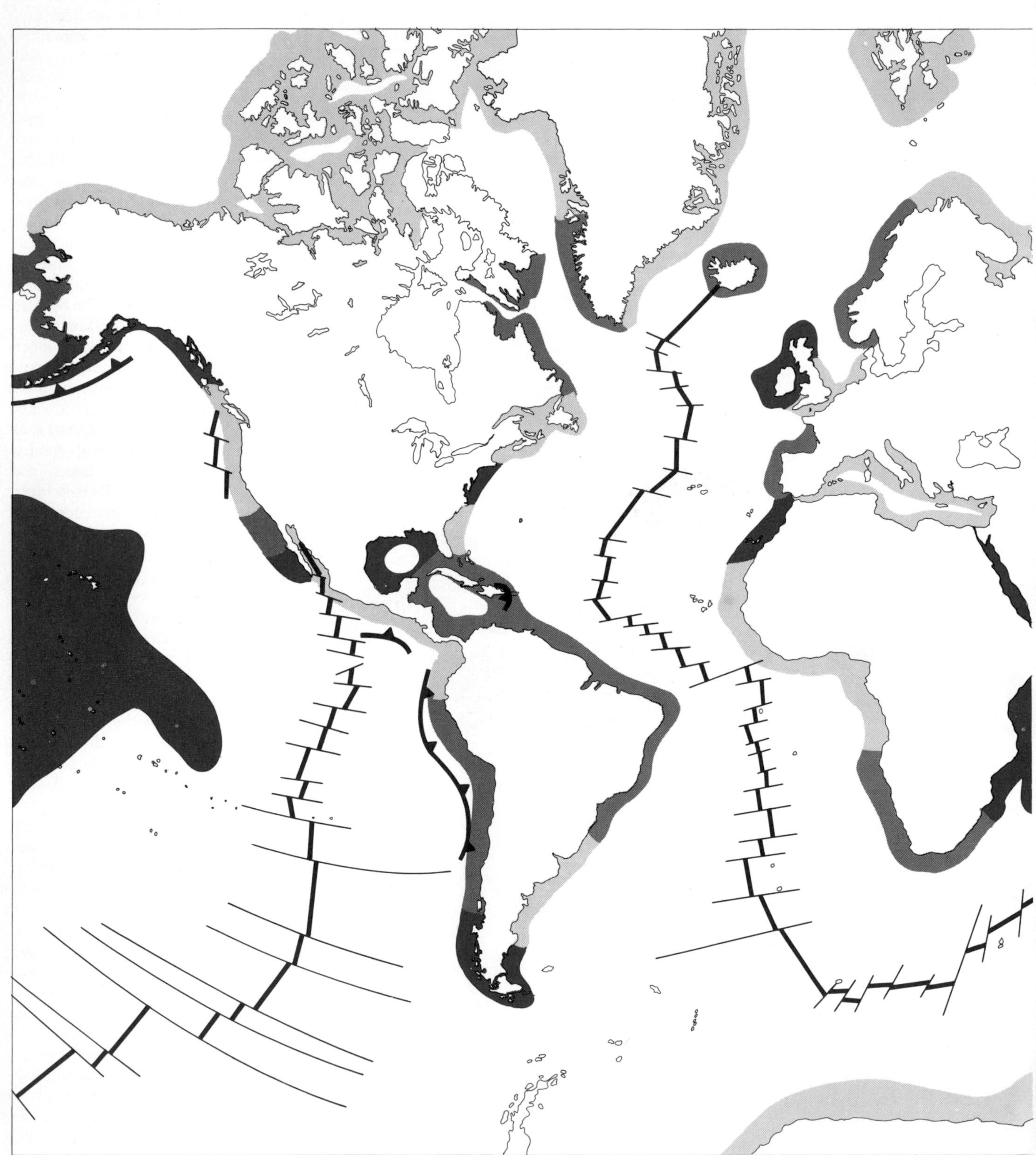

PRINCIPAL SHALLOW-WATER MARINE PROVINCES at present are indicated by the colored areas. The dominant north-south chains of provinces along the continental coastlines are created by the present high latitudinal gradient in ocean temperature

what higher diversity. It seems likely that the late Precambrian environment was quite unstable and that there may well have been a supercontinent in existence, or at least that the continents then were collected into a more compact assemblage than at present. In the late Precambrian period one finds the first unequivocal records of invertebrate life, including burrowing forms that were probably coelomic, or hollow-bodied, worms. In the Cambrian four continents may have existed although they were not arranged in the present pattern. During the Cambrian a skeletonized fauna appears that is at first almost entirely surface-dwelling and that includes chiefly detritus-feeding and suspension-feeding forms, probably with some browsers.

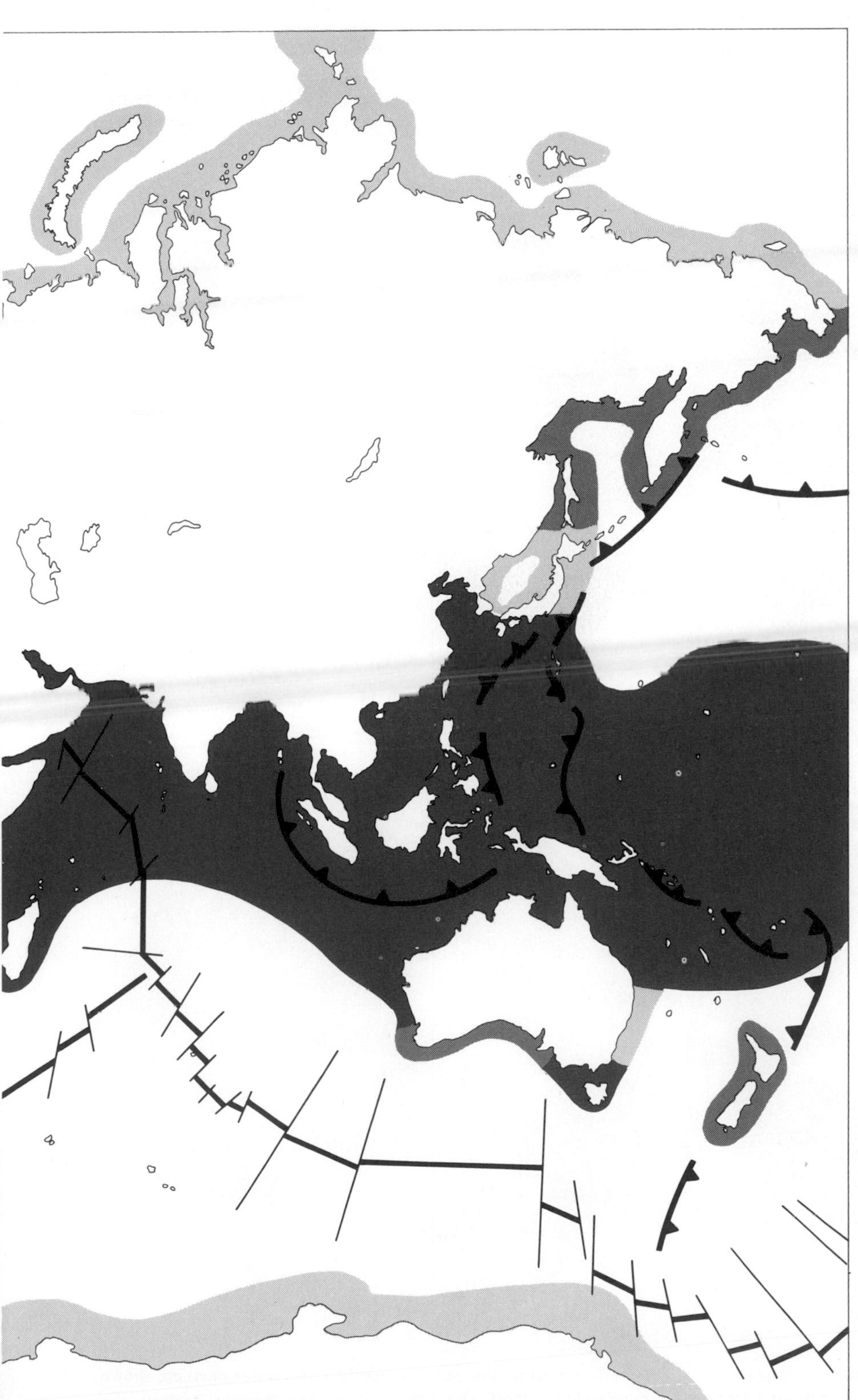

and by the undersea barriers formed by spreading ridges. The vast Indo-Pacific province (*darkest color*) spills out onto scattered islands as indicated. There are 31 provinces shown.

It seems possible, therefore, that the late Precambrian species were adapted to highly unstable conditions and became diversified chiefly as a bottom-living, detritus-feeding assemblage. The coelomic body cavity, evidently a primitive adaptation for burrowing, was developed and diversified into a variety of forms, perhaps as many as five basic ones: highly segmented worms that lived under the ocean floor and were detritus feeders; slightly segmented worms that lived attached to the ocean floor and were suspension feeders; slightly segmented worms that lived attached to the ocean floor and were detritus feeders; "pseudosegmented" worms that lived on the ocean floor and were detritus feeders or browsers, and nonsegmented worms that lived under the ocean floor and fed by means of an "introvert." In addition to these coelomates there were a number of coelenterate stocks (such as corals, sea anemones and jellyfishes) and probably also flatworms and other noncoelomate worms.

From the chiefly wormlike coelomate stocks higher forms of animal life have originated; many of them appear in the Cambrian period, when they evidently first became organized into the groups that characterize them today. Animals with skeletons appeared in the fossil record at that time. Presumably the invasion of the sea-floor surface by coelomates and the origin of numerous skeletonized species accompanied a general amelioration of environmental conditions as the continents became dispersed; the skeletons themselves can be viewed as adaptations required for worms to lead various modes of life on the surface of the sea floor rather than under it. The sudden appearance of skeletons in the fossil record therefore is associated with a generalized elaboration of the bottom-dwelling members of the marine ecosystem. Later, free-swimming and underground lineages developed from the skeletonized ocean-floor dwellers, with the result that skeletons became general in all marine environments.

The correlation of major events in the history of life with major environmental changes inferred from plate-tectonic processes is certainly striking. Even though details of the interpretation are still provisional, it seems certain that further work on this relation will prove fruitful. Indeed, the ability of geologists to determine past continental geographies should provide the basis for reconstructing the historical sequence of global environmental conditions for the first time. That sequence can then be compared with the sequence of organisms revealed in the fossil record. The following tentative account of such a comparison, on the broadest scale and without detail, will indicate the kind of history that is emerging; it is based on the examples reviewed above and on similar considerations.

Before about 700 million years ago bottom-dwelling, multicellular animals had developed that somewhat resembled flatworms. As yet no fossil evidence for their evolutionary pathways exists, but evidence from embryology and comparative anatomy suggests that they arose from swimming forms, possibly larval jellyfish, which in turn evolved from primitive single-celled animals.

Approximately 700 million years ago, perhaps in response to the onset of fluctuating environmental conditions brought about by continental clustering, a true coelomic body cavity was evolved to act as a hydrostatic skeleton in roundworms; this adaptation allowed burrowing in soft sea floors and led to the diversification of a host of worm architectures as that mode of life was explored. Burrows of this type are still preserved in some late Precambrian rocks. As the environment later became more stable, several of the worm lineages evolved more varied modes of life. The changes in body plan necessary to adapt to such

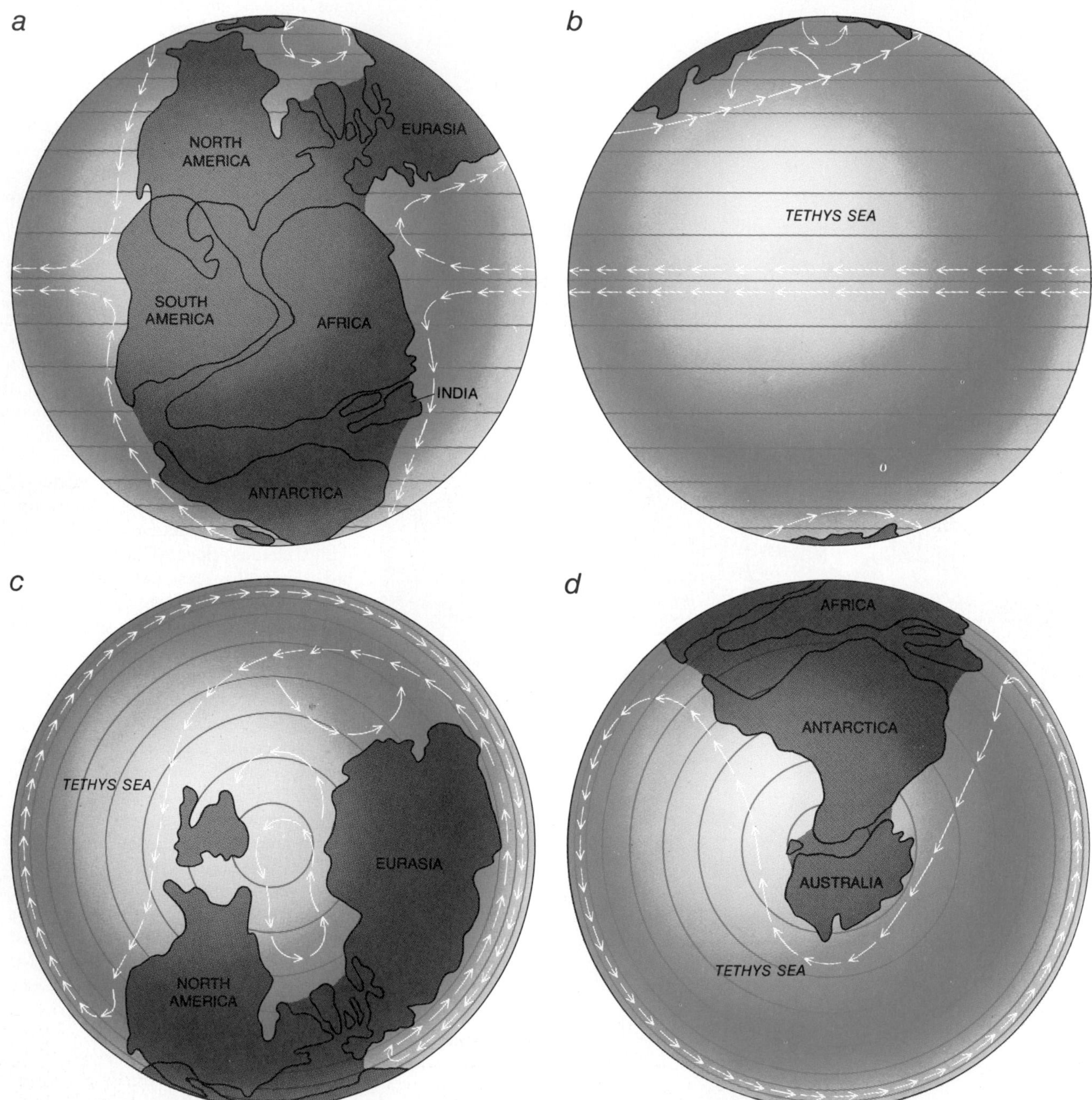

ANCIENT OCEAN CURRENTS in the vicinity of Pangaea, the single "supercontinent" that is believed to have existed near the beginning of the Triassic period some 225 million years ago, are indicated here in two equatorial views (*a, b*) and two polar views (*c, d*). Owing to a combination of geographic and environmental factors, including the predominantly warm-water currents shown, one would expect the continuous shallow-water margin that surrounded Pangaea to have been populated by comparatively few but widespread species. Such low species diversity combined with low provinciality is precisely what the fossil record indicates.

a life commonly involved the development of a skeleton. There were evidently three or four main types of worms that are represented by skeletonized descendants today. One type was highly segmented like earthworms, and presumably burrowed incessantly for detrital food; these were represented in the Cambrian period by the trilobites and related species. A second type was segmented into two or three coelomic compartments and burrowed weakly for domicile, afterward filtering suspended food from the seawater just above the ocean floor; these evolved into such forms as brachiopods and bryozoans. A third type consisted of long-bodied creepers with a series of internal organs but without true segmentation; from these the classes of mollusks (such as snails, clams and cephalopods) have descended. Probably a fourth type consisted of unsegmented burrowers that fed on surface detritus and gave rise to the modern sipunculid worms. These may also have given rise to the echinoderms (which include the sea cucumber and the spiny sea urchin), and eventually to the chordates and to man. Although the lines of descent are still uncertain among these primitive and poorly known groups, the adaptive steps are becoming clearer.

The major Cambrian radiation of the underground species into sea-floor surface habitats established the basic evolutionary lineages and occupied the major marine environments. Further evolutionary episodes tended to modify these basic animals into more elaborate structures. After the Cambrian period shallow-water marine animals became more highly specialized and richer in species, suggesting a continued trend toward resource stabilization. Suspension feeders proliferated and exploited higher parts of the water column, and predators also became more diversified. This trend seems to have reached a peak (or perhaps a plateau) in the Devonian period, some 375 million years ago. The characteristic Paleozoic fauna was finally swept away during the reduction in diversity that accompanied the great Permian-Triassic extinctions. Thus the rise of the Paleozoic fauna accompanied an amelioration in environmental conditions and increased provinciality, whereas the decline of the fauna accompanied a reestablishment of severe, unstable conditions and decreased provinciality. The subsequent breakup and dispersal of the continents has led to the present biosphere.

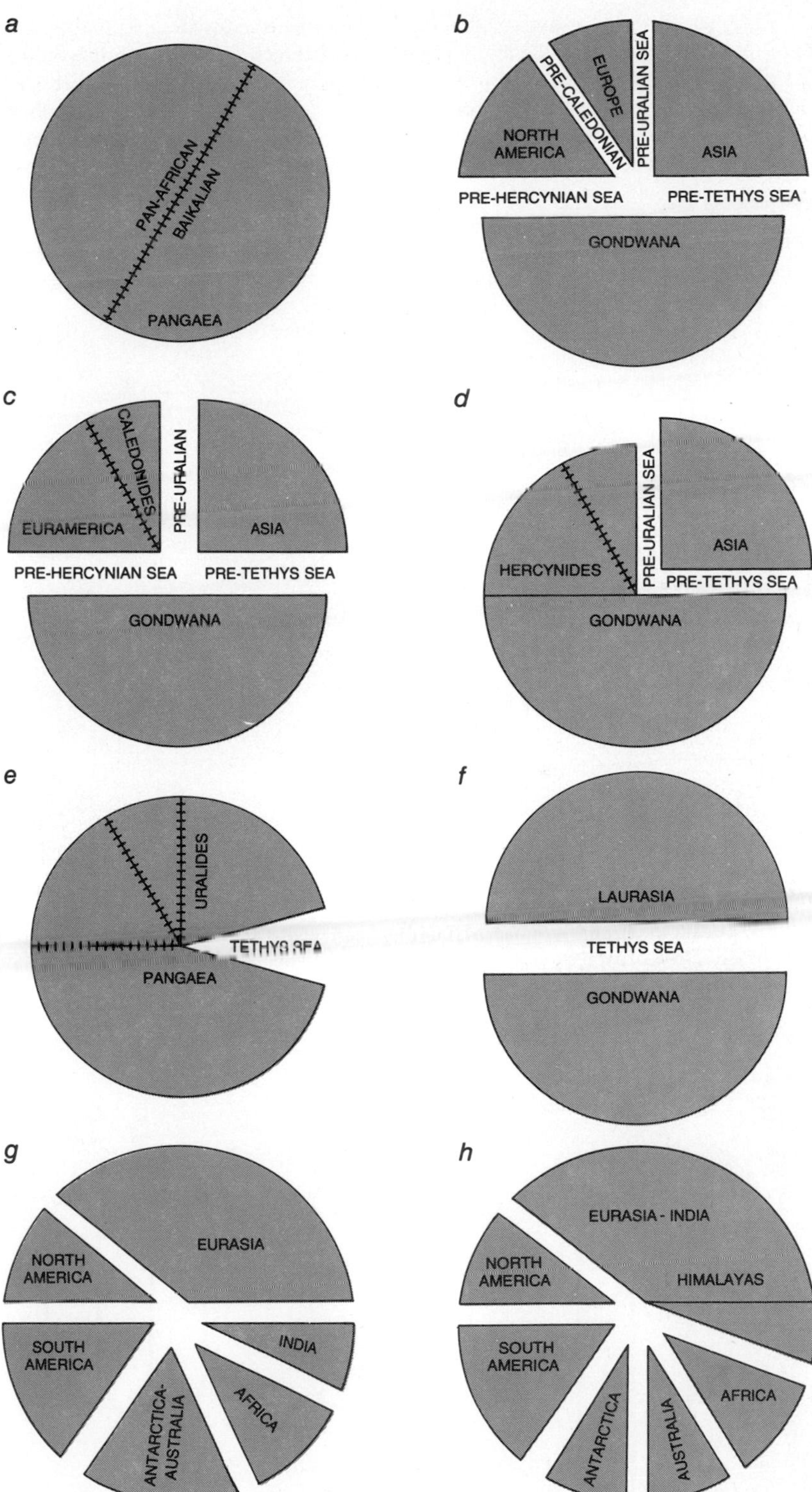

SIMPLIFIED DIAGRAMS are employed to suggest the relative configuration of the continents and the oceans during the past 700 million years. The late Precambrian supercontinent (*a*), which probably existed some 700 million years ago, may have been formed from previously separate continents. The Cambrian world (*b*) of about 570 million years ago consisted of four continents. The Devonian period (*c*) of about 390 million years ago was distinguished by three continents following the collapse of the pre-Caledonian Ocean and the collision of ancient Europe and North America. In the late Carboniferous period (*d*), about 300 million years ago, Euramerica became welded to Gondwana along the Hercynian belt. In the late Permian period (*e*), about 225 million years ago, Asia was welded to the remaining continents along the Uralian belt to form Pangaea. In early Mesozoic time (*f*), about 190 million years ago, Laurasia and Gondwana were more or less separate. In the late Cretaceous period (*g*), about 70 million years ago, Gondwana was highly fragmented and Laurasia partially so. The present continental pattern (*h*) shows India welded to Eurasia.

Today we live in a highly diverse world, probably harboring as many species as have ever lived at any time, associated in a rich variety of communities and a large number of provinces, probably the richest and largest ever to have existed at one time. We have been furnished with an enviably diverse and interesting biosphere; it would be a tragedy if we were to so perturb the environment as to return the biosphere to a low-diversity state, with the concomitant extinction of vast arrays of species. Of course, natural processes might eventually recoup the lost diversity, if we waited patiently for perhaps a few tens of millions of years. Alternatively we can work to preserve the environment in its present state and therefore to preserve the richness and variety of nature.

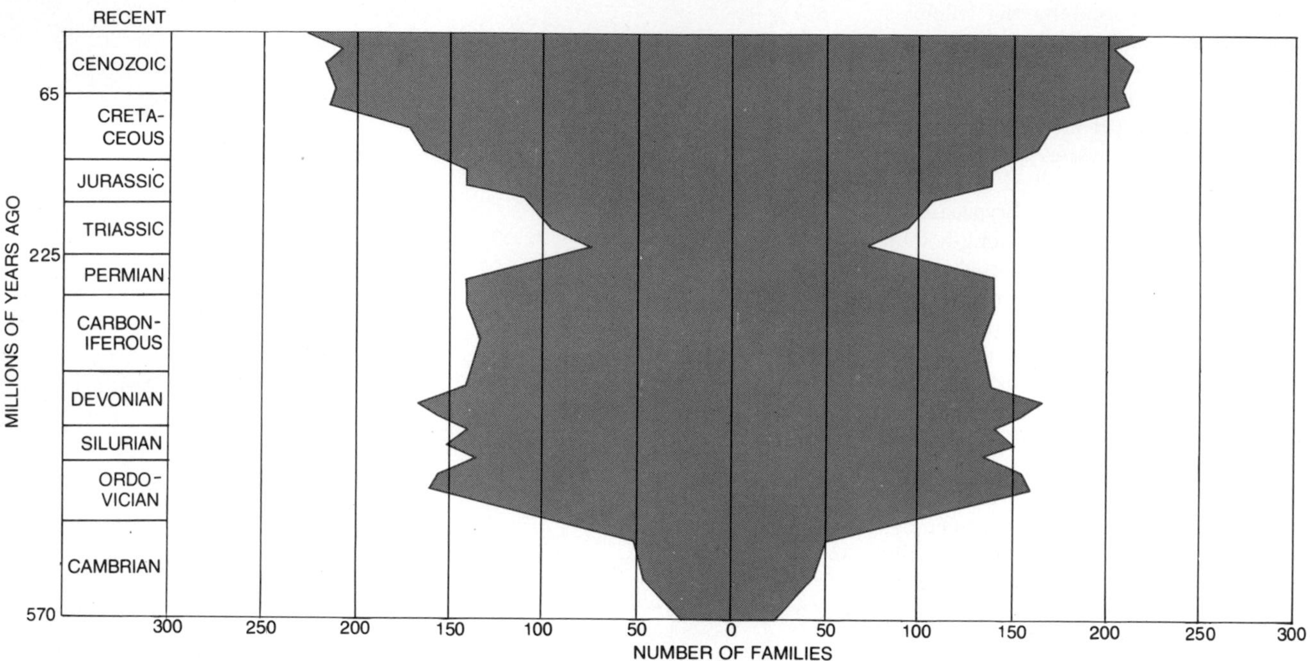

FLUCTUATIONS in the number of families, and hence in the level of diversity, of well-skeletonized invertebrates living on the world's continental shelves during the past 570 million years are plotted by geologic epoch in this graph. Time proceeds upward.

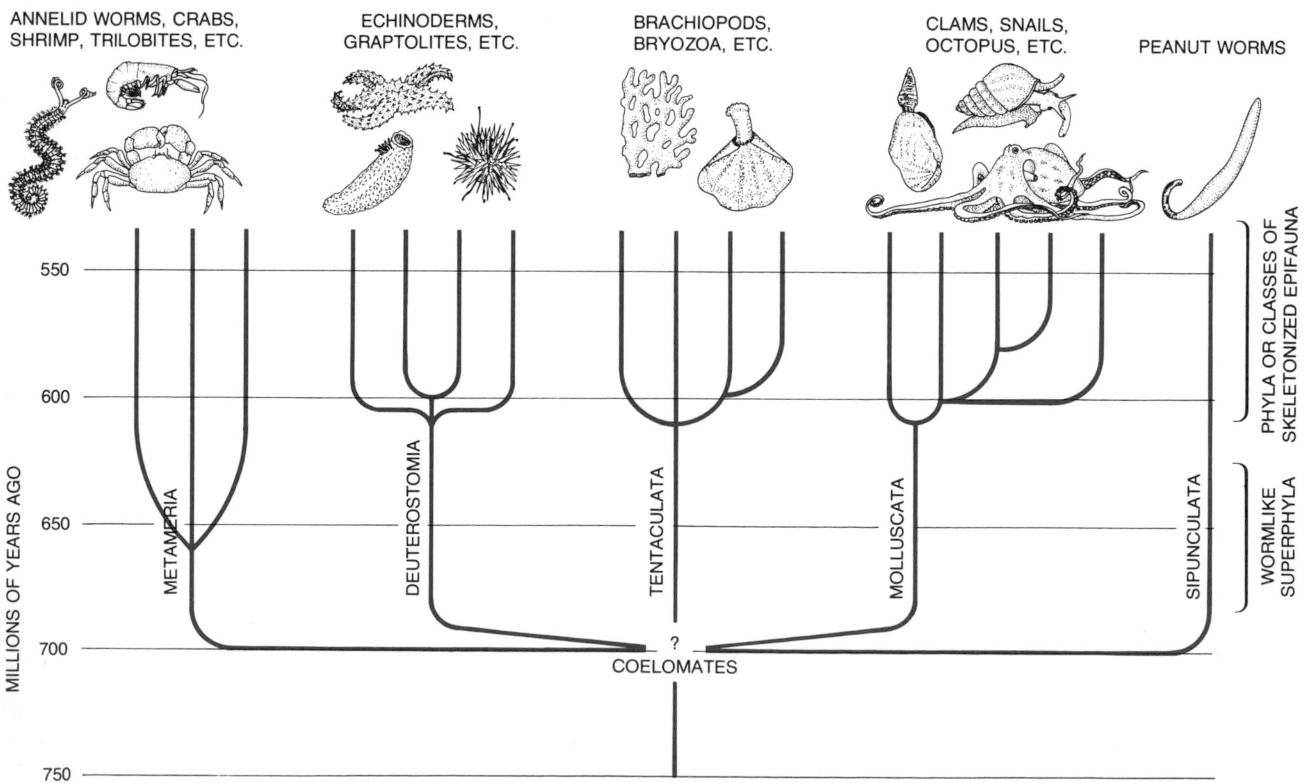

PHYLOGENETIC MODEL of the evolution of coelomate, or hollow-bodied, marine organisms is based on inferred adaptive pathways. The late Precambrian lineages were chiefly worms, which gave rise to epifaunal (bottom-dwelling), skeletonized phyla during the Cambrian period. The organisms depicted in the drawings at top are modern descendants of the major Cambrian lineages.

V

PHYSICAL MARINE RESOURCES

PHYSICAL MARINE RESOURCES

V

INTRODUCTION

The sea contains a wealth of fundamental physical resources—materials, energy, and space—which has been used by mankind since time immemorial. Unlike the land, which could be owned and fought over, the sea had the marvelous property that it belonged to everyone, and whatever resources it contained were abundant enough for everyone. The seas were open to all commerce, the grains of beach sand were proverbially countless, the supply of salt was limitless, and the ability of the sea to absorb waste was infinite.

Times have changed. There are many more of us and we each consume, at least on average, much more energy and material, and produce ever more waste. Moreover, humans are increasingly urban and their wastes are correspondingly concentrated in the very waters where they want to swim and fish. Likewise, new resources, such as oil, are not distributed uniformly everywhere, new ships are too big to go everywhere, and long-lived poisons, such as DDT, permeate the surface waters of the world ocean. We are crowding each other in the once free and open sea, and an international scramble is underway to determine who will control what, under whose auspices, and, ultimately, with what force.

A sketchy history of human interest in manganese nodules may serve as an example of how and why attitudes have changed toward physical marine resources just as they have for living ones. Concretions of ferromanganese oxides were dredged in many places by the *Challenger* expedition. They aroused considerable interest because of their many curious properties. They were black and resembled potatoes or, as the British scientists reported, cricket balls. Many were layered like onions and some had strange nuclei such as sharks' teeth or the earbones of whales. Sir John Murray and his colleagues agreed that these nodules were precipitated from sea water, but they disagreed about whether the chemicals came from continental erosion or from submarine vulcanism. There was a hint of things to come in the discovery that some nodules contained nickel and copper, but the concentrations were of no commercial interest at the time. Murray went on to discover similar nodules in nondepositional environments in the very shallow sea lochs of Scotland not far from the offices of the Challenger Society in Edinburgh. This was not unlike the experience of the Dutch marine geologists who explored the waters of the East Indies, partly because of the rich oil and gas fields on the islands that then were colonies of the Netherlands. It was only after Indonesia became independent that the Dutch discovered a gas field bigger than any in the Indies under Groningen, the home of one of the most eminent of the marine geologists.

Deep-sea dredging continued to find manganese nodules, but they remained scientific curiosities, largely neglected after Murray's death. Interest was revived by the new expeditions and the new science of the 1950s. Two discoveries were made that strongly influenced the future transformation of nodules from curiosities to ore deposits. First, bottom photographs disclosed that in some places nodules were abundantly sprinkled on the sediment surface and elsewhere they formed a continuous surface like a cobbled pavement. They were far more abundant than anyone had thought. The other discovery depended on advances in analytical chemistry since Murray's time. My colleague Edward Goldberg and other chemists found a wide range of elements in all nodules, and concentrations in some of more than 1% of the metals nickel, cobalt, and copper. Another change had occurred since Murray's day. The older and richer mines on land had been exhausted, and new finds of high-grade ore were unable to meet the demands of an expanding industrial society. Consequently, much lower grades of ore were being mined, and the lowest grades on land were similar to the highest on the sea floor. A fascinating scientific problem was exposed by the ability of geochemists to measure the rate of accumulation of the layers of manganese nodules and also the sedimentation rate of the red clay and pelagic oozes on which the nodules rested. The sediment was piling up at least 100 times faster than the nodules were growing, and yet the nodules were on top. To this day there is no general agreement about the explanation of this paradox. Very likely the matter is complex. However, it appears that some nodules are at the surface because they are rolled by the intermittent and accidental nudges of benthonic organisms. Rolling on an accumulating surface automatically moves an object upward. This does not explain how nodules can form continuous pavements. It may be that these are formed when sediment is eroded, thereby leaving a lag gravel of nodules like the pavement of stones that forms on the surface of a wind-deflated desert.

Intrigued by the scientific puzzles and appreciating the importance of new mineral resources, oceanographers, chiefly in Russia and the United States, began systematic study of manganese nodules in the late fifties. These programs were rejuvenated and vastly expanded a decade later with the start of the International Decade of Ocean Exploration, which still continues. The investigations have uncovered an extraordinary number of new puzzles, and it is beginning to appear that manganese nodules may have some of the most complicated histories of any objects on the surface of the earth.

While the scientists found puzzles, many of the more enlightened mining companies became interested enough in manganese nodules to support research. Then, gradually, they began to explore with their own ships, seeking the best ore deposits. Corporations in many of the major industrial nations, especially those importing large quantities of ores, joined in national and international consortia and prepared for the day when deep-sea mining would be economical. Promising ores were discovered and mapped in the eastern equatorial Pacific. These occurred in patchy but locally continuous pavements and contained more than 1% each of nickel and copper. The consortia dredged large quantities of nodules and used them to develop extraction methods for the metals and then to make test runs in small pilot plants.

The corporations also planned and built parts of the systems that would bring the ores up to the surface. These consist of various combinations of ships, barges, chain dredges, air-lift devices like giant vacuum cleaners, and so on. For some years the most promising indication of the imminence of deep-sea mining was that the secretive billionaire Howard Hughes was putting his personal fortune into building a mining system. Details of the system were closely held industrial secrets, but there could be no doubt that it was being built and that the capital expenditures were so great it would have to be put to immediate use. The industry was electrified. Eventually the system was used, not to dredge nodules but to attempt to lift a sunken Russian submarine, and the true story was leaked. It wasn't Howard Hughes investing his money, it was the CIA investing our money.

Meanwhile and nonetheless, other businessmen continued to develop mining systems and the time approached when they would begin to stake claims. From the beginning it had been realized that this would present problems. Who had the authority to receive claims and who would protect the rights of the claimant? Initially this was not a very serious problem because it would not arise for some time and, anyway, it appeared that there were enough nodules for everyone. Because of this lack of urgency, a priceless opportunity to solve the problem may have been lost in 1958 when the United Nations held its first conference on the law of the sea, in Geneva. That meeting was largely concerned with the definition of the continental shelf, because the resources there were within reach of then-existing technology. With regard to deeper resources, it was agreed that each country owned them as far out and as deep as it could continuously develop them. The United States under President Truman had already declared unilaterally that it owned the minerals of its continental shelf, so the treaty appeared merely to confirm the status quo.

However, technology constantly improves. A decade later the United States had in *Glomar Challenger* a ship capable of drilling in the deep sea. Wherein lay governmental responsibility? Certainly the federal agencies were sensitive to the fact that they might have some responsibilities. In fact, one agency proposed that DSDP could not drill a hole 2000 kilometers west of California, in 5 kilometers of water, unless the ship was outfitted with the same blowout preventers that were required on the continental shelf. This implied a jurisdiction extending anywhere that the 1958 treaty indicated. Fortunately, this bureaucratic nonsense was squelched because there wasn't the slightest possibility of natural gas at the site. However, the implied jurisdiction was not questioned. Did it then extend to other activities of U.S. citizens in the deep sea? Apparently not, because other agencies of the federal government proved unwilling to help industry start mining by assuming jurisdiction over deep-sea mining claims.

Something had to be done, so on 14 November 1974 an American corporation, Deepsea Ventures, Inc., filed with the State Department a *Notice of Discovery and Claim of Exclusive Mining Rights, and Request for Diplomatic Protection and Protection of Investment.* The document specified the boundaries of the claim in the usual fashion by giving the geographical coordinates of the corners. However, the corners were not located by land coordinates of township and range, but by latitude and longitude in the open sea. To this date the State Department has not granted the request, but it has been considering the future of deep-sea mining in the course of the seemingly endless U.N. conference on the law of the sea.

This conference has now met for long periods on three continents and has yet to reach any public agreement on anything, perhaps because there has never been a vote on a matter of substance. The conference is providing superb examples of the difficulties of defining an equitable solution when the sea is used for incompatible or conflicting purposes. One bone of contention is the ownership of the manganese nodules. As near as one can tell in the absence of open voting, the rich, resource-depleted, industrial nations generally want the nodules to belong to whoever can mine them. The poor, developing nations that lack resources prefer to have the U.N. own the nodules and tax the miners. The developing nations that export nickel, cobalt, and copper prefer no sea mining at all.

There the matter stands. Over a century the nodules have successively aroused the interest and the vexation of scientists, industrialists, and diplomatists. Will nothing about them ever be resolved?

Unlike nodules, most potential resources that are known to exist in the sea are being utilized. These are discussed in "The Physical Resources of the Ocean," by Edward Wenk, Jr. It was published in 1969. As it forecast, the world has since grown more crowded and complex. However, the types of marine resources now being used differ hardly at all from the ones Wenk describes. What have changed are the amounts used, which, almost without exception, have increased in the world as a whole. Indeed, in United States waters even these changes have been small. Long Island Sound is not the site of nine nuclear power stations, as Wenk expected from plans current at the time he wrote; nor are the continental shelves peppered with drilling platforms. During the interval, the National Environmental Protection Act and the Coastal Zone Management Act have come to influence a wide range of large-scale activities in the United States. Together, and over a period of time, they may have had more of an effect in slowing or stopping some types of industrial actions than has been generally realized. If so, however, the effect may be largely transient. It has taken industry and government agencies some years to obtain legal interpretations of their obligations under the new laws, but they now know how to do such things as prepare an environmental impact report that will be accepted in a court of law. Meanwhile, environmentalists have become more satisified by the power of the laws, and have learned how to ensure that they are enforced.

The article on physical resources calls attention to the most critical problem confronting the development and utilization of marine resources, and the one that is troubling the U.N. conference on the law of the sea, namely, the problem of multiple use. This problem is familiar on land. It is widely acknowledged that elementary schools should not be built under the flight patterns of busy airports and that resort hotels are not compatible with wilderness areas. However, the problems are much worse at sea because the uses are so diverse, traditional, incompatible, and ungoverned by policing agencies with clear jurisdiction. Worst of all, perhaps, is the fact that we are too ignorant of the environment to know whether different uses do in fact conflict. As the sea grows more crowded we shall need to predict, even before they begin, whether certain actions will conflict with others. In the United States such predictions are already required by the new environmental protection laws.

Coastal areas are even more subject to the overlapping claims of a multitude of different users than the open sea is. Prior to the enactment of the Coastal Zone Management Act, people, businesses and governmental agencies made incompatible uses of coastal resources on the basis of "first come, first served." Some people built high-rise hotels where others wanted bird sanctuaries. Some wanted quiet fishing where others wanted to water ski. Some wanted the jobs that would be brought by a new refinery while others wanted to preserve the view from the homes they had bought for a long-planned retirement. The new law called for a halt in the initiation of most new coastal activities until the states had established equitable rules and priorities. That halt is ending. Meanwhile people are adjusting, as they usually do, to the crowded reality of competing interests. Within a mile of my office at Scripps are beaches for swimmers, beaches for surfers, beaches for nudists, kelp beds used for sports fishing, and an underwater nature preserve where shellfish and plants are protected, and there is a constant flow of joggers along the beach and swimmers across it without anybody bumping.

The following two articles deal with other seemingly incompatible uses of the sea that can both go on for another reason, namely, that the sea is so vast. If it were not, we could not dump in pollution and remove salt. "Fresh Water from Salt," by David Jenkins, discusses a practice that has come into widespread use in recent decades. The expansion has been particularly impressive around the Persian Gulf because the countries there have unlimited oil and gas for energy and hardly any water. However, many other arid coastal regions also are finding desalination of seawater worth the cost. John Isaacs has observed that if we think of water in terms of what we are willing to pay for the other things we drink, desalinization seems very cheap.

"The Disposal of Waste in the Ocean," by Willard Bascom, addresses some of the scientific problems connected with this use of the oceans. For example, environmentalists may ask for laws that forbid cities to dump certain wastes, say poisonous heavy metals, into the sea. Legislators would want advice about whether such laws are necessary or reasonable. After careful study, scientists may find that rivers dump much more of the metals into the sea than cities do; thus, the laws would have little benefit and would be unreasonable even if the cost were small. They may find that the metals are killing vast beds of valuable shellfish around cities; thus, moderate costs would be balanced by benefits. It is conceivable that they would discover that the metals are poisoning human swimmers, so that the pollution must be eliminated regardless of cost. The article also discusses the problems that arise in assessing cause and effect when humans change an environment that is already changing naturally. Humans have a difficult time in evaluating their impact on nature—an impact that only recently has grown to terrifying levels. They spread DDT to kill insects in North America and are rightfully accused of poisoning penguins in Antartica. Alternatively, they fish for sardines off California and are denounced for greedy overfishing, when the sardines have apparently just swum elsewhere, as they have many times in ages past. Scientists are barely able to evaluate some kinds of natural fluctuations, let alone understand and predict them. As Bascom observes, additional scientific research will be required.

The final article is an example of the new practical insights that can be gained from basic research. Peter Rona, in "Plate Tectonics and Mineral Resources," extends the plate tectonics revolution into an explanation of the origin of ores and the location of mineral deposits. Thus we come full circle. *Challenger* set to sea a century ago and began the harvest of ideas that have at last returned to change our understanding of the land.

The Physical Resources of the Ocean

by Edward Wenk, Jr.
September 1969

They include not only the oil and minerals of the bottom and the minerals dissolved in seawater but also seawater itself and the shoreline carved by the action of the sea

Men have caught fish in the ocean and extracted salt from its brine for thousands of years, but only within the past decade have they begun to appreciate the full potential of the resources of the sea. Three converging influences have been responsible for today's intensive exploration and development of these resources. First, scientific oceanography is generating new knowledge of what is in and under the sea. Second, new technologies make it feasible to reach and extract or harvest resources that were once inaccessible. Third, the growth of population and the industrialization of society are creating new demands for every kind of raw material.

The ocean's resources include the vast waters themselves, as a processing plant to convert solar energy into protein [see the article "The Food Resources of the Ocean," by S. J. Holt, beginning on page 210], a storehouse of dissolved minerals and fresh water, a receptacle for wastes, a source of tidal energy and a medium for new kinds of transportation. They also include the sea floor, sediments and rocks below the waters as sites of fossil fuel and mineral deposits; the seacoast as a unique resource that is vulnerable to rapid, irrevocable degradation by man.

Because the oceans are so wide and so deep, statistics on their gross resource potential are impressive. It is important to understand, however, that the immediate significance of these resources and their long-term relevance to society involve both exploration and development, and development depends on economic, social, legal and political considerations. One special feature of marine resources that may at first retard development may in the long run promote it: the fact that almost without exception sea-floor resources are in areas not subject to private ownership (although the resources will be largely privately developed). More than 85 percent of the ocean bottom lies beyond the present boundaries of national jurisdictions, and in the areas that are subject to national control the resources are considered common property. This circumstance may uniquely invoke a balancing of public and private interests, disciplined resource management and enhanced international cooperation.

OFFSHORE OIL PLATFORM in the photograph on the opposite page is in Alaska's Cook Inlet, 60 miles southwest of Anchorage. Wells are drilled from derricks set over the massive legs, 14 feet in diameter. The plume of flame is burning natural gas, a waste product in this case. Oil and gas account for more than 90 percent of the value of minerals now being retrieved from the oceans.

The 350 million cubic miles of ocean water constitute the earth's largest continuous ore body. Dissolved solids amount to 35,000 parts per million, so that each cubic mile (4.7 billion tons of water) contains about 165 million tons of solids. Although most chemical elements have been detected (and probably all are present) in seawater, only common salt (sodium chloride), magnesium and bromine are now being extracted in significant amounts. The production of salt (which can be traced back to Neolithic times and resulted in the first U.S. patent) is currently valued at $175 million per year worldwide. Magnesium, the third most abundant element in the oceans, is by far the most valuable mineral extracted from seawater in this country, with annual production worth about $70 million. Although the ocean contains bromine in concentrations of only 65 parts per million, it is the source of 70 percent of the world's production of this element, which is used principally in antiknock compounds for gasoline. The economic recovery of other chemicals from seawater is questionable because of extraction costs. In a cubic mile of seawater the value of 17 critical metals (including cobalt, copper, gold, silver, uranium and zinc) is less than $1 million at current prices; a plant to handle a cubic mile of water per year would have to process 2.1 million gallons per minute every minute of the year, and operating it would cost significantly more than the value of all its products.

One of the potential resources of seawater that has been most difficult to extract economically is water itself—fresh water. As requirements for water for domestic use, agriculture and industry rise sharply, however, desalting the sea becomes increasingly attractive. More than 680 desalination plants with a capacity of more than 25,000 gallons of fresh water per day are now in operation or under construction around the globe, and the growth rate is projected at 25 percent per year over the next decade. The cost of desalting has been decreased by new technology to less than 85 cents per 1,000 gallons, but this is still generally prohibitive in the U.S., where the cost of 1,000 gallons of fresh water is about 35 cents. In water-deficient areas or where the local water supply is unfit for consumption, however, desalted water is competitive. This accounts for the presence of more than 50 plants in Kuwait, 22 on Ascension Island and the 2.6-million-gallon facility at Key West, the first U.S. city to obtain its water supply directly from the ocean. Considerably lower costs will be attained within the next decade where large-scale desalting operations are combined with nuclear-fueled power plants to take advantage of their output of waste heat.

Once upon a time man could safely utilize the waters of the sea as a recep-

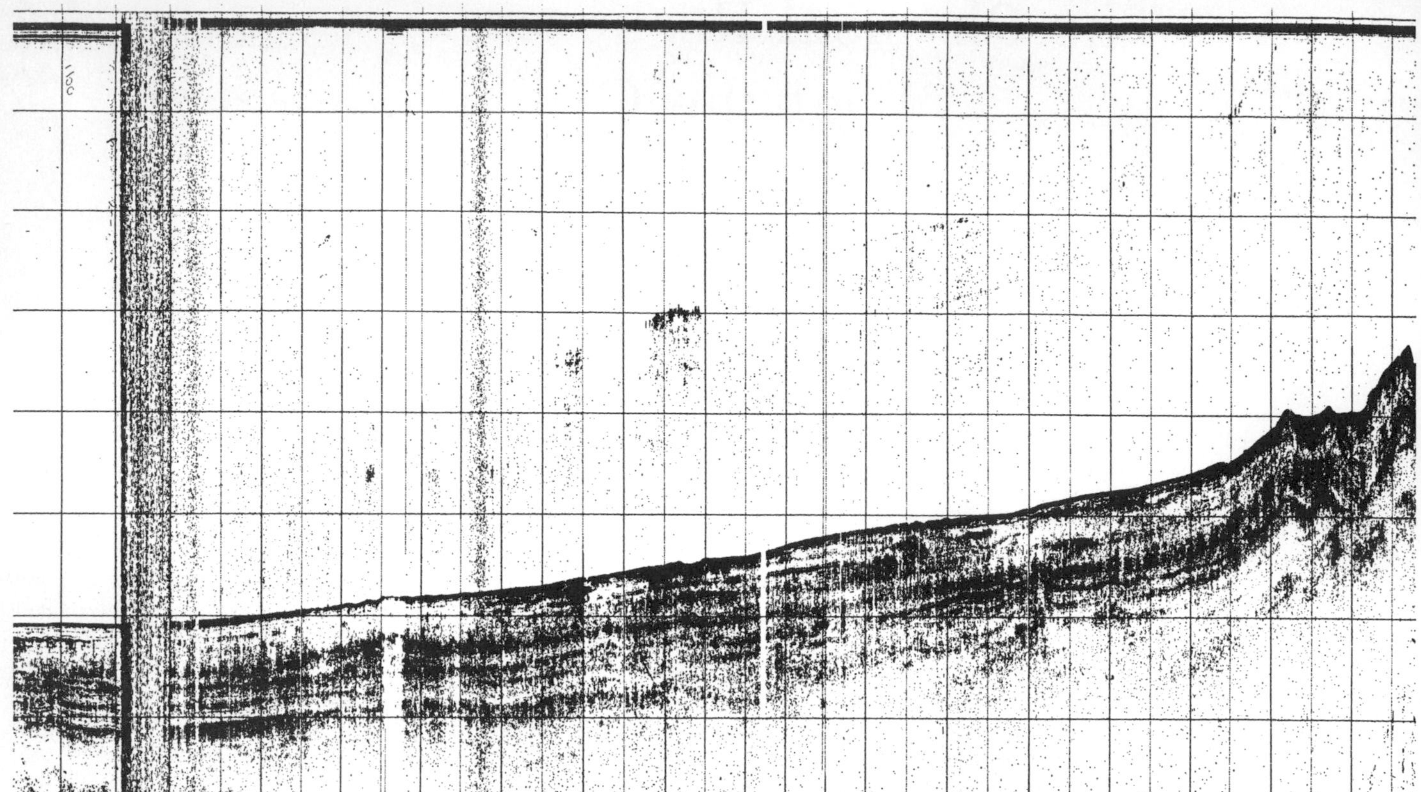

CONTINENTAL RISE, which may be rich in resources, is evident in this seismic profile made off Liberia by the Global Ocean Floor Analysis and Research Center of the U.S. Naval Oceanographic Office. The hard, straight line across the top of the record is a water reflection. The abyssal plain (*left*) is about 15,000 feet below sea level. From this plain a thick apron of land-derived layered sediments comprising the continental rise slants gently up to the toe of the continental slope. The continental slope, which is here

tacle for sewage and other effluents from municipalities and industries, confident that the wastes would rapidly be diluted, dispersed and degraded. With the growth of population and the concentration of coastal industry that is no longer possible. The sheer bulk of the material disposed of and the presence of new types of nondegradable waste products are a special threat to coastal waters—the same waters that, as we shall see, are subject to increasing demands from a wide range of competing activities. In addition pollutants are now beginning to concentrate at an alarming rate far from shore in the open ocean. Since tetraethyl lead was introduced into gasoline 45 years ago, lead concentrations in Pacific Ocean waters have jumped tenfold. Toxic DDT residues have been detected in the Bay of Bengal, having drifted with

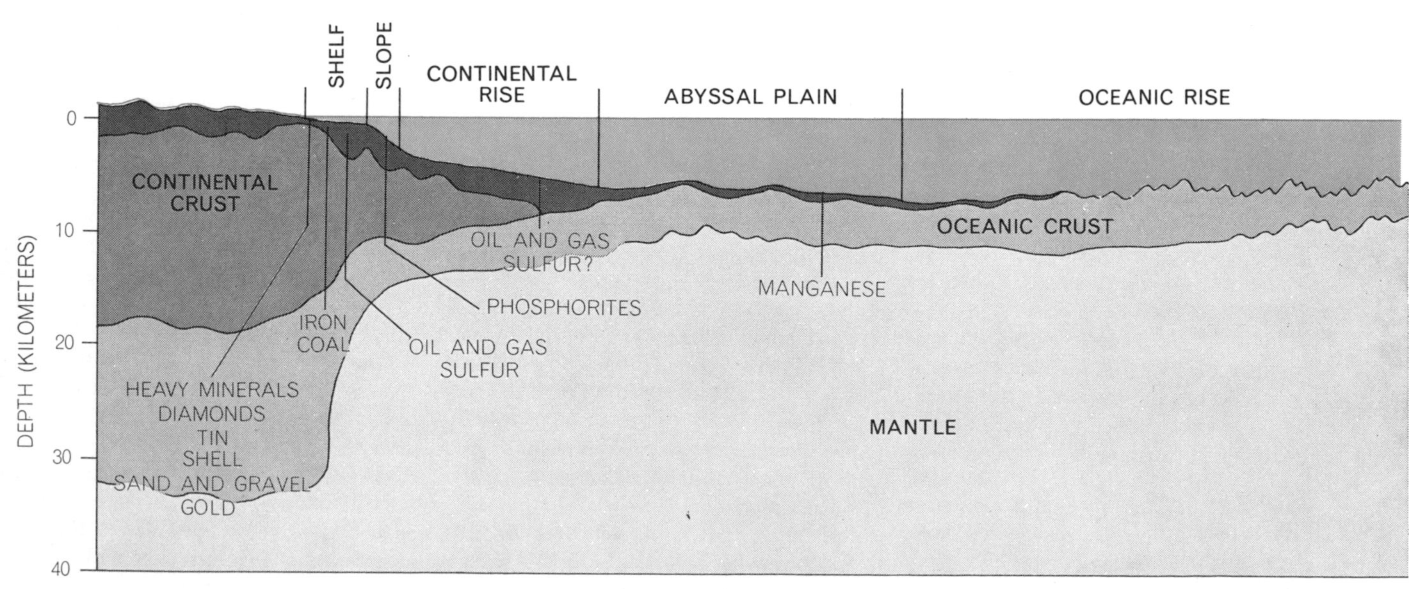

OCEAN-FLOOR RESOURCES that are known or believed to exist in the various physiographic provinces are indicated on a schematic cross section of a generalized ocean basin extending from a continent out to a mid-ocean ridge. Some of these resources are now

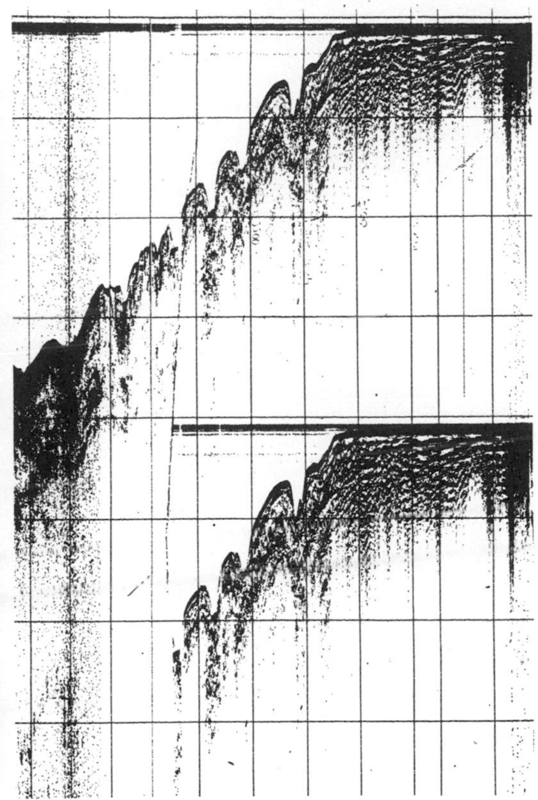

marked by large sedimentary ridges, ascends more steeply to the shallow continental shelf. The "multiple" (*right*) is in effect an echo of the structures shown above it.

the wind from as far away as Africa. And man-made radioactivity from nuclear fallout can be isolated in any 50-gallon water sample taken anywhere in the ocean.

The mineral resources of the seabed, unlike those of the essentially uniform overlying waters, occur primarily in scattered, highly localized deposits and structures on top of and within the sediments and rocks of the ocean floor. They include (1) fluids and soluble minerals, such as oil, gas, sulfur and potash, that can be extracted through boreholes; (2) consolidated subsurface deposits, such as coal, iron ore and other metals found in veins, which are so far mined only from tunnels originating on land, and (3) unconsolidated surface deposits that can be dredged, such as heavy metals in ancient beaches and stream beds, oyster shell, sand and gravel, diamonds, and "authigenic" minerals: nodules of manganese and phosphorite that have been formed by slow precipitation from seawater. Economic exploitation has so far been confined to the continental shelves in waters less than 350 feet deep and within 70 miles of the coastline.

Oil and gas represent more than 90 percent by value of all minerals obtained from the oceans and have the greatest potential for the near future. Offshore sources are responsible for 17 percent of the oil and 6 percent of the natural gas produced by non-Communist countries. Projections indicate that by 1980 a third of the oil production—four times the present output of 6.5 million barrels a day—will come from the ocean; the increase in gas production is expected to be comparable. Subsea oil and gas are now produced or are about to be produced by 28 countries; another 50 are engaged in exploratory surveys. Since 1946 more than 10,000 wells have been drilled off U.S. coasts and more than $13 billion has been invested in petroleum exploration and development. The promise of large oil reserves has stimulated industry to invest more than $1.7 billion since mid-1967 to obtain Federal leases off Louisiana, Texas and California that guarantee only the right to search for and develop unproved reserves. To date more than 6.5 million acres of the outer continental shelf off the U.S. have been leased, which is half of the acreage offered, resulting in lease income to the Federal Government of $3.4 billion. With more than 90 percent of the most favorable inland areas explored and less than 10 percent of the U.S. shelf areas surveyed in detail, the prospects are encouraging for additional large oil finds off U.S. coasts.

Sulfur, one of the world's prime industrial chemicals, is found in the cap rock of salt domes buried within continental and sea-floor sediments. The sulfur is recovered rather inexpensively by melting it with superheated water piped down from the surface and then forcing it up with compressed air. Only 5 percent of the explored salt domes contain commercial quantities of sulfur, and offshore production has been limited to two mines off Louisiana that supply two million tons, worth $37 million, a year. Now a critical shortage of sulfur and the recent discovery by the deep-drilling ship *Glomar Challenger* of sulfur-bearing domes in the deepest part of the Gulf of Mexico have stimulated an intensive search for offshore sulfur deposits.

Undersea subfloor mining can be traced back to 1620, when coal was ex-

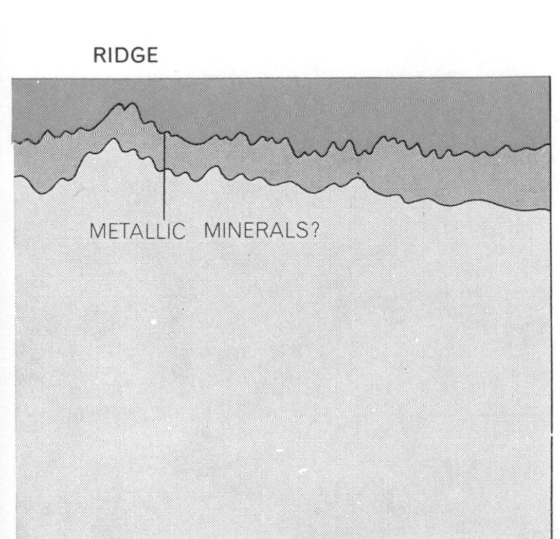

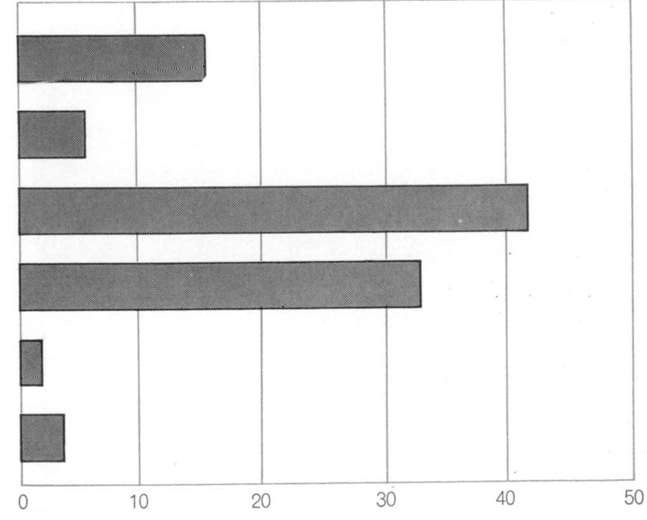

being exploited but others may not be economic for years. Sedimentary layers (*black*) are the most likely site of recoverable raw materials. The chart (*right*) shows what percent of the ocean floor's 140 million square miles of area is occupied by each province.

tracted in Scotland through shafts that were driven seaward from an offshore island. To date 100 subsea mines with shaft entries on land have recovered coal, iron ore, nickel-copper ores, tin and limestone off a number of countries in all parts of the world. Coal extracted from as deep as 8,000 feet below sea level accounts for almost 30 percent of Japan's total production and more than 10 percent of Britain's. With present technology subsea mining can be conducted economically as far as 15 miles offshore, given mineral deposits that are worth $10 to $15 per ton and occur in reserves of more than $100 million. The economically feasible distance should increase to 30 miles by 1980 with the development of new methods for rapid underground excavation. Eventually shafts may be driven directly from the seabed

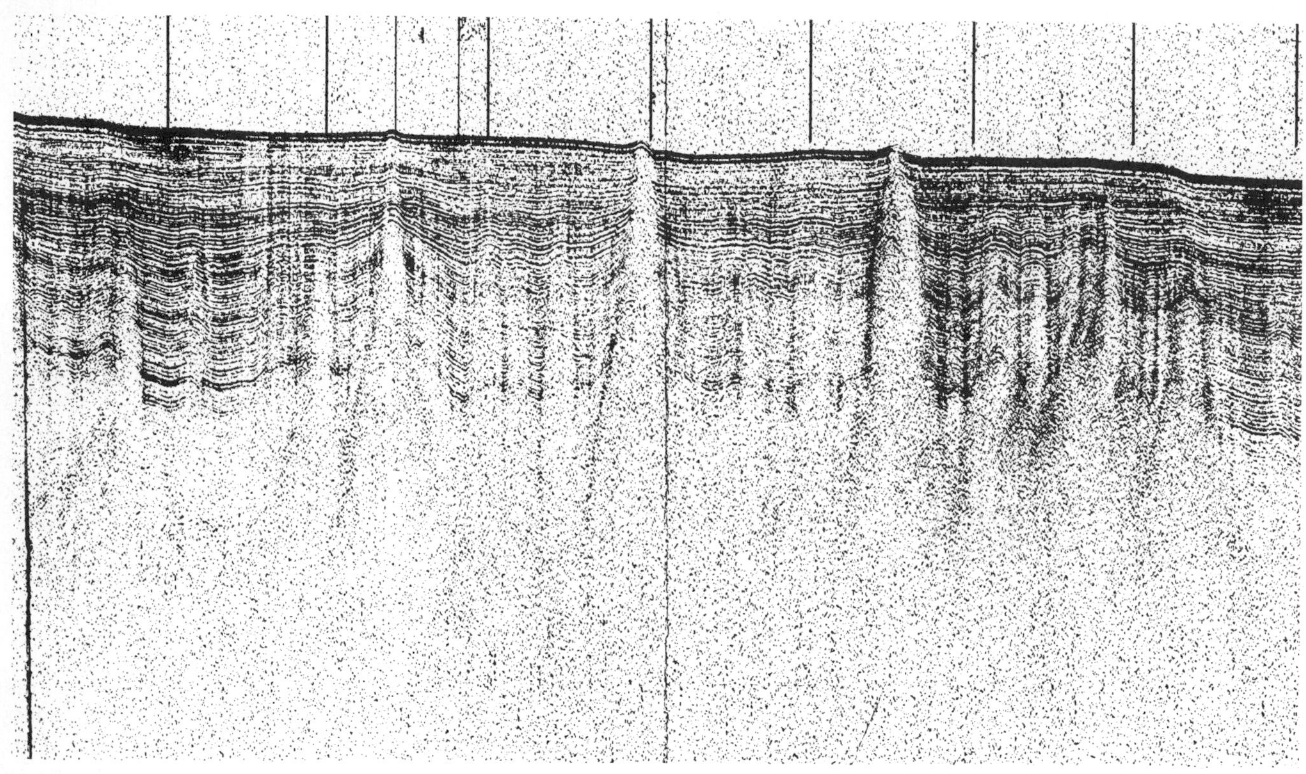

POSSIBILITY OF OIL in the deep-sea floor was revealed by this record from some 250 miles northwest of the Cape Verde Islands. The record, like the one on the preceding pages, was made by the research ship *Kane* of the Naval Oceanographic Office. The tall narrow structures appear to be salt domes, along the flanks of which surface-seeking oil is often trapped in tilted sedimentary layers.

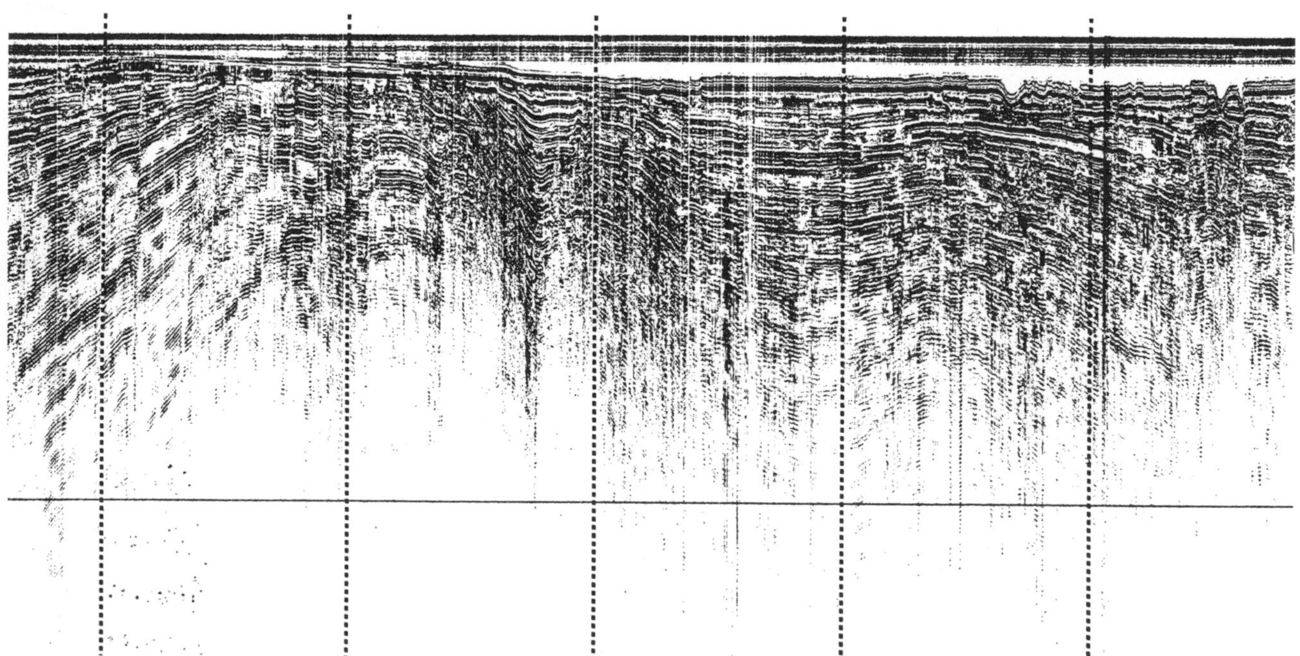

FOLDED SEDIMENTARY LAYERS are shown in this *Kane* record made on the continental shelf north of Trinidad. The band at top is from water reflections. The record shows anticlines (arches) and synclines (troughs); oil is often trapped in crests of anticlines.

if ore deposits are located in ocean-floor rock far from land.

Seventy percent of the world's continental shelves consist of ancient unconsolidated sediments from which are dredged such commodities as sand, gravel, oyster shell, tin, heavy-mineral sands and diamonds. Dredging is an attractive mining technique because of low capital investment, quick returns and high profits and the operational mobility offered by floating dredges. So far it has been limited to nearshore waters less than 235 feet deep and protected from severe weather effects. As knowledge of resources in deeper water increases, industry will undoubtedly upgrade its dredging technology.

Of the many potentially valuable surface deposits, sand and gravel are the most important in dollar terms, and only these and oyster shells are now mined off the U.S. coast. Some 20 million tons of oyster shells are extracted from U.S. continental shelves annually as a source of lime; sand and gravel run about 50 million cubic yards. As coastal metropolitan areas spread out, they cover dry-land deposits of the very construction materials required to sustain their expansion; in these circumstances sea-floor sources such as one recently found off New Jersey, which is thought to contain a billion tons of gravel, become commercially valuable.

In the deeper waters of the continental shelves, on the upper parts of the slopes and on submarine banks and ridges widespread deposits of marine phosphorite nodules are found at depths between 100 and 1,000 feet. The best-known large deposits are off southern California, where total reserves are estimated at 1.5 billion tons, and off northwestern Mexico, Peru and Chile, the southeastern U.S. and the Union of South Africa. The only major attempt at mining was made in 1961, when a company leased an area off California, but that lease was returned unexploited to the Federal Government four years later. With large land sources generally available to meet the demand for phosphates for fertilizer and other products, offshore exploitation of this resource is not likely to occur soon, except possibly in phosphate-poor countries.

The only known minerals on the floor of the deep ocean that appear to be of potential economic importance are the well-publicized manganese nodules, formed by the precipitation from seawater of manganese oxides and other mineral salts, usually on a small nucleus such as a bit of stone or a shark's tooth.

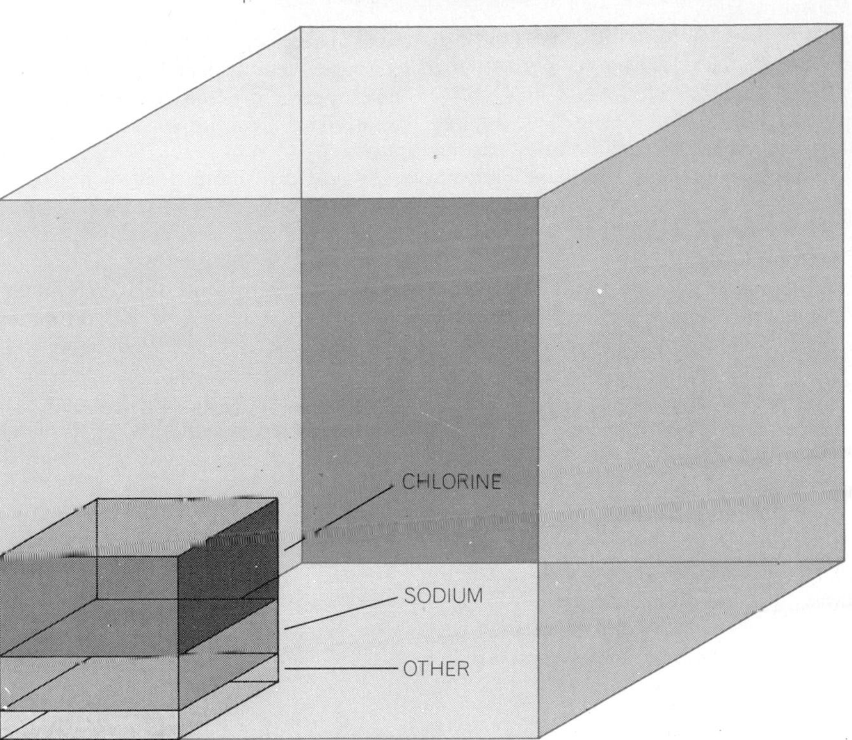

SEAWATER contains an average of 35,000 parts per million of dissolved solids. In a cubic mile of seawater, weighing 4.7 billion tons, there are therefore about 165 million tons of dissolved matter, mostly chlorine and sodium (*gray cube*). The volume of the ocean is about 350 million cubic miles, giving a theoretical mineral reserve of about 60 quadrillion tons.

ELEMENT	TONS PER CUBIC MILE	ELEMENT	TONS PER CUBIC MILE
CHLORINE	89,500,000	NICKEL	9
SODIUM	49,500,000	VANADIUM	9
MAGNESIUM	6,400,000	MANGANESE	9
SULFUR	4,200,000	TITANIUM	5
CALCIUM	1,900,000	ANTIMONY	2
POTASSIUM	1,800,000	COBALT	2
BROMINE	306,000	CESIUM	2
CARBON	132,000	CERIUM	2
STRONTIUM	38,000	YTTRIUM	1
BORON	23,000	SILVER	1
SILICON	14,000	LANTHANUM	1
FLUORINE	6,100	KRYPTON	1
ARGON	2,800	NEON	.5
NITROGEN	2,400	CADMIUM	.5
LITHIUM	800	TUNGSTEN	.5
RUBIDIUM	570	XENON	.5
PHOSPHORUS	330	GERMANIUM	.3
IODINE	280	CHROMIUM	.2
BARIUM	140	THORIUM	.2
INDIUM	94	SCANDIUM	.2
ZINC	47	LEAD	.1
IRON	47	MERCURY	.1
ALUMINUM	47	GALLIUM	.1
MOLYBDENUM	47	BISMUTH	.1
SELENIUM	19	NIOBIUM	.05
TIN	14	THALLIUM	.05
COPPER	14	HELIUM	.03
ARSENIC	14	GOLD	.02
URANIUM	14		

CONCENTRATION of 57 elements in seawater is given in this table. Only sodium chloride (common salt), magnesium and bromine are now being extracted in significant amounts.

They are widely distributed, with concentrations of 31,000 tons per square mile on the floor of the Pacific Ocean. Although commonly found at depths greater than 12,500 feet, nodules exist in 1,000 feet of water on the Blake Plateau off the southeastern U.S. and were located last year at a depth of 200 feet in the Great Lakes.

The nodules average about 24 percent manganese, 14 percent iron, 1 percent nickel, .5 percent copper and somewhat less than .5 percent cobalt. Since ore now being mined from land deposits in a number of countries averages 35 to 55 percent manganese, it may be the minor constituents of the nodules, particularly copper, cobalt and nickel, that first prove to be attractive economically. Many experts think the key to profitable exploitation is the solution of a difficult metallurgical separation problem created by the unique combination of minerals in the nodules.

Few discoveries have created more excitement among earth scientists than the location, by different expeditions in 1964, 1965 and 1966, of three undersea pools of hot, high-density brines in the middle of the Red Sea. The brines contain minerals in concentrations as high as 300,000 parts per million—nearly 10 times as much solid matter as is commonly dissolved in ocean water—and overlie sediments rich in such heavy metals as zinc, copper, lead, silver and gold. Similar deposits may be characteristic of other enclosed basins associated, as is the Red Sea, with rift valleys.

As this decade ends resource exploration is advancing on many fronts. Chromite has been found by Russian oceanographers in sea-floor rifts in the Indian Ocean, and zirconium, titanium and other heavy minerals have been detected in sediments from extensive areas off the Texas coast. Methane deposits sufficient to supply Italy's needs for at least six years have been confirmed in the Adriatic Sea. New oil fields of economic value have been discovered off Mexico, Trinidad, Brazil, Dahomey and Australia. Surveys of the Yellow Sea and the East China Sea indicate that the continental shelf between Taiwan and Japan may contain one of the richest oil reserves in the world. It is now becoming clear that the continental rises, which lie at depths ranging from about 5,000 to 18,000 feet and contain a far larger total volume of sediments than the shelves, may hold significant petroleum reserves. Within the past year the *Glomar Challenger* has drilled into oil-bearing sediments lying under 11,700 feet of water in the Gulf of Mexico, and seismic surveys have revealed what appear to be typical oil-bearing structures under the deep ocean-basin floor [*see upper illustration on page 260*].

As on land, resource development of a frontier requires a mixture of public and private entrepreneurship. Historically basic exploration has been sponsored by government; this broad-ranging exploration reveals opportunities that are followed up by detailed privately funded surveys. This pattern is likely to persist, and as the International Decade of Ocean Exploration gets under way a wide variety of new opportunities for marine resource development will surely come to light.

Limitations on the exploitation of the oceans stem partly from lack of knowledge about the distribution of resources and the state of the art of undersea technology. The major limits, however, are set by venture economics, the motivating factor for the profit sector. That factor is influenced by the availability of competing land deposits, by extraction technology and the legal situation and, most critically, by market demand. On the basis of projections of world population and gross national products to the year 2000, which indicate respective in-

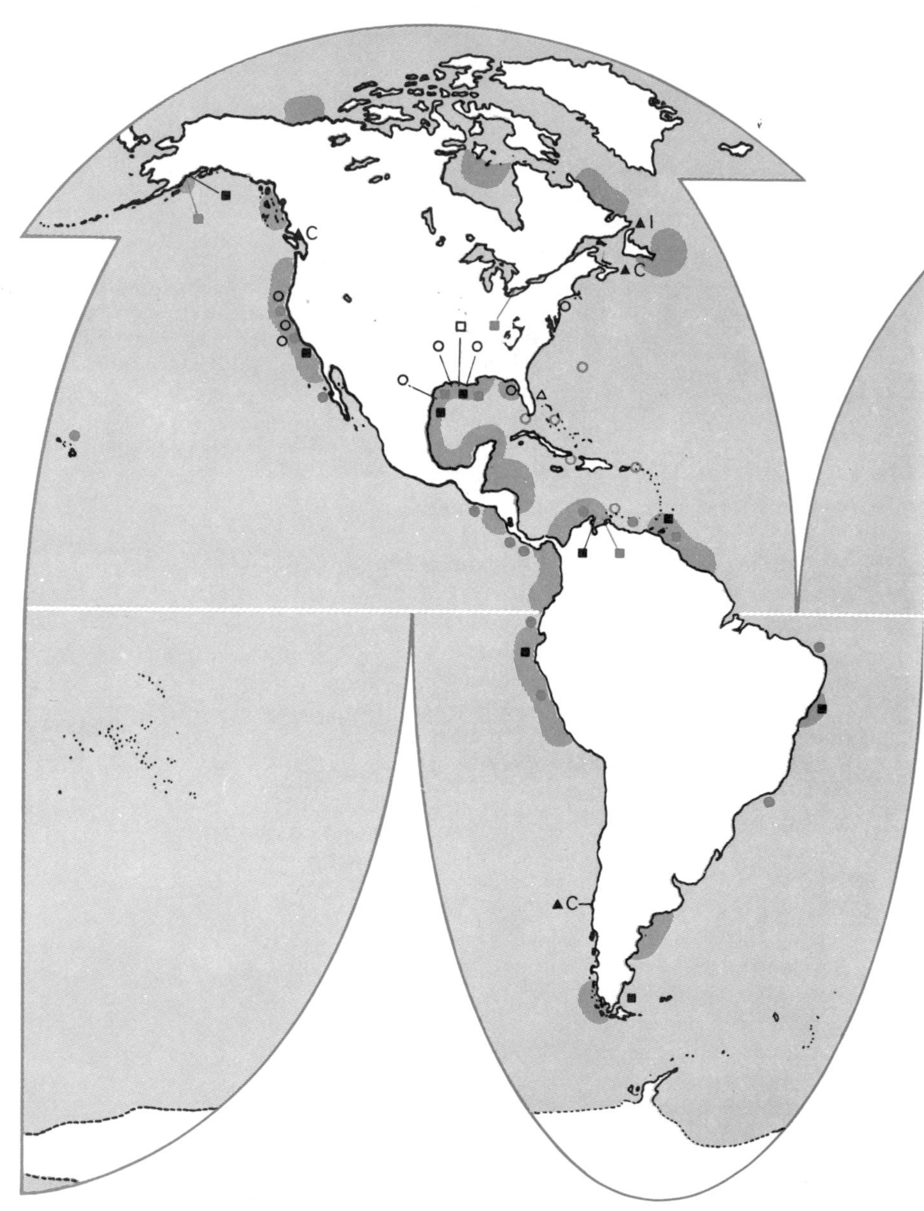

CURRENT PRODUCTION of major ocean resources (except sand, gravel and shell) is mapped with areas of oil and gas exploration. Data come from U.S. Geological Survey, *Oil &*

creases of almost 100 and 500 percent over 1965, a sharp rise in total resource demand can be anticipated, and with it a greater role for the sea.

Other major impediments to the rapid development of ocean resources arise from social and legal constraints. Damage to beaches and wildlife from oil leaks, as in the Santa Barbara Channel, and uncertainty about the effect of dredging on marine organisms have brought public awareness that offshore development may have detrimental consequences. The public, the owner of the resources, is demanding greater safeguards, questioning the wisdom of resource development in areas where it may threaten the environment. In deeper waters seabed development comes up against the potent issue of ownership. There are major questions about the boundaries of national jurisdictions and about the jurisdiction over the seabed beyond such boundaries [see "The Ocean and Man," by Warren S. Wooster; SCIENTIFIC AMERICAN Offprint 888].

The coastal margin—the ribbon of land and water where people and oceans meet and are profoundly influenced by each other—has only recently come to be recognized and treated as a valuable and perishable resource. It is actually a complex of unique physical resources: estuaries and lagoons, marshes, beaches and cliffs, bays and harbors, islands and spits and peninsulas.

In the year 2000 half of the estimated 312-million population of the U.S. will live on 5 percent of the land area in three coastal urban belts: the megalopolises of the Atlantic, the Pacific and the Great Lakes. Along with the people will come an intensification of competing demands for the limited resources of the narrow, fragile coastal zone. To make matters worse, the coastal resource is shrinking under the pressure of natural forces (hu-

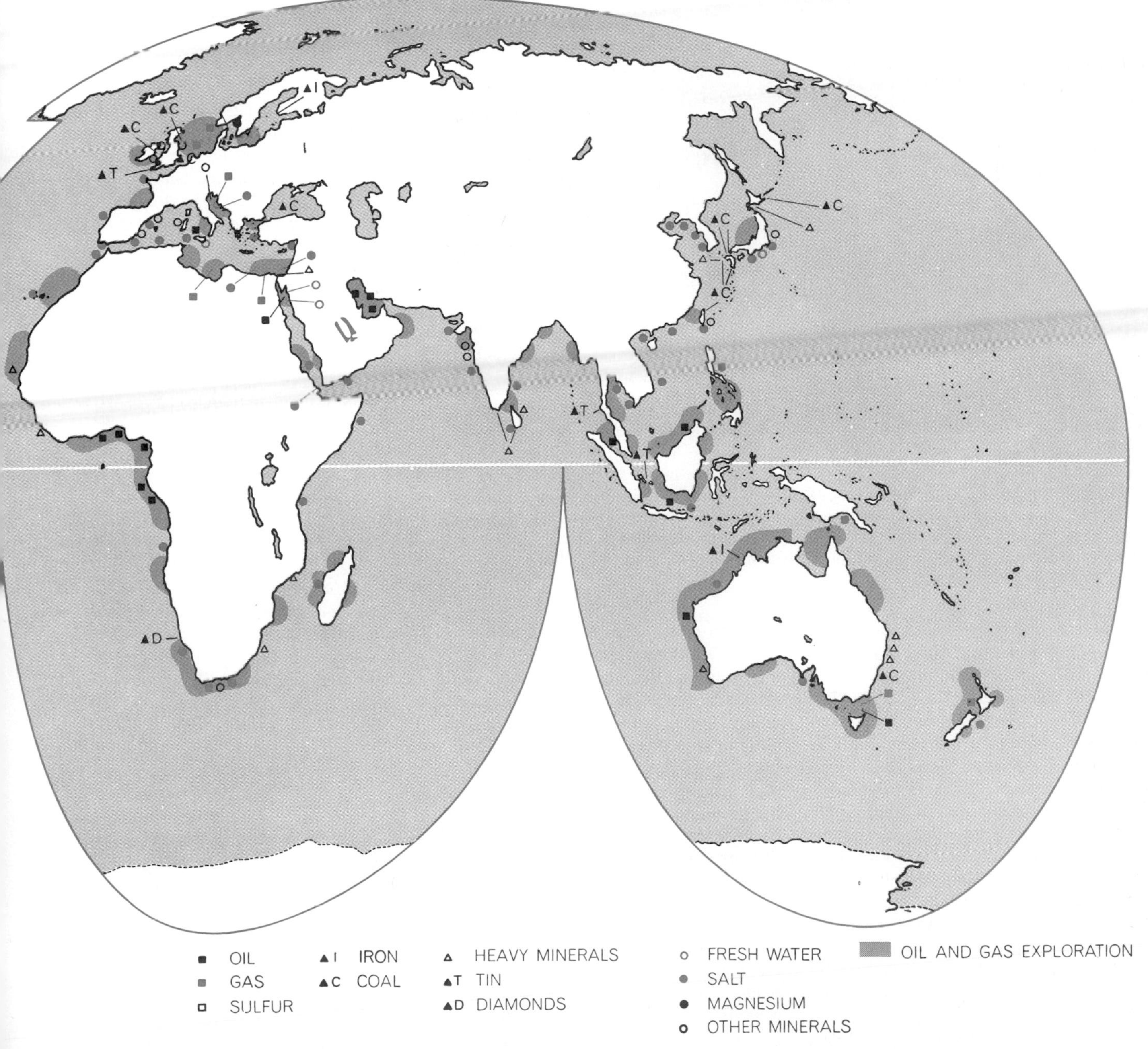

***Gas Journal*, the magazine *Offshore* and other sources. Oil, gas and sulfur are produced by drilling; coal and iron ore from mines driven from dry land; heavy minerals, tin and diamonds by dredging; fresh water, salt, magnesium and other minerals from seawater.**

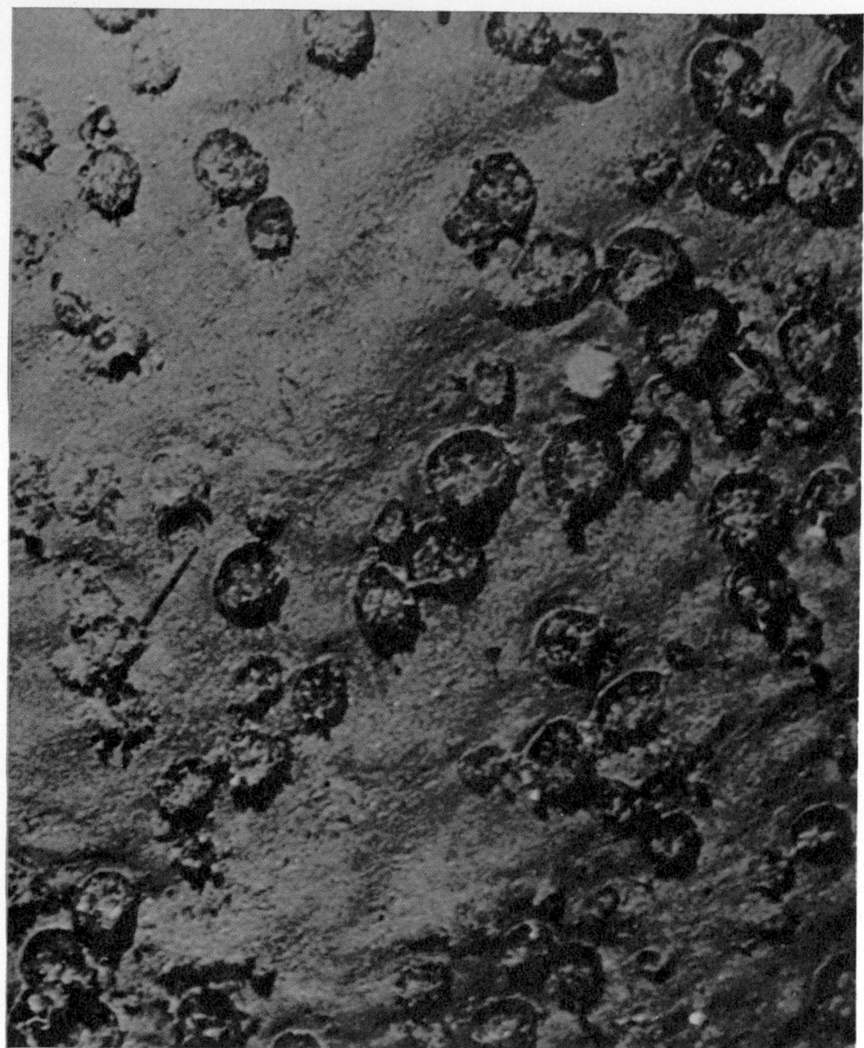

MANGANESE NODULES, formed by precipitation from seawater, are generally found on the deep-sea floor. These nodules were photographed on the Blake Plateau off the southeastern U.S., less than 3,000 feet deep, by a prospecting ship operated by Deepsea Ventures, Inc. They average two inches in diameter, about a quarter-pound in weight. The manganese content is between 15 and 30 percent, the nickel and copper content about 1 percent each.

ricanes have caused $5 billion in damage to the U.S. economy in the past 15 years) and human exploitation and neglect.

More than a tenth of the 10.7 million square miles of shellfish-producing waters bordering the U.S. is now unusable because of pollution. Dredging, drainage projects and even chemical mosquito-control programs are having devastating effects on fish and other aquatic life. The amount of industrial waste reaching the oceans will increase sevenfold within the decade. Whereas 14 nuclear-powered generating plants are operating in the U.S. today, more than 100 are scheduled by 1975, with nine planned for Long Island Sound alone. Thermal pollution from the discharge of hot water is therefore a potential threat to coastal waters as well as inland lakes and rivers.

In the competition for the zone's resources among different uses—industrial and housing development, ports, shipbuilding, recreation, commercial fisheries and waste disposal—natural wetlands and estuarine open spaces are losing out. Of the tidal wetlands along the Atlantic coast from Maine to Delaware, 45,000 acres were lost between 1955 and 1964. An inventory shows that 34 percent of that area was dried up by being used as a dumping ground for dredging operations; 27 percent was filled for housing developments; 15 percent went to recreational developments (parks, beaches and marinas) and 10 percent to bridges, roads, parking lots and airports; 7 percent was turned into industrial sites and 6 percent into garbage and trash dumps. (In Maryland 176 acres of submerged land in Chesapeake Bay were sold recently for $100 an acre and, after being filled with dredged bay-bottom muck, were subdivided into lots selling for between $4,000 and $8,000 each.)

With the demand for marine recreation growing with the coastal population, pressure is increasing on the one-third of the coastal zone that has recreational potential. Only about 6.5 percent of this is now in public ownership, yet in order to meet the projected demand it is considered essential that about 15 percent be accessible to the public. The mere fact that coastal land with recreational potential exists, moreover, is far from meaning that it will ever be put to recreational use. Swimming, boating and skin diving are often incompatible with competing alternative uses, many of which appear to have equally valid claims. In the face of conflicts between public and private, and long-term and short-term, benefits, how and by whom will the ultimate decisions be made on the proper utilization of coastal land?

Management of the coastal zone is unwieldy because the environment is almost hopelessly fragmented by political subdivisions: 24 states, more than 240 counties, some 600 coastal cities, townships, towns and villages and numerous regional authorities and special districts with their own regulatory powers. Superimposed on the many public jurisdictions there is another tapestry of private ownership. Because the states hold coastal resources in trust out to the three-mile limit, the Federal Government has a restricted role in resolving disputes, but it may be able to exert leadership in defining the issues.

Thoreau once admonished: "What is the use of a house if you haven't got a tolerable planet to put it on." Unless rational alternatives among competing uses are evaluated, the trend will continue to be toward single-purpose uses, motivated by short-term advantages to individuals, industry or local governments. Such exploitation may actually dissipate resources. Private beach development restricts public access; dredging and filling downgrade commercial fishing; offshore drilling rigs limit freedom of navigation. Each single-purpose use may seem justifiable on its own, but the overall effect of piecemeal development can be chaos.

In this technological age man can do many more of the things he wants to do. The oceans place before him a vast store of little-developed material resources; the tools of science and technology are at his disposal. This combination of a new frontier, new knowledge and new technical capability may be unique in the human experience. We are accumulating the basic information with which to define the ecological base from which we operate, to understand the natural forces at work and to predict the consequences of each insult to the environment. With this new comprehension it will soon be possible to develop the engineering with which to harvest mineral wealth, maintain water quality, inhibit beach erosion, create modern ports and harbors—and to establish the criteria for making necessary choices among courses of action and the law and institutions to effectuate them. In time we may even be able to correct mistakes that were made long ago in ignorance or that occur in the future because of man's stupidity, neglect or greed.

TIDAL WETLANDS, an important coastal resource, are disappearing rapidly. The top photograph shows Boca Ciega Bay, near St. Petersburg on the west coast of Florida, as it was in 1949. The bottom photograph shows the same area filled and developed, in 1969.

Fresh Water from Salt

by David S. Jenkins
March 1957

The thirst of civilization now presses hard upon the naturally available supply of fresh water. It is easy enough to desalt sea water, but it must be done cheaply, abundantly and soon

In the face of our increasing control over nature, it is ironic that we have steadily been losing ground with regard to one of mankind's most vital needs—water. Over much of the civilized world, water shortage is a grave and growing problem.

To be sure, water has always been a major concern of man. The children of Israel recovered their faith in God only when Moses smote the rock and produced water. Egypt rose and fell with the flow of the Nile and even today is placing its hopes for the future on plans to develop the resources of that great river. Few things have more powerfully influenced the course of the human race than the perennial search for fresh water.

But in today's world the need for water has become acute in many areas. In some arid countries the per capita consumption of water, thanks to improved sanitation, has suddenly risen from two or three quarts per day to 20 or 30 or more. Underdeveloped countries seeking to raise their standard of living by industrialization and irrigation find themselves with huge new needs for water supplies. Even our own water-favored country is beginning to be concerned, with many communities already facing shortages. The problem is widespread, not confined to localities such as the drought-stricken Southwest. Since 1900 the U. S. has increased its consumption of water almost sevenfold. By 1975 our water requirement will have nearly doubled again. We shall then be using about 27 per cent of the total supply of natural fresh water in our rivers, lakes, springs and wells. Many areas will have reached the limit of their local resources. The remaining 73 per cent of the total supply, largely stream water, will probably be prohibitively expensive to collect, store and distribute to the places where it is needed.

There are two major steps we can take. The first, and most important for the years immediately ahead, is to reduce our lavish waste of water. Among other things, we can reduce the pollution of our streams, recover used water by purification, capture floodwaters and manage the industrial use of water more carefully. Some industries have already shown what can be accomplished along this line. For instance, the Kaiser steel mill at Fontana, Calif., has reduced the consumption of water in manufacturing steel from the average of 65,000 gallons per ton to only 1,400 gallons per ton, by recirculating the water it uses.

But in the last analysis we must also increase the water supply itself. There is one way this can be achieved on a very large scale: by converting salt water. There is plenty of water in the oceans, and man's ingenuity is certainly equal to the task of converting sea water and other saline waters into fresh water at a reasonable cost.

Five years ago the U. S. Congress, recognizing the gravity of the situation, passed a Saline Water Conversion Act which authorized a program of research and development. Under the administration of the Department of the Interior a coordinated campaign of investigation is now under way in the U. S., with cooperation from abroad. Government agencies, private industries and other institutions are engaged in more than a score of laboratory investigations, a few of which have progressed to pilot-plant tests. The salt-water conversion methods under investigation include a number of well-known processes and several completely new ideas.

Fresh water is commonly defined as water containing less than 1,000 parts per million of dissolved salts. But how fresh the water needs to be depends on the use to be made of it. Drinking water, according to the U. S. Public Health Service standards, should have no more than 1,000 and preferably less than 500 parts per million. In general salinity of water for agricultural irrigation should be no more than 1,200 parts per million, the allowable concentration depending on the specific salts it contains. For some industrial purposes, such as cooling and flushing, unrefined sea water will do; on the other hand, in high-pressure boilers it may be necessary to have almost pure water containing not more than two or three parts per million of salt. Thus the economic feasibility of sea-water conversion depends on the use to which the water is to be put: for some purposes the cost may be reasonable, for others not. For the guidance of the research program a survey of the various industries' water requirements is urgently needed.

The salinity of the waters available for conversion varies greatly. The oceans are

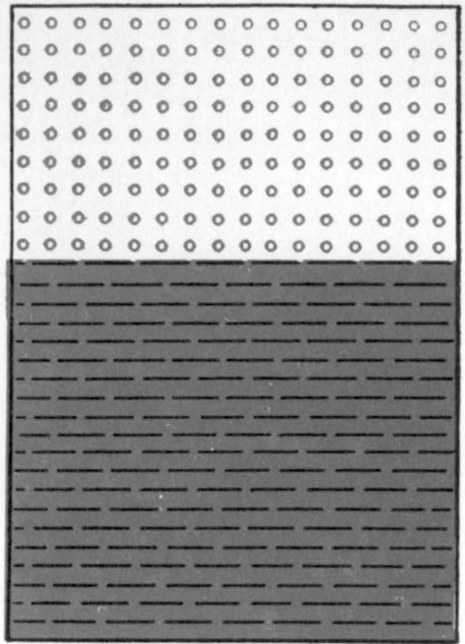

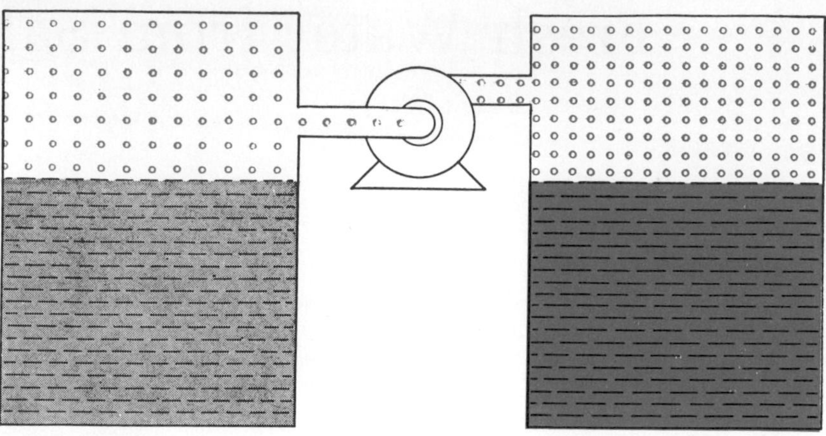

THEORETICAL ENERGY needed to separate water molecules from salt ions can be calculated from experiment shown here. At the same pressure and temperature, more water molecules will go into the vapor phase from fresh water (*left*) than from salt water (*center*). By compressing the vapor from the salt water to the same density or vapor pressure as that of the fresh water vapor, it can be just made to condense to fresh water (*right*). The energy needed is equal to the energy that binds the water molecules to salt ions.

fairly uniform, averaging about 35,000 parts per million of dissolved salt. But in the Persian Gulf it is nearly 40,000 parts per million; in Chesapeake Bay about 15,000; in the Baltic Sea only 7,000. Any water less salty than the oceans but with more than 1,000 parts per million is called brackish.

Common salt, sodium chloride, accounts for most of the saltiness of sea water. However, sea water contains small amounts of many other salts—some 44 dissolved elements in all [*see table on page 272*].

What is required to desalt water? The basic facts are simple enough. A salt dissolved in water is separated into ions —*e.g.*, in the case of sodium chloride: the positively charged sodium ion and the negatively charged chlorine ion. The ions are bound to water molecules by their electric charges. The problem, then, is to pull the water molecules and the ions apart.

We can make a calculation of how much energy this takes. Consider two sealed flasks, one partly filled with pure water, the other with sea water. At room temperature, say, a certain amount of the water in each flask evaporates into the unfilled part of the vessel, and this establishes a certain equilibrium vapor pressure. The vapor pressure in the container of sea water is lower than in the flask of pure water, because its water molecules, being bound to salt ions, do

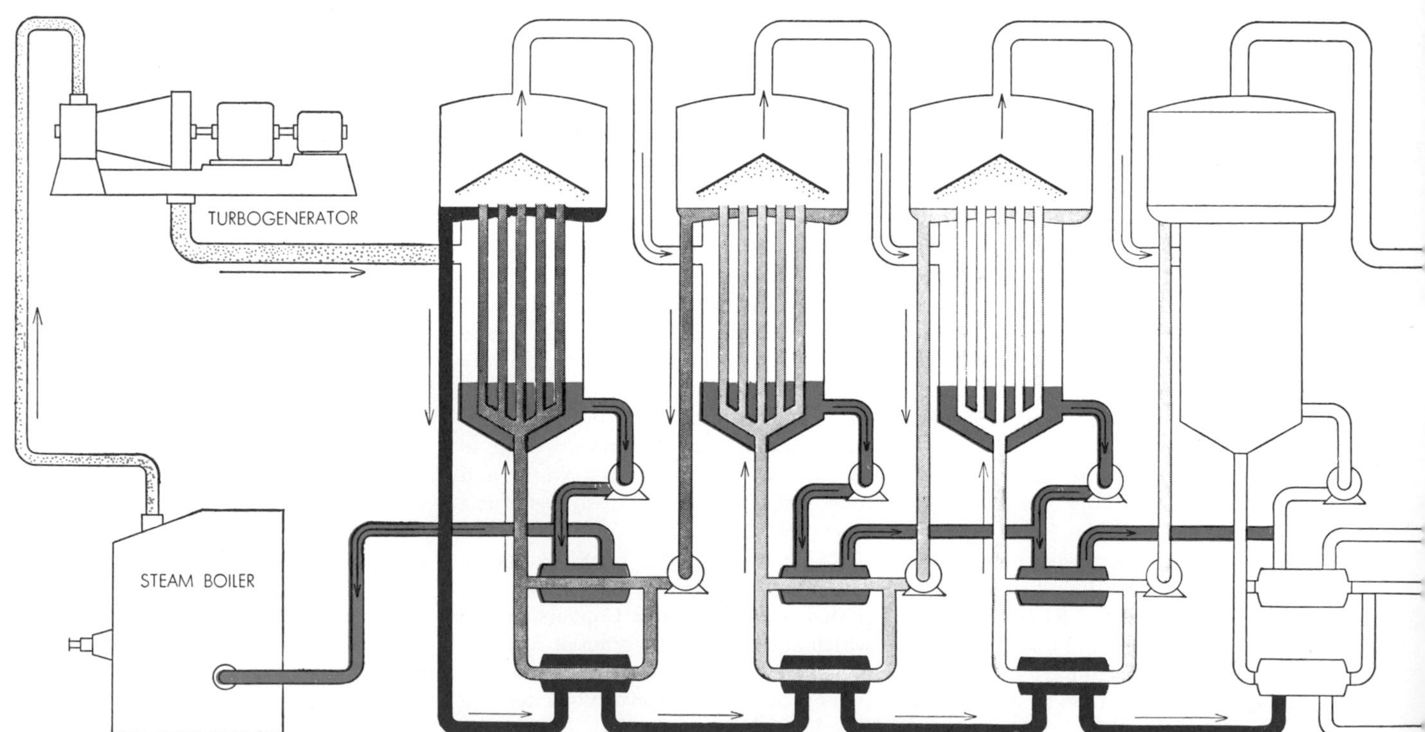

MULTI-STAGE STILL uses the latent heat released by condensation of water vapor at each stage to evaporate the brine flowing in from the next stage. The first stage (*left*) provides a condenser for the steam from a turbogenerator. The steam (*colored dots*) is condensed to water by the cooler salt water (*gray*) flowing in through the evaporator tubes. The salt water, heated by the condensation of the steam, separates into water vapor (*colored dots*) and droplets of brine (*black dots*) in the steam chest above the evaporator.

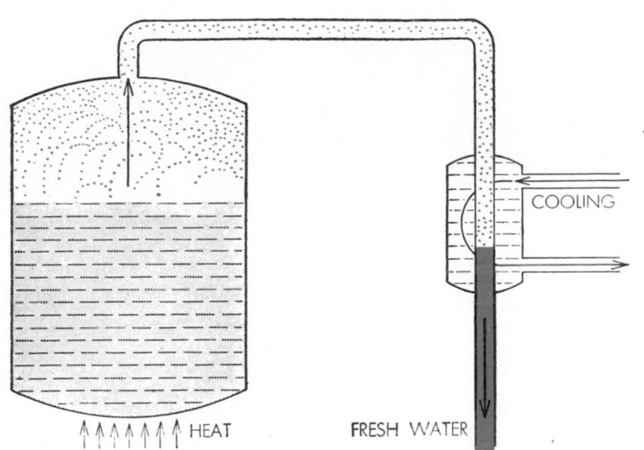

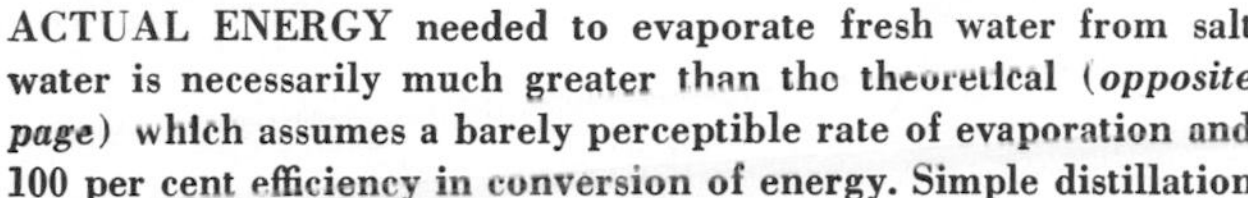

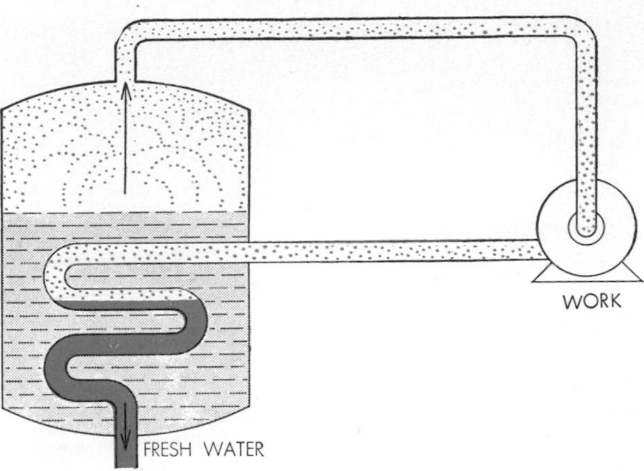

ACTUAL ENERGY needed to evaporate fresh water from salt water is necessarily much greater than the theoretical (*opposite page*) which assumes a barely perceptible rate of evaporation and 100 per cent efficiency in conversion of energy. Simple distillation (*left*) may require 1,000 times as much energy. Compression distillation (*right*) is more efficient. Compression of the vapor raises its temperature; the superheated vapor is piped through the boiler where it condenses, yielding its latent heat to evaporate more brine.

not evaporate as easily. Now the extra energy needed to separate water molecules from the ions can easily be measured: it is just equal to the energy we have to supply to compress the vapor from the sea-water bottle so that its pressure is the same as that of the vapor in the bottle of pure water [*see diagrams on opposite page*]. For sea water of average saltiness this energy amounts to 2.8 kilowatt hours per 1,000 gallons.

But this is merely the minimum amount needed to tip the scale of vapor-pressure equilibrium so that evaporation can proceed—at a barely perceptible rate. To raise the rate of evaporation to a useful level requires a great deal more energy. Furthermore, there are fundamental limitations on the efficiency of conversion of energy to useful work in any process or machine; consequently a considerable part of the energy we feed into the machine is unavoidably wasted. In practice, to separate water from salt by evaporation in a simple still takes 1,000 times the amount of energy given above as the theoretical thermodynamic requirement—that is, about 2,800 kilowatt hours of energy (in the form of heat) per 1,000 gallons of water. But there are, of course, much more efficient processes than simple, single-stage distillation. It is estimated that some of the processes now under study may reduce the energy requirement to about four or five times the thermodynamic minimum,

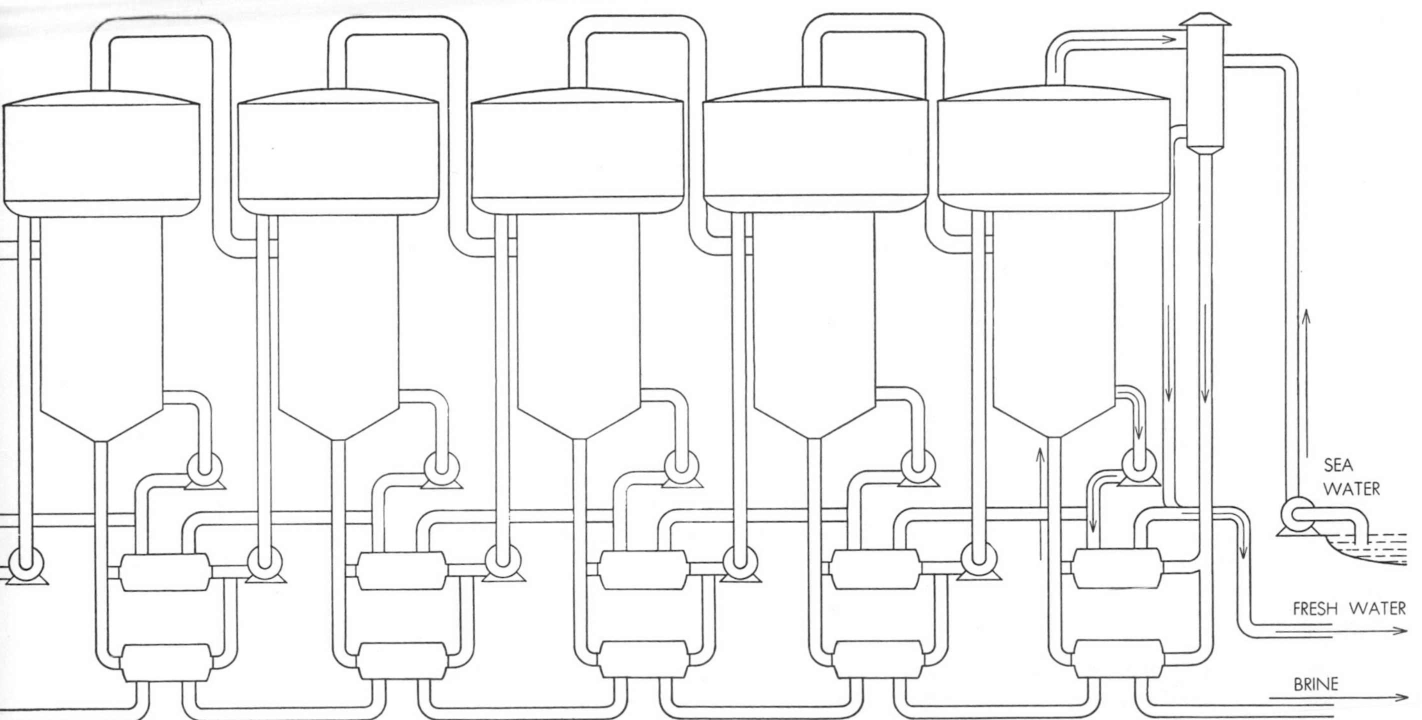

Here the brine droplets are stopped by the conical baffle plate and fall to the bottom of the steam chest, draining out at left. The water vapor rushes out at the top and to the right into the second stage. There it gives up its heat to evaporate the salt water flowing in from the third stage and condenses to fresh water. Because the temperature of the salt water at each stage is lower, the boiling point must be brought lower; this is accomplished by reducing the pressure at each stage, as indicated by increased size of steam chests from left to right.

EXPERIMENTAL SOLAR STILLS designed by Maria Telkes are shown on rooftop at New York University. In unit at left water vapor rising from shallow pool of sea water condenses on underside of inclined glass or plastic roof. Units at right are "sloping stills," in which sea water is flowed through a black wick tilted at right angles to sun's rays with considerable increase in output.

or about 10 to 15 kilowatt hours per 1,000 gallons.

Let us see what devices we can employ to enhance the efficiency of distillation. The simple still evaporates water to steam at atmospheric pressure. But now if we compress the steam to a few pounds per square inch above atmospheric pressure, the temperature of the steam will rise slightly, and we can use this added heat to evaporate more water. In other words, we have increased the yield of distilled water without feeding more heat to the system, merely spending a little energy to drive a mechanical compressor. This method, called "compression distillation," reduces the total energy requirement from 2,800 kilowatt hours to about 200 kilowatt hours per 1,000 gallons of water. The idea is more than 100 years old: it was first patented by a Frenchman, Pierre Pelletan, in 1840. The U. S. armed services used it extensively during World War II for supplying water to troops in areas lacking ready fresh water.

In the past three years interest in compression distillation has been heightened by an exciting new system. It was devised by Kenneth C. D. Hickman, a collaborator in the governmental research program. In essence what Hickman has added is a simple device for increasing phenomenally the rate of heat transfer to the water: namely, spreading it out in a thin film. The salient feature of his device is a rotating drum, shaped something like a child's musical top [*see diagram on page 274*]. Salt water at a temperature of 125 degrees Fahrenheit is sprayed on the inside surface of the drum. The centrifugal force of the drum's rotation spreads the water over this surface as a very thin, turbulent film. Some of the water evaporates (the unevaporated brine is constantly drawn off through a scoop). The water vapor leaves the drum via a pipe where a blower compresses it slightly, raising its temperature. The warmed vapor then circulates to the outside surface of the drum; there it condenses and gives up its latent heat; the drum shell transmits this heat to the film of water on its inside surface, speeding evaporation. The condensed vapor is collected as distilled water.

This system is recommended not only by its simplicity and low power requirement but also by another great advantage: the low operating temperatures (125 to 150 degrees F.). In distillation processes using much higher temperatures, the sea-water salts are deposited on metal surfaces as scale. Scale formation, which impedes the transfer of heat to the water, is the greatest single enemy of efforts to bring down the cost of distillation. In Hickman's apparatus little scale forms, because of the low operating temperatures. The main limitation of the rotary compression still is that such a still obviously must be limited in size.

There are other highly promising attacks on the problem of improving the efficiency of distillation. One of the most hopeful is the multi-stage still. In this system the latent heat released by the condensation of the evaporated water at each stage is used for the next stage, providing a chain effect. For example, in the first stage sea water is evaporated to steam at atmospheric pressure or higher; the steam passes to coils in a second evaporator, condenses there and is collected as distilled water; in condensing it releases its latent heat to evaporate sea water in that container, and so on through a number of stages [*see diagram on pages 268 and 269*]. Vacuum pumps keep each successive evaporator under lower pressure, so that its water boils at a lower temperature. Such a system could operate on the exhaust (waste) steam from an electricity-generating plant. The main problem is to prevent the formation of scale, and the method is promising enough to justify the considerable research being conducted on that aspect. If scale can be eliminated, a 20-stage still may be feasible, and it might produce fresh water at between 30 and 40 cents per 1,000 gallons.

There is a comparatively new distillation method which uses a sudden reduction of pressure instead of heat to evaporate water. At a given temperature, the amount of water vapor that air can hold depends on the air pressure. If salt water is fed into a closed chamber in which the pressure is lower than outside, part of the water will "flash" to steam. This method is being used extensively in multi-stage systems. In French West Africa engineers are attempting to develop a flash evaporator which will operate on the temperature differences between upper and lower levels of the ocean, using the colder water to chill and condense the vapor from the flash chamber. The U. S. Department of the Interior and the University of California have been working on similar systems.

Various other distillation processes have been studied, including distillation at "supercritical" temperatures and pressures. It is not possible to discuss here all the distillation ideas that are under study. But even this brief description of the work in progress must make clear that the prospects for distilling salt water to fresh at a reduced cost are bright.

The sun, which showers us with a vast abundance of energy free of charge, is responsible for our natural supply of fresh water by its evaporation of the seas. Is there any way to harness solar energy to provide us with more? Some ingenious solar stills have been proposed.

The simplest form of solar still is a pan

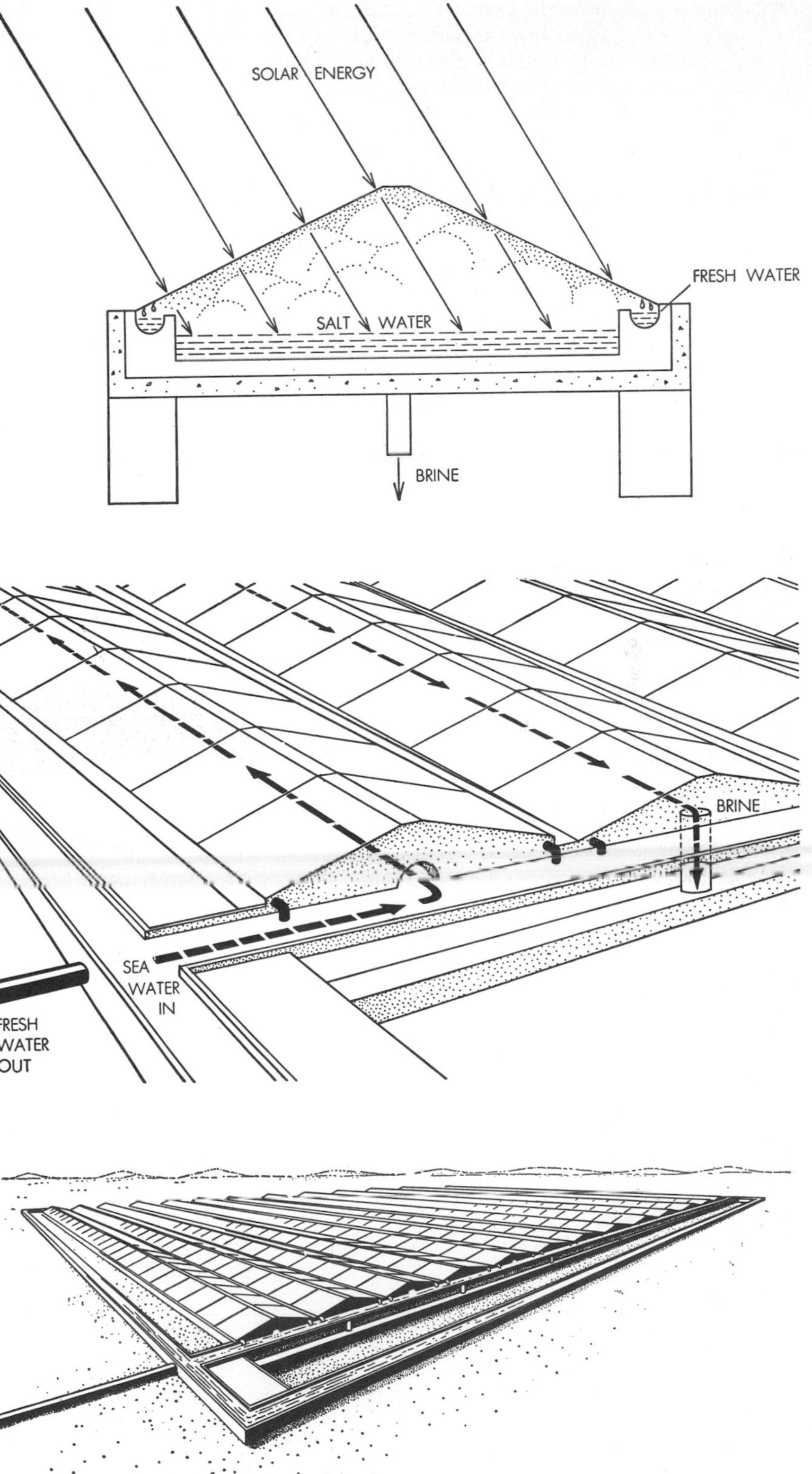

SOLAR STILL of type under development by Department of the Interior is diagrammed. Water vapor evaporating from salt water condenses on underside of glass or plastic plates and drains into gutter on either side of unit (*top diagram*). Incoming sea water (*middle diagram*) flows into first unit and returns through second. Outlet pipes for fresh water and brine are immersed in incoming sea water in order to preheat it by heat-exchange effect. Drawing at bottom shows the plan of a large installation which might cover many acres.

containing a shallow layer of salt water (say about an inch) and covered with a sloping glass plate. The glass is transparent to the sun's radiation but holds in the heat reradiated within the pan. Water evaporated from the bottom condenses on the glass, trickles down its sloping surface and is collected in a trough. This type of still, using only about half of the incoming solar energy, can produce little more than a pint of fresh water per day per square foot of area, even in the hot, clear climate of Arizona.

Some economy can be achieved by reducing the cost of the equipment. Several manufacturers have recently produced transparent plastic films which can replace glass at much less expense. One of them is a fluorocarbon called Teflon, reported to resist all forms of weathering. E. I. du Pont de Nemours and Company has designed an arrangement in which the Teflon canopy is supported by inflating it to slightly higher than atmospheric pressure, eliminating the need for a supporting frame.

Several radically new designs for solar stills are now under serious study. In Denver George O. G. Lof, a consulting engineer, is investigating for the Department of the Interior a still in which the ground acts as a storage bank for the sun's heat. A basin containing a foot of water is placed directly on the ground, so that solar heat absorbed by the water is transmitted to the ground. This heat reservoir then continues to evaporate water when the sun is not shining. If the loss of heat by radiation at night is not too great, it has been estimated that this type of still may produce up to a fifth of a gallon of fresh water per day per square foot at something like 50 cents per thousand gallons.

Maria Telkes of New York University has designed an interesting 10-stage still. It operates without machinery or any energy requirement except solar heat (or comparatively low-temperature heat from some other source). The apparatus is a sandwich-like arrangement of alternate absorbing and condensing layers. A black wick in sheet form, soaked with salt water, absorbs the sun's heat. The evaporated water condenses in the next layer, gives up heat to warm the next wick, and so on. This arrangement produces five or six times as much water as a single-stage solar still per square foot of area exposed to the sun.

Solar stills of various types are being developed in the U. S., North Africa, Australia, Spain, Italy and elsewhere.

EXPRESSED AS SALTS	PARTS PER MILLION PARTS SEA WATER (APPROXIMATE)
SODIUM CHLORIDE ($NaCl$)	27,213
MAGNESIUM CHLORIDE ($MgCl_2$)	3,807
MAGNESIUM SULFATE ($MgSO_4$)	1,658
CALCIUM SULFATE ($CaSO_4$)	1,260
POTASSIUM SULFATE (K_2SO_4)	863
CALCIUM CARBONATE ($CaCO_3$)	123
MAGNESIUM BROMIDE ($MgBr_2$)	76
TOTAL	35,000
EXPRESSED AS IONS	
CATIONS	
SODIUM (Na^+)	10,722
MAGNESIUM (Mg^{++})	1,297
CALCIUM (Ca^{++})	417
POTASSIUM (K^+)	382
TOTAL	12,818
ANIONS	
CHLORIDE (Cl^-)	19,337
SULFATE (SO_4^-)	2,705
BICARBONATE (HCO_3^-)	97
CARBONATE (CO_3^{--})	7
BROMIDE (Br^-)	66
TOTAL	22,212

SALTS IN SEA WATER are of many varieties, the principal being shown here. Sea water contains 44 principal elements, including gold in the amount of .000006 parts per million.

Now let us turn from distillation to other methods of separating salt from water. In recent years the ion-exchange method has been used by industry for special purposes, such as refining brackish water. But it appears that ion-exchange systems will not become sufficiently economical for large-scale desalting of sea water.

In the ion-exchange process for treating water, the salt water is washed through resins or other material where its salt ions are replaced by unobjectionable ions. Ion exchange has been employed for softening water, for purifying water for special industrial purposes and for desalting brackish well water in the Sahara Desert and elsewhere.

Now the ion-exchange principle has been applied to form selective membranes which can separate ions from water. Ion-exchanges within the membrane make it impermeable either to positive or to negative ions. In the case of a membrane impermeable to positive ions but not to negative ones, an electric current will drive the negative ions through the membrane while it repels the positive ones. If a current is applied in a tank of salt water divided into compartments by a series of membranes, alternately permeable to positive and to negative ions, the salt ions collect in alternate compartments and the water in

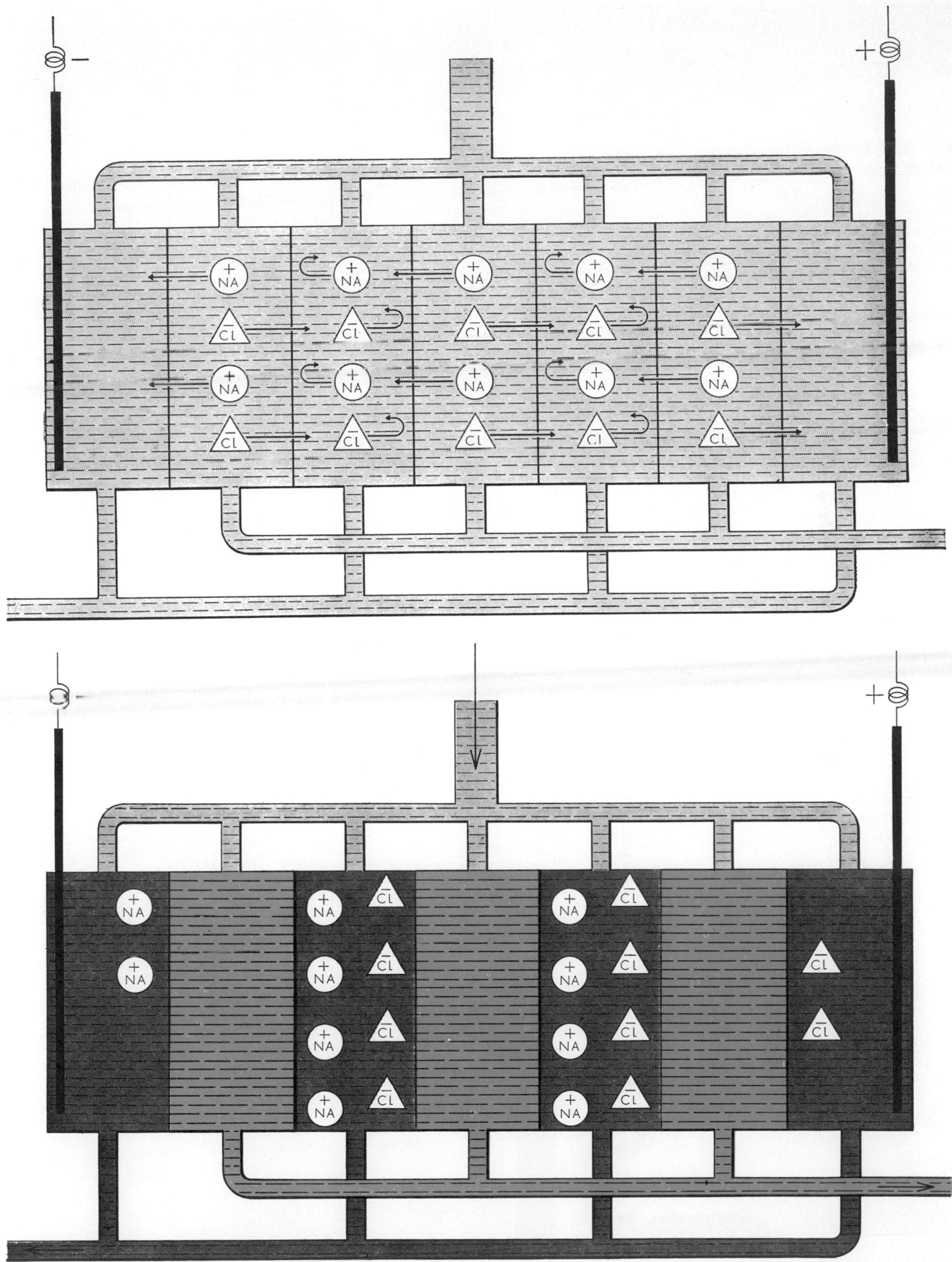

ION-EXCHANGE SEPARATION of fresh water from salt employs membranes which are alternately permeable to the sodium or chlorine ion and impermeable to the other. By applying an electric current across the system (*top diagram*), the sodium ions are attracted toward one end of the system and the chlorine ions toward the other. The ions are thus concentrated in alternate cells, leaving desalted water in the cells between (*bottom diagram*). The brine can then be drawn off via one pipe and desalted water via another.

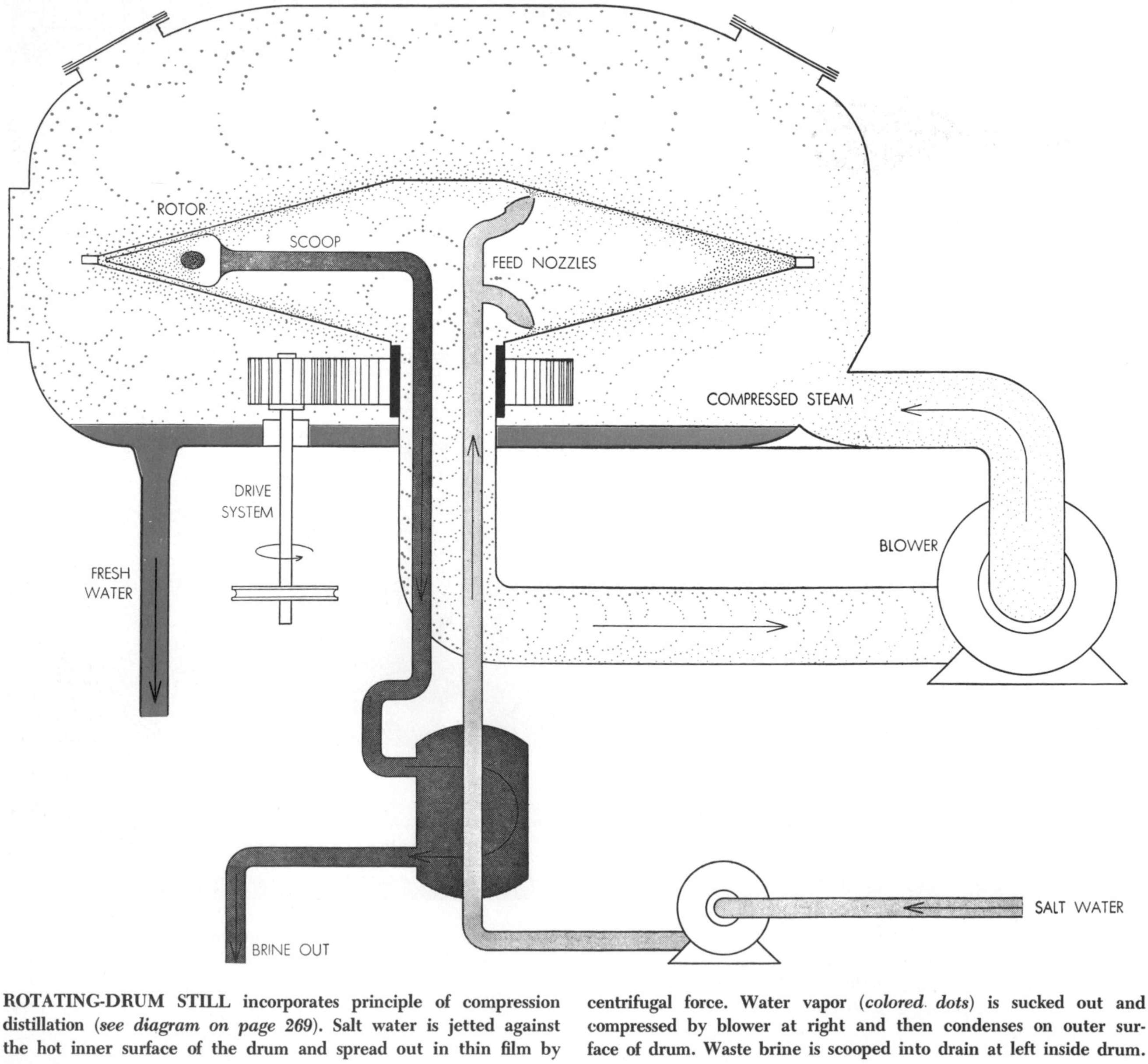

ROTATING-DRUM STILL incorporates principle of compression distillation (*see diagram on page 269*). Salt water is jetted against the hot inner surface of the drum and spread out in thin film by centrifugal force. Water vapor (*colored dots*) is sucked out and compressed by blower at right and then condenses on outer surface of drum. Waste brine is scooped into drain at left inside drum.

the intervening ones is desalted [*see diagram on page 273*].

This process, called electrodialysis, is being developed by research groups in the U. S., the Netherlands, England and the Union of South Africa. Because of the electric power requirement, it does not look economically promising for converting sea water, but it offers good prospects for desalting brackish waters.

George W. Murphy, now at the University of Oklahoma, has proposed using the electrical charge of ions from a strong brine, instead of an electric current, as the driving power to push ions through the membrane. Research on this possibility is being done at the Southern Research Institute in Birmingham.

Another promising membrane method is based on the phenomenon of osmosis. As every student of chemistry knows, if a salty solution is divided from a less salty one by an osmotic membrane, which is impermeable to salt but not to water, water passes through the membrane into the more salty solution, tending to equalize the salinity on both sides of the membrane. But this process can be reversed by applying to the more concentrated solution a mechanical pressure greater than the osmotic pressure acting on the water (which amounts to 350 pounds per square inch between fresh water and sea water). That is to say, the "reverse osmosis" forces water through the membrane out of the salty solution, while the membrane holds back the salt ions. Charles E. Reid of the University of Florida has demonstrated that with membranes made of cellulose acetate, 90 to 95 per cent of the salt can be removed from sea water in one pass.

Two other processes which offer promise are being developed. One is separation of salts from water by freezing. The Carrier Corporation, under contract with the Department of the Interior, is conducting research on a very attractive combination of freezing and evaporation, and similar developments have been reported by Israel and Yugoslavia. The other promising process is separation of water by dissolving it in organic solvents which do not dissolve salts.

We are already converting salt water to fresh for some purposes. In five to 15 years we should be able to convert it at reasonable cost for a much larger number of uses, with industrial uses first. Undoubtedly man will find not one but many solutions of the problem, and will begin manufacturing water by a number of devices and on a scale ranging from small household stills to large municipal and industrial plants.

MULTI-ROTOR STILL made up of eight rotating drums of type shown on opposite page, each eight feet in diameter, is under development at Badger Manufacturing Company in Cambridge, Mass. Rotary drum achieves high rate of heat transfer and thus high efficiency.

24 The Disposal of Waste in the Ocean

by Willard Bascom
August 1974

Contrary to some widely held views, the ocean is the plausible place for man to dispose of some of his wastes. If the process is thoughtfully controlled, it will do no damage to marine life

No one would dispute the wisdom of protecting the sea and its life against harm from man's wastes. An argument can be made, however, that some of the laws the U.S. and the coastal states have adopted in recent years to regulate the wastes that can be put into the oceans are based on inadequate knowledge of the sea. It is possible that a great effort will be made to comply with laws that will do little to make the ocean cleaner.

This discussion of waste disposal will be limited to disposal in the ocean; it will not take up disposal in lakes, rivers, estuaries, harbors and landlocked bays. Indeed, part of the problem is that insufficient distinction has been drawn between the ocean and the other bodies of water, whose chemistry, circulation, biota and utilization differ from those of the ocean in many ways. It is not sensible to try to write one set of water-quality specifications that will cover all bodies of water. My concern here is only with the quality of ocean water and marine life along the U.S. Atlantic, Gulf and Pacific coasts. The scientific findings in those areas apply, however, to nearly any other coastal waters that are exposed to ocean waves and currents.

Some of the changes that human activities have wrought in the ocean environment are already irreversible. For example, rivers have been dammed, so that they release much smaller quantities of fresh water and sediment. Ports have been built at the mouth of estuaries, changing patterns of flow and altering habitats. On the other hand, certain abuses of the ocean have already been stopped almost completely by the U.S. Nuclear tests are no longer conducted in the atmosphere, so that radioactive material is no longer distributed over the land and the sea; the massive dumping of DDT has been halted, and the reckless development of coastal lands has been restrained by laws calling for detailed consideration of the impact on the environment.

Between these extremes is a broad realm of uncertainty. Exactly how clean should the ocean be? How unchanged should man try to keep an environment that nature is changing anyway? The problem is to decide what is in the best interest of the community and to achieve the objective at some acceptable cost. At the same time it is necessary to guard against the danger that excessive demands made in the name of preserving ecosystems will lead to action that is both useless and expensive.

Waste disposal automatically suggests pollution, which is a highly charged word meaning different things to different people. A definition is needed for evaluating accidental and deliberate inputs into the ocean. Athelstan F. Spilhaus of the National Oceanic and Atmospheric Administration, who has written extensively on pollution, defines it as "anything animate or inanimate that by its excess reduces the quality of living." The key word is excess, because most of the substances that are called pollutants are already in the ocean in vast quantities: sediments, salts, dissolved metals and all kinds of organic material. The ocean can tolerate more of them; the question is how much more it can tolerate without damage.

One approach to the question was suggested by the National Water Commission in its report of June, 1973, to the President and Congress: "Water is polluted if it is not of sufficiently high quality to be suitable for the highest uses people wish to make of it at present or in the future." What are "the highest uses" that can be foreseen for ocean waters, particularly those near the shore? They are probably water-contact sports, the production of seafood and the preservation of marine life.

Water-contact sports are occasionally

MONTEREY BAY in California appears in a deliberately overexposed aerial photograph on the opposite page. The overexposure through a filter is a technique that shows details of turbidity in the water caused by mud, organisms or waste. At bottom right a harbor projects into the bay. The light spot of water along the shore above and slightly to the left of the harbor is a sewage outfall. The dark brown spots are beds of kelp, and the reddish purple is a "red tide" consisting of large numbers of the marine organism *Gonyaulax*.

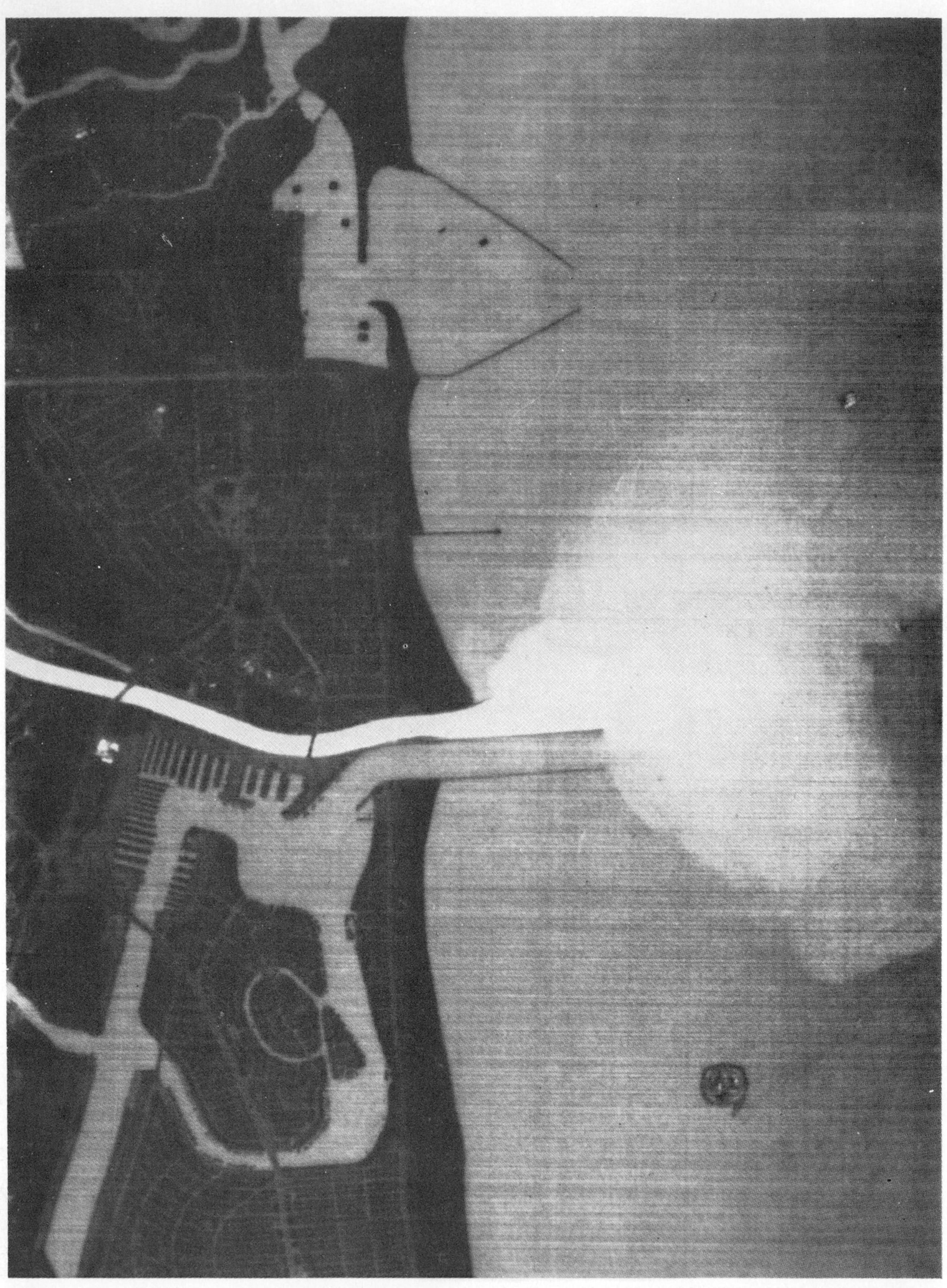

THERMAL EFFLUENT from generating plants of the Southern California Edison Co. and the Los Angeles Department of Water and Power appears in a thermograph of the San Gabriel River, which is the white strip at center, and San Pedro Bay. The intake temperature of cooling water from the bay was 18.9 degrees Celsius. On discharge into the river it was 26.7 degrees C. At the point where the river enters the bay the water temperature was 24.4 degrees C. The two plants jointly generate 900 megawatts of power.

inhibited by pollutants on the seacoasts of the U.S. Where such conditions exist they should be corrected at once. Even where coastal waters are clean the community must be alert to keep them so.

To maintain the ocean waters at an acceptable level of quality it is necessary to consider the main inputs of possible pollutants resulting from human activity. One of them is fecal waste (75 grams dry weight of solids per person per day), which after various degrees of treatment ends up in the ocean as "municipal effluent." Wastes also flow from a host of industrial activities. They are usually processed for the removal of the constituents that are most likely to be harmful, and the remaining effluent is discharged through pipes into the ocean. Dumping from barges into deep water offshore is a means of disposing of dredged materials, sewage sludge and chemical wastes. Thermal wastes include the warmed water from coastal power plants and cooled water from terminals where ships carrying liquid natural gas are berthed. In addition ships heave trash and garbage overboard and pump oily waste from their ballast tanks and bilges.

Such are the intentional discharges, but pollutants reach the ocean in other ways. Aerial fallout brings minute globules of pesticide sprayed on crops, particles of soot from chimneys and the residue of the exhaust of automobiles and airplanes. Painted boat bottoms exude small amounts of toxicants intended to discourage the growth of algae and barnacles. Forest fires put huge amounts of carbon and metallic oxides into the air and thence into the sea. Oil spills from ship collisions and blowouts during underwater drilling operations add an entire class of compounds.

Moreover, natural processes contribute things to the sea that would be called pollutants if man put them there. Streams add fresh water, which is damaging to marine organisms such as coral, and they also bring pollutants washed by rain from trees and land. Volcanic eruptions add large quantities of heavy metals, heat and new rock. Oil has seeped from the bottom since long before man arrived.

Finally, the ocean is neither "pure" nor the same everywhere. It already contains vast amounts of nearly everything, including a substantial burden of metals at low concentration and oxygen at relatively high concentration, plus all kinds of nutrients and chemicals. It has hot and cold layers, well stratified by the thermocline (the boundary between the warm, oxygen-rich upper layer and the cold, oxygen-poor depths). Waves and currents keep the water constantly in motion. It is against this complex background that man must measure the effects of his own discharges.

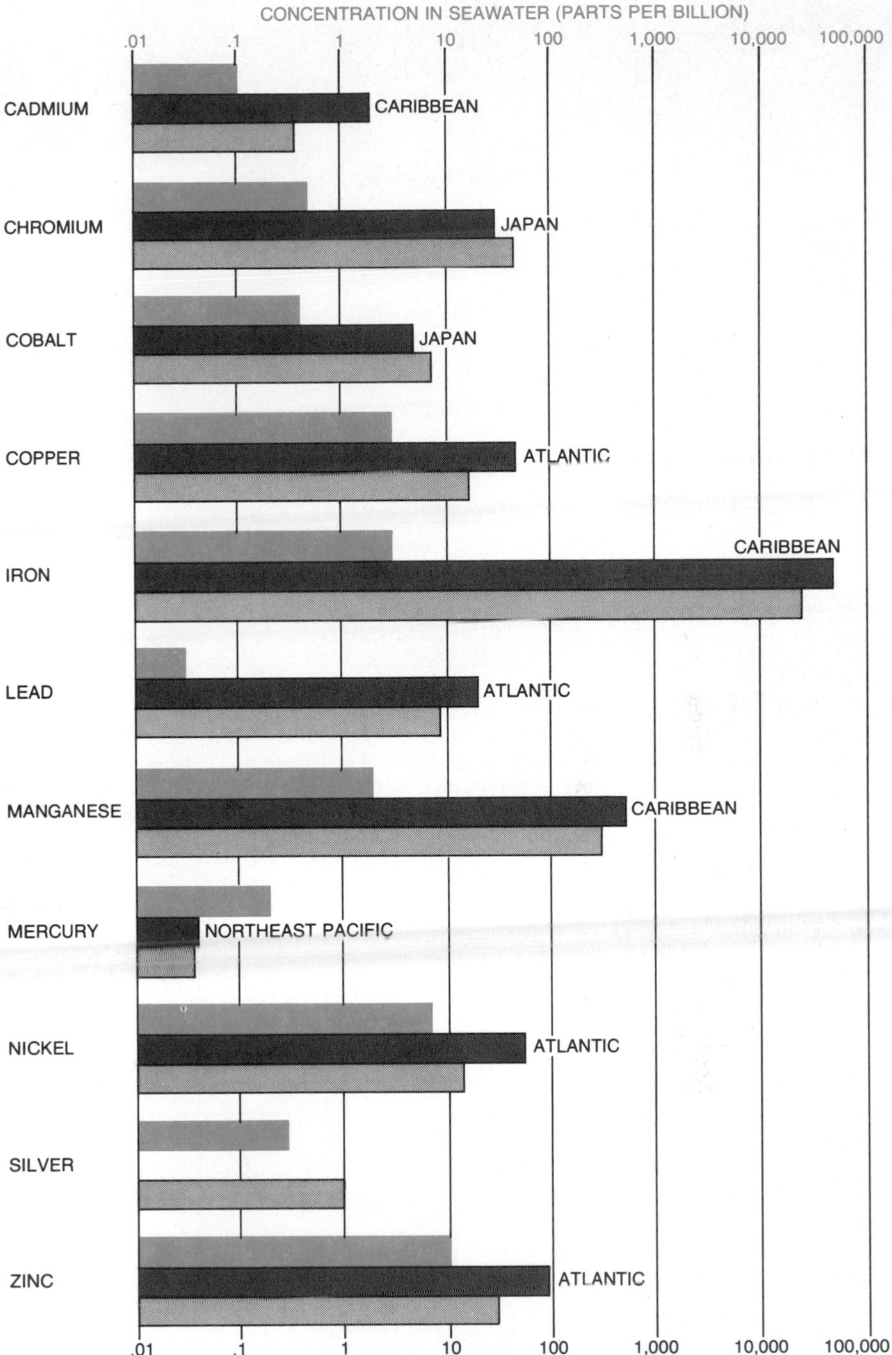

TRACE METALS in seawater (*color*) and top 10 centimeters of sediment (*gray*) are charted. Seawater figures are a worldwide average. The darker gray bars show concentrations at several sampling sites and the lighter bars the average from five sites along California coast.

Even if there were no people living on seacoasts, it would be impossible to predict accurately the kind and quality of marine life because of the natural variability of the ocean. The biota shifts constantly because the temperature and the currents change. Great "blooms" of plankton develop rapidly when conditions have become exactly right and then die off in a few days, depleting the oxygen in the water on both occasions. Within a single year the population of such organisms as salps, copepods and euphausids can change by a factor of 10. When the waters off California become warmer as the current structure shifts, red "crabs" (which look more like small lobsters and are of the genus *Pleuroncodes*) float by in fantastic numbers, fol-

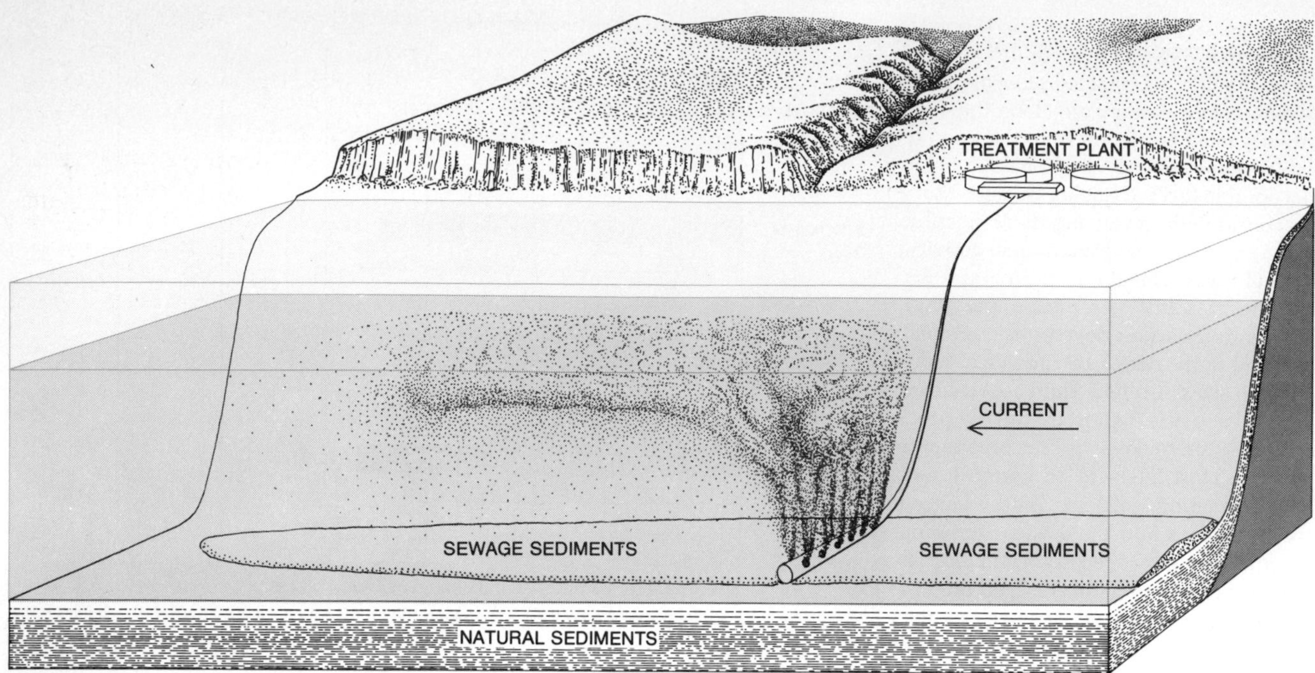

OCEANIC DISPOSAL of waste from a sewage-treatment plant is portrayed. The system is for a plant with a capacity of about 100 million gallons per day. Effluent from the plant flows a distance of from two to five miles through an outfall pipe that is from six to 12 feet in diameter. For about a quarter of a mile at its end the pipe has dozens of six-inch discharge ports. The mostly liquid material it discharges rises to the thermocline, which is the boundary between deep, cool water and the warmer surface layer. Prevailing current mixes material and moves it to one side or the other, depending on wind and tide. Some solid particles settle on the bottom.

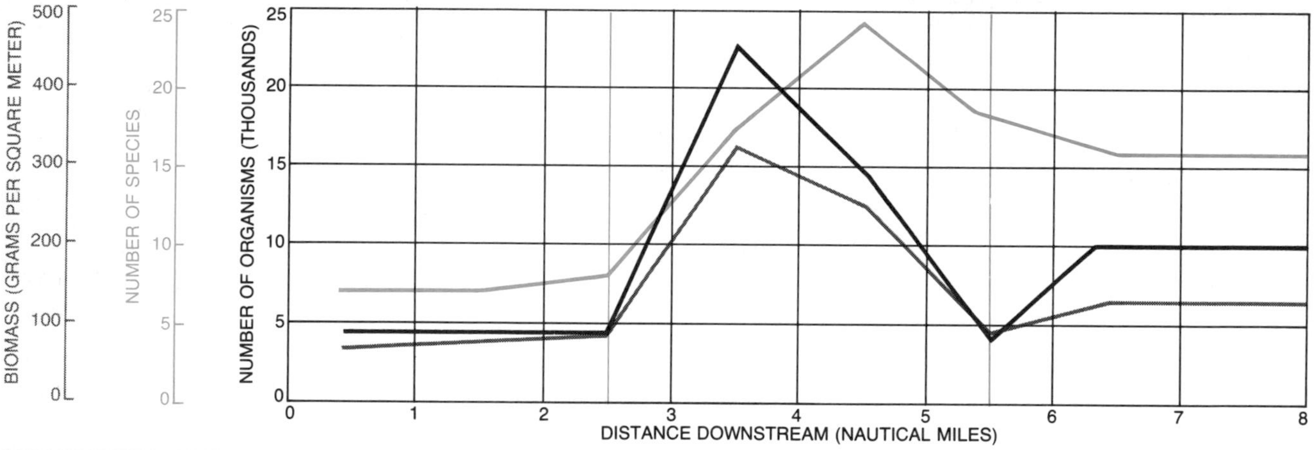

EFFECT OF OUTFALL on a community of polychaete worms is depicted. For about three miles downstream from the outfall the worm population is reduced. For the next three miles it is above normal. Thereafter it is about the same as in uninfluenced seabed.

lowed by large populations of bonito and swordfish. They came in 1973 as they did in 1958 and 1963, but the water soon turned cold again and the fish departed, leaving windrows of dead *Pleuroncodes* along the beaches.

The investigator's problem is to learn enough about the major natural changes so that he can tell whether or not human activities have any effect, either positive or negative. It is a signal-to-noise problem; here the changes one is trying to detect are often only a tenth of the natural biological and oceanic background variations. Both types of variation are hard to quantify.

In the case of the sardine, however, a record of the natural changes has been preserved below the floor of the Santa Barbara Basin off the coast of southern California. The bottom of the basin is anaerobic, that is, lacking in oxygen and so supporting little life. The particles that sift down to form sediments are undisturbed by burrowing creatures and therefore remain exactly as they land, in thin strata, layer on layer, one per year. The years can be counted backward, and the count can be confirmed with the lead-210 dating technique.

Some years ago John D. Isaacs of the Scripps Institution of Oceanography, who is also director of the California Marine Life Program, discovered in work with Andrew Soutar that each layer contains identifiable fish scales. Each layer showed a more or less constant number of anchovy and hake scales, but the sardine scales were present erratically, indicating major changes in the population. When the sardines disappeared about 1950, human activities were blamed. The geologic record clearly shows, however, that the sardines had come and gone many times before man arrived. Someday they will return.

It is obvious that some of man's wastes can be damaging to sea life; indeed, products such as DDT, chlorine and ship-bottom paint have been specifically designed to protect man against insects, bacteria and barnacles. Ionic solutions of certain metals are also known to be toxic at some level, as are numerous other substances. The problem is to determine what level is harmful, remembering that some of the substances are actually required for life processes. For example, copper is beneficial or essential for a number of organisms, including crabs, mollusks and oyster larvae. Other marine animals seem to require nickel, cobalt, vanadium and zinc.

Oceanographers would like to be able to demonstrate cause and effect in the ocean, that is, to show that some specific level of a metal does not harm marine life. Proving the absence of damage, however, including long-term and genetic effects, is difficult. Only on fairly rare occasions has it been possible to directly link a specific oceanic pollutant with biological damage. Examples include the finding by Robert Risebrough of the University of California at Berkeley that the decline of the brown pelican off California was attributable to DDT, which inhibits the metabolism of calcium and so makes the shells of the eggs so thin that the mother pelican breaks her own eggs by sitting on them. After patient scientific detective work the source was found to be a single chemical plant in the Los Angeles area. As a result of the work the plant was required to stop discharging DDT wastes into the ocean, and the brown pelican is now returning to California.

From what I have said so far it can be seen that the question of what is a pollutant or what amount of a substance represents pollution is not always easy to answer. Let me now try to put the main kinds of waste in proper perspective.

Municipal sewage containing human fecal material is the type of waste one usually thinks of first. It is certainly a natural substance; indeed, as "night soil" it has long been in demand as a fertilizer in many countries. Since it is not appreciably different from the fecal material discharged by marine animals, is there any reason to think it will be damaging to the ocean, even without treatment? Isaacs has pointed out that the six million metric tons of anchovies off southern California produce as much fecal material as 90 million people, that is, 10 times as much as the population of Los Angeles, and the anchovies of course comprise only one of hundreds of species of marine life.

Two aspects of municipal sewage do require attention. One of them is disease microorganisms. Human waste contains vast numbers of coliform bacteria; they are not themselves harmful, and they die rapidly in seawater (90 percent of them in the first two hours), but they are routinely sampled along public beaches because they indicate the level of disease microorganisms. When there are no endemic diseases in the city discharging the waste (the normal condition in the U.S.), there will be none in the water. It should be noted, however, that the assumption that disease microorganisms die off at the same rate as coliform bacteria is being questioned. It is necessary to guard against the possibility that such organisms will survive in bottom muds long enough to be stirred up by a major storm.

The usual way of reducing the bacterial count is to add chlorine to waste water that is about to be discharged. This approach seems reasonable, since chlorine is commonly added to drinking water and swimming pools to kill bacteria and algae. The trick is to add just the right amount, so that the chlorine exactly neutralizes the bacteria and no excess of either enters the ocean.

The other problem with sewage is one of aesthetics. People do not like to look at discolored water or oily films. A greater effort to reduce effluent "floatables" (tiny particles of plastic, wood, wax and grease) will help to reduce such effects. It will also reduce the number of bacteria reaching the shore, since many of them are attached to the particles.

Petroleum products are perhaps the most controversial marine pollutants. They are seen as small, tarlike lumps far out to sea and on beaches, as great slicks and as brown froth. From two to five million metric tons of oil enter the ocean annually. At least half of it is from land-based sources such as petroleum-refinery wastes and flushings from service stations. Significant quantities of oil enter the marine environment from airborne hydrocarbons. A considerable amount of oil must enter the ocean as natural seepage from the bottom, but it is obviously difficult to estimate how much.

Oil pollution from ships is the most serious problem. Oceangoing vessels shed oil in three ways: by accidents such as collisions; during loading and unloading, and by intentional discharge, which includes the pumping of bilges, the discharge of ballast by tankers and the cleaning of oil tanks by tankers. The ballast component is the worst.

After a tanker unloads its cargo of oil it takes on seawater (about 40 percent of the full load of oil) so that it will not ride too high in the water and be unmanageable. Any oil that remains in the tanks mixes with the water and is discharged with it when the ballast is pumped out in preparation for reloading the vessel with oil. The discharge of oil can be reduced in two ways. One is to wash the tank with water and stow the water aboard in a "slop tank," where the oil slowly separates from it. Then the water is discharged and the next load of oil is put on top of the oil that remains in the slop tank. This practice, which is described as being 80 percent effective, is followed in tankers carrying about 80 percent of the oil now transported at sea. The other stratagem is to build segregated ballast spaces into the double bottoms of new tankers, which reduces the discharge by 95 percent.

A system of international controls could virtually eliminate such discharges. There is an extra incentive for international controls because wherever oil is discharged, and by whatever ship, there is no telling to what shore it will be carried by winds, waves and currents. Substantial progress toward this kind of agreement has been made recently.

Ships are also responsible for most of the littering of the ocean and its shores. Waste consisting of paper, plastics, wood, metal, glass and garbage is customarily thrown overboard. The heavier material sinks quickly, littering the bottom; paper products disintegrate or become waterlogged and sink slowly, and the foods are soon consumed by marine scavengers. The wood, sealed containers and light plastics float ashore.

The estimated yearly litter from ships is about three million metric tons, much of which seems to come from the fishing fleets. The litter that Americans see and are annoyed by comes mostly from the land by way of streams or is thrown into harbors or tossed overboard from pleasure craft. Littering is an aesthetic problem rather than an ecological one, but it certainly reduces the quality of living. It can be curbed by the force of public opinion.

Dumping is a word with a specific meaning; it should not be confused with littering or with discharges from pipes. Dumping means carrying waste out to sea and discharging it at a designated site. Barges carrying solids simply open bottom doors and drop their cargo. Barges carrying liquids generally pump the material out through a submerged pipe into the turbulent wake of the vessel. Still other barges dump wastes enclosed in steel drums or other containers.

Much of the material is dredge spoil sucked up from harbor bottoms by hopper dredges to deepen ship channels. Some 28 million tons of this material were dumped into the Atlantic in 1968. Next in quantity in the New York area is relatively clean material removed from excavations for buildings; then comes sewage sludge, and finally industrial waste such as acids and other chemicals.

The amount of sludge dumped annually into New York Bight is about 4.6 million tons. Much of it is sewage sludge, which is a slurry of solid waste formed by sedimentation in primary sewage treatment or by secondary treatment in the activated-sludge process. For ocean discharge the material is thickened by settling or centrifuging to from 3 to 8 percent solids. Much of the solid material is silt, but complex organic materials and heavy metals are also present.

In some parts of the country the sludge is not dumped but is discharged into the ocean through special pipes. In others it is buried in landfill or spread as fertilizer, although the metals it contains may cause problems later. A broad spectrum of industrial effluents (solvents from pharmaceutical production, waste acid from the titanium-pigment industry, caustic solutions from oil refineries, metallic sodium and calcium, filter cake, salts and chlorinated hydrocarbons) are dumped intermittently at certain sites under Government license.

What damage is done to marine organisms by materials of this kind? The turbidity created by dumping is usually dispersed within a day. Dumped dredge spoil buries bottom-dwelling animals under a thin blanket of sediment, but many of them dig out and the others are replaced by recolonization in about a year. Sewage sludge is high in heavy metals, which may be toxic, particularly when they combine with organic materials to create a reducing (oxygen-poor) environment in which few animals can live. Sludge can also have a high bacterial count. It is clear that much industrial waste could be harmful to marine life and should not be dumped into the sea.

The entire matter of dumping needs more study. With reliable data it will be possible to retain the option of disposal at sea for some materials, such as dredge spoil, and to reject it for others, such as chemicals. Deep-water sites could be set aside for dumping on the same logic that applies to city dumps, namely that it is a suitable use for space of low value where few animals could be harmed.

Thermal waste is discharged into the sea by power plants because the sea is a convenient source of cooling water. The temperature of the water on discharge is typically 10 degrees Celsius higher than it was on intake. The difference is within the range of natural temperature variations and so is not harmful to most adult marine animals. The eggs, larvae and young animals that live in coastal waters, however, are sucked through the power plant with the cooling water. They are subjected to a sudden rise in temperature and decrease in pressure that is likely to be fatal. For this reason and others it would seem logical to put new power plants offshore. There they could draw deeper and cooler water from a level that is not rich in living organisms. For a nuclear plant the hazards of a nuclear accident would be reduced; for an oil- or coal-fired plant fuel could be delivered directly by ship, and the shoreline could be reserved for nonindustrial purposes.

Some industries discharge substantial quantities of heavy metals and complex organic compounds into municipal waste-water systems whose effluent reaches the ocean. Certain of the metals (mercury, chromium, lead, zinc, cadmium, copper, nickel and silver) are notably toxic and so are subject to stringent regulation. The most dangerous substances, however, are synthetic organic compounds such as DDT and polychlorinated biphenyls. The discharge of these substances as well as the heavy metals must be prevented. The best way to do so is by "source control," meaning the prevention of discharge into the sewer system. Each plant must be held responsible for removing and disposing of its own pollutants.

Other waste substances that generate controversy are those with nutrient value. Since they are decomposed by bacteria, oxygen is required. This biological oxygen demand is commonly measured in units that express how much oxygen (in milligrams per liter) will be required in a five-day period.

There is good reason to restrict the amount of nutrient material that is discharged into lakes and rivers, where oxygen is limited and a reducing environment can be created. The ocean is another matter. It is an essentially unlimited reservoir of dissolved oxygen, which is kept in motion by currents and is constantly being replenished by natural mechanisms.

It is nonetheless possible to overwhelm a local area of the ocean with a huge discharge of nutrient material that may form a deposit on the sea floor if the local conditions are not carefully considered. The materials must be presented to the ocean in the right places and at reasonable rates. Among the ways of achieving that objective are the use of discharge pipes that lead well offshore and have many small diffuser ports and, if the volume of discharge is exceptionally large, the distribution of the effluent through several widely dispersed pipes.

Problems caused by the addition of nitrogen and phosphate to inland waters, which they overfertilize, do not apply to the ocean. There they could be helpful by producing the equivalent of upwelling, the natural process that brings nutrients from deep water to the surface waters where most marine organisms are found. As Isaacs has pointed out, "the sea is *starved* for the basic plant nutrients, and it is a mystery to me why

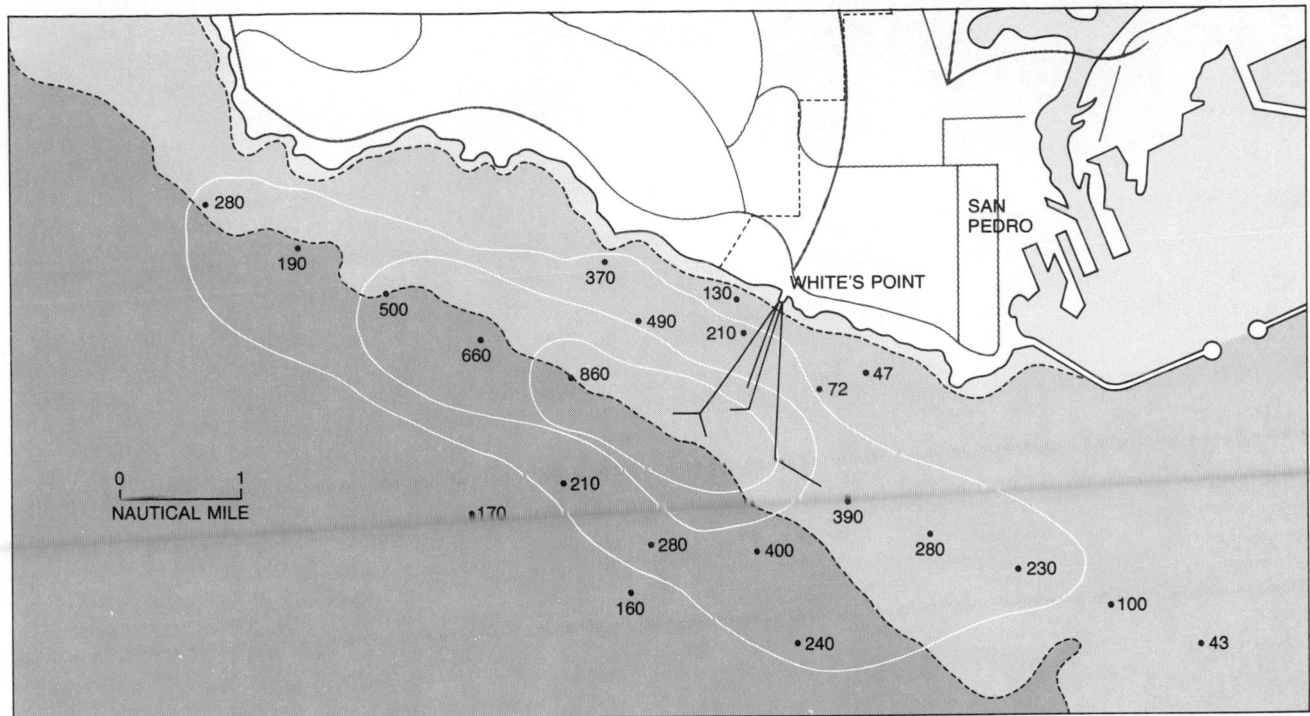

CONCENTRATION OF CHROMIUM in the upper sediments adjacent to a major industrial outfall off San Pedro, Calif., is charted in parts per million. Four outfall lines enter the sea from White's Point. The smallest area surrounded by an elliptical contour line is the area where the concentration is 800 parts per million or more, and the larger contoured areas have concentrations above 500 and 200 parts per million respectively. A depth of three fathoms is shown by the broken line close to shore. Farther out is a 50-fathom line. Other heavy metals discharged into the ocean make similar patterns based on current structure and the slope of the bottom.

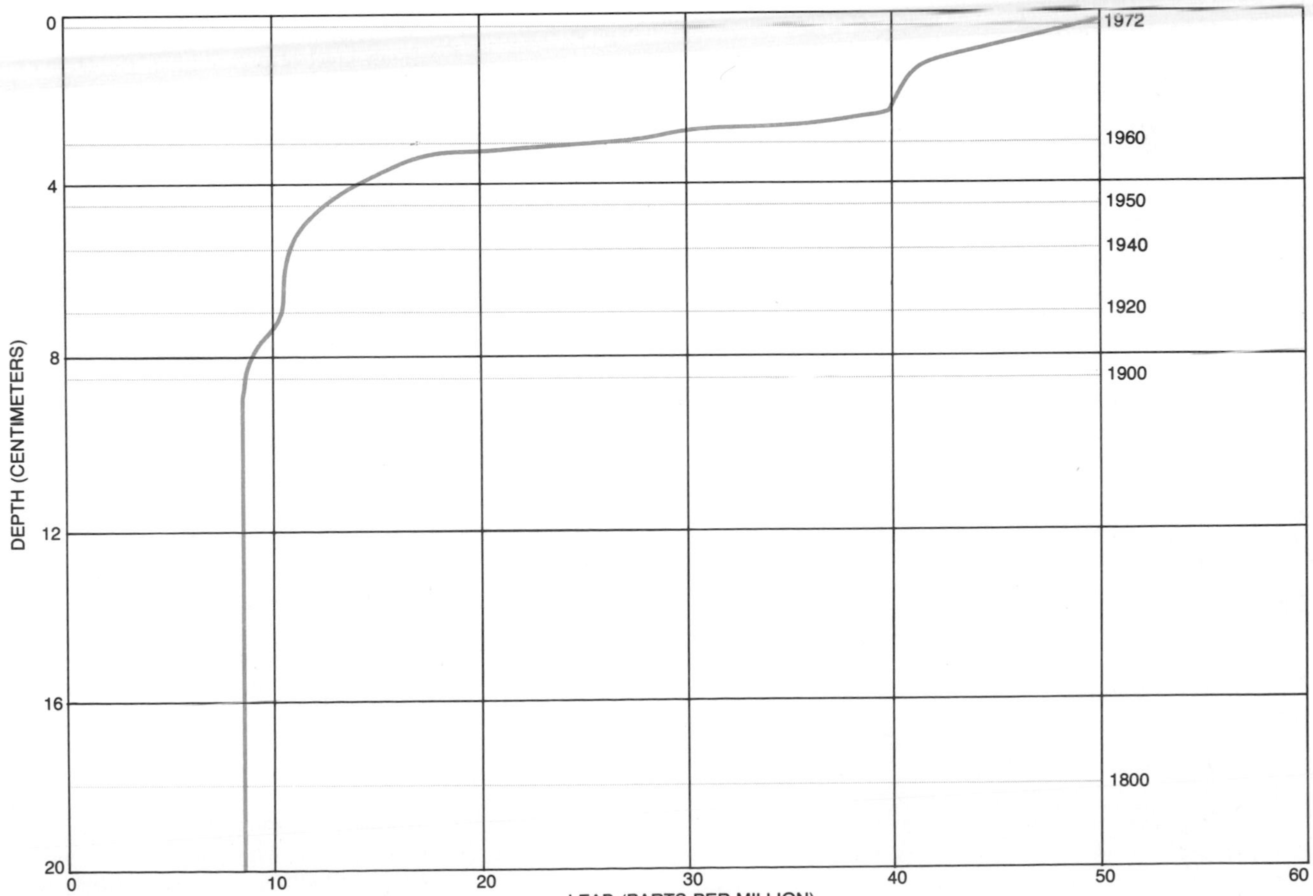

DEPTH (CENTIMETERS)
0
4
8
12
16
20
1972
1960
1950
1940
1920
1900
1800
0 10 20 30 40 50 60
LEAD (PARTS PER MILLION)

CONCENTRATION OF LEAD in the sediments of the San Pedro Basin shows an increase in recent years because of airborne lead that originates mainly from automobile emissions. The figures are based on samples of the sediments obtained by means of coring.

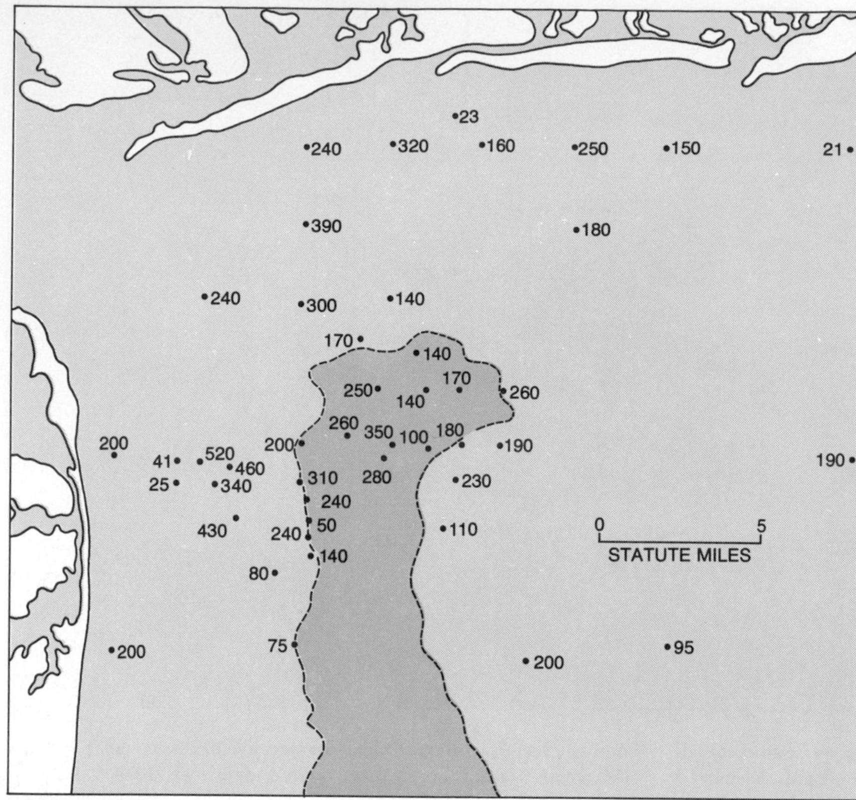

EFFECT OF DUMPING on sediments of the New York Bight is indicated by the concentration of chromium in parts per million in and around the dumping area. A bight is an open bay formed by a bend in a coast. Material transported to the bight to be dumped includes sewage sludge and rubble. The broken contour line represents a depth of 30 meters.

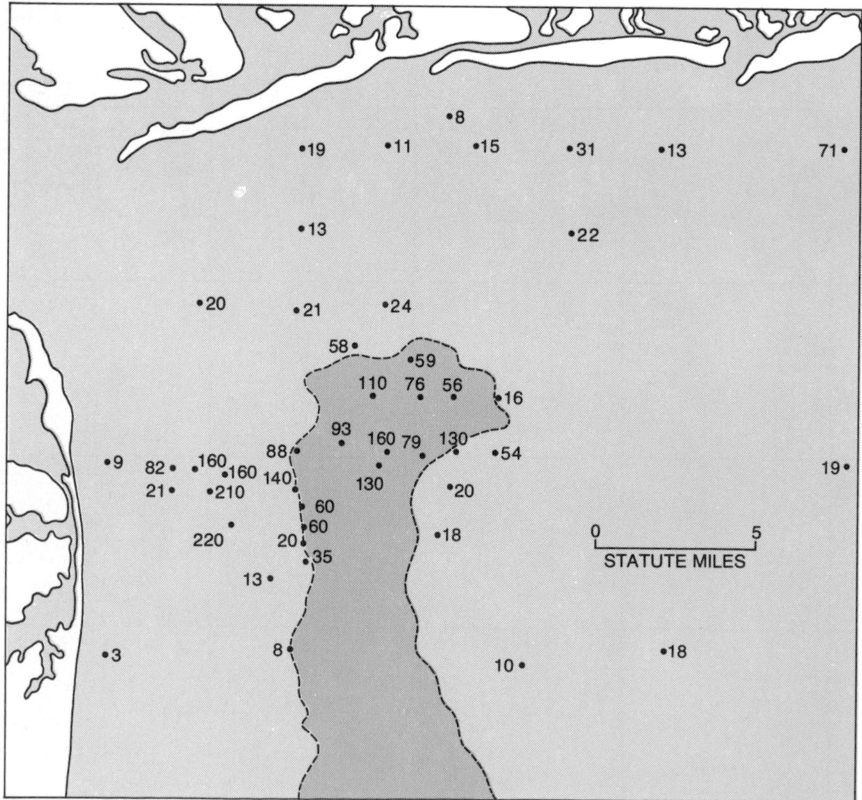

COPPER CONCENTRATION in sediments of the dumping area of the New York Bight is shown in parts per million. Data on chromium and copper in the New York Bight are based on work by Grant Gross of the oceanography section of the National Science Foundation.

we should be concerned with their thoughtful introduction into coastal seas in any quantity that man can generate in the foreseeable future."

Once possible pollutants reach the ocean it is necessary to keep track of where they go, the extent to which they are altered or diluted and what animals they affect. In order to obtain this information many of the techniques of oceanography are brought into play. Currents are measured above and below the thermocline; other instruments measure the temperature, salinity and dissolved oxygen. Water is sampled at various depths, and so are bottom sediments. It is also useful to directly monitor any changes in plant and animal communities with divers or television cameras.

A good indicator of change is the response of the polychaete worms in the bottom mud. Close to an effluent-discharge point the number of species may be as low as from four to 10 per sample and the total weight of worms as low as 50 grams per square meter. A short distance away the number of species may be 40 or more and the total weight 700 grams per square meter. At greater distances the figures drop off to normal: about 25 species and 300 grams per square meter. This local enrichment shows that worms thrive at some optimum level of organic material. Laboratory tests by Donald Reish and Jack Word of California State University at Long Beach show that worms have a similar optimum for toxic metals such as zinc and copper.

Man must do something with his wastes, and the ocean is a logical place for some of them. No single solution will be sensible for all kinds of waste or all locations, but the following suggestions may help to protect both the land and the sea in the long run. (1) Clearly define what is ocean, separating it from inland freshwaters and from harbors and shallow bays, and make laws that are appropriate for each environment. (2) Avoid the assumption that anything added to the ocean is necessarily harmful and consider instead what substances might cause damage and eliminate excesses of them. (3) Rigorously prohibit the disposal in the ocean of all man-made radioactive materials, halogenated hydrocarbons (such as DDT and polychlorinated biphenyls) and other synthetic organic materials that are toxic and against which marine organisms have no natural defenses. (4) Set standards based on water quality (after reasonable mixing) that are compatible with what is known about the threshold of damage to marine life, providing a safety factor of at least 10. (5) Work to obtain international cooperation in prohibiting ships from disposing of litter or oil and from pumping bilges. (6) Set aside ocean areas of deep water and slow current where certain materials can be dumped with minimal damage. (7) Require each discharger to make studies to demonstrate how his specific effluent will influence the adjacent ocean. (8) Support additional research on the effect of pollutants on the ocean and its life. (9) Anticipate pollutants that may become serious as technology produces new chemical compounds in greater quantities.

A more rational basis is needed for making decisions about how to treat wastes and where to put them. No oceanographer wants damaging waste in the ocean where he works or on land where he lives. Since the waste must go to one place or the other, however, one would prefer the choice to be based on a knowledge of all the factors. Unemotional consideration of which materials can be introduced into the sea without serious damage to marine life will result in both an unpolluted ocean and a large saving of national resources.

SEWAGE-TREATMENT PLANT serves the City of Los Angeles, discharging into the Pacific Ocean about 235 million gallons per day of primary treated effluent and 100 million gallons per day of secondary effluent. The discharge pipe is 12 feet in diameter and nearly five miles long. At the discharge end it is in 197 feet of water. The plant separately discharges sludge, consisting of about 1 percent solids, through a seven-mile pipe to a depth of more than 300 feet. The sludge is discharged at the brink of a marine canyon.

MARINE ORGANISMS grow on the outfall pipe from the Los Angeles sewage-treatment plant. At left are anemones about three feet high and at right is a gorgonian. The location is near the discharge point of the outfall pipe at a depth of approximately 200 feet.

25

Plate Tectonics and Mineral Resources

by Peter A. Rona
July 1973

The concepts of continental drift and sea-floor spreading provide clues to the location of economically important minerals such as oil and metals. These clues have already led to promising deposits

A scientific revolution is in progress that over the past five years has already changed our understanding of the earth as profoundly as the Copernican revolution changed medieval man's understanding of the solar system. The Copernican revolution entailed a fundamental change in man's world view from an earth-centered planetary system to a sun-centered one and led to the development of modern astronomy and the exploration of space. The current scientific revolution entails a fundamental change in man's world view from a static earth to a dynamic one and presages comparable benefits. Some of the benefits may even be economic. The implications of the new global tectonics for mineral resources, particularly the mineral resources of the ocean floor, are only now beginning to emerge.

At present the only undersea mineral resources that certainly have economic value are the vast oil and gas reserves found under many continental shelves and continental slopes, gravel, sand, shells and placer deposits on the continental shelves, various other minerals buried under the continental shelves in specific relation to adjacent continental deposits, and fields of manganese nodules that blanket large areas of the deep-sea floor. Even this limited knowledge is remarkable in the light of the difficulty that was encountered in obtaining it. Consider how much we would know about the mineral deposits of the continents if our sampling procedure were limited to flying in a balloon at an altitude of up to six miles and suspending a bucket at the end of a cable to scrape up loose rocks from the surface of the land. What are the chances that we would find the major known ore bodies, which generally underlie areas of less than a square mile?

Yet this farfetched analogy accurately describes man's present capacity for sampling the sediments and rocks of the ocean bottom, utilizing a variety of coring, drilling and dredging devices lowered from ships through the water column over an area twice as large as that of the continents. Averaged over the world's oceans, the distribution of ocean-floor rocks that have been sampled to date is only about three dredge hauls per million square kilometers!

In recent years every major discovery of a hidden mineral resource has been anticipated by a theoretical vision. For example, once field geologists realized that there was a definite association between the type of sedimentary structure termed an anticline and accumulations of oil, they knew where to drill and the rate of discovery of oil deposits accelerated accordingly. In the same way the right conceptual framework can be used to extend man's limited direct knowledge of resources of the ocean basin toward a realistic appraisal of their potential. The test of the value of such a conceptual framework is how well it explains what one sees and predicts what one does not see.

The old conceptual framework of a static earth held that the continents and ocean basins were permanent features that had existed in their present form since early in the 4.5-billion-year history of the earth. Only the most accessible continental mineral deposits were discovered, largely by trial and error, with little understanding of why or where they existed. The recent change to a conceptual framework based on a dynamic-earth model, in which continents are constantly moving and ocean basins are opening and closing, is leading toward a better understanding of the global distribution of mineral deposits in both space and time.

The basis of the new conceptual framework is the theory of plate tectonics, the essentials of which have already been reported in SCIENTIFIC AMERICAN [see "Plate Tectonics," by John F. Dewey; Offprint 900]. "Tectonics" is a geological term pertaining to earth movements. The movements in question involve the lithosphere, the rigid outer shell of the earth, which is of the order of 60 miles thick. The lithosphere, which behaves as if it were floating on an underlying plastic layer, the asthenosphere, is segmented into about six primary slabs, or plates, each of which may encompass a continent and part of an adjacent ocean basin [*see top illustration on page 289*].

The boundaries of the lithospheric plates are delineated by narrow earthquake zones where the plates are moving with respect to each other. Three types of boundary are recognized. One type, called a convergent plate boundary, is where two adjacent plates move

TROODOS MASSIF on the island of Cyprus, the site of economically important mineral deposits that originated at a divergent tectonic-plate boundary, stands out clearly as the dark-colored mountainous region in the middle of the satellite photograph on the opposite page. The photograph was made recently from an altitude of nearly 600 miles by a multispectral camera system on board the first Earth Resources Technology Satellite (ERTS I). Region is believed to be a slice of oceanic lithosphere that was formed by the process of sea-floor spreading from a submerged mid-oceanic ridge and was subsequently thrust upward.

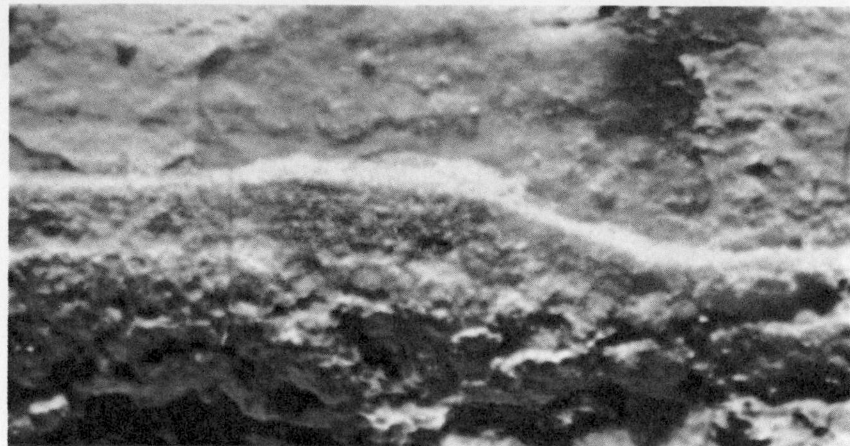

SMALL VEIN OF PURE COPPER was discovered in a core sample of sedimentary rock obtained by the Deep Sea Drilling Project some 350 miles southeast of New York City. The copper vein, the horizontal reddish structure in this longitudinal section of a piece of the original core, is about half an inch long. It was found in sediment about 65 feet above the volcanic basement rocks under the lower continental rise at a water depth of 17,000 feet.

METAL-RICH CORE, collected from the Atlantis II Deep, one of the hot-brine pools located along the axial valley of the Red Sea at a depth of about 6,600 feet below sea level, represents the most concentrated submarine metallic sulfide deposits known. The muddy sediments containing the sulfide minerals fill the Red Sea basins to a thickness estimated at between 65 and 330 feet. The deposits are saturated with (and overlain by) salty brines considered to be the hydrothermal solutions from which the sulfide minerals were precipitated. The photograph was made by David A. Ross of the Woods Hole Oceanographic Institution.

together and collide or where one plate plunges downward under the other plate and is absorbed into the interior of the earth.

The second type of boundary, called a divergent plate boundary, is where two adjacent plates move apart because new lithosphere is added to each plate by the process of sea-floor spreading. The new lithosphere, which moves more or less symmetrically to each side of the divergent plate boundary, acts like a conveyor belt, carrying the continents apart in the motion that has become known as continental drift. The dual existence of convergent boundaries where lithosphere is destroyed and divergent boundaries where lithosphere is created implies that the diameter of the earth is not changing radically.

The third type of tectonic-plate boundary is the parallel plate boundary, where two adjacent plates move edge to edge along their common interface.

Hydrothermal mineral deposits, that is, mineral deposits formed by precipitation from solutions, constitute a major part of our useful metallic ores on the continents. Economically the most important types of hydrothermal deposit are the sulfides, in which various metals combine with sulfur to precipitate from the hydrothermal solution. About a year ago Frederick Sawkins, a geologist at the University of Minnesota, pointed out that most of the sulfide deposits of the world are located along present or former convergent plate boundaries where an oceanic lithospheric plate plunges under the margin of a continent (including the continental shelf) or under a chain of volcanic islands. The processes that concentrate the sulfide deposits along convergent plate boundaries, which are at present only partly understood, involve mineralizing solutions that emanate from the plunging lithospheric plate, which melts as it is absorbed into the interior of the earth.

Metallic sulfide deposits along convergent plate boundaries include the Kuroko deposits of Japan, the sulfide ore bodies of the Philippines and the deposits extending along the mountain belts of western North America and South America (the Coast Ranges, the Rockies and the Andes) and from the eastern Mediterranean region to Pakistan. Gold-bearing deposits are not sulfides but often accompany sulfide minerals. The majority of gold deposits in Alaska, Canada, the southeastern U.S., California, Venezuela, Brazil, West Africa, Rhodesia, southern India and

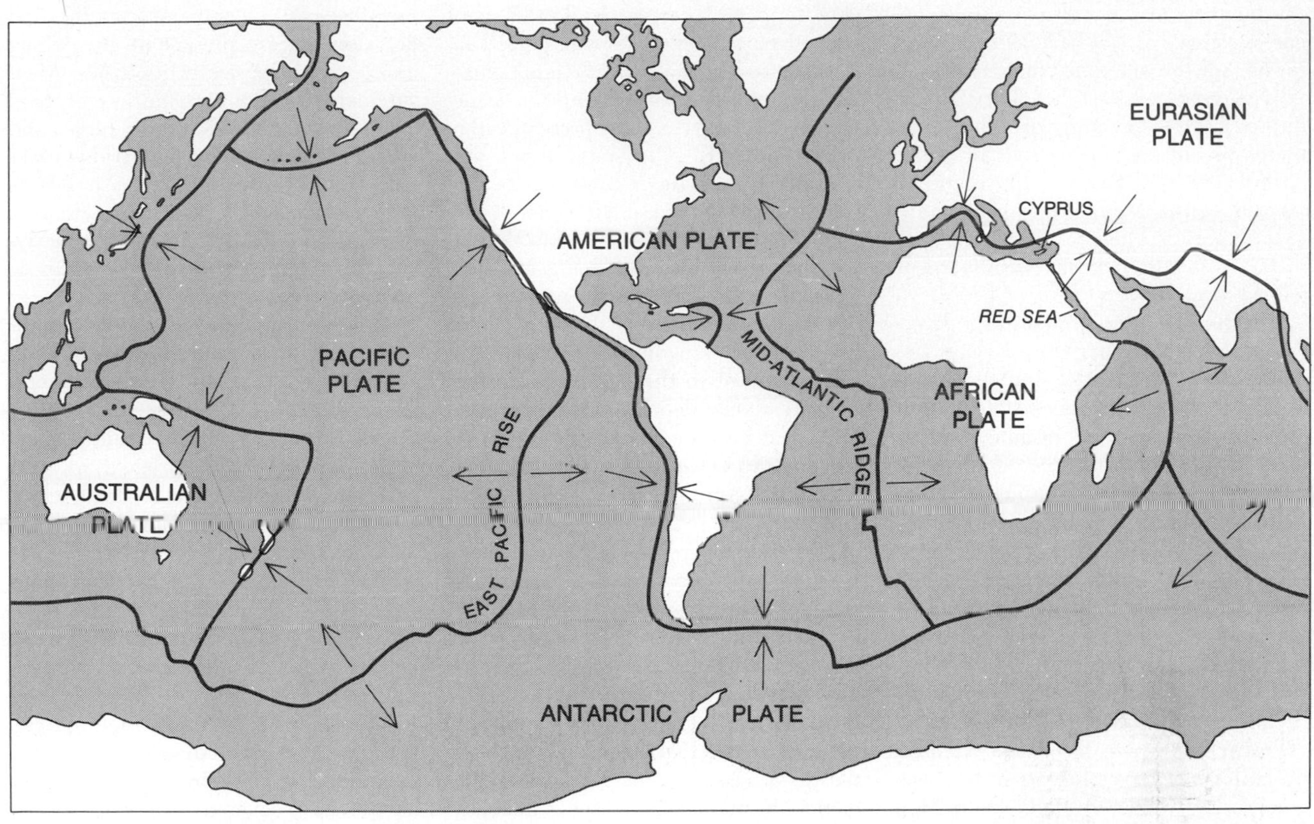

SIX PRINCIPAL TECTONIC PLATES of the lithosphere, the rigid outer shell of the earth, are delineated by the heavy color lines on this world map. The paired arrows indicate whether a plate boundary is convergent or divergent (*see illustration below*).

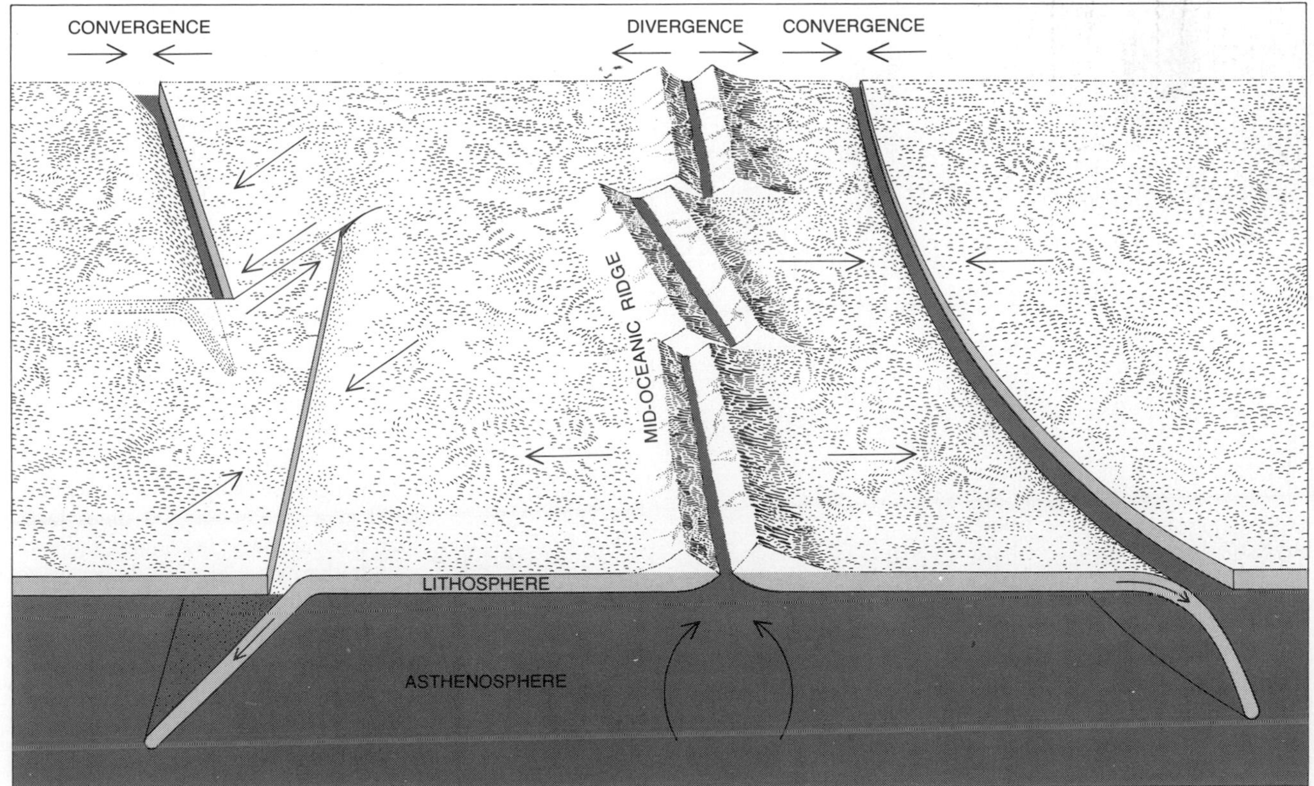

TWO TYPES OF PLATE BOUNDARY are illustrated schematically in this block diagram. The 60-mile-thick lithospheric plates move outward like conveyor belts from the mid-oceanic ridges (divergent plate boundaries) and plunge downward under the deep-sea trenches (convergent plate boundaries). The third major type of plate boundary, not shown here, is the parallel plate boundary.

southeastern and western Australia occur in rocks that can be associated with former convergent plate boundaries.

Divergent plate boundaries are formed by the spreading of lithospheric plates in the central portions of ocean basins. The Red Sea and the island of Cyprus in the Mediterranean Sea provide important clues to the potential of metallic sulfide deposits at divergent plate boundaries.

The Red Sea, the product of a divergent plate boundary developing between the African plate and the Eurasian plate, provides an accessible natural laboratory for the study of mineral processes associated with divergent plate boundaries. About five years ago the richest submarine metallic sulfide deposits known were found in three rather small basins along the center of the Red Sea at a depth of about 6,600 feet below sea level. The sulfide minerals are disseminated in sediments that fill the basins to a thickness estimated at between 65 and 330 feet. The top 30 feet or so of sediment, which has been explored by coring the largest of the basins, has a total dry weight of about 80 million tons, with average metal contents of 29 percent iron, 3.4 percent zinc, 1.3 percent copper, .1 percent lead, .005 percent silver and .00005 percent gold. The deposits are saturated with (and overlain by) salty brines carrying the same metals in solution as those present in the sulfide deposits. The salty brines are considered to be the hydrothermal solutions from which the sulfide minerals are precipitated. It remains controversial whether the brines are being charged with minerals from volcanic sources under the Red Sea or from sediments with high copper, vanadium and zinc contents adjacent to the basins where the metallic sulfide deposits are found [see "The Red Sea Hot Brines," by Egon T. Degens and David A. Ross; SCIENTIFIC AMERICAN, April, 1970].

The Red Sea represents the earliest stage in the growth of an ocean basin: the stage where a divergent plate boundary rifts a continent in two. The most advanced growth stage of a divergent plate boundary is the mid-oceanic-ridge system, a 47,000-mile undersea mountain chain that extends through all the major ocean basins and girdles the globe. The mid-oceanic-ridge system has not been adequately sampled to determine whether or not concentrations of metallic sulfides comparable to the Red Sea deposits are present at sites along its crest or in basins in its flanks. Measurements of the distribution of heat emanating from mid-oceanic ridges and of the chemical alteration of ridge rocks indicate that seawater forms a hydrothermal solution by penetrating fissures, dissolving minerals from rocks underlying the ridges and precipitating those minerals in concentrated deposits.

A limited amount of sampling indicates that hydrothermal processes are actively concentrating metals from volcanic sources underlying mid-oceanic ridges. Sediments on active mid-oceanic ridges are generally enriched in iron, manganese, copper, nickel, lead, chromium, cobalt, uranium and mercury, with trace amounts of vanadium, cadmium and bismuth. The concentrations typical of sediments covering widespread areas on mid-oceanic ridges are not economic, but much higher concentrations exist locally.

Metallic sulfides are found in rocks dredged from the Indian Ocean Ridge. In addition small veins of pure copper have been recovered by the Deep Sea Drilling Project at several sites. At the crest of the Ninety East Ridge near the Equator in the Indian Ocean, for example, veins of copper are found in volcanic rocks overlain by 1,440 feet of sediment at a water depth of 7,380 feet. Some 350 miles southeast of New York City a small vein of pure copper and clusters of copper crystals have been discovered in sediment about 65 feet above the volcanic basement rocks under the lower continental rise at a water depth of 17,000 feet [*see top illustration on page 288*].

A specimen of manganese 1.7 inches thick recently dredged from a water depth of about 12,000 feet in the median valley of the Mid-Atlantic Ridge by the Trans-Atlantic Geotraverse of the National Oceanic and Atmospheric Administration has particular significance. The composition, form and thickness of this manganese sample, which accumulated at a rate about 100 times faster than the manganese in nodules, indicates a hydrothermal origin and demonstrates that hydrothermal mineral deposits are actively accumulating at certain divergent plate boundaries in ocean basins. Because the sea floor is supposed to originate by spreading from mid-oceanic ridges, a mineral deposit on a mid-oceanic ridge would be expected to extend in a linear zone from the ridge across the ocean basin to the adjacent continental margin if the depositional process is a

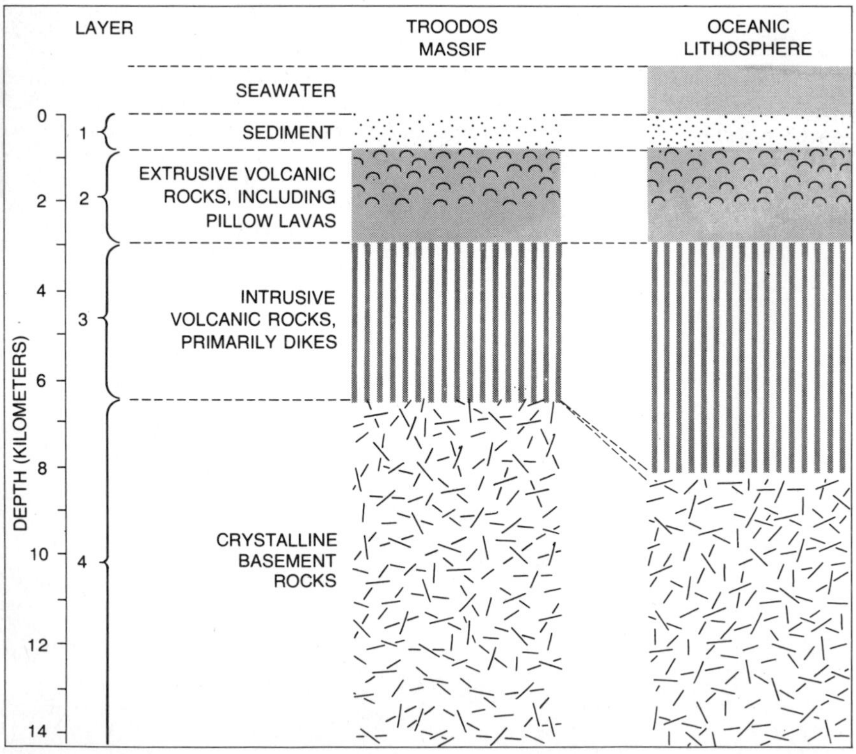

CLOSE CORRESPONDENCE between the layered sequence of rocks in the Troodos Massif (*left*) and that of the oceanic lithosphere (*right*) is evident from this comparison. The geological structure of the Troodos Massif was determined directly from rock outcrops; the structure of the oceanic lithosphere was determined indirectly by seismic-refraction techniques. The sulfide ore bodies of the Troodos Massif are in the upper portion of layer made up of extrusive volcanic rocks. Pillow shapes form when volcanic lava cools on the sea floor.

continuous one [*see illustration on next page*].

At this point in man's exploration of the oceans it would seem to be too much to expect that it would be possible to make detailed observations on an economically important metallic sulfide deposit that originated at a divergent plate boundary on a submerged mid-oceanic ridge. Yet such a deposit is known and has been extensively studied. The Troodos Massif on the island of Cyprus is interpreted as being a slice of oceanic lithosphere that was formed by the process of sea-floor spreading from a mid-oceanic ridge and was subsequently thrust upward to its present position [*see illustration on page 287*]. The composition and layered sequence of rocks that constitute the Troodos Massif are the

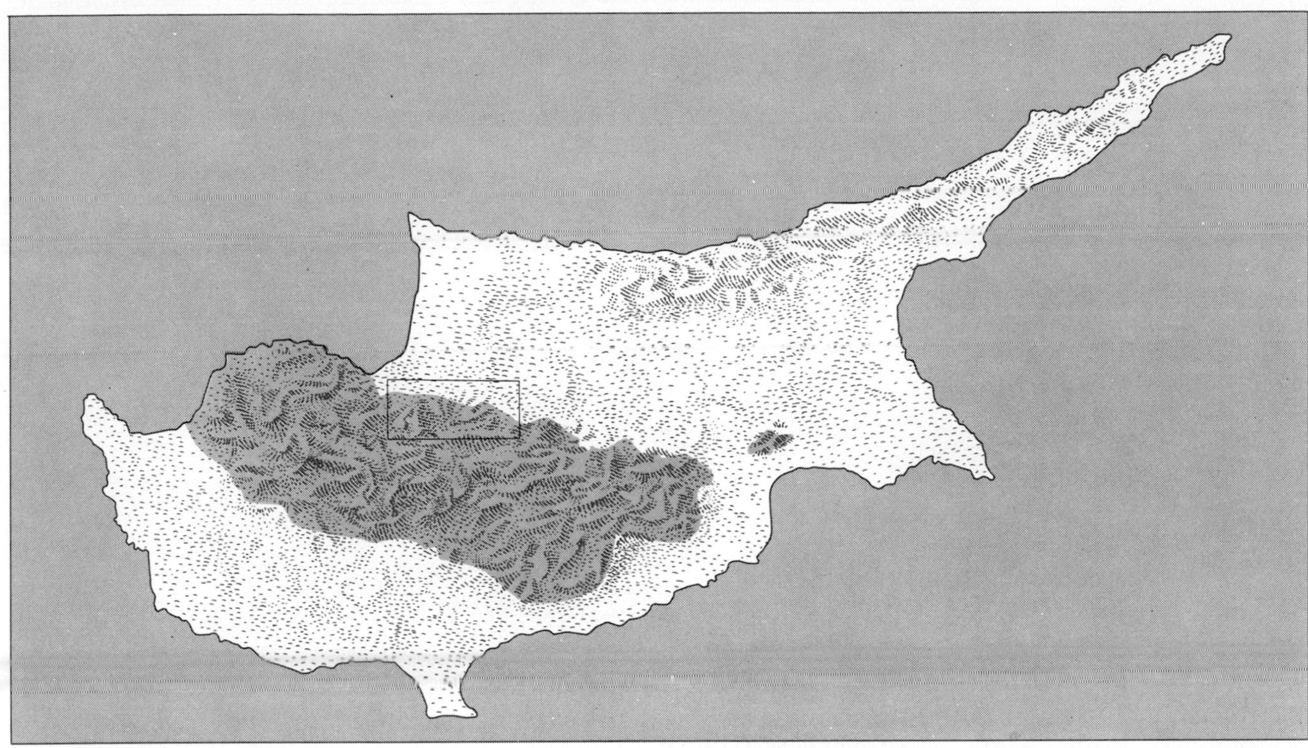

ISLAND OF CYPRUS has been famous for its mineral wealth since Phoenician times. The principal ore bodies are in the uppermost volcanic layers of the Troodos Massif, the total extent of which is indicated by the dark-colored area. The hatched area represents sediments, including alluvium. A geological map of a portion of the Troodos igneous complex (*small rectangle*) is shown below.

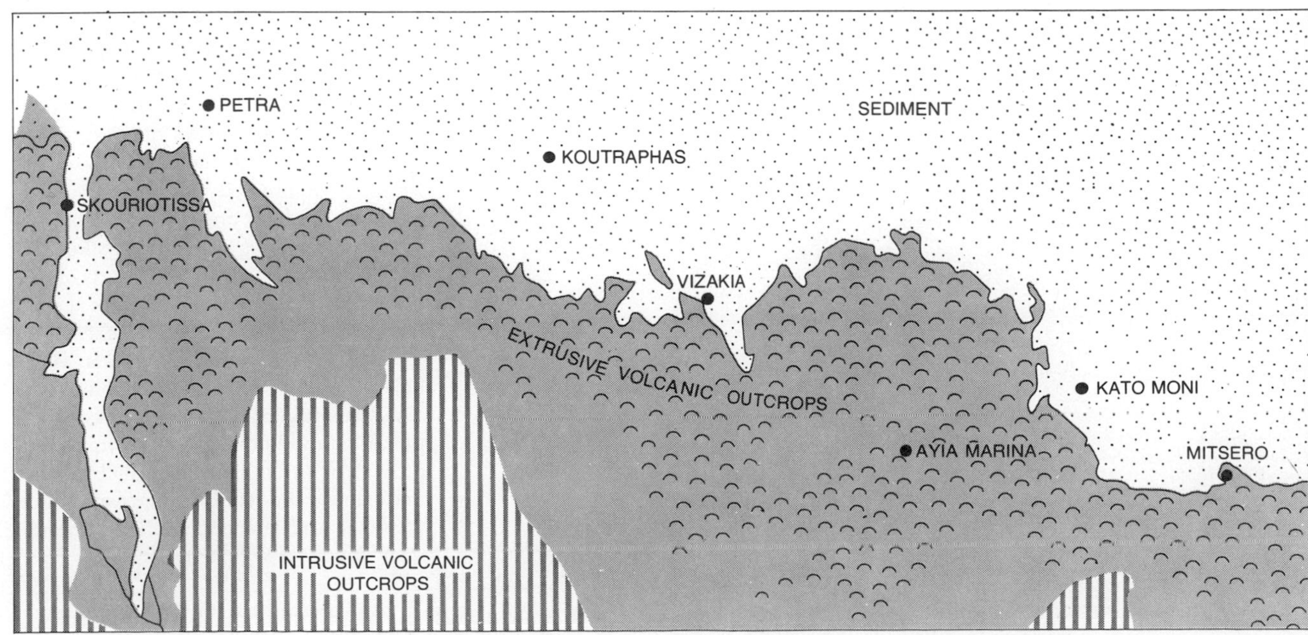

GEOLOGICAL MAP of a region that lies along the northern fringe of the Troodos Massif is based on studies that were undertaken by the Geological Survey of Cyprus. The map shows outcrops of extrusive volcanic rock that incorporate bodies of metallic sulfide ore.

same as those known to underlie the seabed.

Cyprus has long been famous for its mineral wealth. The mining of copper (for which the island is named) was an important industry in Roman and even in Phoenician times. The brilliant green stains of copper sulfides on ancient mine tailings have attracted modern prospectors. Between 1965 and 1970 the average annual exports amounted to about a million tons each of iron pyrites, chromite and gypsum, about 150,000 tons of copper pyrites and 100,000 tons of copper concentrates. The estimated value of the mineral products exported from Cyprus in 1970 amounted to $30 million.

The principal ore bodies are in the uppermost volcanic layers of the Troodos Massif. It has been uncertain whether the Troodos sulfide-ore bodies originated before the upthrust of the Troodos Massif or afterward. In the first instance the ore bodies would be representative of the seabed. In the second the ore bodies would be attributed to special conditions unrelated to the seabed. The sulfide deposits are clearly related to the volcanic rocks in which they occur. Recent studies reveal that iron-rich and manganese-rich sediments interlayered with the volcanic rocks and associated with the ore bodies of the Troodos Massif are chemically identical with those metal-enriched sediments found on active mid-oceanic ridges, indicating that both the sediments and the ore bodies were formed on the sea floor by hydrothermal processes.

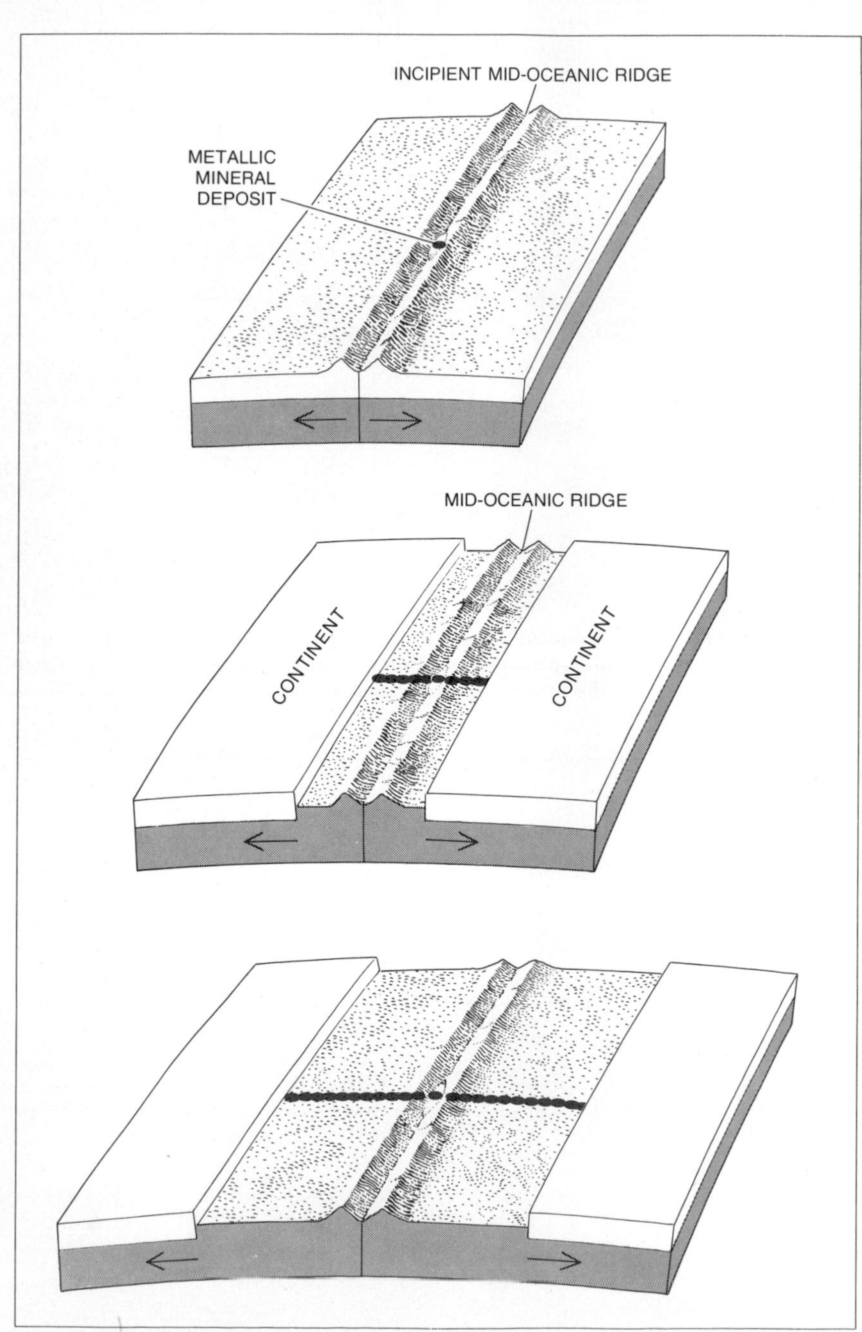

HYDROTHERMAL MINERAL DEPOSIT (*color*) formed in a hot-brine pool on the axis of a mid-oceanic ridge would be expected to extend in a linear zone from the ridge across the ocean basin to the adjacent continental margins as the ocean basin progressively widens (*from top to bottom*) as a consequence of sea-floor spreading from the mid-oceanic ridge.

The Troodos ore bodies may provide the first firm evidence on the nature of metallic sulfide deposits in ocean basins. The Skouriotissa ore body, for example, is roughly elliptical in plan view, measures approximately 2,000 feet long by 600 feet wide and is lens-shaped in cross section. Its estimated mass is six million tons. The average composition of the ore is 2.25 percent copper (ranging to greater than 5 percent), 48 percent sulfur and 43 percent iron.

The Mavrovouni ore body is also roughly elliptical in plan view, measures approximately 1,000 feet long by 600 feet wide and forms a lens that attains a thickness of 800 feet in cross section. Its estimated mass is greater than 15 million tons. The average composition of the ore is 4.2 percent copper, 48 percent sulfur, 43 percent iron, .4 percent zinc, .25 ounce per ton gold and .25 ounce per ton silver.

Sediments underlying the Skouriotissa ore body, presumably a disintegration product of the pyrite in the ore body, contain 2.12 ounces of gold per ton and 12.96 ounces of silver per ton. Exposed patches of metallic oxides indicate the presence of the ore bodies under the mountainous surface of the Troodos Massif. The Skouriotissa ore body is exploited by underground shafts and the Mavrovouni ore body by strip-mining.

What kind of target for exploration would a Troodos ore body make if it were submerged under thousands of feet of water on the crest or flank of a mid-oceanic ridge? It is unlikely that any of the present exploration methods would be capable of detecting the ore body. The resolution of present geophysical exploration methods will have to be improved in order to detect such an ore body under the sea. Both the exploration methods and the engineering development involved will be costly.

The prerequisites for the accumulation of petroleum consist of a source of organic matter to generate the petroleum, a natural reservoir to contain it and a trap to concentrate its fluid and gas constituents. Petroleum is hydrocarbons derived from the remains of plants and animals. As the progenitor of the petroleum, the organic matter must accumulate in an environment where it is preserved. The preservation of organic matter is favored by an environment that is toxic to life (so that the organic matter is not consumed as food) and deficient in oxygen (so that the hydrocarbon is not decomposed). How do conditions favorable to the accumulation of petroleum relate to convergent and divergent plate boundaries?

Convergent plate boundaries where the oceanic portion of a lithospheric plate plunges under the margin of a continent are characterized by the presence of a deep-sea trench running along their length. A system of deep-sea trenches runs along the entire western margin of North America and South America where the Pacific lithosphere is plunging under the continents. In addition to a deep-sea trench, chains of volcanic islands are present along some convergent plate boundaries; they are located between the trench and the continent. There are many such chains of volcanic islands at the western margin of the Pacific, including the Aleutians, the Kuriles, Japan, the Ryukyus, the Philippines and Indonesia. Other such chains are the Marianas, the South Sandwich Islands and the West Indies. The island chains divide an ocean basin into smaller basins partially enclosed between the islands and the adjacent continent; such basins include the Bering Sea, the Sea of Okhotsk, the Sea of Japan, the Yellow Sea, the East China Sea and the South China Sea.

Both the marginal trenches and the volcanic-island chains create a habitat that is favorable for the accumulation of petroleum in several respects. First, the trenches and island chains act as barriers that catch sediment and organic matter from the continent and the ocean basin. Second, the shape of the trenches and the small ocean basins acts to restrict the circulation of the ocean, so that oxygen is not replenished in the seawater and organic matter is preserved. Third, the accumulation of sediments and the geological structures that develop as a result of the deformation of the sediments by tectonic forces provide reservoirs and traps for the accumulation of petroleum. According to Hollis D. Hedberg of Princeton University, "these marginal semienclosed basins constitute some of the most promising areas in the world for petroleum accumulation."

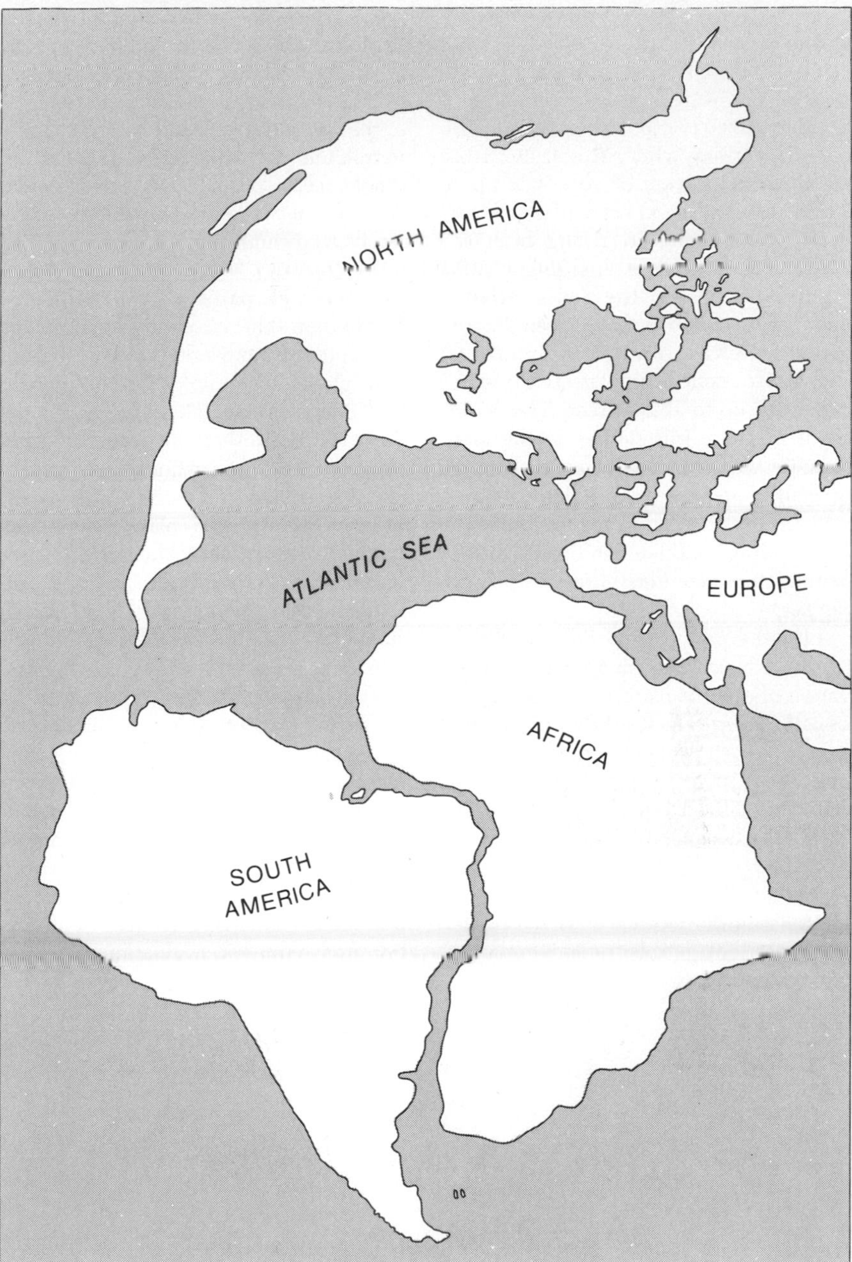

AT AN EARLY STAGE of continental drift the Atlantic was a sea with its circulation restricted by the surrounding continents. As in the present Red Sea, conditions in the Atlantic Sea favored the preservation of organic matter and the deposition of rock salt, leading to the formation of petroleum accumulations under the present continental margins.

The development of divergent plate boundaries may also create a habitat favorable for the accumulation of oil, a finding that would open immense possibilities for petroleum resources in the deep ocean basin. When a divergent plate boundary develops under a continent, the continent is rifted in two and the continental fragments are carried apart on a conveyor belt of new lithosphere generated at the divergent plate boundary. As the two continental fragments move apart, a sea forms between them. The surrounding continents act as barriers to restrict the circulation of the sea. As a result organic matter is preserved and, if the evaporation of the seawater exceeds its replenishment, layers of rock salt are deposited along with the organic matter. As the continental fragments continue to move apart and to subside along with the adjacent sea floor, the restricted sea becomes an open ocean. The layers of organic matter and salt are buried under sediments. The organic matter subsequently develops into petroleum (by processes that are only partly understood) and the salt forms into dome-shaped masses that act to trap the petroleum.

The Red Sea is an example of a restricted sea formed at an early stage of development of the divergent plate boundary along which Arabia is rifting from Africa. Layers of rock salt up to 17,000 feet thick and organic muds have been found under it. Along both the eastern and western margins of the North Atlantic and the South Atlantic apparent salt domes have been discovered extending seaward from continental shelves to continental rises in water depths of up to 16,500 feet. The occurrence of these salt domes in the deep Atlantic indicates that at an early stage of continental drift the Atlantic was a sea with its circulation restricted by the surrounding continents in their positions at that time [*see illustration on preceding page*].

Like the present Red Sea, conditions in the Atlantic Sea favored the preservation of organic matter and the deposition of rock salt. As the Atlantic widened in response to the symmetric creation of new lithosphere by sea-floor spreading from the Mid-Atlantic Ridge, the Atlantic Sea became an ocean and the organic matter and salt were buried under sediments, forming the present margins of the Atlantic Ocean. It is reasonable to expect that petroleum accumulations will extend seaward under the continental shelf, the continental slope and the continental rise to water depths of about 18,000 feet along large portions of both the eastern and western margins of the North Atlantic and South Atlantic. Petroleum may likewise be found in other ocean basins that have grown through the stage of a restricted sea by sea-floor spreading.

In short, the patterns of mineral distribution that are emerging from the conceptual framework provided by the new global tectonics will clearly help to guide man's search for new mineral deposits. Hydrothermal processes have concentrated the majority of known metallic sulfide ore bodies along convergent lithospheric plate boundaries originally at continental margins. Hydrothermal processes are also active at divergent plate boundaries from initial stages (represented by the metallic sulfide deposits accumulating in the Red Sea) to advanced stages (represented by the metal concentration in sediments on mid-oceanic ridges and by possible metallic sulfide deposits of the Troodos Massif type). The Troodos Massif metallic sulfide ore bodies provide an actual example of the type of deposits that can be expected in sea-floor rock generated by mid-oceanic ridges. The confirmation and economic evaluation of metallic sulfide deposits of the Troodos Massif type in ocean basins await technological advances in marine exploration methods.

With regard to petroleum, convergent plate boundaries create conditions that form accumulations in small ocean basins and deep-sea trenches marginal to continents. Divergent plate boundaries, on the other hand, create conditions that favor the development of oil accumulations extending from the continental shelf into the deep ocean basin under the continental rise.

The global patterns of mineral distribution that are emerging from such models can be expected to accelerate the discovery of resources not only on the seabed but also on the continents.

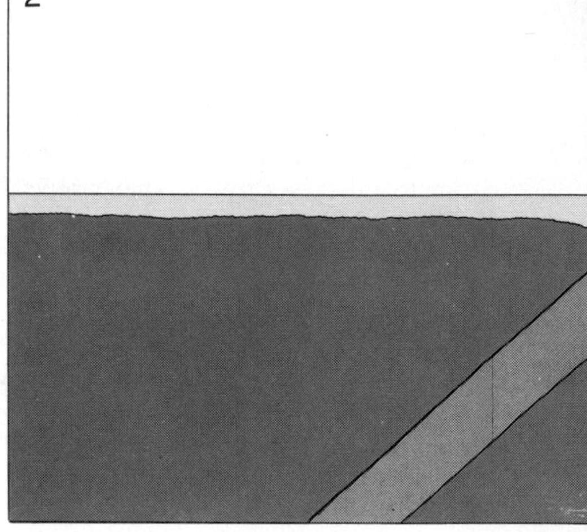

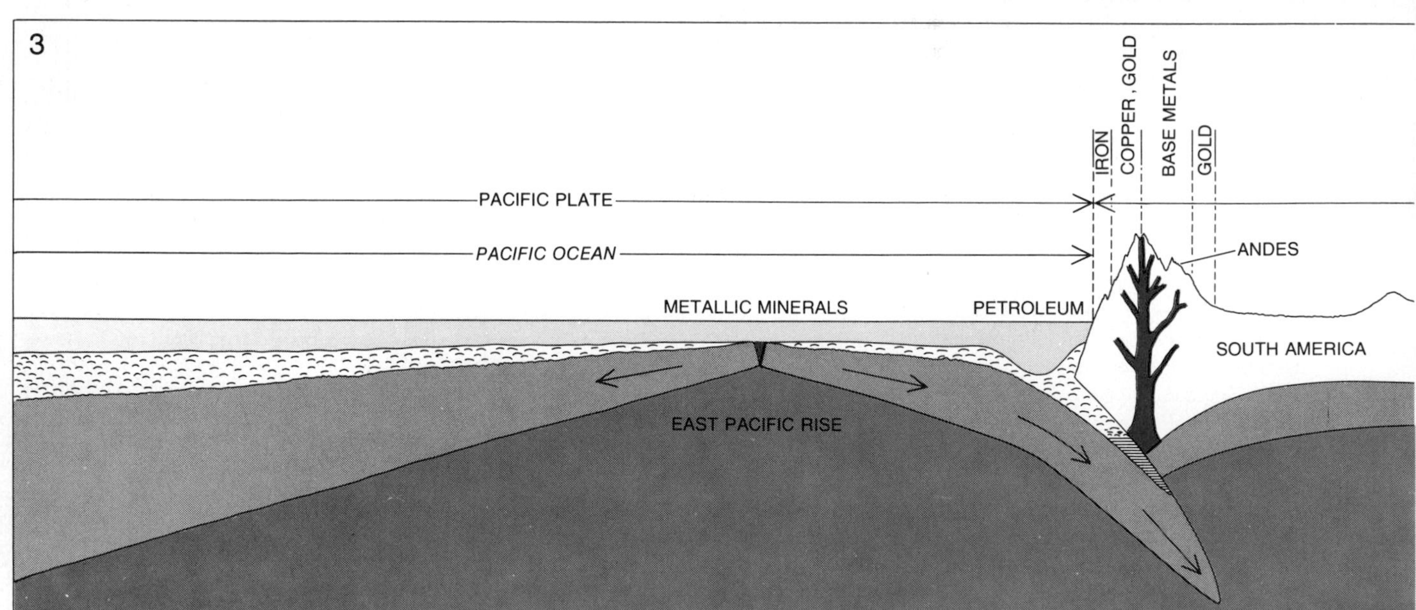

ROLE OF PLATE BOUNDARIES in the accumulation of mineral deposits is exemplified in this sequence of cross-sectional views of the development of the South Atlantic Ocean. The position of Africa is assumed to be stationary throughout the sequence of cross sections. In stage *1* a single ancestral continent, called Pangaea, is rifted into two continents (South America and Africa) about a divergent plate boundary. In stage *2* the oceanic crust created by the process of sea-floor spreading from the divergent plate boundary (a precursor of the Mid-Atlantic Ridge) rafts South America westward and is compensated for by the consumption of oceanic crust at a trench (a convergent plate boundary) that develops to the west of South America. Thick layers of rock salt, organic matter and metallic minerals accumulate in the Atlantic Sea during this early stage of continental drift. In stage *3* continued sea-floor spreading

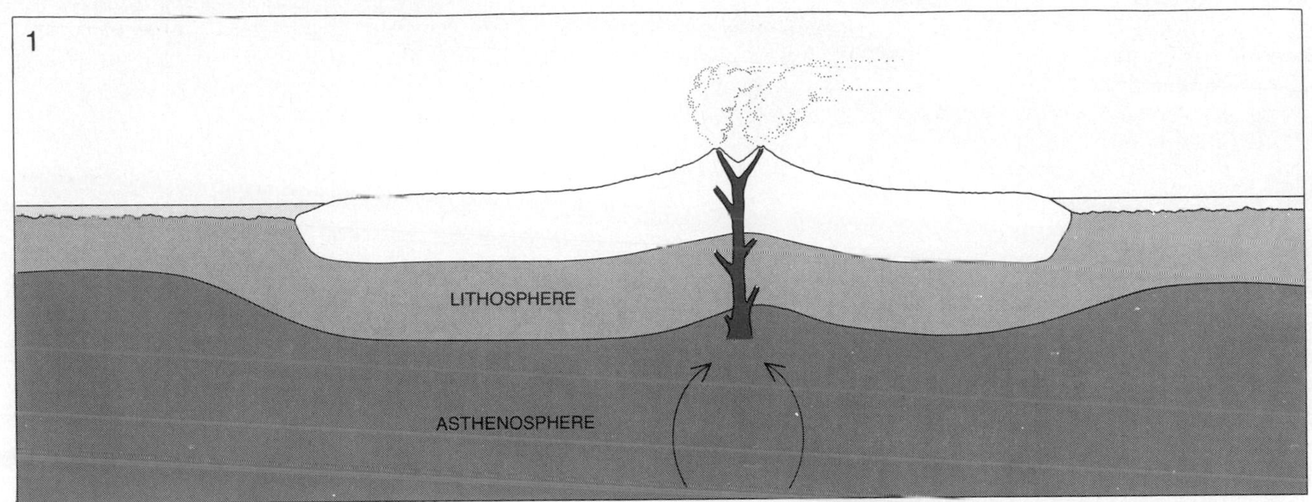

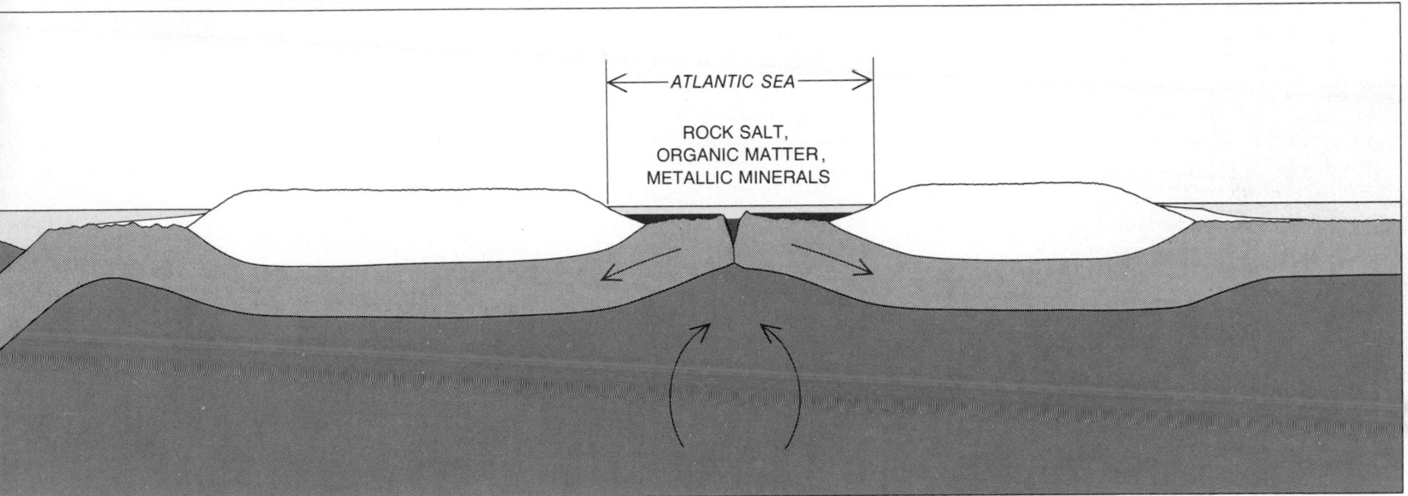

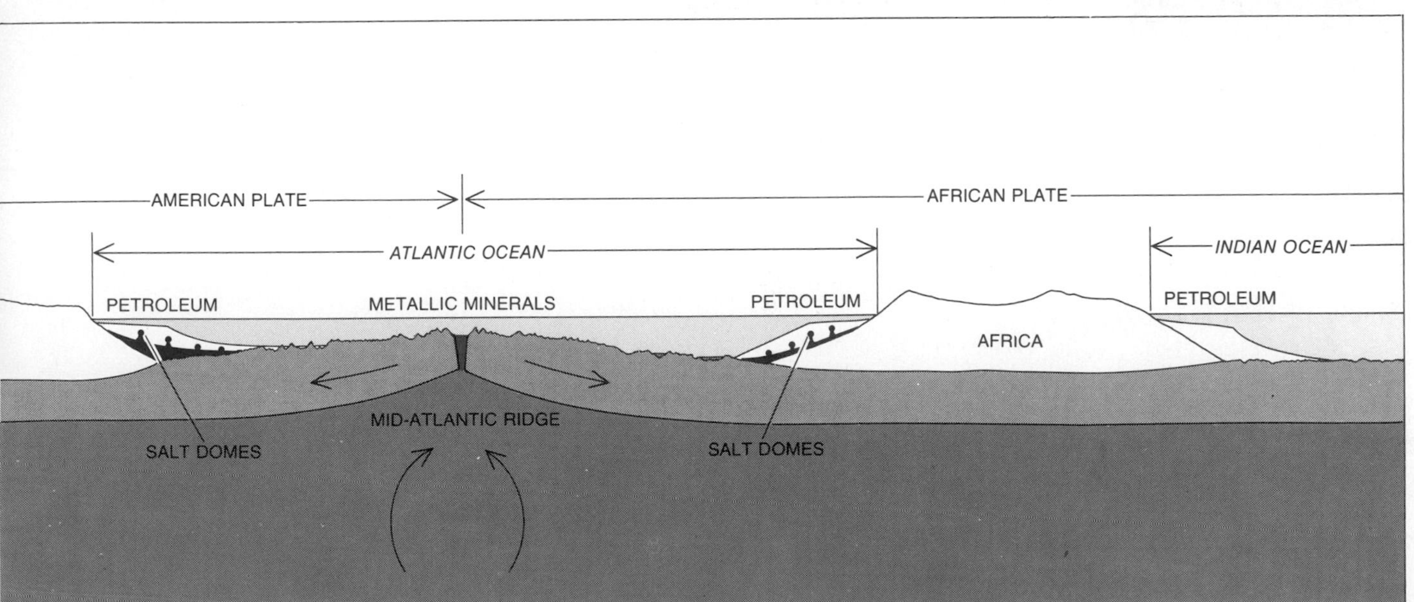

from the Mid-Atlantic Ridge widens the Atlantic into an ocean, rafts South America westward over the trench, reversing the inclination of the trench and producing the Andes mountain chain as a consequence of the deformation that develops at the convergent plate boundary along the western margin of South America. Metallic minerals that are melted from the Pacific plate as it plunges under South America ascend through the overlying crustal layers and are deposited in them to form the metal-bearing provinces of the Andes. Meanwhile in the Atlantic Ocean metallic minerals continue to accumulate about the Mid-Atlantic Ridge. Salt originating in the thick layers of rock salt that have been buried under the sediments of the continental margins rises in large, dome-shaped masses that act to trap the oil and gas that are generated from the organic matter that was preserved in the former Atlantic Sea.

BIBLIOGRAPHIES

I THE HISTORY OF OCEANOGRAPHY

1. The Voyage of the *Challenger*

Report on the Scientific Results of the Voyage of H.M.S. Challenger During the Years 1873–1876. Sir C. Wyville Thomson and Sir John Murray. Macmillan & Co., 1884.

2. Technology and the Ocean

Marine Science Affairs—a Year of Broadened Participation: The Third Report of the President to the Congress on Marine Resources and Engineering Development. U.S. Government Printing Office, January, 1969.

The Submersible as a Scientific Instrument. A. Conrad Neumann in *Oceanology International,* Vol. 3, No. 5, pages 39–43; July–August, 1968.

Relict Sediments on Continental Shelves of World. K. O. Emery in *The American Association of Petroleum Geologists Bulletin,* Vol. 52, No. 3, pages 445–464; March, 1968.

II MARINE GEOLOGY

3. The Continental Shelves

Ancient Oyster Shells on the Atlantic Continental Shelf. Arthur S. Merrill, K. O. Emery and Meyer Rubin in *Science,* Vol. 147, No. 3656, pages 398–400; January 22, 1965.

The Atlantic Continental Margin of the United States during the Past 70 Million Years. K. O. Emery in *The Geological Association of Canada Special Paper No. 4, Geology of the Atlantic Region,* pages 53–70; November, 1967.

Characteristics of Continental Shelves and Slopes. K. O. Emery in *Bulletin of the American Association of Petroleum Geologists,* Vol. 49, No. 9, pages 1379–1384; September, 1965.

Reversals of the Earth's Magnetic Field. Sir Edward Bullard in *Philosophical Transactions of the Royal Society of London: Series A, Mathematical and Physical Sciences,* Vol. 263, No. 1143, pages 481–524; December 12, 1968.

Sea-Floor Spreading and Continental Drift. Xavier Le Pichon in *Journal of Geophysical Research,* Vol. 73, No. 12, pages 3661–3697; June 15, 1968.

Seismology and the New Global Tectonics. Bryan Isacks, Jack Oliver and Lynn R. Sykes in *Journal of Geophysical Research,* Vol. 73, No. 18, pages 5855–5899; September 15, 1968.

4. The Origin of the Oceans

The History of the Earth's Crust. Edited by Robert A. Phinney. Princeton University Press, 1968.

5. The Deep-Ocean Floor

History of Ocean Basins. H. H. Hess in *Petrologic Studies: A Volume in Honor of A. F. Buddington,* edited by A. E. J. Engel, Harold L. James and B. F. Leonard. The Geological Society of America, 1962.

A New Class of Faults and Their Bearing on Continental Drift. J. Tuzo Wilson in *Nature,* Vol. 207, No. 4995, pages 343–347; July 24, 1965.

Rises, Trenches, Great Faults and Crustal Blocks. W. Jason Morgan in *Journal of Geophysical Research,* Vol. 73, No. 6, pages 1959–1982; March 15, 1968.

Sea Floor Spreading, Topography, and the Second Layer. H. W. Menard in *Transactions American Geophysical Union,* Vol. 48, No. 1, page 217; March, 1967.

Spreading of the Ocean Floor: New Evidence. F. J. Vine in *Science,* Vol. 154, No. 3755, pages 1405–1415; December 16, 1966.

6. The Floor of the Mid-Atlantic Rift

Atlantic Ocean Floor: Geochemistry and Petrology of Basalts from Legs 2 and 3 of the Deep Sea Drilling Project. Fred A. Frey, Wilfred B. Bryan and Geoffrey Thompson in *Journal of Geophysical Research,* Vol. 79, No. 35, pages 5507–5527; December 10, 1974.

FAMOUS: A Plate Tectonic Study of the Genesis of the Lithosphere. J. R. Heirtzler and X. Le Pichon in *Geology,* Vol. 2, No. 6, pages 273–274; 1974.

Flow of Lava into the Sea. James G. Moore, R. L. Phillips, Richard W. Grigg, Donald W. Peterson and Donald A. Swanson in *Geological Society of America Bulletin,* Vol. 84, No. 2, pages 537–546; February, 1973.

Understanding the Mid-Atlantic Ridge: A Comprehensive Program. Edited by J. R. Heirtzler. National Academy of Sciences, 1972.

7. The Evolution of the Pacific

Deep Ocean Drilling with Glomar Challenger. M. N. A. Peterson and N. T. Edgar in *Oceans,* Vol. 1, No. 5, pages 17–32; June, 1969.

Diachronous Deposits: A Kinematic Interpretation of Post Jurassic Sedimentary Sequence on the Pacific Plate. B. C. Heezen, I. D. MacGregor, H. P. Foreman, G. Forristal, H. Hekel, R. Hesse, R. H. Hoskins, E. J. W. Jones, A. Kaneps, V. A. Krasheninnikov, H. Okada and M. H. Ruef in *Nature,* Vol. 241, No. 5384, pages 25–32; January 5, 1973.

200,000,000 Years beneath the Sea: The Story of the Glomar Challenger. Peter Briggs. Holt, Rinehart and Winston, Inc., 1968.

8. When the Mediterranean Dried Up

Initial Cruise Reports of the Deep Sea Drilling Project. W. B. F. Ryan, K. J. Hsü, *et al.* U.S. Government Printing Office, Washington, 1972.

The Tectonics and Geology of the Mediterranean Sea. William B. F. Ryan, Daniel J. Stanley, J. B. Hersey, Davis A. Fahlquist and Thomas D. Allan in *The Sea,* Vol. IV, Parts II and III, edited by A. E. Maxwell. John Wiley & Sons, Inc., 1971.

III THE SEA AND ITS MOTIONS

9. Why the Sea Is Salt

Chemical Oceanography. Edited by J. P. Riley and G. Skirrow. Academic Press, 1965.

The Composition of Sea-Water, Section I: Chemistry, in *The Sea: Ideas and Observations on Progress in the Study of the Seas, Vol. II,* edited by M. N. Hill. Interscience Publishers, 1963.

Marine Chemistry: The Structure of Water and the Chemistry of the Hydrosphere. R. A. Horne. Wiley-Interscience, 1969.

The Ocean as a Chemical System. Lars Gunnar Sillén in *Science,* Vol. 156, No. 3779, pages 1189–1197; June 2, 1967.

The Oceans: Their Physics, Chemistry, and General Biology. H. U. Sverdrup, Martin W. Johnson and Richard H. Fleming. Prentice-Hall, Inc., 1961.

10. Ocean Waves

Breakers and Surf: Principles in Forecasting. Hydrographic Office Publication No. 234, 1944.

The Oceans: Their Physics, Chemistry and General Biology. H. U. Sverdrup, Martin W. Johnson and Richard H. Fleming. Prentice-Hall, Inc., 1942.

Practical Methods for Observing and Forecasting Ocean Waves by Means of Wave Spectra and Statistics. Willard J. Pierson, Jr., Gerhard Neumann and Richard W. James. Hydrographic Office Publication No. 603, 1958.

11. The Atmosphere and the Ocean

Descriptive Physical Oceanography. G. L. Pickard. Pergamon Press, 1968.

ENCYCLOPEDIA OF OCEANOGRAPHY. Edited by Rhodes W. Fairbridge. Reinhold Company, 1966.

THE GULF STREAM: A PHYSICAL AND DYNAMICAL DESCRIPTION. Henry Stommel. University of California Press and Cambridge University Press, 1965.

THE INFLUENCE OF FRICTION ON INERTIAL MODELS OF OCEANIC CIRCULATION. R. W. Stewart in *Studies on Oceanography,* edited by Kozo Yoshida. University of Washington Press, 1965.

AN INTRODUCTION TO PHYSICAL OCEANOGRAPHY. William S. von Arx. Addison-Wesley Publishing Company, Inc., 1962.

12. The Circulation of the Abyss

MEASUREMENTS OF DEEP CURRENTS IN THE WESTERN NORTH ATLANTIC. J. C. Swallow and L. V. Worthington in *Nature,* Vol. 179, No. 4571, pages 1183–1184; June 8, 1957.

A NEUTRAL-BUOYANCY FLOAT FOR MEASURING DEEP CURRENTS. J. C. Swallow in *Deep-Sea Research,* Vol 3, No. 1, pages 74–81; October, 1955.

A THEORY OF ICE AGES. Maurice Ewing and William L. Donn in *Science,* Vol. 123, No. 3207, pages 1061–1066; June 15, 1956.

13. The Top Millimeter of the Ocean

ELECTRIFICATION OF THE ATMOSPHERE BY PARTICLES FROM BREAKING BUBBLES. D. C. Blanchard in *Progress in Oceanography: Vol. 1,* edited by M. Sears. Pergamon Press, 1963.

JOURNAL OF GEOPHYSICAL RESEARCH. Vol. 77, No. 27; September 20, 1972.

14. The Microstructure of the Ocean

AN ANALYSIS OF THE STIRRING AND MIXING PROCESSES IN INCOMPRESSIBLE FLUIDS. Carl Eckart in *Journal of Marine Research,* Vol. 7, No. 3, pages 265–275; 1948.

A NEW CASE OF CONVECTION IN THE PRESENCE OF COMBINED VERTICAL SALINITY AND TEMPERATURE GRADIENTS. J. S. Turner and Henry Stommel in *Proceedings of the National Academy of Sciences of the United States of America,* Vol. 52, No. 1, pages 49–53; July, 1964.

OCEANIC FINE STRUCTURE. Thomas R. Osborn and Charles S. Cox in *Geophysical Fluid Dynamics,* Vol. 3, No. 4, pages 321–345; July, 1972.

THE VERTICAL MICROSTRUCTURE OF TEMPERATURE AND SALINITY. M. C. Gregg and C. S. Cox in *Deep-Sea Research and Oceanographic Abstracts,* Vol. 19, No. 5, pages 355–376; May, 1972.

WAVE-INDUCED SHEAR INSTABILITY IN THE SUMMER THERMOCLINE. J. D. Woods in *Journal of Fluid Mechanics,* Vol. 32, Part 4, pages 791–800; June 18, 1968.

15. Beaches

MARINE GEOLOGY. Ph. H. Kuenen. John Wiley & Sons, Inc., 1950.

PROCEEDINGS OF THE SIXTH CONFERENCE ON COASTAL ENGINEERING. Edited by J. W. Johnson. Council on Wave Research—The Engineering Foundation, 1958.

SHORE PROCESSES AND SHORELINE DEVELOPMENT. Douglas Wilson Johnson. John Wiley & Sons, Inc., 1959.

WAVES, TIDES, CURRENTS AND BEACHES: GLOSSARY OF TERMS AND LIST OF STANDARD SYMBOLS. Robert L. Wiegel. Council of Wave Research—The Engineering Foundation, 1953.

IV MARINE LIFE AND LIVING RESOURCES

16. The Nature of Oceanic Life

BIOLOGY OF SUSPENSION FEEDING. C. B. Jorgensen. Pergamon Press, 1966.

OCEANS: AN ATLAS-HISTORY OF MAN'S EXPLORATION OF THE DEEP. Edited by G. E. R. Deacon. Paul Hamlyn, 1962.

THE OPEN SEA: FISH AND FISHERIES. Alister C. Hardy. Houghton Mifflin Company, 1959.

THE OPEN SEA: THE WORLD OF PLANKTON. Alister C. Hardy. Houghton Mifflin Company, 1957.

17. Active Animals of the Deep-Sea Floor

DEEP-SEA PHOTOGRAPHY. Edited by John B. Hersey. Johns Hopkins University Press, 1968.

GIANT AMPHIPOD FROM THE ABYSSAL PACIFIC OCEAN. Robert R. Hessler, John D. Isaacs and Eric L. Mills in *Science,* Vol. 175, No. 4022, pages 636–637; February 11, 1972.

ROLE OF BIOLOGICAL DISTURBANCE IN MAINTAINING

DIVERSITY IN THE DEEP SEA. P. K. Dayton and R. R. Hessler in *Deep-Sea Research and Oceanographic Abstracts, Vol. 19, No. 3,* pages 199–208; March, 1972.

DYNAMIC INTERACTIONS. Milner B. Schaefer in *Transactions of the American Fisheries Society,* Vol. 99, No. 3, pages 461–467; 1970.

18. The Food Resources of the Ocean

FISHERIES BIOLOGY: A STUDY IN POPULATION DYNAMICS. D. H. Cushing. University of Wisconsin Press, 1968.

LIVING RESOURCES OF THE SEA: OPPORTUNITIES FOR RESEARCH AND EXPANSION. Lionel A. Walford. The Ronald Press Company, 1958.

MARINE SCIENCE AND TECHNOLOGY: SURVEY AND PROPOSALS. REPORT OF THE SECRETARY-GENERAL. United Nations Economic and Social Council, E/4487, 1968.

THE STATE OF WORLD FISHERIES: WORLD FOOD PROBLEMS, No. 7. Food and Agriculture Organization of the United Nations, 1968.

WORK OF FAO AND RELATED ORGANIZATIONS CONCERNING MARINE SCIENCE AND ITS APPLICATIONS. FAO Fisheries Technical Paper No. 74. Food and Agriculture Organization of the United Nations, September, 1968.

19. The Anchovy Crisis

THE CLUPEOID RESOURCES OF TROPICAL SEAS. Alan R. Longhurst in *Oceanography and Marine Biology—an Annual Review,* edited by Harold Barnes, Vol. 9, pages 349–385; 1971.

THE FISH RESOURCES OF THE OCEAN. Edited by J. A. Gulland, pages 136–145. Fishing News Ltd., Surrey, England, 1971.

MEN, BIRDS AND ANCHOVIES IN THE PERU CURRENT—

20. Marine Farming

AQUACULTURE, ITS STATUS AND POTENTIAL. J. H. Ryther and G. C. Matthiessen in *Oceanus,* Vol. 14, No. 4, pages 2–14; February, 1969.

THE FOOD RESOURCES OF THE OCEAN. S. J. Holt in *Scientific American,* Vol. 221, No. 3, pages 178–194; September, 1969.

THE OYSTERS OF LOCMARIAQUER. Eleanor Clark. Pantheon Books, 1964.

PHOTOSYNTHESIS AND FISH PRODUCTION IN THE SEA. John H. Ryther in *Science,* Vol. 166, No. 3901, pages 72–76; October 3, 1969.

WHALE CULTURE—A PROPOSAL. Gifford B. Pinchot in *Perspectives in Biology and Medicine,* Vol. 10, No. 1, pages 33–43; Autumn, 1966.

21. Plate Tectonics and the History of Life in the Oceans

DYNAMICS IN METAZOAN EVOLUTION. R. B. Clark. Oxford University Press, 1964.

EVOLUTIONARY PALEOECOLOGY OF THE MARINE BIOSPHERE. J. W. Valentine. Prentice-Hall, Inc., 1973.

GLOBAL TECTONICS AND THE FOSSIL RECORD. James W. Valentine and Eldridge M. Moores in *The Journal of Geology,* Vol. 80, No. 2, pages 167–184; March, 1972.

A REVOLUTION IN THE EARTH SCIENCES: FROM CONTINENTAL DRIFT TO PLATE TECTONICS. A. A. Hallam. Oxford University Press, 1973.

V PHYSICAL MARINE RESOURCES

22. The Physical Resources of the Ocean

ENCOURAGING DEVELOPMENT OF NONLIVING RESOURCES. *Marine Science Affairs—a Year of Broadened Participation: The Third Report of the President to the Congress on Marine Resources and Engineering Development.* U.S. Government Printing Office, January, 1969.

ENHANCING BENEFITS FROM THE COASTAL ZONE. *Marine Science Affairs—a Year of Broadened Participation: The Third Report of the President to the Congress on Marine Resources and Engineering Development.* U.S. Government Printing Office, January, 1969.

THE MINERAL RESOURCES OF THE SEA. John L. Mero. Elsevier Publishing Company, 1965.

MINERAL RESOURCES OF THE WORLD OCEAN: PROCEEDINGS OF A SYMPOSIUM HELD AT THE NAVAL WAR COLLEGE, NEWPORT, RHODE ISLAND, JULY 11–12, 1968. Edited by Elisabeth Keiffer. Graduate School of Oceanography, University of Rhode Island, Occasional Publication No. 4, 1968.

PETROLEUM RESOURCES UNDER THE OCEAN FLOOR. National Petroleum Council, 1969.

USES OF THE SEAS. Edited by Edmund A. Gullion. Prentice-Hall, Inc., 1968.

23. Fresh Water from Salt

DEMINERALIZATION OF SALINE WATERS. Oscar L. Chapman, Goodrich W. Lineweaver and David S.

Jenkins. United States Department of the Interior, 1952.

Fresh Water for the Future. E. R. Gilliland in *Industrial and Engineering Chemistry,* Vol. 47, No. 12, pages 2410–2422; December, 1955.

Fresh Water from the Sea. Thomas K. Sherwood in *The Technology Review,* Vol. 57, No. 1, pages 15–20; November, 1954.

Saline Water Conversion. *Annual Report of the Secretary of the Interior for 1954. Annual Report of the Secretary of the Interior for 1955.* U.S. Department of the Interior, January, 1955; January, 1956.

The Water Conversion Problem: Energy Requirements and Energy Standards. K. C. D. Hickman in *Industrial and Engineering Chemistry,* Vol. 48, No. 4, pages 7A–20A; April, 1956.

We Will Drink the Sea. William Grigg in *Science News Letter,* Vol. 68, No. 6, pages 90–91; August 6, 1955.

24. The Disposal of Waste in the Ocean

Marine Environmental Quality: Suggested Research Programs for Understanding Man's Effect on the Oceans. National Academy of Sciences, 1971.

Marine Pollution Monitoring: Strategies for a National Program. National Oceanic and Atmospheric Administration of the U.S. Department of Commerce, 1972.

Wastes Management Concepts for the Coastal Zone: Requirements for Research and Investigation. National Academy of Sciences and National Academy of Engineering, 1970.

25. Plate Tectonics and Mineral Resources

Exploration Methods for the Continental Shelf. Geology, Geophysics, Geochemistry. P. A. Rona. National Oceanic and Atmospheric Administration Technological Report ERL 238-AOML 8, U.S. Government Printing Office, 1972.

Hydrothermal Manganese in the Median Valley of the Mid-Atlantic Ridge. Martha R. Scott, Robert B. Scott, Andrew J. Nalwalk, P. A. Rona and Louis W. Butler in *EOS: American Geophysical Union Transactions,* Vol. 54, No. 4, page 244; April, 1973.

Sulfide Ore Deposits in Relation to Plate Tectonics. F. Sawkins in *Journal of Geology,* Vol. 80, pages 377–397; 1972.

INDEX